2010 8th International Conference on Advanced Semiconductor Devices & Microsystems

(ASDAM 2010)

Smolenice, Slovakia
25 – 27 October 2010

IEEE Catalog Number: CFP10469-PRT
ISBN: 978-1-4244-8574-1

Copyright © 2010 by the Institute of Electrical and Electronic Engineers, Inc
All Rights Reserved

Copyright and Reprint Permissions: Abstracting is permitted with credit to the source. Libraries are permitted to photocopy beyond the limit of U.S. copyright law for private use of patrons those articles in this volume that carry a code at the bottom of the first page, provided the per-copy fee indicated in the code is paid through Copyright Clearance Center, 222 Rosewood Drive, Danvers, MA 01923.

For other copying, reprint or republication permission, write to IEEE Copyrights Manager, IEEE Service Center, 445 Hoes Lane, Piscataway, NJ 08854. All rights reserved.

***This publication is a representation of what appears in the IEEE Digital Libraries. Some format issues inherent in the e-media version may also appear in this print version.**

IEEE Catalog Number: CFP10469-PRT
ISBN 13:   978-1-4244-8574-1

Additional Copies of This Publication Are Available From:

Curran Associates, Inc
57 Morehouse Lane
Red Hook, NY 12571 USA
Phone: (845) 758-0400
Fax: (845) 758-2633
E-mail: curran@proceedings.com
Web: www.proceedings.com

ORGANIZERS

Microelectronics Department,
Faculty of Electrical Engineering and Information Technology,
Slovak University of Technology, Bratislava

*****

Promos Foundation

*****

Institute of Electrical Engineering,
Slovak Academy of Sciences, Bratislava

COSPONSORSHIP

Slovak University of Technology in Bratislava
Faculty of Electrical Engineering and Information Technology

Promos Foundation

The IEEE Electron Devices Society

GENERAL CHAIRMAN

Daniel Donoval (*Slovak University of Technology, Bratislava, Slovakia*)

PUBLICATION CHAIRMAN

Juraj Breza (*Slovak University of Technology, Bratislava, Slovakia*)

INTERNATIONAL PROGRAMME COMMITTEE

Duncan Allsopp (*University of Bath, United Kingdom*)
Stefan Bengtsson (*Chalmers University, Gothenburg, Sweden*)
Juraj Breza (*Slovak University of Technology in Bratislava, Slovakia*)
Michel Brillouet (*CEA Grenoble, France*)
Carles Cane (*CNM Barcelona, Spain*)
Sylvain Delage (*Alcatel –Thales, France*)
Frantisek Dubecky (*Slovak Academy of Sciences, Bratislava, Slovakia*)
Karol Fröhlich (*Slovak Academy of Sciences, Bratislava, Slovakia*)
Zsolt J. Horvath (*Hungarian Academy of Sciences, Budapest, Hungary*)
Miroslav Husak (*Czech University of Technology, Prague, Czech Republic*)
Hiroshi Iwai (*Tokyo Institute of Technology, Japan*)
Peter Kordos (*Slovak Academy of Sciences, Bratislava, Slovakia*)
Jaroslav Kovac (*Slovak University of Technology in Bratislava, Slovakia*)
Jozef Kristofik (*Academy of Sciences of the Czech Republic, Prague,Czech Republic*)
Stefan Kubicek (*IMEC Leuven, Belgium*)
Tibor Lalinsky (*Slovak Academy of Sciences, Bratislava, Slovakia*)
Gaudenzio Meneghesso (*University of Padova, Italy*)
Androula Nassiopoulou (*IMEL/NCSR Demokritos, Athens, Greece*)
Jozef Osvald (*Slovak Academy of Sciences, Bratislava, Slovakia*)
Marek Tlaczala (*University of Wroclaw, Poland*)
Vladimir Tvarozek (*Slovak University of Technology in Bratislava, Slovakia*)
Frantisek Uherek (*Slovak University of Technology in Bratislava, Slovakia*)
Gerhard Wachutka (*Technical University of Munich, Germany*)
Miroslav Zeman (*Technical University of Delft, Netherlands*)

FOREWORD

The organizers and hosts warmly welcome you to the International Conference on Advanced Semiconductor Devices and Microsystems, ASDAM 2010. This is the eighth in the series of conferences held at the Smolenice Castle. Microelectronics, and today nanoelectronics with better understanding of the physical behaviour of semiconductor devices and structures, numerous novel technologies, materials, interfaces and deep-submicrometer dimensions continue to be the most rapidly developing technology. This also brings about a swift expansion in the field of microsystems and a wide variety of their applications.

We receive great pleasure presenting you the Conference Proceedings which contains 5 invited lectures and 78 contributed papers selected carefully by the International Program Committee from more than 100 contributions from 18 countries world wide. The invited talks given by renowned speakers, contributed oral presentations and poster sessions along with many interesting personal meetings during coffee breaks, lunches and common evenings in the beautiful Castle of Smolenice will give all of us excellent chances for fruitful discussions and dissemination of recent results.

We believe that ASDAM 2010, as demonstrated by the interest and presence of participants from about 20 countries, is a forum for regular European meetings bringing together leading experts from the world, particularly from the countries of Europe. Frequent and intense international collaboration and contacts of the Department of Microelectronics of the Slovak University of Technology in Bratislava as well as the geographical location of Bratislava in the heart of Europe, on the border of Slovakia, Austria and Hungary, very close to the International Airport of Vienna, support this idea as well.

It is our pleasure to thank all authors and participants for the effort they made to prepare their contributions. We are very grateful to everybody who helped to make this Conference a reality, especially the members of the programme and organizing committees, the conference secretariat, and all those who support the conference, accepted invitations to present talk or who submitted papers for consideration.

We appreciate the technical co-sponsorship of the IEEE Electron Devices Society.

On behalf of the Organizing Committee we would like to wish all of you a great deal of success in your scientific discussions, exchange of information, and a pleasant and memorable stay in Smolenice.

Juraj Breza
Publication Chairman

Daniel Donoval
General Chairman

vi

CONTENTS

Organizers   *iii*

International Programme Committee   *iv*

Foreword   *v*

Contents   *vi*

Invited Talk 1

Nanoscaled SiGe based MOSETs   1
M. Östling

Structures & Devices 1

Single photon detection by means of SiGe-quantum dot arrays   9
J. Moers, N. P. Stepina, J. Gerharz, E.S. Koptev, A.I. Nikiforov, A.V. Dvurechenskii, D. Grützmacher

Model for Evaluation of Terahertz Plasma Resonances in HEMT-Based Devices with Grating Gate   13
I. Khmyrova, R. Yamase, N. Watanabe

Ultra High Frequency Performance in All Ternary $In_{0.52}Al_{0.48}As-In_{0.53}Ga_{0.47}As-In_{0.52}Al_{0.48}As$ DHBT   17
R. Knight, J. Sexton, M. Missous

Structures & Devices 2

Light emitting diode with 2D PhC structure in the surface analysed by NSOM   21
L. Suslik, D. Pudis , J. Skriniarova, J. Kovac , J. Kovac, jr., I. Kubicova, I. Martincek, J. Jakabovic, J. Novak

100mV noise performances of Te-doped Sb-HEMT   25
A. Noudeviwa, A. Olivier, Y. Roelens, F. Danneville, N. Wichmann, N. Waldhoff, L. Desplanque, X. Wallart, S. Bollaert

Development of Advanced Gunn Diodes and Schottky Multipliers for High Power THz sources   29
F. Amir, C. Mitchell, M. Missous

3-D Simulation of a 45 nm Partially Depleted Silicon on Insulator (SOI) Transistor with Diamond-Shaped Body Contact   33
A. Daghighi, A. Farajzadeh

Electrical and optical properties of ZnO/Si photodiodes with embedded CdTe and CdSe/ZnS nanoparticles   37
J. Hotovy, J. Kovac, J. Skriniarova, I. Novotny, J. Jakabovic, J. Kovac jr.

Poster Session 1 Technology & Devices

Fabrication of Novel High Frequency and High Breakdown InAlAs-InGaAs pHEMTs 41
M. Mohamad Isa, D. Saguatti, G. Verzellesi, A. Chini, K. W. Ian, M. Missous

A Width-Dependent Body-Voltage Model to Obtain Body Resistance in PD SOI MOSFET Technology 45
A. Daghighi, A. Asgari-Khoshooie

Reactive ion etching of AlxGa1-xN/GaN heterostructure using Cl$_2$ BCl$_3$/Ar gas plasma 49
W. Oleszkiewicz, J. Gryglewicz, B. Paszkiewicz, R. Paszkiewicz, A. Szyszka, M. Ramiączek - Krasowska, A. Stafiniak, , M. Tłaczała

Noise in the InAlN/GaN HEMT transistors 53
K. Rendek, A. Satka, J. Kovac, D. Donoval

Inter-digitated AlGaN/GaN Schottky diode for monolithic integration 57
B. Paszkiewicz, R. Paszkiewicz, M. Wosko, M. Tlaczala

Preparation and properties of ZnO nanomaterials for sensoric applications 61
T. Brath, D. Buc, M. Caplovicova, L. Caplovic, M. Predanocy, V. Hrnciar

Influence of thickness on transparency and sheet resistance of ITO thin films 65
M. Mazur, D. Kaczmarek, J. Domaradzki, D. Wojcieszak, S. Song, F. Placido

Thermoelectrical properties of TiO2:(Co, Pd) and TiO2:Nb thin films 69
E. Prociow, M. Mazur, J. Domaradzki, D. Wojcieszak, D. Kaczmarek, T. Gawor

Effect of Substrate Temperature on Oblique-Angle Sputtered ZnO:Ga Thin Films 73
I. Novotny, D. Kuluuva, S. Flickyngcrova, V. Tvarozek, L. Spiess, P. Schaaf, M. Netrvalova, P. Sutta

Nanocrystalline Diamond/amorphous Composite Carbon Films Prepared by PECVD Technology for Photocathode Application 77
J. Huran, N. I. Balalykin, G. D. Shirkov, P. Bohacek, A. P. Kobzev, A. Valovic

N-doped Nanocrystalline Silicon Carbide Films Prepared by PECVD Technology 81
P. Bohacek, J. Huran, A. Valovic, A. P. Kobzev, V. N. Shvetsov, M. Kucera, L. Malinovsky

Material optimization of the alignment marks for the EBDW lithography 85
L. Matay, R. Andok, V. Barak, A. Ritomsky, A. Konecnikova, I. Kostic, S. Partel, P. Hudek

Patterning of nanometer structures by using direct-write e-beam lithography for the sensor development 89
P. Durina, M. Stefecka, T. Roch, J. Noskovic, M. Trgala, A. Pidik, I. Kostic, A. Konecnikova, L. Matay, P. Kus, Plecenik

Temperature dependence of the pyroelectric behaviour in GaN/AlGaN 93
A. Laposa, J. Jakovenko, M. Husak

Off-state stress investigation of InAlN/GaN HFETs with different AlN buffer layer 97
M. Florovic, J. Kovac, H. Behmenburg, P. Kordos, J. Skriniarova, D. Donoval

Growth and structural properties of GaAs on Al pseudo-substrates for ultrafast optoelectronics 101
Z. Sofer, D. Sedmidubsky, M. Mikulics

Invited Talk 2

GaN Power Electronics 105
B. Lu, D. Piedra, T. Palacios

GaN Based Structures & Devices 1

AlGaN/GaN HEMT on Si (111) substrate for millimeter microwave power applications 111
S. Bouzid, V. Hoel, N. Defrance, H. Maher, F. Lecourt, M. Renvoise, D. Smith, J. C. De Jaeger

Characterisation of electrical properties of AlGaN/GaN Schottky diode at very high temperature 115
A. Chvala, D. Donoval, R. Sramaty, J. Marek, J. Kovac, P. Kordos, J. Skriniarova

On the Identification of Trap Location in AlGaN/GaN HEMTs during Electrical Stress 119
M. Tapajna, R. J. T. Simms, Y. Pei, U. K. Mishra, M. Kuball

GaN Based Structures & Devices 2

Study of temperature distribution in the channels of AlGaN/GaN HEMT devices by μ-Raman characterization techniques 123
J. Kovac jr., S. K. Jha, E. V. Jelenković, O. Kutsay, M. Pejović, C. Surya, J. A. Zapien, I. Bello, R. Srnanek, J. Kovac, S. Flickyngerova

Modelling and optimisation of a sapphire/GaN-based diaphragm structure for pressure sensing in harsh environments 127
M. J. Edwards, S. Vittoz, R. Amen, L. Rufer, P. Johander, C. R. Bowen, D. W. E. Allsopp

HEMT-SAW Structures for Chemical Gas Sensors in Harsh Environment 131
I. Ryger, T. Lalinsky, G. Vanko, M. Tomaska, I. Kostic, S. Hascik, M. Vallo

Investigation of Deep Energy Levels in Heterostructures based on GaN by DLTS 135
L. Stuchlikova, J. Sebok, J. Rybar, M. Petrus, M. Nemec, L. Harmatha, J. Benkovska, J. Kovac, J. Skriniarova, T. Lalinsky, R. Paskiewicz, M. Tlaczala

Molecular dynamics and Electrical Simulation of a Novel GaN/4H-SiC Hetero-structure Optically Triggered Vertical NPN Device 139
S. Bose, S. K. Mazumder

Analysis of structure geometry and interface charge on electrical characteristics of InAlN/GaN HEMTs 143
J. Marek, D. Donoval, J. Kovac, M. Molnar, A. Chvala, P. Kordos

Invited Talk 3

GaN for THz Sources 147
M. Marso

GaN Based Structures & Devices 3

Preparation and properties of AlGaN/GaN MOS-HFETs with atomic layer deposited Al_2O_3 as gate oxide 155
R. Stoklas, D. Gregusova, M. Blaho, P. Kordos, M. Tajima, T. Hashizume

Comparison of AlGaN/GaN HFETs and MOSHFETs in prospect of oscillator design 159
A. Fox, M. Mikulics, B. Strang, M. Marso, D. Grützmacher, P. Kordos

Role of the gate-to-drain distance in the performance of the normally-off InAlN/GaN HEMTs 163
J. Kuzmik, Ostermaier, G. Pozzovivo, B. Basnar, W. Schrenk, J.-F. Carlin, M. Gonschorek, E. Feltin, N. Grandjean, Y. Douvry, Ch. Gaquière, J.-C. De Jaeger, G. Strasser, D. Pogany, E. Gornik

Influence of interface states on C-V characteristics of AlGaN/GaN heterostructures 167
J. Osvald

Materials & Technology 1

Effects of soft-UV irradiation on organic thin film transistors with different gate dielectrics 171
N. Wrachien, A. Cester, G. Meneghesso, J. Kovac, J. Jakabovic, D. Donoval

Design, preparation and properties of spin-LED structures based on InMnAs 175
P. Telek, S. Hasenöhrl, J. Soltys, I. Vavra, M. Drzik, J. Novak

Study of ZnO Films Grown with Different Dopants - Physical Properties and Their Comparison 179
L. Prusakova, M. Netrvalova, P. Sutta

Synthesis and Doping of Zinc-Oxide Thin Films by RF Sputtering and Ion Implantation 183
M. Milosavljević, D. Perusko, V. Milinović, P. Gaspierik, I. Novotny, V. Tvarozek

MO CVD growth of ZnO with different growth rate 187
D. Nohavica, P. Gladkov, J. Grym, Z. Jarchovsky

Poster Session 2 Characterization & Sensors

Optimization of Position of Piezoresistive Elements on Substrate Using FEM Simulations 191
P. Kulha

A new model of trap assisted band-to-band tunnelling 195
M. Mikolasek, J. Racko, L. Harmatha, O. Gallo, J. Reznak, F. Schwierz, R. Granzner

Simulation Study of Conduction-state Charge Imbalance in High Voltage Super-junction Power MOSFET 199
K. Pravin N

Electrode configuration for EMG measurements 203
E. Vavrinsky, K. Rendek, M. Daricek, M. Donoval, F. Horinek, M. Horniak, D. Donoval

Semi-insulating GaAs radiation detectors: PICTS study of neutron-induced defects 207
F. Dubecky, M. Ladziansky, D. Kindl, V. Necas

Wireless Sensor System for Overhead Line Ampacity Monitoring 211
J. Frolec, M. Husak

Wireless Sensor Network Control System 215
M. Husak, A. Boura, J. Jakovenko

Detection of soft X-rays using semi-insulating GaAs detector 219
B. Zatko, F. Dubecky, P. Bohacek, V. Necas, L. Ryć

Use of Barometric Sensor for Vertical Velocity Measurement 223
M. Husak, J. Jakovenko

Monitoring of Car Driver Physiological Parameters 227
E. Vavrinsky, V. Tvarozek, V. Stopjakova, P. Solarikova, I. Brezina

Broadband amplitude-stabilized oscillator 231
J. Foit, J. Novak

Potentiality of the Inductive Powering for Measurement in the Enclosed Systems 235
A. Boura, M. Husak

Influence of Conductor Systems on the Crosstalks in Integrated Circuits 239
J. Novak, J. Foit, V. Janicek

Simulation of a planar Micro Ion Mobility Spectrometer for Security Applications 243
R. Cumeras, I. Gràcia, E. Figueras, L. Fonseca, J. Santander, M. Salleras, C. Calaza, N. Sabaté, C. Cané

Characterization of high permittivity $GdScO_3$ films prepared by liquid injection MOCVD 247
M. Jurkovic, K. Husekova, K. Cico, E. Dobrocka, M. Nemec, J. Fedor, K. Fröhlich

Biomedical signal amplifier for EMG wireless sensor system 251
K. Rendek, M. Daricek, E. Vavrinsky, M. Donoval, D. Donoval

Resistive switching in $RuO_2/TiO_2/RuO_2$ MIM structures for non-volatile memory application 255
B. Hudec, M. Hranai, K. Husekova, J. Aarik, A. Tarre, K. Fröhlich

Invited Talk 4

Micro-power converters for energy harvesting devices 259
E. Sangiorgi, A. Romani, M. Tartagni

Sensors & Microsystems 1

A monolithic micro fuel cell based on a functionalized porous silicon membrane 263
N. Torres-Herrero, J. Santander, N. Sabaté, C. Cané, T. Trifonov, A. Rodriguez, R. Alcubilla

Experimental Analysis and Modeling of the Mechanical Impact during the Dynamic Pull-In of RF-MEMS Switches 267
M. Niessner, J. Iannacci, G. Schrag, G. Wachutka

Sensors & Microsystems 2

Hybrid photonic/plasmonic ZnO/Au composites for sensing applications 271
J. A. Zapien, L. Yu, C. H. To, C. Limiao, J. Kovac jr., I. Bello, S. T. Lee

Constitutive Equation of the Dipole Layer in Hydrogen-sensing Metal-Oxide-Semiconductor Structures 275
F. Srobar, O. Prochazkova

Radiation Effects on CMOS Image Sensors due to X-Rays 279
J. Tan, B. Büttgen, A. J. P. Theuwissen

First Measurement on the DEPFET Mini-Matrix Particle Detector System 283
J. Scheirich, C. Oswald, P. Kodys

Invited Talk 5

SEM techniques for characterization of GaN nanostructures and devices 295
A. Satka, J. Kovac, J. Priesol, A. Vincze, F. Uherek, M. Michalka

Characterization of Materials & Structures

Study of optical and electrical properties of sputtered indium oxide films 297
M. Predanocy, I. Fasaki, M. Wilke, I. Hotovy, I. Kosc, L. Spiess

Characterization and optical properties of TiO2 prepared by pulsed laser deposition 301
O. Kadar, F. Uherek, J. Chlpik, J. Remsa, J. Bruncko, A. Vincze, M. Jelinek

New InP Based pHEMT Double Stage Differential to Single-ended MMIC Low Noise Amplifiers for SKA 305
N. Ahmad, S. Arshad, M. Missous

SIMS depth profile characterisation of InAlN/GaN structures 309
A. Vincze, J. Kovac, H. Behmenburg, R. Srnanek, F. Uherek, D. Donoval, M. Heuken

Modeling & Simulation

Physics-Based Modeling of Electromagnetic Parasitic Effects in Interconnects and Busbars 313
G. Wachutka, P. Böhm

Compact Model Extraction from Quantum Corrected Statistical Monte Carlo Simulation of Random Dopant Induced Drain Current Variability   317
U. Kovac, C. Alexander, G. Roy, B. Cheng, A. Asenov

Monte Carlo Simulations of Channel Scaling to Ultimate Limit in Si and $In_{0.3}Ga_{0.7}As$ Bulk MOSFETs   321
A. Islam, K. Kalna

Analytical Modelling of InGaP/GaAs HBTs   325
G. Dutta, S. Basu

Materials & Technology 2

Structure and optical properties of the hydrogen diluted a-Si:H thin films prepared by PECVD with different deposition temperatures   329
M. Netrvalova, M. Fischer, J. Mullerova, M. Zeman, P. Sutta

Structural and chemical analysis of self-aligned titanium silicide formed by furnace annealing   333
E. Barbarini, S. Guastella, F. Pirri

The Compound Oxides Based on TiO_2 and NiO Thin Films for Low Temperature Gas Detection   337
I. Kosc, I. Hotovy, M. Kompitstas, R. Grieseler, M Wilke, V. Rehacek, M. Predanocy, T. Kups, L. Spiess

RuO_2/TiO_2 based MIM capacitors for DRAM application   341
B. Hudec, K. Husekova, J. Aarik, A. Tarre, A. Kasikov, K. Fröhlich

Author index   345

Nanoscaled SiGe based MOSETs

Mikael Östling

School of Information and Communication Technology, KTH Royal Institute of Technology
Electrum 229, SE-164 40 Kista, Sweden,
E-mail: ostling@kth.se

This paper presents an overview of the technological challenges facing the future scaling of device dimensions needed to meet the performance scaling in accordance with Moore's law. A number of performance boosters have to be introduced in order to keep up with the expected performance gain in each new technology node. The introduction of strain engineering is an important feature as well as the implementation of high-k dielectrics. From the 32 nm node and forward there is an urgent search for a fundamental breakthrough to achieve low access resistance to the drain and source areas. This paper will focus to a large extent on this latter area and discuss metallic source/drain (MSD) contacts in nanoscaled MOSFET technology. MSD contacts offer extremely low S/D parasitic resistance, abruptly sharp junctions between S/D and channel and preferably low temperature processing. Recently great efforts have been achieved on Pt- and Ni-silicide implementation. A conclusion is that MSD MOSFETs are competitive candidates for future generations of CMOS technology.

1. Introduction

Today's ever increasing performance requirements can only be met by a vast number of technology boosters, i.e. substrate engineering, local strain, new dielectric materials, metal source/drain and new device architecture. MOSFETs on strained substrates have enhancements in mobility by more than 100% compared to non-strained devices. Introduction of high-k dielectrics have been successfully achieved to scale beyond 1 nm EOT and dopant segregated MSD contacts have been shown possible to implement by conventional process technology. In this extended abstract very few details will be shown but a list of references is presented for further reading.

2. The impact of strain on MOS performance

Strain engineering was adopted in silicon technology in the early nineties. The first application was silicon-germanium heterojunction bipolar transistors. For these devices the bandgap engineering using a SiGe base yielded a favorable current gain at high base doping levels. The growth technology was well developed and high frequency performance and noise properties could be well-controlled and in some case the increased current gain could be traded for reduced noise.

The second generation of strained devices was the strained channel MOSFETs, with the main objective to increase charge carrier mobility [1,2]. Here, the strain is applied to the silicon material by either deposition of stressor films or by growth of strain relaxed SiGe buffer layers, with high Ge concentration. The strain in general increases the transconductance of a MOSFETs, which is beneficial for speed performance but also translates to a lower input referred noise for a given technology, e.g. gate length. The silicon dioxide to channel interface is sensitive to the strain level and it is suggested that some defects (recombination centers) are reduced due the change in lattice constant and surface energy [3].

In a recent study [4] it was shown that mobility enhancement factors of up to 2.1 could be achieved for relaxed SiGe buffer layers with a Ge content of 27% overgrown by a thin strained Si layer of 10 nm. For a Ge content of 15 % an enhancement factor of 1.7 was demonstrated. The strained technology is today implemented in the leading edge products.

One drawback from thick relaxed buffer layers is that the poor thermal conductivity and hence the difficulties to the dissipated heat in the highly scaled MOSFETs. In order to reduce this problem it is possible to grow very thin buffer layers and create virtual substrates. In figure 1a) a thin buffer layer structure is shown and the subsequent figure 1b) shows the resulting mobility in the strained Si layer [5].

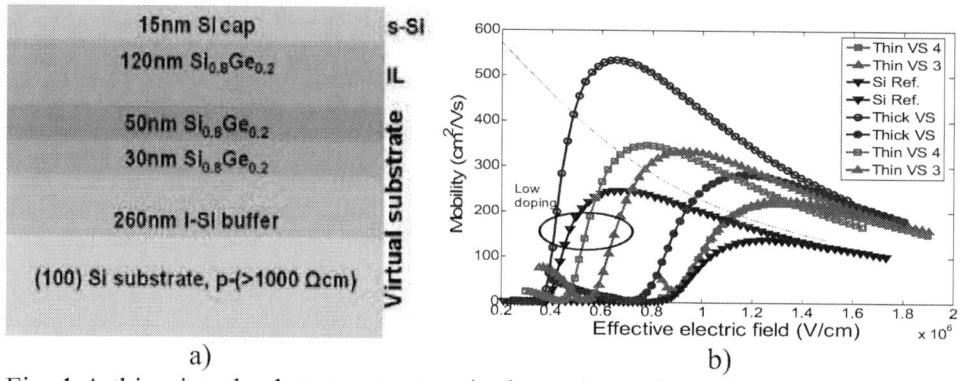

a)   b)

Fig. 1 A thin virtual substrate structure is shown in a schematic cross-section a). The resulting mobility is shown b).

Other techniques to increase the charge carrier mobility are also explored, i.e. the possibility of other substrate orientation and replacing the channel material with III-V materials with higher inherent mobility.

3.   High-k dielectrics for high performance MOSFETs

Scaling of MOSFETs requires dielectric thicknesses below 1 nm electrical equivalent thickness (EOT). The ultimate replacement for silicon dioxide is still subject for intense investigations by both academia and industry. The first replacement for SiO_2 was a dielectric based on HfO_2 introduced by Intel [6]. Implementation of the new dielectric materials is strictly dependent on process compatibility with Si processing and hence reduces the possible choices considerably. In Fig 2 a chart of various dielectric materials is shown. A high dielectric constant is not the only physical property that will solve the problem to control the electric vector field in transistor channels at required leakage currents. It has been identified by the high k gang [7] that the offset value between the energy bands of silicon and the gate insulator material is decisive for limiting current leakage by tunneling. The relation between the k-value and the energy offset delta E_c can be approximated by a hyperbolic function as shown in Fig. 2. For the 22 nm bulk node a rough condition for a material fulfilling the 22 nm bulk CMOS node is delta E_c x $k \approx 70$ eV [8].

978-1-4244-8574-1/10 $26.00 © 2010 IEEE   2

A major obstacle with implementing high-k dielectrics is that the mobility in the MOS channel is clearly deteriorated due to more imperfections in the dielectric material compared to SiO_2. If a metal gate electrode such as TiN is used studies have shown that the remote phonon scattering is substantially reduced [9].

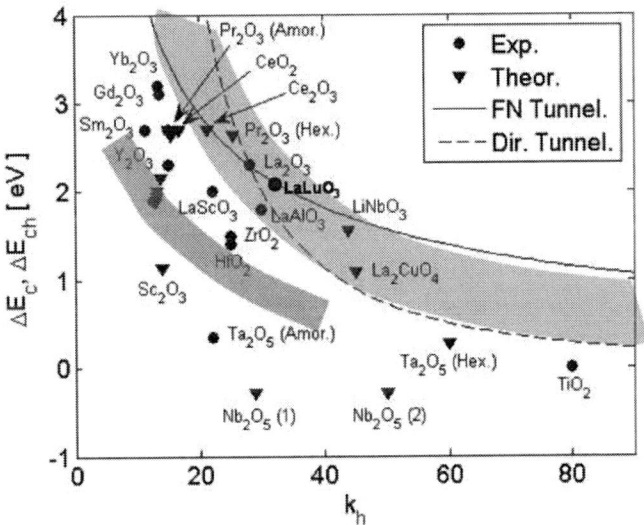

Fig 2. Energy offset values versus k-values for different oxides. The upper shadowed area represents the requirements for the 22 nm LSTP *bulk* CMOS node, the lower shadowed area represents the requirements for the 22 nm LSTP *fully depleted* double gate SOI node. [8]

3. Advanced device architecture and the implementation of Metallic Source/Drain Schottky Barrier Contacts

In a recent PhD thesis by Luo [10] a comprehensive discussion on the MSD technique is carefully presented together with some device geometry aspects. Future scaling calls for more advanced device structures, *i.e.*, multi-gate MOSFETs such as FinFETs [11] in Fig. 2.5, Ω-FETs [12] and gate-all-around FETs [13], have also been employed to effectively control short channel effects (SCE). The multi-gate MOSFETs have been studied extensively as the ultimate solution for extremely scaled devices. Instead of controlling the channel over one surface for planar devices, multi-gate devices control the channel from many surfaces, leading to greatly improved SCEs and device performance. Making devices on Ultra Thin Body (UTB)-SOI substrates also presents a way to control SCEs [14,15]. UTB-SOI substrates also reduces the extrinsic capacitance. With downscaling of MOSFETs, the intrinsic capacitance decreases; however, the non-scalable extrinsic capacitances are becoming more comparable to their intrinsic counterparts, which offsets the desired performance enhancement. With UTB-SOI substrates, the extrinsic parasitic capacitance is eliminated due to the excellent isolation from the bulk silicon, improving power consumption at matched performance. The circuit delay time, $\tau \sim CV/I$, is thus reduced, resulting in the improved switching speed for CMOS in digital circuits.

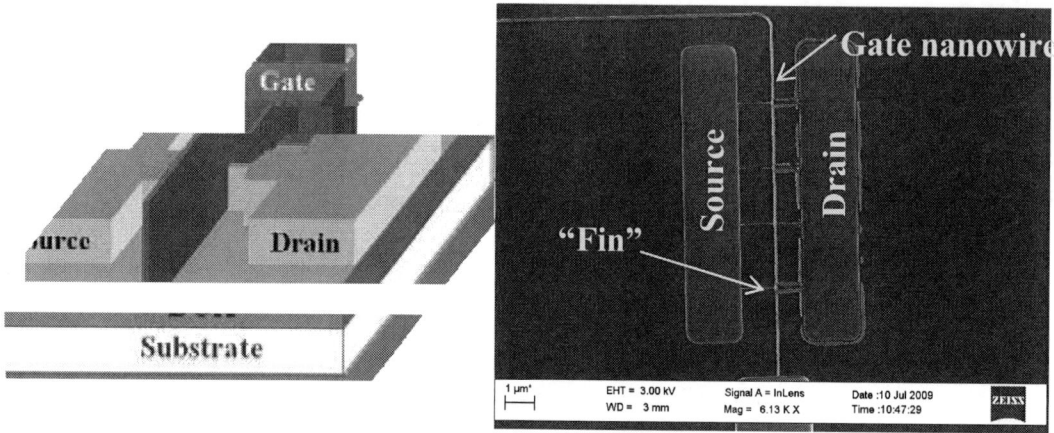

Fig 3. Schematic drawing of a FinFET structure and a top-view scanning electron microscopy (SEM) image of FinFET fabricated at KTH.

Considering ordinary UTB-SOI MOSFETs with highly doped S/D, one of the most challenging problems is the parasitic series S/D resistance due to the ever-reducing top Si layer. An attractive alternative is to use metallic S/D contacts [16, 17] to circumvent the high series resistance problem with conventional UTB-SOI MOSFETs. With this technology, the highly doped S/D regions are replaced with metallic S/D as shown in Fig. 2.6. Metal silicides are of particularly interest for the metallic S/D materials due to the established self-aligned silicidation process and the low resistivity. This metallic silicide S/D architecture provides an elegant solution to the high series resistance problem, since the contacts between connecting metal and metallic silicides provide very low contact resistivity as well as low sheet resistance. In addition, low thermal budget of metal silicide S/D also facilitates the integration of high-k, metal gate and strained channel. In table 1 the ITRS requirements on the source and drain resistances are depicted.

Year of production	2015	2016	2017	2018	2019	2020	2021
Technology node (nm)	22	20	17.7	15.7	14.0	12.5	11.1
Physical gate length (nm)	17	15	14	12.8	11.7	10.7	9.7
Max contact resistivity for FDSOI ($10^{-8}\ \Omega \cdot cm2$)	4.0	2.0	1.0	0.8	0.7	0.6	0.5
Max S/D extension sheet resistance for FDSOI (Ω/sq)	730	752	809	890	983	1104	1199

Table 1. ITRS 2009: requirements on source/drain for FDSOI substrate

Among various metal silicides, PtSi and NiSi are exceptionally favorable due to its low resistivity, low Si consumption and low thermal budget (500-600 ^{o}C) during silicidation [18]. Low temperature processing is of special importance for the integration of high-k and metal gates in a CMOS process flow [19].

Although the parasitic S/D resistance of a MSD MOSFET is greatly reduced, the intrinsic large Schottky barrier heights (SBH) of most silicides, however, limit the devices performance. Simulation studies indicate that SBH has to be smaller than 0.1 eV in order to outperform doped S/D devices [20]. Furthermore, to simplify CMOS fabrication, one silicide with low SBHs for both n- and p-type polarities is favorable. Since not a single silicide fulfills this requirement, the SBH engineering technique is therefore needed.

One of SBH engineering techniques, i.e., silicide as dopant source (SADS), was employed to tune the SBHs of PtSi and NiSi by inducing segregated dopants at silicide/Si interface [21-23]. A schematic SADS sequence is depicted below in Fig.4.

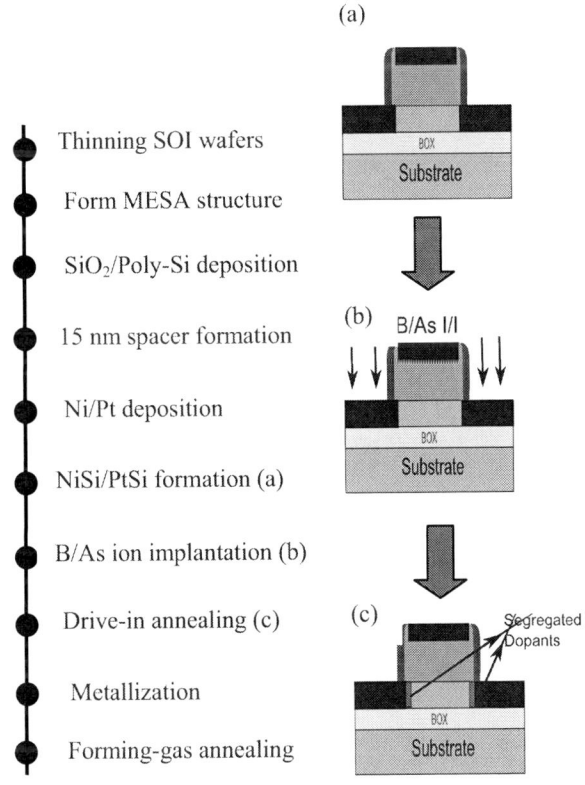

Figure 4. Process flow for the fabrication of MSD SB-MOSFETs (left), and a low-temperature SADS technique to tune SBHs for PtSi and NiSi (right). A 15 nm thin spacer is formed in order to minimize excess channel resistance.

Thin NiSi films often suffer from poor morphological stability and start to agglomerate at 600 °C as a consequence of the low melting point of NiSi. In order to combat the poor morphological stability, the addition of C and N presents a potential solution. In addition, adding Pt to NiSi leading to the formation of the ternary monosilicide alloy $Ni_{1-x}Pt_xSi$ has been shown to be an effective solution [24]. However, for CMOS devices fabricated on SOI substrates requires thinner Si layers than 5 nm in order to attain the predicted device/circuit performance. It has been observed that especially NiSi can experience agglomeration problems below 10 nm thickness. However, recently it was found that Ni-silicide films exhibit two different morphological stability characteristics depending on the deposited thickness of Ni (t_{Ni}). For $t_{Ni} \geq 4$ nm, polycrystalline NiSi films form and agglomerate at lower temperatures for thinner films like PtSi films; however, for $t_{Ni}<4$ nm, epitaxially aligned Ni-silicide films readily grow and exhibit extraordinary morphological stability up to 800°C [25] (Fig. 5). Adding Pt into Ni also improves the morphological stability of $Ni_{1-x}Pt_x$ silicide films.

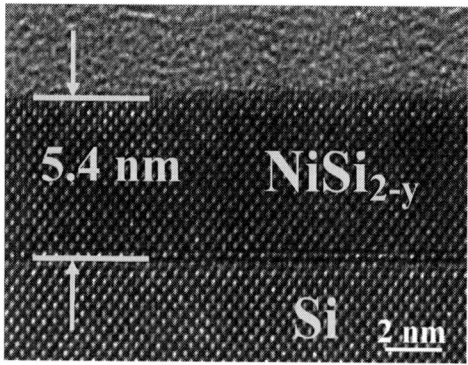

In Fig. 5 cross-sectional and top-view TEM images of the ultrathin Ni-silicide film with $t_{Ni}=2$ nm formed at 750 °C are shown. Almost perfect epitaxially aligned Ni-silicide layer is observed.

Conclusions

This paper has discussed three areas of major challenges for future scaled MOSFETs. The first challenge concerned the future need for implementing strain in devices to obtain higher mobility. Several strain engineering solutions are available and even further technologies are underway. The second big challenge is the development of high-k dielectrics where a vast number of material choices exist but the ultimate solution will be a compromise based on both high-k properties but also process compatibility. The third challenge discussed in the paper is that of minimizing access resistance to source/drain contacts. A promising technique called dopant segregation in metal silicide Schottky barrier contacts was discussed and suggested as a promising contact technology

Acknowledgement

The author greatly acknowledge his research team members and funding from the Swedish Foundation for Strategic Research, the Swedish Research Council, and the EU FP7 NoE NANOSIL.

References

[1]   Scott E. Thompson et al, *IEEE Trans. Electron Devices, vol. 51, no. 11, 2004*

[2]   M. Chu, Y. Sun, U. Aghoram, and S. E. Thompson, *Annual Review of Materials Research, Vol. 39: 203-229 (August 2009).*

[3]   A. Stesmans, P. Somers, V. V. Afanas'ev, C. Claeys, and E. Simoen, *Applied Physics Letters, vol. 89, pp. 152103*

[4]   F. Driussi, D. Esseni, L. Selmi, P.-E. Hellström, G. Malm, J. Hållstedt, M. Östling, T. J. Grasby, D. R. Leadley, and X. Mescot, *Solid-State Electronics, vol. 52, pp. 498-505, 2008.*

[5] P.-E. Hellström, J. Edholm, M. Östling., S. Olsen, Anthony O'Neill, K. Lyutovich, M. Oehme, and E. Kasper, "Strained-Si NMOSFETs on Thin 200 nm Virtual Substrates," presented at *International Solid State Device Research Symposium*, Bethesda, MD, USA, 2005.

[6] R. Chau et al, *IEEE Electron Device Lett, vol. 25, no. 6, 2004*

[7] http://www.high-k-gang.eu

[8] O.Engström, B. Raeissi, S. Hall, O. Buiu, M.C.Lemme, H. D. B. Gottlob, P. K. Hurley and K. Cherkaoui, *Solid State Electron. 51, 622 (2007)*

[9] R. Chau, *IEEE Electron Device Lett., vol. 25,, pp. 408-410*

[10] Jun Luo, PhD Thesis, KTH, "Integration of metallic source/drain contacts in MOSFET technology" *TRITA-ICT/MAP AVH Report 2010:06, ISBN 978-91-7415-680-5, (2010)*

[11] X. J. Huang, W.-C. Lee, C. Kuo, D. Hisamoto, L. Chang, J. Kedzierski, E. Anderson, H. Takeuchi, Y.-Kyu. Choi, K. Asano, V. Subramanian, T.-J. King, J. Bokor and C. M. Hu, *IEDM Tech. Dig., pp. 67–70, 1999*

[12] J. P. Colinge, *Solid-state electronics, vol. 48, no. 6, pp. 897-905*

[13] N. Singh, A. Agarwal, L. K. Bera, T. Y. Liow, R. Yang, S. C. Rustagi, C. H. Tung, R. Kumar, G. Q. Lo, N. Balasubramanian, and D.-L. Kwong, *IEEE Electron Device Lett., vol. 27, no. 5, pp. 383-386, May. 2006*

[14] K. K. Young, *IEEE Trans. Electron Devices,* vol. 36, no. 2, pp. 399-402, Feb. 1989

[15] . C. S. Woo, K. W. Terrill, and P. K. Vasudev, *IEEE Trans. Electron Devices,* vol. 37, no. 9, pp. 1999-2006, Sept. 1999

[16] A. M. Waite, N. S. Lloyd, P. Ashburn, A. G. R. Evans, T. Ernst, H. Achard, S. Deleonibus, Y. Wang, P. Hemment, *Proc. ESSDERC,* pp. 223-226, 2003

[17] C. Mazure, J. Fitch, and C. Gunderson, *IEDM Tech. Dig.,* pp. 853–856, 1992

[18] S.-L. Zhang and M. Östling, "Metal silicides in CMOS technology: past, present and future trends," *Critical reviews in solid state and materials science,* vol. 28, no. 1, pp. 1-129

[19] G. Larrieu, E. Dubois, R. Valentin, N. Breil, F. Danneville, G. Dambrine, J. P. Raskin and J. C. Pesant, *in IEDM Tech. Dig.,* p.147 (2007)

[20] S. Xiong, T.-J. King and J. Bokor, *IEEE Trans. Electron Devices,* 52, p.1859 (2005)

[21] Z. Zhang, Z. J. Qiu, R. Liu, M. Östling, and S.-L. Zhang, IEEE Electron Device Lett. 28, 565 (2007).

[22] Z. J. Qiu, Z. Zhang, M. Östling, and S. -L. Zhang, IEEE Trans. Electron Devices, 55, 396 (2008).

[23] J. Luo, Z.-J. Qiu, Z. Zhang, M. Östling and S.-L. Zhang, J. Vac. Sci. & Technolo. B, vol. 28, issue. 1, 2010.

[24] D. Mangelinck, J. Y. Dai, J. S. Pan, and S. K. Lahiri, *Appl. Phys. Lett.,* vol. 75, no. 12, pp. 1736-1738, 1999

[25] J. Luo, Z.-J. Qiu, C. L. Zha, Z. Zhang, D. P. Wu, J. Lu, J. Åkerman, M. Östling, L. Hultman, and S.-L. Zhang, Appl. Phys. Lett., vol. 96, no. 3, pp. 031911-1-031911-3, 2010.

978-1-4244-8574-1/10 $26.00 © 2010 IEEE

Single photon detection by means of SiGe-quantum dot arrays

J. Moers[1,2], N. P. Stepina[3], J. Gerharz[1,2], E.S. Koptev[3], A.I. Nikiforov[3],
A.V. Dvurechenskii[3], D. Grützmacher[1,2]

[1] Jülich-Aachen Research Alliance, JARA, Fundamentals of Future Information Technology
[2] Institute of Bio- and Nanosystems, Forschungszentrum Jülich, 52425 Jülich, Germany
[3] Institute of Semiconductor Physics, 630090 Novosibirsk, Russia

The fabrication process of single-photon detectors (SPD) based on high-density arrays of Ge quantum dots (QD) is proposed. The design of the SPD exploits the phenomenon that the contribution of an individual QD to the hopping transport through this high-density Ge QDs array crucially depends on the occupation of the dot with carriers. A change in the conductance of the array can be induced by changing the charge state of one QD by illumination. For sub-micron sized devices step-like variations of the conductance due to the absorption of single photons are expected to be noticeable. The fabrication process combines molecular beam epitaxy (MBE) with means of semiconductor technology as E-Beam lithography and reactive ion etching. The process flow developed is described and measurement results are shown, confirming the suitability of the device design for single photon detection.

1.   Introduction

Emission and detection of single photons is required for a broad range of future device applications: in quantum information and cryptography, medical diagnosis and imaging, chemical analysis and materials characterization [1]. As the privacy protection in quantum cryptography requires no more than one photon to be present in each laser pulse, high demands are placed on Single Photon Detection (SPD). It should have high quantum efficiency of detection, low noise, and sufficient computational speed. Conventionally, single photons can be detected by the multiplication of a charge carrier generated by a photon in a photomultiplier tube (PMT) or avalanche photodiode (APD). At present, the best SPDs for practical use in this area are GaInAs/InP APDs [2,3]. To count single photons, the APD are set to operate in the Geiger mode [4], in which one photon is capable of causing an avalanche of charge carriers. However, this usually leads to a considerable increase in dark noise and the probability of the so-called afterpulses arising from APD operation.

Our concept of a Si based SPD employing hopping transport in densely packed Ge quantum dot (QD) arrays has the potential to overcome these drawbacks. The QD-arrays show a conductance, which is governed by hopping transport: the current will follow percolation paths through the array by hopping from QD to QD. The hopping probability between two dots depends on the charge state of the dots, hence changing the charge in one dot will alter the hopping probability and leads to rearrangement of percolation path. Illuminating the QD-array using light with a wavelength between 1.3 µm and 1.55 µm will extract electrons from one dot by photon absorption. The generated electron can be captured by another dot, leaving one dot with an additional positive charge and one dot with less positive charge. As a result a redistribution of carriers among different QDs occurs. Changing the potential landscape leads to new conductive paths, hence the conductance of the array changes; it can either be an increase or a decrease of conductance. This change can be interpreted as the detection of the photon.

978-1-4244-8574-1/10 $26.00 © 2010 IEEE

In [5] it was shown that changing the global current through this array does not necessarily requires the alteration of the charge state in a QD which belongs to the percolation path. So changing the charge state in any QD of the array will change the current through the whole array.

2. Device fabrication

Figure 1 shows a schematic cross sectional view of the device. Stranskii-Krastanov-growth of Ge-islands on a silicon (100) substrate is exploited to form those arrays of Ge-QD with high enough densities ($1\text{-}4\times10^{11}\text{cm}^{-2}$). They can successfully be embedded in the crystal matrix of the

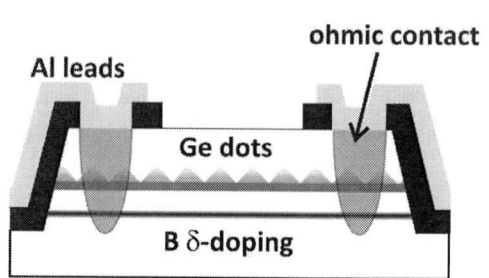

Figure 1: Schematic cross sectional view of the SPD: A high density array of Ge-QD is grown by MBE. The conductance through this Ge-QD is governed by hopping transport, which is crucially depended on the charge state of every dot. The charging/discharging of one dot generated by single photon absorption will change the conductance of the whole device.

elemental semiconductor system Si-Ge and can easily be doped with carriers [6]. Yakimov et al. showed [7] that the conductance through the dot array is crucial dependent on the filling factor of the dots. Therefore a δ-doped layer is introduced beneath the QD-layer to supply a defined number of holes in each dot. The device process starts with the fabrication of high density Ge-QD array by Molecular Beam Epitaxy (MBE). It is grown on a 100 mm (001) p-Si substrate with a resistivity of 20 Ωcm. The growth temperature for 10 monolayers (ML) of Ge was 300°C and the growth rate was 0.2 ML/s. The Ge-QD-layer was capped with a 40 nm Si layer. As a result, the areal density of the dots of 4×10^{11} cm^{-2} was obtained. A Boron δ-doped silicon layer was inserted in 5 nm distance to the QD-layer. The amount of doping was chosen to adjust the number of holes per dot to ensure significant changes in the conductance of the device by one single charging/discharging of one island. After epitaxy the wafer were cut into dies of 2 cm × 2 cm.

Optical contact lithography with exposure wavelength λ of 250 nm was used to define adjustment markers for further optical and electron beam lithography. For e-beam markers

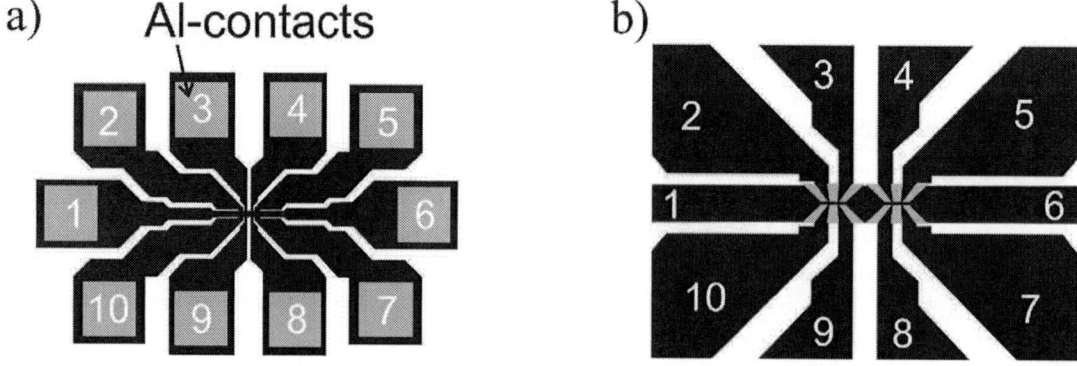

Figure 2: a) broad mesa-structure defined by optical lithography, b) enlargement of the inner part of a). The e-beam defined structures are shaded grey. There are two specimens on each mesa: 4-point measurements can be performed by feeding the current between contact 1 and contact 6, while the voltage can be measured between contact 2 and 3 or 10 and 9 for one sample and between contact 4 and 5 or 8 and 7 for the other sample.

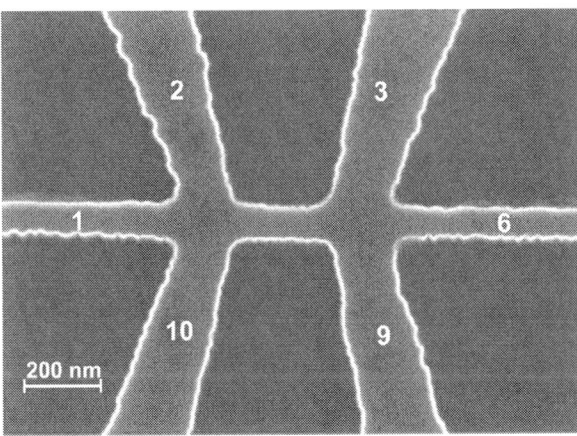

Figure 3: SEM top view image of an active structure with active length L_A = 200 nm (between contacts 2 and 3 or 10 and 9, respectively) and width W_A = 70 nm (between contact 2 and 10 or 3 and 9, respectively). The contacts are denoted in the same scheme as in figure 2b.

square holes with vertical sidewalls, a width of 10 μm and a depth of 600 nm etched by reactive ion etching were used. The broad mesa structure of the device (figure 2a) was defined by optical lithography, too. The structure has 10 contacts, so that in each structure two devices in series could be integrated. The pattern transfer is done by reactive ion etching using an anisotropic Ar/SF$_6$-plasma with low BIAS to prevent surface damage. The resulting etching depth was 96 nm, so the Ge-QD layer and Boron δ-doped layer outside the mesa were removed.

After mesa-etch the sample was cleaned applying a standard RCA clean. A second optical lithography defines the contact areas. Before evaporation of 500 nm of Al the residues of the resist in the exposed areas are removed by oxygen plasma and the native oxide layer was removed with a dip in 1% HF for 60 s. Immersion of the evaporated samples in solvent structures the Al, so it only remains in the contact areas (Al-pads in figure 2a).

Afterwards the active device area is defined by e-beam lithography using a VISTEC EBPG 5000+ high end e-beam tool. High resolution ZEP 520A-7 resist was spun on the samples at 4000 rpm resulting in a thickness of 210 nm. The e-beam was used to shrink down the lateral dimension of the broad mesa to active sizes of width W_A and length L_A varying from 70 nm to 500 nm (figure 2b); the structure is transferred to the substrate by means of reactive ion etching again using the same plasma as for the broad mesa. Figure 3 shows a SEM top view image of the resulting structure. On each mesa two specimens are designed on which two-point as well as four point measurements can be performed (cnf. figure 2b). After removing the e-beam resist in DMAc at room temperature for 20 minutes the contacts are annealed at 480°C for 30 minutes under N$_2$ atmosphere to ensure good ohmic behaviour.

3. Electrical measurements

Time resolved 4-point measurements of the conductance of the device at 4.2K were performed. A fiber coupled laser with 1.55 μm wavelength and initial laser power of 1 mW was used for illumination. To get extremely low light intensity the initial laser power was attenuated up to 60 dB. The conductance traces of the device show step like changes due to single QD charging and discharging events (Figure 4 and inset in figure 4). When the light is off, there is a constant conductance. For the instant the light is switched on, the charging and discharging of the dots begins and hence the conductance of the array is decreased or increased for each event. When the light is switched off again, the rearrangement of the carriers in the dots stops, and hence the changes in conductance stop, too. The number of changing events per time unit depends linearly on intensity of illumination, which is a prerequisite for single photon detection [8 and references there in].

4. Conclusions and outlook

A process flow to fabricate Single Photon Detectors on basis of high density Ge quantum dot arrays is proposed. The process and the detector design are compatible to CMOS technology allowing the integration of the detectors in advanced electronics for signal processing. Electrical measurements on obtained structures show their potential for single photon detection. In future work the influence on different cleaning procedures and plasma etching conditions on the sidewall of the active area and hence on the channel conductance will be investigated further.

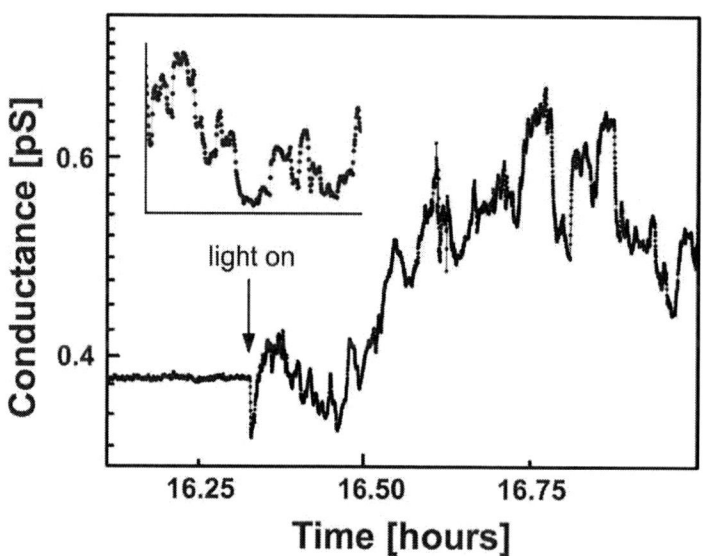

Figure 4: Measured conductance trace of the SPD. The step-like changes are due to charging/discharging of single Ge-QDs under illumination. The inset is an enhanced part of the curve.

References

[1]   R. H. Kingston, *Optical Sources, Detectors and Systems*, Academic, London, 1995
[2]   A. Rochas, C. Guillaume-Gentil, J.-D. Gautier, A. Pauchard, G. Ribordy, H. Zbinden, Y. Leblebicib, L. Monat, *Proc. SPIE* 6583, 65830F (2007)
[3]   R. T. Thew, D. Stucki, J. -D. Gautier, H. Zbinden, and A. Rochas, *Appl. Phys. Lett.* **91**, 201114 (2007)
[4]   N. Gisin, G. Ribordy, W. Tittel, and H. Zbinden, *Rev. Mod. Phys.* 74, 145 (2002)
[5]   Ratno Nuryadi, Yasuhiko Ishikawa, Michiharu Tabe, *Phys. Rev. B* **73**, 045310 (2006)
[6]   A.I. Nikiforov, V.A. Cherepanov, O.P. Pchelyakov, A.V. Dvurechenskii and A.I. Yakimov, *Thin Solid Films,* 2000 (380) 158
[7]   A. I. Yakimov, A. V. Dvurechenskii, A. I. Nikiforov, and A. A. Bloshkin, *JETP Letters*, 77, 376 (2003)
[8]   N. P. Stepina, E.S. Koptev, A.V. Dvurechenskii, I.V. Osinnyukh, A.I. Nikiforov,D. Gruetzmacher, J. Moers, J. Gerharz: 18th Int. Symp. "Nanostructure:Physics and Technology" St Peterburg, Russia, 2010, p255

Model for Evaluation of Terahertz Plasma Resonances in HEMT-Based Devices with Grating Gate

Irina Khmyrova, Ryosuke Yamase, and Norikazu Watanabe

University of Aizu,
Aizu-Wakamatsu, 965-8580, Japan
e-mail: khmyrova@u-aizu.ac.jp

We propose simple analytical model for calculation of spatial distribution of the sheet electron density in the channel of high-electron mobility transistor (HEMT) periodically modulated by the bias voltage applied to grid-grating gate. The contribution of ungated regions of two-dimensional electron gas (2DEG) channel is taken into account. The developed model allows to evaluate resonant frequencies of plasma oscillations excited in such periodically modulated 2DEG channel. The proposed model can be useful in the interpretation of experimentally obtained data and optimization of the grating-gated HEMT-like structures for THz applications.

1. Introduction

An interest in terahertz (THz) region of electromagnetic spectrum is driven by strong demand in compact detectors and sources for such applications as sensing and imaging in radioastronomy, defence, biomedicine. Dyakonov and Shur [1] have proposed to use the excitation of electron plasma oscillations in the channel of high-electron mobility transistor (HEMT)-like structures with submicrometer gate as an approach to realize such systems. Two-dimensional electron gas (2DEG) channel region beneath the gate contact can serve as a resonant cavity for excited plasma waves with resonant frequencies being dependent on the length of the gate contact. The resonant frequency in such a system can be tuned also by variation of the gate bias voltage. Recently THz detection and generation have been observed in double- and single-quantum well HEMT-like structures with grid-grating gate [2-6]. Grid-grating gate has been introduced [2] to enhance the coupling of THz radiation as its wavelength exceeds the dimensions of the HEMT. Resonant plasma frequencies observed

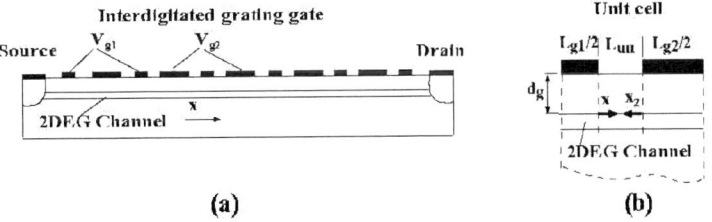

Fig.1. Schematic structure of the interdigitated HEMT (a)
and unit cell (b).

experimentally have been compared with those predicted theoretically by an ideal model [6]. Indeed, realistic devices can differ significantly from those represented by ideal model, in particular, they contain rather large ungated portions of the 2DEG channel. We believe that realistic model accounting for the contribution of ungated regions on such performance of grating-gated HEMT structures as resonant plasma frequencies can be useful in the interpretation of experimental results as well as to gain insight into some phenomena observed experimentally, for example, spectra broadening.

2. Spatial distribution of sheet electron density

The paper deals with the development of the analytical model for evaluation of resonant frequencies of plasma oscillations excited in the 2DEG channel of the HEMT –like structure schematically shown in Fig. 1a. 2DEG channel is formed at the heterointerface InAlAs/InGaAs with δ-doped wide-gap InAlAs layer. Interdigitated gate fabricated on top surface consists of (N+1) gate fingers G1 with length L_{g1} and N gate fingers G2 with length L_{g2}. 2DEG channel is periodically modulated by the bias voltages V_{g1} and V_{g2} applied to gate fingers G1 and G2, respectively. In realistic grating gated HEMT structures [3-6] with gate-to-channel separation of about 50-70 nm ungated regions are affected by fringing field when bias voltage is applied to gate fingers G1 and G2. Due to periodical modulation of the 2DEG channel one can limit consideration by a single unit cell between the centers of two adjacent gate fingers G1 and G2 shown in Fig. 1b. The model developed for a single-gate HEMT without source-drain current [7] has been generalized to the case of a double-gate unit cell.

We derived 2DEG sheet electron density in such unit cell $\Sigma_{uc} = \dfrac{\Sigma_1 + \Sigma_2}{2}$,

where $\Sigma_{1,2} = \dfrac{\varepsilon_0 \varepsilon}{e d_g \left(1 + \exp \xi_{1,2}\right)} \left(V_{g1,2} - V_{th} + V_{p2,1} \exp \xi_1\right)$,

$V_{p1,2} = \dfrac{d_d}{d_g \left(1 + \exp \xi_{1,2}\right)} \left(V_{g1,2} - V_{th} + V_p \exp \xi_{1,2}\right)$,

V_{th} is threshold voltage, $V_p = e\Sigma_d d_d / \varepsilon_0 \varepsilon$, e is electron charge, d_d and Σ_d are the depth of the doping plane and dopant concentration, respectively; ε_0 and ε are dielectric constants of vacuum and wide-gap semiconductor layer. Coordinates x_1 and x_2 are related to parameters ξ_1 and ξ_2 and to each other: $x_{1,2} = \dfrac{1}{\pi}\left(1 + \xi_{1,2} + \exp \xi_{1,2}\right)$ and $x_2 = \dfrac{L_{un}}{d_g} - x_1$, d_g is the thickness of the layer separating the gate contact and 2DEG channel .

Fig. 2 shows spatial distribution of 2DEG sheet electron density in the unit cell Σ_{un} calculated at different bias voltages V_{g1} and V_{g2}, gate lengths $L_{g1} = 100$ nm and $L_{g2} = 300$ nm and length of ungated region $L_{un} = 100$ nm. One can see from Fig. 2 that the sheet electron density in the ungated regions is significantly affected by fringing field. 2DEG channel regions beneath the gates are subjected to its influence as well. Fig.3 demonstrates evolution of sheet electron density spatial distribution in the unit cell with the increase of the length of ungated region from 100 to 300 nm at fixed bias voltages $V_{g2} = -2.5$ V and $V_{g1} = -1$ V (dashed line) and $V_{g1} = -2.75$ V (solid line). Non-saturated source-drain current [8] flowing along periodically modulated 2DEG channel of the HEMT with grid-grating gate can also affect the spatial distribution of the sheet electron density. To account for the contribution of the source-drain current previously developed model [9] has been used. Fig. 4 illustrates the modification of the 2DEG sheet electron density spatial distribution by the source-drain current in the HEMT with interdigitated gate with all fingers of length $L_g = 100$ nm and gate bias voltage $V_{g1} = V_{g2} = -1$ V. Length of ungated regions $L_{un} = 200$ nm.

978-1-4244-8574-1/10 $26.00 © 2010 IEEE

3. Frequencies of plasma resonances

To derive the expressions for evaluation of resonant frequencies of plasma oscillations we generalize the formalism used for a single-gate HEMT [7] on the case of periodically modulated 2DEG HEMT channel under consideration. The phase velocity of the plasma wave propagating along the unit cell can be expressed as

$$v_{uc} = \sqrt{\frac{e}{2m^*}\left(V_{g1} + V_{g2} - 2V_{th} + V_{p1}\exp\xi_2 + V_{p2}\exp\xi_1\right)}.$$

Due to spatial nonuniformity of the sheet electron density the calculation of the phase change of the wave traveling along the unit cell requires integration

$$\theta_{uc} = \omega \int_{x_l}^{x_r} \frac{dx}{v_{uc}}, \text{ where } x_l = L_{g1}/2d_g \text{ and } x_r = \left(L_{un} + L_{g2}/2\right)/d_g, \omega \text{ is angular frequency.}$$

Calculating total phase change in the structure with interdigitated (N+1) fingers G1 and N fingers G2 and applying the condition for the standing wave existence we may evaluate resonant frequency of plasma oscillations in the following form:

$$f_{fr} \propto \frac{\pi}{4}\left(2N\theta_{uc}\right)^{-1}.$$

Dependences of the fundamental resonant frequencies of plasma oscillation versus gate bias voltage V_{g1} calculated at different voltages V_{g2} for 5 gate fingers G2 and 6 gate fingers G1 are shown in Fig. 5. The following parameters for the InAlAs/InGaAs structure have been used: ε = 12.7, $\Sigma_d = 5 \times 10^{12}$ cm^{-2}, $d_g = 50$ nm, $d_d = 35$ nm, $V_{th} = -2.756$ V.

4. Conclusions

Analytical model to evaluate spatial distribution of the sheet electron density in the 2DEG channel of the realistic HEMT-like structure with grid-grating gate is developed. The developed model takes into account the possible contribution of ungated 2DEG channel regions as well as nonsaturated source-drain current which can flow along the channel. Resonant frequencies of plasma oscillations in the HEMT structure with periodically modulated 2DEG channel are evaluated. The developed model can be useful to interpret the data obtained experimentally and for optimization of the grating-gated HEMT structures for THz applications.

References

[1] M. Dyakonov and M. Shur, *IEEE Trans. Electron Devices,* **43**, 1640, 1996.
[2] X. G. Peralta, S.J. Allen, M. C.Wanke, N. E. Harff, et al., *Appl. Phys. Lett.*, **81**, 1627, 2002.
[3] E. A. Shaner, M. Lee, M. C. Wanke, A.D. Grine, J. L. Reno, and S. J. Allen, *Appl. Phys. Lett.*, **87**, 193507, 2005.
[4] T. Otsuji, Y. M. Meziani, M. Hanabe, T. Ishibashi, T. Uno, and E. Sano, *Appl. Phys. Lett.*, **89**, 263502, 2006.
[5] Y. M. Meziani ,T. Otsuji, ,M. Hanabe, T. Ishibashi, T. Uno, and E. Sano, *Appl.Phys. Lett.*, **90**, p. 215328, 2007.
[6] H. Saxena, R. E. Peale, and W. R. Buchwald, *J. Appl. Phys.*, **105**, 113101, 2009.
[7] T. Nishimura, N. Magome, I. Khmyrova. T. Suemitsu, W. Knap, and T. Otsuji,

Jpn. J. Appl. Phys., **48**, 04C096, 2009.

[8]   D. Veksler, F. Teppe, A. P. Dmitriev, V. Yu. Kacharovskii, W. Knap, and M. Shur, *Phys. Rev.* **B 73**, 125328, 2006.

[9]   I. Khmyrova, in *Proceedings of the IEEE COMCAS 2009 Conference*, Tel Aviv, Israel, 2009, 1F4-2.

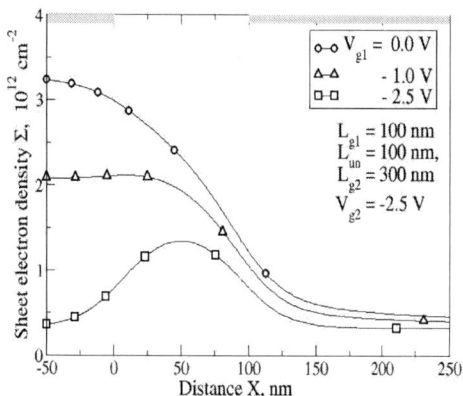

Fig. 2. Spatial distribution of the sheet electron density in the unit cell at different V_{g1}.

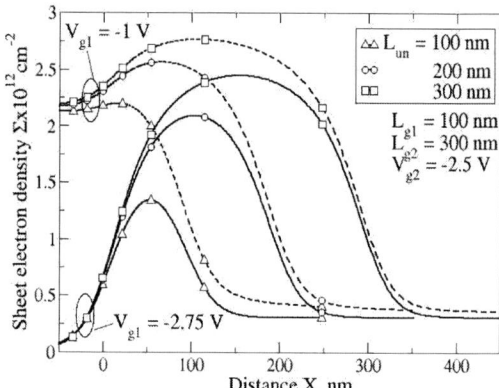

Fig. 3. Sheet electron density spatial distribution in the unit cell at different length of ungated region L_{un}.

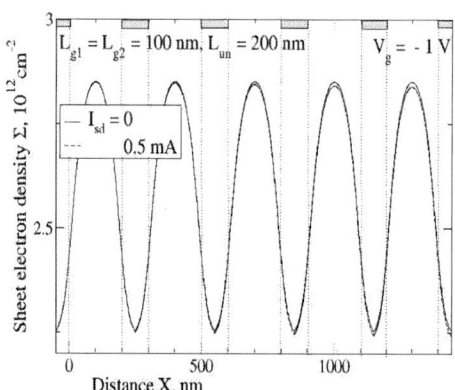

Fig. 4. Sheet electron density in the interdigitated HEMT without source-drain current (solid line) and with source-drain current I_{sd} = 0.5 mA (dashed line).

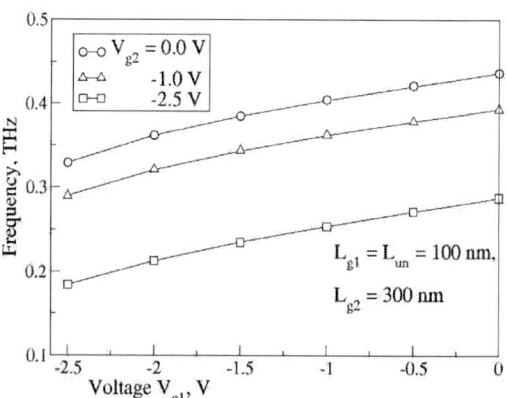

Fig. 5. Plasma resonances versus gate bias V_{g1} at different bias voltages V_{g2}.

ULTRA HIGH FREQUENCY PERFORMANCE IN ALL TERNARY

$In_{0.52}Al_{0.48}As$-$In_{0.53}Ga_{0.47}As$-$In_{0.52}Al_{0.48}As$ DHBT

R.Knight, J.Sexton and M.Missous

Department of Electrical and Electronic Engineering, University of Manchester,
Sackville Street, Manchester, M13 9PL
e-mail: Robert.knight@postgrad.manchester.ac.uk

A new all ternary $In_{0.52}Al_{0.48}As$-$In_{0.53}Ga_{0.47}As$-$In_{0.52}Al_{0.48}As$ double heterojunction bipolar transistor (DHBT) showing complete elimination of current blocking has been grown using solid source molecular beam epitaxy (MBE). The development of the DHBT required careful epitaxial design trade-offs which culminated in an optimum structure achieving ground breaking RF performance in excess of 100GHz and excellent DC performance for emitter dimensions of 1x5 μm^2. These new DHBTs which only use ternary alloys lead to simplified device growth and fabrication options.

1. Introduction

THE operation frequency of InP-based single heterojunction bipolar transistors (SHBT) and DHBT's have been pushed steadily higher as the demand on high-speed communication and high power microwave applications increases. The excellent material characteristics of InGaAs and InAlAs, namely, the high mobility, saturation velocity and the wide band gap (1.48eV) InAlAs; make them attractive for high frequency applications. $In_{0.52}Al_{0.48}As$-$In_{0.53}Ga_{0.47}As$ SHBTs are well established in high speed communication application with reported high cutoff frequencies f_t up to 204GHz [1]. However SHBTs suffer from low breakdown characteristics due to the higher avalanche multiplication factors of the low band gap collector material [2]. Ideally, high-power microwave or precision mixed-signal applications require a large breakdown and/or higher early voltage. These characteristics are readily achievable by utilising a double heterojunction structure in which a large band-gap material is placed within the collector. This region effectively prevents the early breakdown characteristics inherent to SHBTs by reducing the detrimental effect of impact ionisation at the base-collector interface.

Little work has been done on InAlAs InGaAs DHBT's because InAlAs DHBTs suffer from a large discontinuity in the collector. The large discontinuity causes a problem know as current blocking. Current blocking in DHBTs with $In_{0.52}Al_{0.48}As$ collectors is much more pronounced as compared to those with InP collectors because of the unfavourable band alignment (as far as HBTs are concerned) [3] at the $In_{0.52}Al_{0.48}As$-$In_{0.53}Ga_{0.47}As$ heterojunction. There are several ways to reduce the effect of current blocking in DHBTs, some of which include: grading the collector, collector current launcher, doping interface dipole (DID) and composite collector (CC) [4-6].

Many researchers [5, 7, 8] investigated current blocking in $In_{0.52}Al_{0.48}As$ or quaternary InAlGaAs collector based DHBTs and used either super lattice and/or grading at the heterojunctions as a partial solution to the problem. In most cases current blocking can only

978-1-4244-8574-1/10 $26.00 © 2010 IEEE

be removed when quaternary alloys are used in the collector since these tend to have lower ΔE_c but this is usually at the expense of a much reduced breakdown value. McAlister et al [9] and others [10-12] used a combination of dipole doping [6] and setback layer in DHBTs of different materials to reduce current blocking. Recent work at the University of Manchester which involved growth of more than ten wafers with different combinations of composite collector (CC) and doping interface dipole (DID) have successfully achieved their optimum combination to eliminate current blocking in pure ternary $In_{0.52}Al_{0.48}As$-$In_{0.53}Ga_{0.47}As$-$In_{0.52}Al_{0.48}As$ DHBT up to a current density of 2mA/μm^2. This paper reports the first high performance $In_{0.52}Al_{0.48}As$-$In_{0.53}Ga_{0.47}As$-$In_{0.52}Al_{0.48}As$ DHBT with no current blocking demonstrating an f_t of 110GHz and f_{max} of 90GHz for a device with emitter area of 1x5μm^2

2. Growth and Device Structure

2.1 Wafer Growth

The epitaxial layers of fabricated HBTs were grown on a RIBER V100+ solid source MBE system on 2"-3" Fe-doped semi-insulating (100) InP substrates. The system utilised an ultra high vacuum (UHV) cham-ber with a base pressure of <10-10 Torr. Py-rolitic boron-nitride crucibles were used for the evaporation of gallium, arsenic, indium, silicon and beryllium. The growth was per-formed at the relatively low temperature of ~420oC in order to confine the high-concentration Be-base dopant. The entire epitaxial structure comprises of $In_{0.52}Al_{0.48}As$ and $In_{0.53}Ga_{0.47}As$ lattice-matched layers to the InP substrate; no qua-ternary alloys are used.

2.2 Device Structure

The epitaxial design as shown in Table 1 has a thin undoped InGaAs spacer layer between emitter and base to reduce Beryllium (Be) out diffusion during growth and to act as a etch stop when fabricating the emitter. The high base doping gives specific contact resistance of 12Ω. The 750 Å spacer and doping interface dipole, doping level of 1($10^{16}cm^{-3}$) and 4($10^{18}cm^{-3}$); between the base and collector reduce the discontinuity enough to eliminate current blocking.

TABLE I
EPITAXIAL STRUCTURE OF $In_{0.52}AL_{0.48}AS$-$In_{0.53}GA_{0.47}AS$-$In_{0.52}AL_{0.48}AS$ DHBTS

Layer	Material	Thickness(Å)	Doping (cm^{-3})
Cap1	$In_{0.53}Ga_{0.47}As$	800	2x10^{19} (n)
Cap2	$In_{0.52}Al_{0.48}As$	800	2x10^{19} (n)
Emitter	$In_{0.52}Al_{0.48}As$	500	5x10^{17} (n)
Spacer	$In_{0.53}Ga_{0.47}As$	100	nid
Base	$In_{0.53}Ga_{0.47}As$	450	4x10^{19} (p)
Setback	$In_{0.53}Ga_{0.47}As$	750	1x10^{16} (n)
p-n dipole doping	$In_{0.53}Ga_{0.47}As$ - $In_{0.52}Al_{0.48}As$	2 x 100	4x10^{18} (pn)
Collector	$In_{0.52}Al_{0.48}As$	1000	1x10^{16} (n)
Collector 2	$In_{0.52}Al_{0.48}As$	500	1x10^{19} (n)
Sub-collector	$In_{0.53}Ga_{0.47}As$	3500	1x10^{19} (n)
Buffer	$In_{0.52}Al_{0.48}As$	300	nid
Substrate	InP		Semi-insulating

nid = no intentional doping

Table 1: DHBT Epitaxial Structure

3. Fabrication

Large emitter-area devices of 20×20 μm^2 and small geometry, RF devices of emitter areas 1x15 μm^2, 1×10 μm^2 and 1×5 μm^2 were fabricated using a triple mesa; self aligned emitter and base process [13]. The ohmic contacts for emitter, base, collector and interconnects are all non-alloyed Ti/Au deposited by thermal evaporation. The emitter was etched in two stages starting with the cap layer then the emitter. An H_2O_2:H_2O:H_3PO_4 chemical wet etch in the ratio of (1:50:3) with an average rate of 800Å/min was used to remove the InGaAs cap layer. The InAlAs emitter was removed with HCl:H2O (3:1)

978-1-4244-8574-1/10 $26.00 © 2010 IEEE

chemical wet etch. The InGaAs layer below the emitter does not react with the HCl:H$_2$O (3:1) wet etch. This etch stop layer ensures no over etching occurs into the base and allows for excellent emitter definition.

The base metal layer patterned a self aligned air bridge structure as seen in figure 1. The base/collector and isolation etches undercut the metal reducing the base collector area thus minimising the base collector capacitance Cjc of the device. The collector region was etched in a similar process to that of the emitter. Stumps were then added to the base and collector layers to planarise the device to the emitter contact. Cyclotene was then spun on and CF$_4$:0$_2$ reactive ion (RIE) etched back to expose the contact stumps. The Ti/Au interconnection metal is deposited on the passivation layer of cyclotene to finish the device.

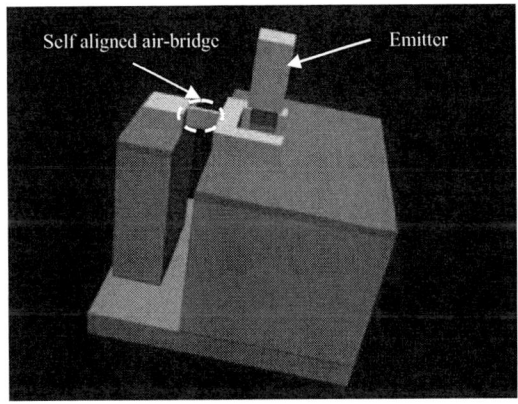

Figure 1: 3D view of the self aligned base air bridge

3. DC and RF Results

The small geometry, RF devices of emitter areas 1x15 μm^2, 1×10 μm^2 and 1×5 μm^2 were measured using a 110GHz VNA high frequency analyser. Figure 2 shows (a) the DC characteristic of a 1×5 μm^2 emitter device showing an I$_b$ of 55μA a collector current of 2mA demonstrating a current gain of 37. The DHBT device shows no signs of current blocking evident from the small (1.2V) knee voltage and has a V$_{CE(offset)}$ of 0.5V.

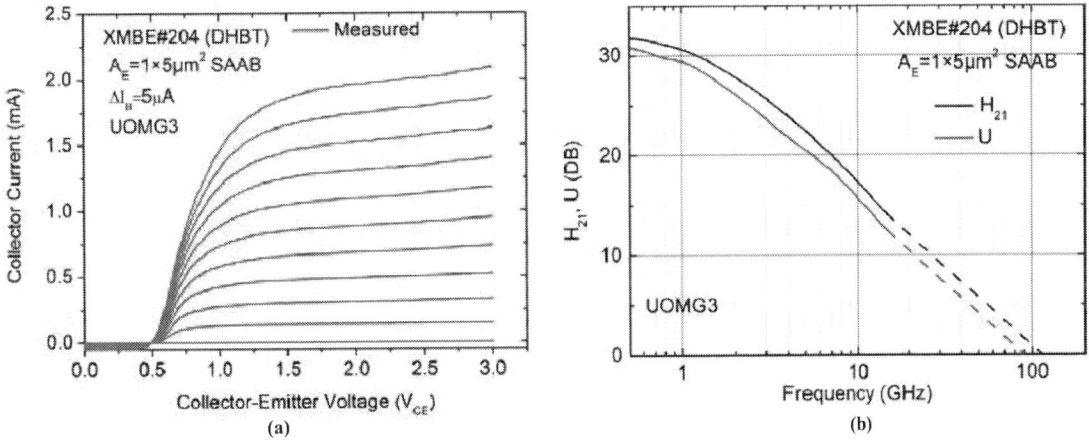

Figure 2: InAlAs/InGaAs DHBT results (a) DC characteristics and (b) f_t H$_{(2,1)}$ and f_{max} U

Figure 2 (b) demonstrates a cut off frequency f_t of 110GHz and a maximum oscillation frequency f_{max} of 90GHz. for a device with emitter area of 1x5μm2 of the ultra high speed all ternary In$_{0.52}$Al$_{0.48}$As-In$_{0.53}$Ga$_{0.47}$As-In$_{0.52}$Al$_{0.48}$As DHBT.

978-1-4244-8574-1/10 $26.00 © 2010 IEEE   19

3. Conclusion

The elimination of current blocking in an all ternary $In_{0.52}Al_{0.48}As/In_{0.53}Ga_{0.47}As$ DHBT has provided a solid foundation for the development of ultra high speed epitaxial structures which has paved the way for ground breaking RF performances. This paper demonstrates the first all ternary $In_{0.52}Al_{0.48}As$-$In_{0.53}Ga_{0.47}As$-$In_{0.52}Al_{0.48}As$ DHBT showing measured f_t and f_{max} of 110GHz and 90GHz respectively from a device size of $1x5\mu m^2$ emitter size. To our knowledge these are the highest ft and f_{max} reported to date in this material system demonstrating that the transport properties of $In_{0.52}Al_{0.48}As$ are sufficiently good to permit high frequency operation contrary to earlier speculations that these should limit operation to few tens of GHz [14].

Acknowledgement

The EMRS-DTC & Selex Galileo are gratefully acknowledged for supporting this project.

References

[1]   Sokolich, M., Charles, H., Thomas, S., Montes, M. and Martinez, R. IEEE Journal of, 2001. **36**(9): p. 1328-1334.

[2]   McKinnon, W.R., Driad, R., S.P. McAlister., Renaud, A. and Z.R. Wasilewski., in *Papers from the eighth canadian semiconductor technology conference*. 1998. Ottawa, Canada: AVS.

[3]   Vurgaftman, I., J.R. Meyer, and L.R. Ram-Mohan, Journal of Applied Physics, 2001. **89**(11): p. 5815-5875.

[4]   Yamada, H., T. Futatsugi., H. shigematsu., T. Tomioka., T. Fujii. and N. Yokoyama. in *Electron Devices Meeting, 1991. IEDM '91. Technical Digest., International*. 1991.

[5]   Iwai, T., H. Shigematsu., H. Yamada., T. Tomioka., K. Joshin. and T. Fujii., Japanese Journal of Applied Physics, 1997. **36**(part1, no.2): p. 648-651.

[6]   Capasso, F., Alfred, Y., C.K. Mohammed. and Philip, W., Applied Physics Letters, 1985. **46**(7): p. 664-666.

[7]   Huang, C.H., T.L. Lee, and H.H. Lin. in *Indium Phosphide and Related Materials, 1993. Conference Proceedings., Fifth International Conference on*. 1993.

[8]   Chanh, N., T. Liu., M. Chen., H. Sun. and D. Rensch. Electron Device Letters, IEEE, 1996. **17**(3): p. 133-135.

[9]   McAlister, S.P., W.R McKinnon., R. Driad. and A.P Penaud., Journal of Applied Physics, 1997. **82**(10): p. 5231-5234.

[10]   Yamahata, S., K. Kurishima., H. Ito. and Y. Matsuoka. in *Gallium Arsenide Integrated Circuit (GaAs IC) Symposium, 1995. Technical Digest 1995., 17th Annual IEEE*. 1995.

[11]   Liu, W. and D.S. Pan, Electron Device Letters, IEEE, 1995. **16**(7): p. 309-311.

[12]   Ishibashi, T. and Y. Yamauchi, Electron Devices, IEEE Transactions on, 1988. **35**(4): p. 401-404.

[13]   Choi, K., et al., Solid-State Electronics. **50**(9-10): p. 1483-1488.

[14]   C.W. Farley, J.A. Higgins, W-J. Ho, B.T. McDermott ,and M.F. Chang, J.Vac.Sci. and Technol. B10(2),1023 (1992)

Light emitting diode with 2D PhC structure in the surface analysed by NSOM

Ľ. Šušlik [1], D. Pudiš [1], J. Škriniarová [2], J. Kováč [2], J. Kováč, jr. [2],
I. Kubicová [1], I Martinček [1], J. Jakabovič [2], J. Novák [3]

[1] Dept. of Physics, University of Žilina, Univerzitná 1, 010 26, Žilina, Slovakia
[2] Dept. of Microelectronics, Slovak University of Technology, Ilkovičova 3, 812 19,
Bratislava, Slovakia
[3] Institute of Electrical Engineering, Slovak Academy of Sciences, 841 04, Bratislava,
Slovakia
e-mail: suslik@fyzika.uniza.sk

We present light emitting diode with two-dimensional photonic crystal structure prepared by interference lithography in the light emitting diode surface with an emission maximum at 845 nm. Applied two-dimensional photonic crystal structure improved light extraction efficiency for more than 30 %. The photonic crystal light emitting diode surface morphology was analyzed by atomic force microscope. The enhanced extraction efficiency of the photonic crystal diode was documented from L(I) dependencies from the near-field studies.

1. Introduction

Photonic crystals (PhC) opened wide platform for a new generation of optical and optoelectronical devices. The planar two-dimensional (2D) photonic crystal structures were firstly applied in the light waveguiding applications [1] and semiconductor lasers with high differential quantum efficiency [2]. The PhC structures have also been investigated as a potential candidate for improving the extraction efficiency in light emitting diodes (LED) [3, 4]. It can be improved by minimizing the total internal reflection in the diode surface. In a conventional unpatterned LED the majority of light emitted from the quantum well (QW) active region becomes trapped in the high index active region layer due to the total internal reflection and emitted photons become confined to waveguide modes. Then a small fraction of emitted light radiates away from the top surface [5].

One of the most promising techniques for the photonic structure fabrication is interference lithography [6]. There are different ways for generation of 2D patterns by interference lithography using a single exposure of an interference pattern generated by multiple beams or using multiple exposures of an interference optical field produced by two beams [6, 7]. This multi exposure technique can be used for fabrication of different one- and 2D photonic structures in photoresist materials as well as in different III-V compounds.

In this work we focus on the fabrication of 2D PhC structures prepared in LED surface. The interference lithography based on the two-beam geometry using multiple exposure process is used for the 2D PhC structure preparation in the semiconductor surface of different period, symmetry and shape of removed regions. The 2D PhC structure is applied in the LED surface in order to improve the light extraction from the LED. The surfaces of the prepared PhC LEDs are experimentally investigated by atomic force microscope (AFM). The near field of the diode is studied by near-field scanning optical microscope (NSOM) and the optical power on current dependence is measured [8].

2. Experimental

The LED samples have been grown by low-pressure MOVPE on (001) oriented n-type GaAs substrates. The structure consists of a 350 nm n-doped GaAs buffer layer, 1 300 nm n-doped $Al_{0.45}Ga_{0.55}As$ confinement layer, the QW active region and 650 nm p-doped upper confinement $Al_{0.60}Ga_{0.40}As$ layer. The active region contains three 9 nm thick GaAs QWs separated by 24 nm thick $Al_{0.2}Ga_{0.8}As$ barriers (Fig. 1a). The structure was covered by 65 nm GaAs cap layer.

The patterning of the 2D PhC structures in the LED surface was realized in few lithographic steps. First, the 600 nm thin standard positive photoresists AZ 5214E was deposited as a mask on the LED structure surface. The photoresist film was spin-coated with post-baking at 103 °C for 50 seconds.

In the next step, the sample was exposed by the 2D interference pattern. The 2D interference pattern was formed by the double exposure process with the sample in-plane rotation between individual exposures $\alpha = 90°$. The period of such 2D interference pattern was set to be $\Lambda = 2.4\ \mu m$ by the laser beam geometry.

After exposure, the sample was developed in the AZ 400K developer for 40 seconds, rinsed in DI water and dried with nitrogen. The open regions were etched in RIE mode in CCl_4/He based plasma in a ROTH & RAU MICROSYS 350 machine. The device fabrication was completed by lithography definition of contact regions and evaporation of top and bottom contacts. The upper Au electrical contact with the circular shape of the interior diameter 80 μm was realized.

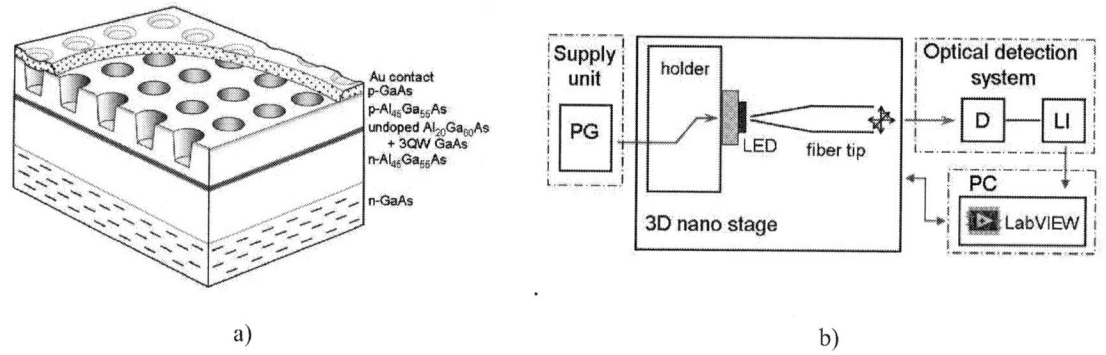

a)   b)

Fig. 1: *a) Schematic illustration of the LED structure with 2D PhC in the surface. b) The schema of the NSOM experiment, PG – pulse generator, D – detector, LI – Lock In.*

The emission properties were analyzed by the spectral characteristics, L(I) measurement, and the near-field pattern investigations of emitting diode. The spectral measurements were taken using Ocean Optics USB 2000 spectrometer. The L(I) dependencies were measured in the integrating sphere by using the calibrated Si detector in an impulse mode. The near field of the 2D PhC diode was investigated using the CDD camera with optical objective and the high-resolution near-field pattern was analyzed by the near-field scanning optical microscope (NSOM) with the metal-coated optical fiber tip as a probe (Fig. 1b) [8].

3. Results and discussion

The 2D PhC structure quality prepared in the LED surface was analyzed by employing AFM. The square 2D structure with the period of 2.45 μm was confirmed by the AFM analysis. The structure line profile documents the 320 nm depth of etched regions in the AlGaAs p-doped confinement layer.

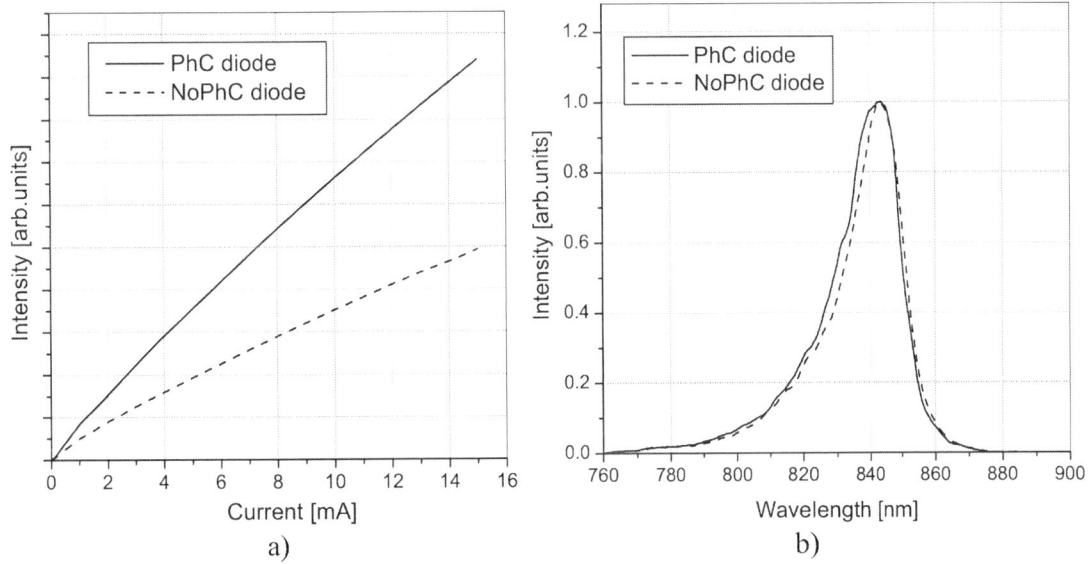

a)

b)

Fig.2: *a) Comparison of LI characteristics and b) the spectral characteristics of the PhC and no-PhC LED.*

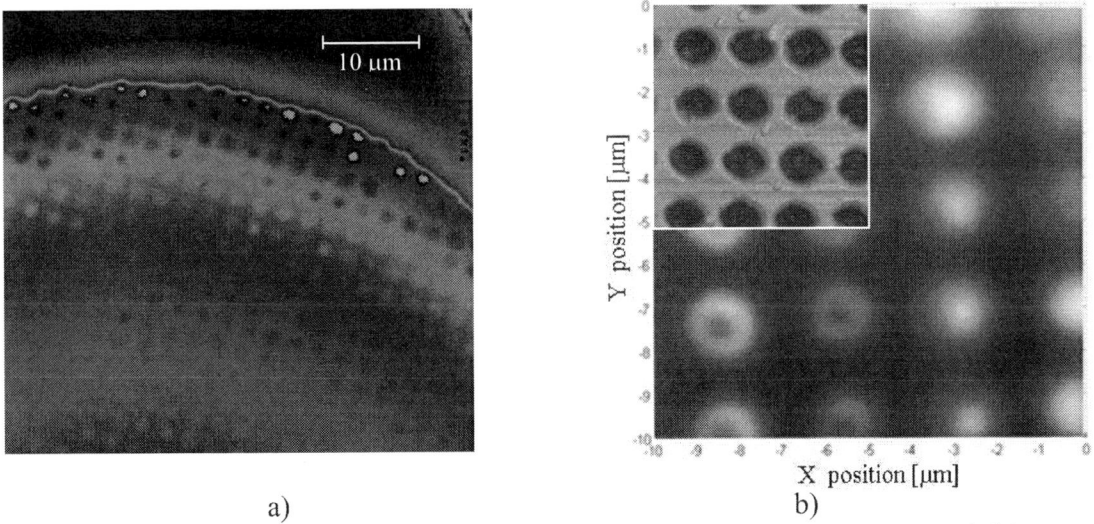

a)

b)

Fig. 3: *a) The near-field pattern of the PhC LED and b) high-resolution near-field pattern with the inset AFM analysis of the LED surface.*

The emission properties analyzed by the L(I) measurement are shown in the Fig. 2a and the spectral measurement in Fig. 2b. The near field of this diode was investigated by the CCD camera optical system (Fig. 3a) and the high resolution NSOM (Fig. 3b). The L(I) characteristics of the PhC and no-PhC LED were taken using integrated sphere and with calibrated Si detector. To minimize the influence of the edge emission, the edges of the diode were covered by wax. The L(I) characteristics document enhanced emission from the PhC LED in comparison with the no-PhC LED. The difference between PhC and no-PhC LED in the LI dependences increases with the current and the relative difference is from 60 to 80 %. The spectral measurements were obtained using the optical fibre with 10 μm core diameter perpendicularly placed to the LED surface. Comparison of the spectra from the PhC and no-PhC LED revealed the weak spectral broadening for the PhC LED, what is probably caused by the more effective irradiation of the guided light in active region. An optimized design of the 2D PhC structure in the surface should lead to the more efficient radiation outcoupling.

From the near-field measurements was found out, that more light is irradiated from the air hole regions in comparison with the surrounding area of the structure. The high-resolution NSOM image clearly revealed enhanced emission from the air hole regions as it is shown in the Fig. 3b. The enhanced emission on the periphery of the air hole region is probably caused by the light scattering on the slight oblique edges of the 2D PhC structure.

4. Conclusion

The 2D PhC structure on the GaAs/AlGaAs LED surface was prepared using interference method with two beam geometry of experiment. An improvement of the light emission for the diode with the PhC structure was documented by L(I) and spectral analysis. The near-field pattern investigated by NSOM documents the enhanced emission from the air hole regions of the 2D PhC structure. The decrease of the period of the PhC structure should lead to more considerable improvement of the emission outcoupling from such emitting structures.

Acknowledgement

The authors would like to thank Dr. S. Haščík for technical assistance. This work was supported by Slovak Academy of Sciences No. VEGA-1/0868/08, 1/0689/09, 1/0683/10.

References

[1]   A. David, C. Meier, R. Sharma, F. S. Diana, S. P. DenBaars, E. Hu, S. Nakamura, C. Weisbuch, H. Benisty, *Appl. Phys. Lett.* **87**, 2005.

[2]   H. Altug, J. Vuckovic, *Opt. Express* **13**, 2005.

[3]   M. L. Hsieh, K. C. Lo. Y. S. Lan, S. Y. Yang, C. H. Lin, H. M. Liu, H.C. Kuo, IEEE Photonic. Tech. L. **20, 2008.**

[4]   A.M. Adawi, R. Kullock, J.L. Turner, C. Vasilev, D.G. Lidzey, A. Tahraoui, P.W. Fry, D. Gibson, E. Smith, C. Foden, M. Roberts, F. Qureshi, N. Athanassopoulou, *Org. Electron* **7**, 2006.

[5]   J. J. Wierer, M. R. Krames, J. E. Epler, N. F. Gardner, M. G. Craford, J. R. Wendt, J. A. Simmons, M. M. Sigalas, *Appl. Phys. Lett.* **84**, 2004.

[6]   N. D. Lai, W.P. Liang, J.H. Lin, C.C. Hsu, C.H. Lin, *Opt. Express* **13**, 2005.

[7]   J. Skriniarová, D. Pudis, I. Martincek, J. Kovác, N. Tarjányi, M. Veselý, I. Turek, *Microelectron. J.* **38**, 2007.

[8]   D. Pudis, I. Martincek, I. Turek, J. Kovac, jr., J. Kovac, V. Gottschalch, B. Rheinländer, M. Dado, *Laser Phys.* **15**, 1623-1628 (2005).

100mV noise performances of Te-doped Sb-HEMT

A.Noudeviwa, A. Olivier, Y.Roelens, F.Danneville, *Member, IEEE,* N. Wichmann, N. Waldhoff, L.Desplanque, X. Wallart, and S.Bollaert

Institut d'Electronique de Microélectronique et de Nanotechnologie (IEMN), UMR CNRS 8520, Université de Lille I, BP 60069, 59652 Villeneuve d'Ascq Cedex, France.(phone:+333.20.19.79.79; fax:+333.20.19.78.92; e-mail: yannick.roelens@iemn.univ-lille1.fr)

In this paper, we present the noise measurement results of InAs/AlSb HEMTs at room temperature under very low drain bias (100mV) at 30GHz. Under these dc bias conditions the transistor exhibit NF_{min}=*1.56 dB and* G_{ass}=*5.3dB @30 GHz for* P_{dc}= *7.3μW/μm. These results are compared to our previous work and the great improvements observed open up the possibility to develop a 100mV electronics at room temperature.*

1. Introduction

During the last two decades, antimonide based heterostructures and related HEMTs have been studied [1] - [2]. InAs/AlSb heterostructure ability to combine high frequency performance and very low power operation makes this technology's transistors to be a potential candidate, featuring low-voltage, low-power consumption for high-speed analog and digital applications. Hence, at room temperature, X-Band Low noise amplifier (LNA) [3]-[5], Ka-band LNA [6], W-band LNA [7] have been demonstrated at V_{ds} below 0.5V. However very low drain bias (under V_{ds} = 100mV) LNA at room temperature haven't been demonstrated and in our previous work[8] we observed that at room temperature the studied device is not usable for LNA at room temperature inspite of his potential use at 77K. In this work we present noise measurements result at room temperature performed on our new devices [9]. These news devices present a better DC and RF performances [9] and huge noise parameters improvements are observed.a rule, the paper should be divided into chapters, each with a heading, so that the reader can follow the logical development of the work.

2. Device structure

The AlSb/InAs HEMT structure was grown in a RIBER Compact 21 MBE reactor equipped with elemental solid sources for group III and V materials on Semi-Isolating InP substrate using an AlSb metamorphic buffer. The structure consists of an AlGaSb/AlSb metamorphic buffer, a 150Å InAs channel layer, a 50Å AlSb spacer, a Te δ-doping, a composite Schottky barrier (60Å AlSb layer, 40Å $Al_{0.5}In_{0.5}As$ layer) and a 50Å Te doped InAs cap layer (figure 1). The AlInAs layer in the composite Schottky barrier avoids exposure of the highly reactive AlSb to air and acts as a hole barrier[9].

978-1-4244-8574-1/10 $26.00 © 2010 IEEE

Cap Layer	InAs (Te 8×10^{18} cm^{-3})	50 Å
Protection layer	Al$_{0.5}$In$_{0.5}$As	40 Å
Barrier layer	AlSb	**60 Å**
δ-doping plane	Te 5×10^{12} cm^{-2}	
Spacer layer	AlSb	50 Å
Channel layer	InAs	150 Å
Barrier layer	AlSb	500 Å
Buffer layer	Al$_{0.8}$Ga$_{0.2}$Sb	2500 Å
buffer	AlSb	7500 Å
	Al$_{0.5}$In$_{0.5}$As	1000 Å
SI Substrate	**InP**	

Fig.1   AlSb/InAs heterostructure

In the following, the geometry of processed HEMTs corresponds to a gate length equal to 120 nm, having 2 gate fingers of 20 μm gate width each and a source-drain distance equal to 1.3μm.

3.   RF performances

The studied HEMT exhibit extrapolated f_T and f_{max} performances of 103GHz and 83GHz respectively at V_{ds} = 100mV [9]. These results are great improvements from our previously reported results at room temperature [8] (see table.1) for antimonide based HEMT. These improvements are related to use of Tellurium in δ-doping plane instead of silicon and the aspect ratio optimization.

	f_T **(GHz)**	f_{max} **(GHz)**
Si-doped[8]	55	52
Te-doped	103	83

table.1: Comparison between 2 technologies f_T and f_{max}

4.   Noise results

This study was to assess AlSb/InAs HEMTs potentialities for very low drain bias applications at room temperature. The four noise parameters (NF_{min} -the minimum noise figure-, R_n –the equivalent noise resistance- and Γ_{opt} -the optimum source reflection coefficient-) were extracted using F_{50} method [10] at room temperature.

First, NF_{min} and associated gain (G_{ass}) at 30 GHz (Ka band) and V_{ds} =100mV are reported as a function of P_{dc} in Fig.2 at room temperature. NF_{min} as low as 1.56 dB is achieved for P_{dc}=7.3 μW/μm (this value is used as reference dc power for numerical data provided in the following). Moreover for this dc power G_{ass} is 5.3dB, indicating that the transistor is potentially usable in Ka Band at room temperature below V_{ds} = 100mV.

	T_{out} **(K)**	R_n **(Ω)**	NF_{min} **(dB)**	G_{ass} **(dB)**
Si-doped[8]	400	107	3.5	0.8
Te-doped	300	37	1.56	5.3

table.2 : the 2 studied devices noise parameters *@30 GHz*

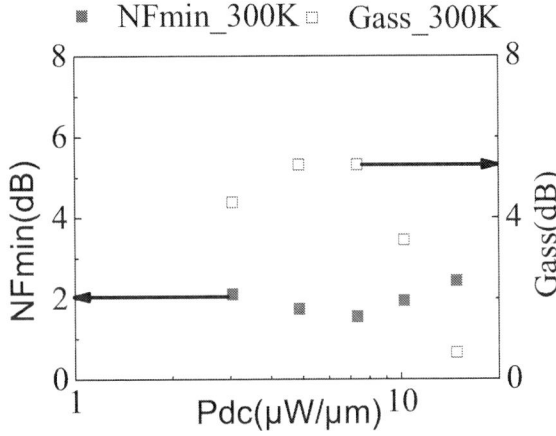

Fig.2 Minimum Noise Figure NF_{min} and associated Gain G_{ass} as a function of power consumption at 30 GHz

Γ_{opt} is reported as function of P_{dc} into the Smith chart (Fig.3). This plot shows an important feature, i.e., the noise matching conditions are not degraded with power dissipation variation.

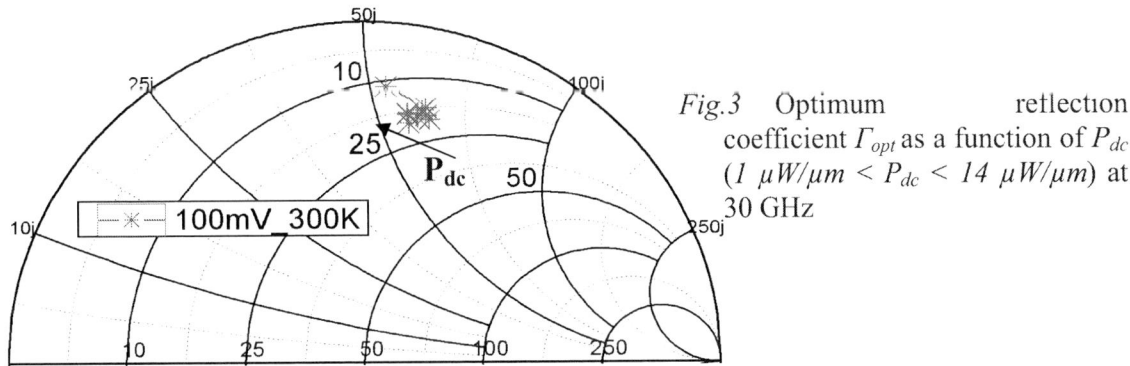

Fig.3 Optimum reflection coefficient Γ_{opt} as a function of P_{dc} ($1~\mu W/\mu m < P_{dc} < 14~\mu W/\mu m$) at 30 GHz

Moreover, in Fig.4, R_n is 3 times lower than observed value in our previous devices [8] in the same conditions (see table.2), indicating a much lower sensitivity to noise mismatching. NF_{min} and R_n improvements with this new device are mainly explained because of the better RF performances (higher f_T) and the lower T_{out} (see analytical expressions in [11]) which decrease from 400K to 300K between our previous device and this one.

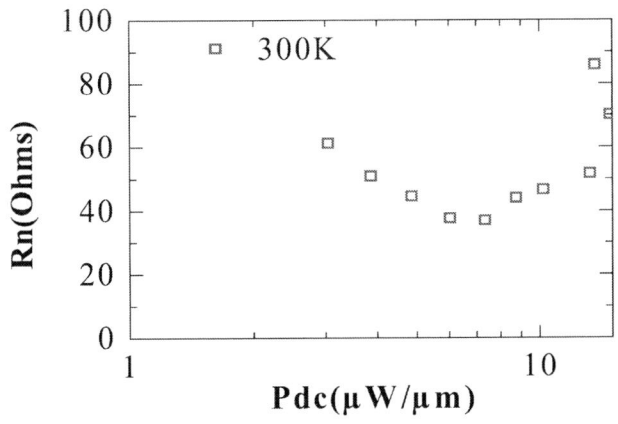

Fig.4 Noise equivalent resistance as a function of power consumption at 30 GHz

5. Conclusion

In this work, trends on AlSb/InAs HEMTs noise performances were investigated at room temperature below V_{ds} = 100mV. When compared to Si doped AlSb/InAs HEMTs noise measurements results at room temperature [8]. The studied device exhibit better results under 100mV drain bias (f_T =103 GHz, f_{max}= 83 GHz, NF_{min}=1.56 dB and G_{ass}=5.3dB @30 GHz, for P_{dc}= 7.3μW/μm). This excellent tradeoff observed between noise figure, Gain, and power consumption, it is expected to realize 100 mV (dc Biased) 300K Sb-HEMTs LNA that will feature very low power consumption.

Acknowledgments

This work was partially supported by CNES (Centre National d'Etudes Spatiales).

References

[1] C. Chang et al., "Electron densities in InAs/AlSb quantum wells" *Journal of Vacuum Science & Technology B: Microelectronics and Nanometer Structures*, vol. 2, Avr. 1984, pp. 214-216.

[2] Tuttle G. et H. Kroemer, "An AlSb/InAs/AlSb quantum well HFT" *IEEE Transactions on Electron Devices*, vol. 34, 1987, p. 2358.

[3] R. Tsai et al., "Metamorphic AlSb/InAs HEMT for low-power, high-speed electronics," *Gallium Arsenide Integrated Circuit (GaAs IC) Symposium, 2003.* pp. 294-297.

[4] B. Buhrow et al., "A low power AlSb/InAs HEMT X-band low noise amplifier," *International Conference on Indium Phosphide and Related Materials* 2005, pp. 617-620.

[5] W. Deal et al., "A Low Power/Low Noise MMIC Amplifier for Phased-Array Applications using InAs/AlSb HEMT," *IEEE MTT-S Microwave Symposium, 2006,* pp. 2051-2054

[6] J. Hacker et al., "An ultra-low power InAs/AlSb HEMT Ka-band low-noise amplifier," *IEEE Microwave and Wireless Components Letters,* vol. 14, 2004, pp. 156-158

[7] J. Hacker et al., "An ultra-low power InAs/AlSb HEMT W-band low-noise amplifier," *IEEE MTT-S Microwave Symposium 2005,* pp. 1029-1032

[8] A.Noudeviwa et al., *"Sb-hemt: towards 100mv cryogenic electronics," IEEE Trans. on Electron Devices, Vol. 57, no. 8,* pp. 1903-1909 *August 2010,*

[9] A.Olivier et al., *" High Frenquency Performance of Tellurium−Doped AlSb/InAs HEMTs at Low Power Supply"* Eumic 2010. *To be published*

[10] G. Dambrine, H. Happy, F. Danneville, and A. Cappy, *"a new method for on-wafer noise measurement"*, IEEE Transactions on microwave theory and techniques, vol. 41, no. 3, pp. 375-381, March 1993

[11] M. Pospieszalski, *"Modeling of noise parameters of MESFETs and MODFETs and their frequency and temperature dependence,"* IEEE MTT-S 1989, vol.1, pp. 385-388.

Development of Advanced Gunn Diodes and Schottky Multipliers for High Power THz sources

F. Amir, C. Mitchell and M. Missous

M&N Group, School of E&EE, University of Manchester, M60 1QD, UK
email: faisal.amir@alumni.manchester.ac.uk and m.missous@manchester.ac.uk

An advanced step-graded Gunn diode (~ 100 GHz fundamental frequency) has been developed using a joint modelling-experimental approach to test GaAs based Gunn oscillators at sub-millimetre wavelengths. These devices are to be used as high power (multi-mW) Terahertz sources in conjunction with multipliers using Schottky diodes as the non-linear elements. The modelled-measured results of low series resistance Schottky diodes with non-alloyed contacts are also discussed.

1. Introduction

The quest continues to close the frequency gap between microwaves and infrared regions of the electromagnetic spectrum, often called the 'THz Gap'. The demand for high-frequency and high output power has warranted focused research in this field. The two semiconductor devices used to achieve this goal, namely Gunn diodes (initial source) and Schottky diodes (non-linear element in the multiplier) have both been modelled in Silvaco using the Virtual Wafer Fabrication (VWF) simulation environment [4] and compared to experimental data.

The Gunn diode modelling work described in this paper follows on from previously reported work [1 - 3] on advanced GaAs Gunn diodes with graded gap AlGaAs hot electron injection. After successful development of 77 GHz (second harmonic) and 125 GHz (second harmonic) models, the computational work focused on increasing the device operating frequency to ~100 GHz fundamental. The main motivation behind this work was to model and test GaAs based Gunn oscillators operating at sub-millimetre waves whilst studying the underlying device physics. The development of such devices, when used with multipliers as high power THz sources, should provide much needed components for use in sub-mm wave imaging applications, providing a sound foundation for future high-frequency high-power multiplier source development in the THz regime.

2. Device physical modelling

The physical models of both the Gunn and Schottky diodes have been developed using the commercial computer aided physical model Silvaco [3, 4]. This is a physics-based platform which allows the electrical and thermal behaviour of a semiconductor device to be simulated under given bias conditions to obtain the DC and time-domain responses.

Models for frequency multipliers are being developed, operating up to ~ 0.6 THz as high-power (multi-mW) and high-efficiency sources. A harmonic balance simulation tool has been created using AWR Microwave Office, specifically for analysis and prediction of GaAs Schottky diode performance given the device's basic electrical characteristics. The tool will be used to study the effects of operating conditions on the multiplier performance, as well as calculating the optimum embedding impedances in the multiplier circuit.

3. Results and Discussion

The simulated DC characteristics for both Gunn diode and Schottky diode are presented here and compared with experimental data. Also presented are time-domain simulations for a novel 100 GHz fundamental Gunn diode that modelled the frequency of oscillation. These simulations also showed the onset of oscillations at low bias voltages. This data did not attempt to predict measured frequency in practical Gunn diodes devices where the frequency is cavity controlled to a degree.

3.1   Schottky Diode Varactor

In the multiplier, a Schottky diode varactor creates harmonic oscillations of the input signal. The Schottky diode epitaxial structure is that of a conventional GaAs Schottky diode, but capped with a heavily silicon doped ($5x10^{18}$ cm^{-3}) graded layer of In$_x$Ga$_{1-x}$As (x=0 to 0.53) and a similarly doped, constant composition In$_{0.5}$Ga$_{0.5}$As layer to form a very low resistance non-alloyed Ohmic contact. These devices were developed using novel process techniques to reduce device parasitics, process complexity and series resistance. These processes have allowed the design of new devices in collaboration with multiplier design requirements across a wide range of chip size and device combination totalling over 700 individual device structures (such as figure 1c). The junction capacitance of such diodes have been extracted from RF testing indicating 4.8 fF at breakdown and a C$_{jo}$ of ~15 fF (figure 1a). These devices gave a theoretical cut-off frequency in excess of 1.1 THz. Initial work on InGaAs 2-DEG Schottky diodes has also begun.

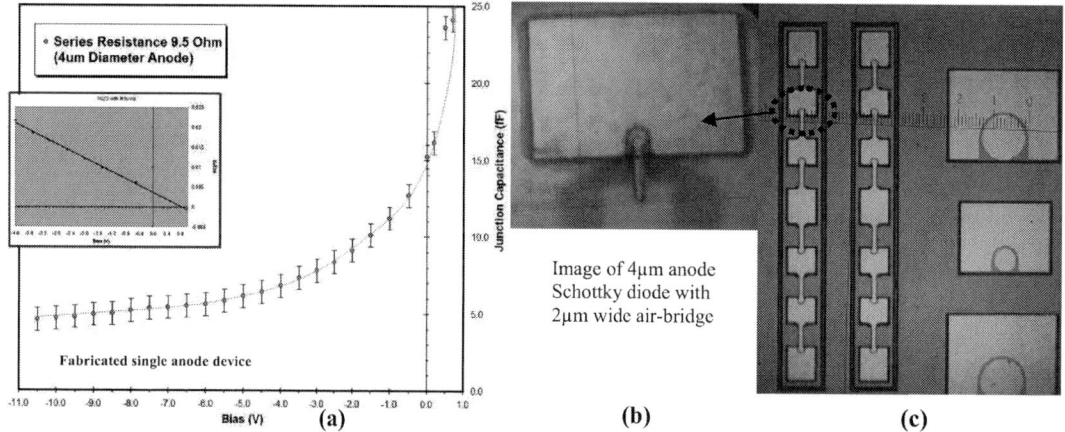

Figure 1   (a) Junction capacitance (bias - 1/C$_j^2$ inset) (b) Schottky test diode (c) 6-anode multiplier circuit

A Schottky diode 2D model has been developed in Silvaco. Figure 2 shows modelled versus measured results matching very well. The model gives a series resistance of 6 Ohms, substantially less than the measured value of 9.5 Ohms. This is attributed to external device resistances which will be incorporated as lumped elements in future models. The model will be used to study underlying device physics and extract parameters such as ideality factor, zero voltage junction capacitance, built in voltage and breakdown voltage. These parameters will be used in harmonic balance simulations created using AWR Microwave Office, which will allow analysing and prediction of the GaAs Schottky diode performance at mm-wave frequencies.

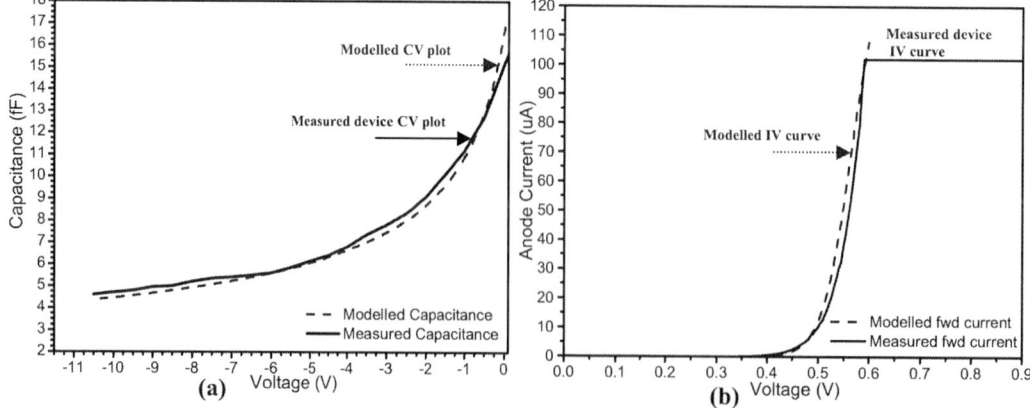

Figure 2   (a) CV plot for a 4 μm diameter anode Schottky diode (b) Forward bias IV curve

3.2   100 GHz fundamental Gunn Diode

The epitaxial structure and conduction band diagram for a GaAs Gunn diode with graded-gap AlGaAs hot electron injection is shown in Figure 3(a). Epitaxial growth of the Gunn diode wafers themselves is performed using an Oxford Instruments V100+ MBE reactor. Figure 3(b) shows the geometry of the 2D model used. To realistically represent the manufactured device which has a cylindrical structure, cylindrical mesh geometry is defined and applied to the 2D model.

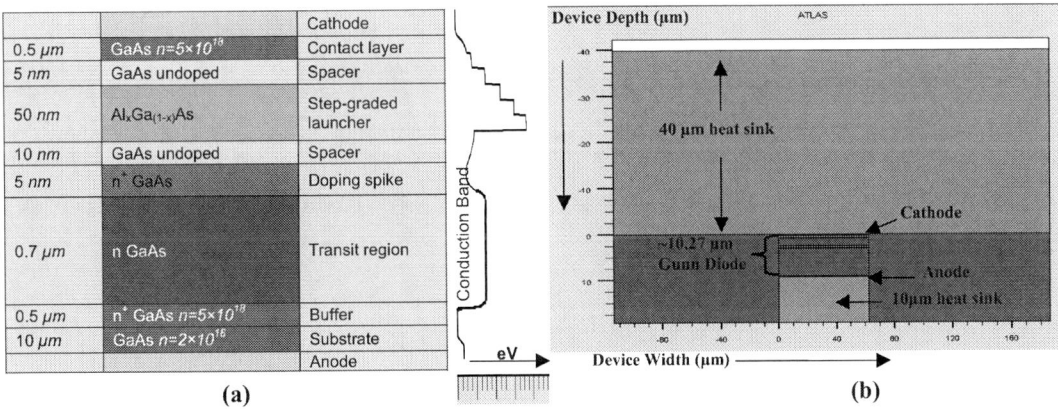

Figure 3   (a) GaAs Gunn diode structure (b) A 2D model showing the electroplated Gold heat sinks.

The simulated forward and reverse-bias IV characteristics of a 0.7 μm transit region length device are shown in Figure 4(a) and match extremely well with measured data thus validating the choice of the physical models and material parameters used. The asymmetry between the forward and reverse bias IV characteristic is due to the effect of the injector on the transit region electric field profile [3]. Figure 4(b) also shows the effects of ambient temperature variation on the simulated and measured results where again a close match is seen. This further validates the model taking into account device self heating effects by using the GIGA simulator [4, 5] within Silvaco.

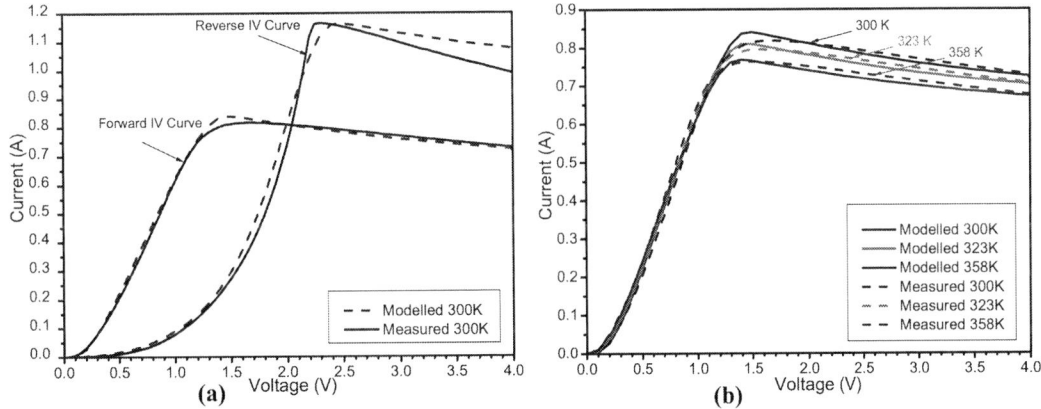

Figure 4   (a) Measured & simulated fwd-rev IV curves (b) Fwd IV characteristics at 300K, 323K & 358K.

The time-domain simulations in figure 5a show stable oscillation at 106 GHz. The manufactured device gave a maximum fundamental power of ~22 mW at 94 GHz the highest ever reported for a GaAs Gunn operating in fundamental mode at this frequency. It can be seen that the simulated time-domain response for the device is noisy at this bias (1.75 V). At higher external bias (2.75 V) the free-running oscillation in the time-domain response (figure 5b) is now lower in frequency (96 GHz) and does not exhibit the noise seen at lower external bias. This is in agreement with [6] regarding the behaviour of free-running Gunn devices as bias voltage is increased.

978-1-4244-8574-1/10 $26.00 © 2010 IEEE

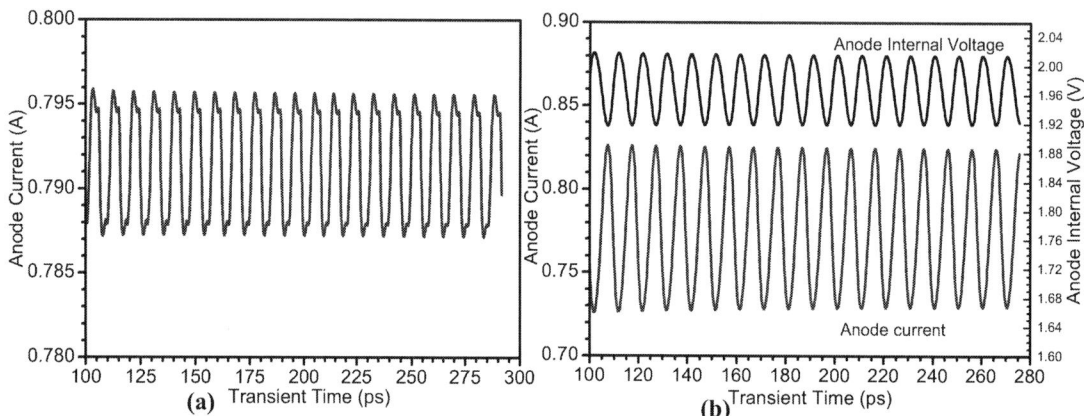

Figure 5 (a) Time-domain stable oscillation at 1.75 V (b) Anode current and anode internal voltage stable oscillations at 2.75 V.

4. Conclusion

Precise methods for accurately determining device performance of Gunn diode and Schottky diode from physical modelling using Silvaco have been discussed. Both models have shown close matches between measured and simulated results validating the choice of physical models and parameters used. The work presented here has provided a sound foundation for future high-frequency high-power multiplier source development in the Terahertz regime.

Future work includes the development of a full functional multiplier to model the interaction between the Gunn diode and a resonant circuit in order to study device operation in a cavity. Harmonic balance simulations along with electromagnetic simulators to design and tune Schottky varactor circuit parameters for optimal circuit performance are on-going. This should then permit the simultaneous design and optimisation of the frequency multiplier.

Acknowledgement

The UK Science and Technology Facilities Council (STFC) and the UK MoD Defence Science and Technology Laboratory (DSTL) are gratefully acknowledged for supporting this programme through the PIPSS scheme. Faisal Amir is funded on a scholarship from National University of Sciences and Technology (NUST, Pakistan). Thanks are due to the IEEE Electron Devices Society for the award of a PhD fellowship to Faisal Amir in 2009. Thanks are also due to Mike Carr (e2v Plc, UK) for providing the experimentally measured data for the Gunn diodes.

References

[1] F. Amir, C. Mitchell, N. Farrington, and M. Missous, "Advanced Gunn Diode as High Power Terahertz Source for a Millimetre Wave High Power Multiplier," Proc. SPIE, vol. 7485, 748-50I, 2009.

[2] F. Amir, C. Mitchell, N. Farrington, and M. Missous, "Advanced Step-graded Gunn Diode for mm-wave Imaging Applications," 5th ESA Workshop on Millimetre Wave Technology and Applications and 31st ESA Antenna Workshop 18-20 May, ESTEC, Noordwijk, The Netherlands, pp. 201-205, 2009.

[3] F. Amir, N. Farrington, T. Tauqeer, and M. Missous, "Physical Modelling of a Step-Graded AlGaAs/GaAs Gunn Diode and Investigation of Hot Electron Injector Performance," Advanced Semiconductor Devices and Microsystems (ASDAM 2008), 12-16 October, Int. Conf., pp.51-54, 2008.

[4] Silvaco International Atlas User's Manual, [Device Simulation Software], Software version 5.16.3.R 6 July, Santa Clara, CA 95054, 2010.

[5] G. K. Wachutka, "Rigorous thermodynamic treatment of heat generation and conduction in semiconductor device modelling," IEEE Transactions on Computer-Aided Design of International Circuits and Systems, vol. 9, pp. 1141-1149 1990.

[6] N. R. Couch, H. Spooner, P. H. Beton, M. J. Kelly, M. E. Lee, P. K. Rees, and T. M. Kerr, "High-performance, graded AlGaAs injector, GaAs Gunn diodes at 94 GHz," IEEE Electron Device Letters July, vol.10, no.7, pp. 288-290, 1989.

3-D Simulation of a 45 nm Partially Depleted Silicon on Insulator (SOI) Transistor with Diamond-Shaped Body Contact

Arash Daghighi[1], Azar Farajzadeh[2]

1- Assistant Professor, Faculty of Engineering, Shahrekord University, Shahrekord, Iran.

E-mail: daghighi-a@eng.sku.ac.ir

2- Young Research Club, Islamic Azad University Najafabad Branch, Isfahan, Iran.

The improvement in output conductance of a 45 nm Partially Depleted (PD) SOI MOSFET with diamond-shaped body contact (DSBC) is shown. The results of 3-D simulations of current drive and body potential for the conventional and DSBC devices demonstrate suppression of floating body effects. DSBC device was compared with conventional body contacted structure and a reduction of small-signal output conductance (g_{ds}) by 24% was observed. The transition frequency of output conductance related to the body resistance (R_{body}) in a DSBC structure is increased by 2.4 times of its value in conventional body contacted transistor. These improvements represent superior intrinsic gain of the DSBC SOI MOSFET in higher frequencies in comparison with conventional body contacted device.

1. Introduction

Silicon-on-insulator (SOI) CMOS offers functionality improvement over bulk CMOS [1]. The presence of buried oxide layer results in a floating body region in partially depleted (PD) SOI MOSFET. Floating body device design complicates device performance and suffers from several floating body modulation effects [2]. Body contacts are used to make the body of PD SOI devices grounded [2]. Body resistance (R_{body}) has a great influence on dc and ac terminal characteristics of the device. Diamond shaped body contact is used as an area efficient body contact to suppress the floating body effects [3]. In this paper, the improvement in output conductance of a 45nm PD SOI MOSFET with diamond-shaped body contact is presented.

2. 3D Simulation Results

Exact inspection of the terminal characteristics of the body contacted PD SOI MOSFET requires full 3-D simulation to take into account the body potential variation along the device width. Fig. 1 shows the 3-D structure of SOI MOSFET with the conventional body contact. As it can be seen, in conventional body contacted device, the p^+ implanted region in the source of the transistor is connected to the body. The DSBC structure is depicted in Fig. 2. In this structure, two p^+ implanted regions with the diamond shape are used [3]. This method of making the body contact has the advantage of increasing the drain current and reduction of the occupied silicon area [3].

Non-isothermal drift-diffusion model was used for 3-D simulations [4]. In order to compute the output conductance, an AC sinusoidal voltage is applied to the drain terminal and the drain current response was measured. Three transistor structures were simulated: (1) a floating body, (2) a conventional body contacted and (3) a DSBC. The graph of I_{DS} vs V_{DS} is shown in Fig. 3. As it can be seen, the kink effect is observed in the output characteristic of the floating body transistor. The kink is eliminated in the transistors with the body contact. As drain voltage increases more than 1 V, the drain current of conventional

978-1-4244-8574-1/10 $26.00 © 2010 IEEE

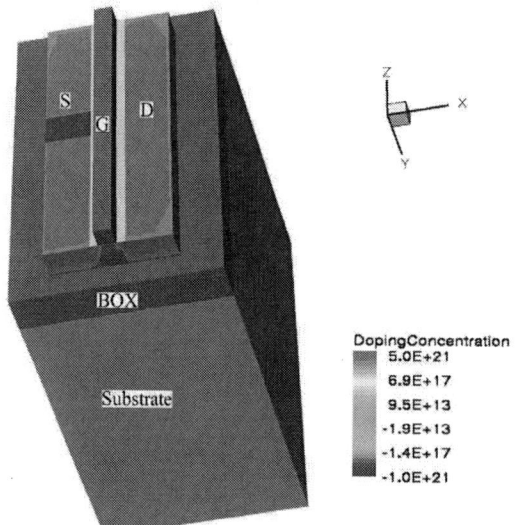

Fig. 1: 3-D structure of a conventional body-tied-source PD SOI MOSFET

body contacted device gradually increases. This is due to the reduced threshold voltage through the elevated body potential.

Body potential variations along the device width for the three structures are shown in Fig. 4. As can be seen, the DSBC device body voltage is lower than the conventional body contacted structure by 48%. Therefore, as a result of superior performance in controlling the body voltage, floating body effects are suppressed in DSBC device.

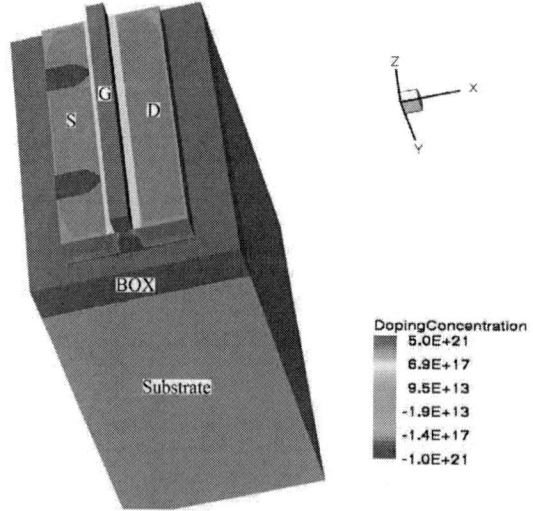

Fig. 2: 3-D structure of a diamond shaped body contacted SOI MOSFET

The key MOSFET characteristics in design of analog integrated circuits in addition to drain current are output conductance and gate trans-conductance. Fig. 5 demonstrates the simplified equivalent circuit of PD SOI transistor seen from the drain terminal while the gate and source are grounded. It can be shown that the frequency dependent relation of output conductance has a pole at [5]:

978-1-4244-8574-1/10 $26.00 © 2010 IEEE

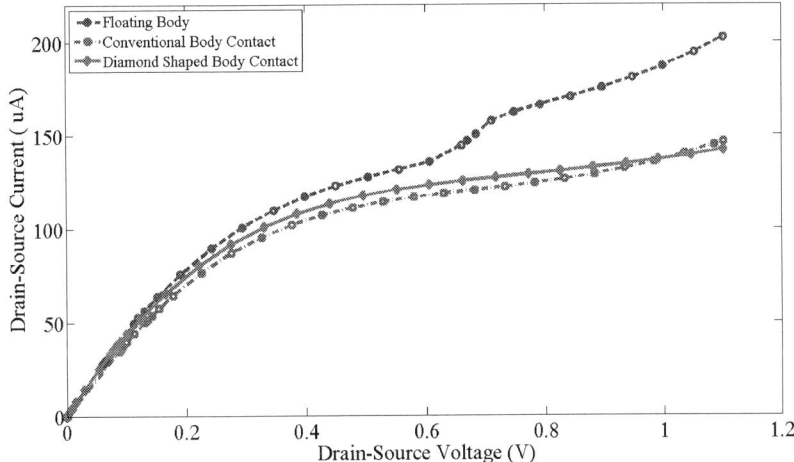

Fig. 3: Drain-source current vs. drain-source voltage for the three structures,
V_{GS}=1.1 V, W=0.4 μm and L=45 nm

$$f_p = \frac{1}{2\pi R_{body}(C_{bd} + C_{bs})} \qquad (1)$$

where C_{bd}, C_{bs} and R_{body} denote body-drain capacitance, body-source capacitance and body resistance, respectively. Eq. 1 shows the direct dependency of the pole frequency (f_p) to the body resistance. Fig. 6 shows output conductance variations with frequency for the DSBC and conventional body contacted devices. As it can be seen, the DSBC device faces a reduction in output conductance by 24% compared to conventional body contacted structure. Reduction in output conductance in these transistors translates to higher gain for analog circuits. The g_{ds} transition for the conventional body contacted device occurs at f_1=6.8 GHz. This transition moves to higher frequencies for the DSBC structure (f_2=16 GHz). The higher frequency for the g_{ds} transition is predicted by Eq. 1 since the body resistance (R_{body}) is smaller for the DSBC device in comparison with the conventional body contacted structure. Therefore, DSBC device can be used for higher gain in wider bandwidths in comparison with the conventional body contacted structures.

3.   Conclusion

Functionality of a 45 nm PD SOI diamond-shaped body contacted device in current drive, body voltage and output conductance was investigated. It was shown that higher intrinsic gain and wider output conductance bandwidth are achievable using DSBC structure.

References
[1] G.G. Shahidi, IBM J. RES. & DEV., Vol. 46, No. 2/3, March/May 2002, pp. 121-131
[2] JP Colinge, Kluwer Academic Publishers, 2004.
[3] A. Daghighi et al, Solid-State Electronics, Vol. 52, pp. 196-204, 2008
[4] DESSIS Manual. ISE Integrated System Engineering, Version 10.0; 2004.
[5] D. Lederer, Semicond. Sci. Technol., 20, pp. 469-472, 2005.

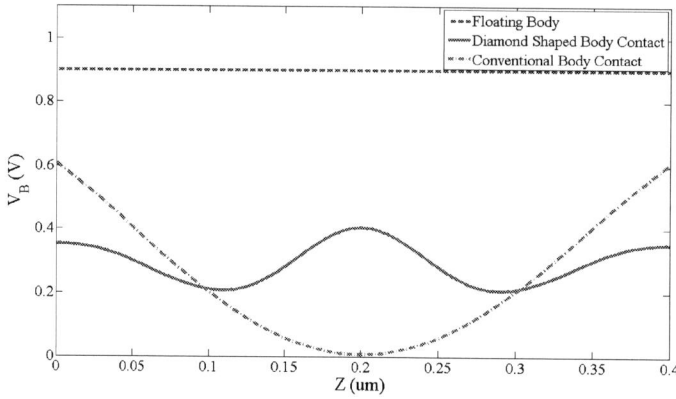

Fig. 4: Body potential distribution along the device width for the three structures, $V_{GS}=1.1V$, $V_{DS}=1.1$ V

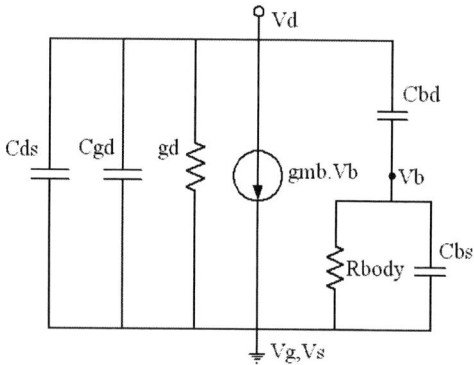

Fig. 5: Equivalent circuit of PD SOI MOSFET seen from drain terminal while the gate and the source are grounded.

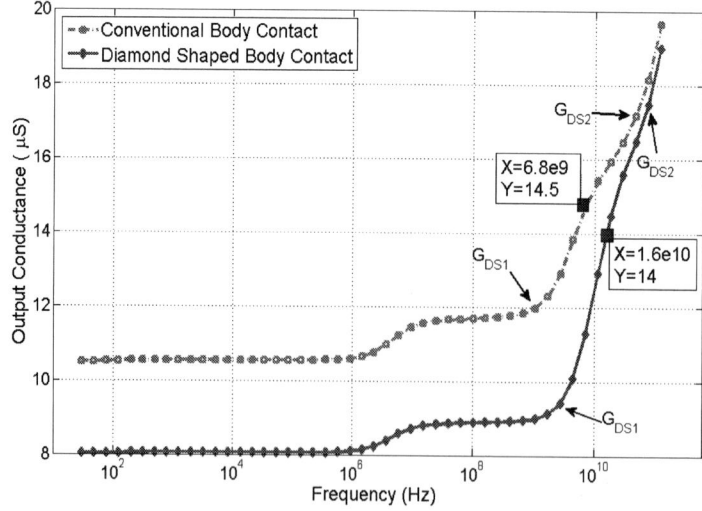

Fig. 6: Variation of output conductance with frequency for the conventional body contact and diamond shaped body contact

Electrical and optical properties of ZnO/Si photodiodes with embedded CdTe and CdSe/ZnS nanoparticles

Hotový, J., Kováč, J., Škriniarová, J., Novotný, I., Jakabovič, J., Kováč, J. jr.

Department of Microelectronics, Slovak University of Technology Bratislava,
Ilkovicova 3, 812 19 Bratislava, Slovakia
e-mail: jurajhotovy@gmail.com

Photodiodes based on ZnO/Si heterostructures were fabricated by sputter deposition of polycrystalline n-ZnO films on p-Si substrates. CdTe and CdSe/ZnS nanoparticles were embedded at the junction in between Si substrate and ZnO thin films. The effect of nanoparticles embedding on electrical and optical properties of ZnO/Si photodiodes has been studied. I-V and photocurrent spectra measurements revealed that embedding of nanoparticles significantly lowers the dark current of ZnO/Si photodiodes and also increases light absorption at around 600 nm for ZnO/Si photodiodes.

1. Introduction

In general, there are two major purposes of nanoparticles embedding in optoelectronic devices. Based on the localized surface plasmon theory, small metal nanoparticles enhance light absorption in thin-film solar cells [1,2] and light emission in LEDs [3]. Light scattering effects of larger metal or dielectric nanoparticles are utilized to increase light trapping inside thin-film solar cells [4,5].

High quality of the interface between ZnO and substrate is the crucial factor towards achieving superior properties of the photodiode. Several research teams have reported formation of the SiO_2 layer (thickness in nm range) in the Si/ZnO interface [6,7]. Therefore, electrons generated by visible photons may face a transport barrier, which suppresses the electron current and thus lowers total photocurrent [6]. In forward bias, the increase of serial resistivity of the diode has been revealed [7]. Another issue is the presence of the interface states caused by defects in the metallurgical junction. Embedding the nanoparticles in between ZnO layer and the substrate is supposed to modify the ZnO/substrate interface. It is a promising method, which might lead to the enhancement of some overall properties of the photodiodes.

2. Experimental

Silicon substrates (<111>; p-type concentration of $2.5\,(10^{18}\ cm^{-3})$) have been used to deposit n-type ZnO thin films by RF sputtering in planar diode Perkin-Elmer Sputtering System 2400 8/L. The solution of nanoparticles was spilled onto the surface of Si substrates (see **Table 1**). The solution of nanoparticles had been prepared by diluting nanopowders (produced by *PlasmaChem Ltd.*) in isopropylalcohol. The nanopowders consist of luminescent inorganic nanocrystals of CdTe and CdSe/ZnS with wavelength of emission maximum at *570* (nm) and *630* (nm), respectively. Wavelength of emission maximum (± 5 nm tolerance is given by the producer) is the function of nanocrystals size. In order to keep the equal preparation conditions, all ZnO thin films were sputtered at the same time at RF power of *600* (W), using $ZnO:Al_2O_3$ target with 2 wt.% Al. (6'' in diameter, with 99.99% purity). The base pressure was 10^{-4} Pa and the pressure of working atmosphere was maintained constant during the sputtering at *1.33* (Pa). In order to fabricate photodiodes lithography processes were employed. Top circle-shaped ohmic contacts were made of vapour-evaporated aluminium. Afterwards the final structures were MESA etched, cut into small pieces and bonded on the sockets.

The crystal structure of the ZnO thin films was identified with a Theta Theta Diffractometer Siemens D5000 with a Goebel mirror in grazing incidence geometry with CuKα radiation at the TU Ilmenau. The surface morphology of the ZnO thin films was observed by atomic force microscopy (AFM) using Park Systems XE-100 under normal air conditions, operating in non-contact mode. Electrical properties of ZnO thin films deposited on the Corning glass substrates were investigated by Hall measurements (Van der Pauw method) at room temperature.

Electrical and optical properties of ZnO/Si photodiodes were studied by I-V and photocurrent spectra measurements using the digital multimeter Keithley2400 and the monochromator SOLAR TII MSDD 1000, respectively. I-V measurement was performed at dark condition and at a lower (L1) and higher (L2) intensity of the halogen light source (in graphs, curves are labelled as *dark*, *L1* and *L2*, respectively). Every publicized characteristic represents behaviour, which is typical for all samples from the particular set of samples.

3. Results and discussion

3.1 Structural and electrical properties of ZnO thin films

The electrical parameters of the samples are summarized in **Table 1**. The structural properties of the ZnO thin films are revealed in the diffractogram (**Fig.1**), acquired by X-ray diffraction (XRD) measurements. The dominance of (002) diffraction peak indicates that the ZnO thin films are polycrystalline with hexagonal structure and have a good c-axis orientation. The values of the full-width at half-maximum of the (002) and (103) diffraction peaks are 0.55° and 1.14°, respectively. The position of (002) peak is shifted to lower 2θ angle of 34.017°; the 2θ angle for the bulk ZnO without stress is equal to 34.431° according to the XRD data of the International Centre for Diffraction Data (ICDD) [8]. The AFM 3D image in **Fig.2** shows the surface morphology of the ZnO thin films over a scale of 2 μm × 2 μm. The surface morphology shows polycrystalline structure, as well as grain size is evident from the picture. The estimated value of RMS roughness was *5.68* (nm).

Table 1 List of samples with the parameters of the ZnO thin films

sample no.	embedded nanoparticles	emission maximum of nanoparticles	Electrical parameters of ZnO thin films		
			resistivity [Ωcm]	concentration [cm^{-3}]	Hall mobility [cm^{2}V^{-1}s^{-1}]
S	-	-			
nano1S	CdTe	630 nm	8.5×10^{-2}	1.2×10^{19}	6
nano2S	CdSe/ZnS	570 nm			

978-1-4244-8574-1/10 $26.00 © 2010 IEEE   38

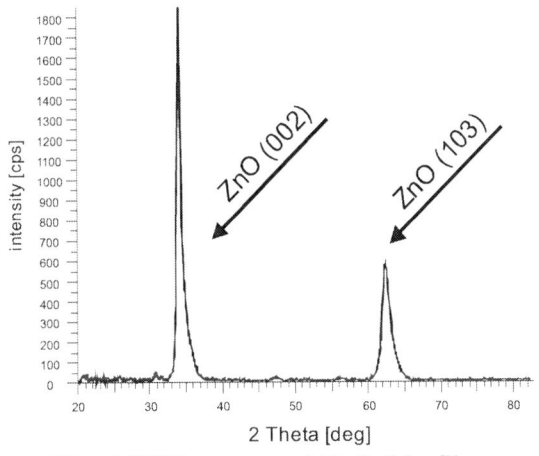

Fig. 1 XRD pattern of ZnO thin films

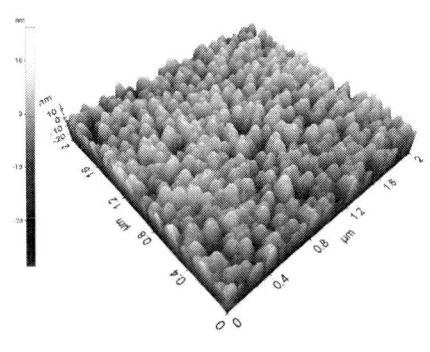

Fig. 2 AFM image of ZnO thin films

3.2 Electrical and optical properties of ZnO/Si photodiodes

All fabricated photodiodes exhibited rectifying behaviour and were light sensitive. The illumination in forward bias influenced also currents of the samples S without nanoparticles (**Fig.3**), as well as the samples nano1S and nano2S with embedded nanoparticles (**Fig.4**). This effect may be explained by change of the ZnO layer photoconductivity. The photodiodes with embedded nanoparticles had considerably lower dark current of 8.75 (10^{-4} A/cm^2) under reverse bias of -5 (V)) in comparison with the photodiodes without nanoparticles (**Fig.5**) of 1 (10^{-2} A/cm^2) at -5 (V). It means that embedding of nanoparticles lowered the dark current density by more than 1 order of magnitude. There are also differences in forward bias. The samples nano1S and nano2S achieved the same values of the forward current density at lower voltages in comparison with the sample S.

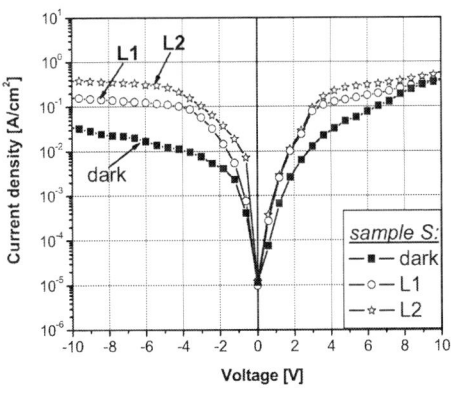

Fig.3 I-V characteristics of the photodiodes without nanoparticles

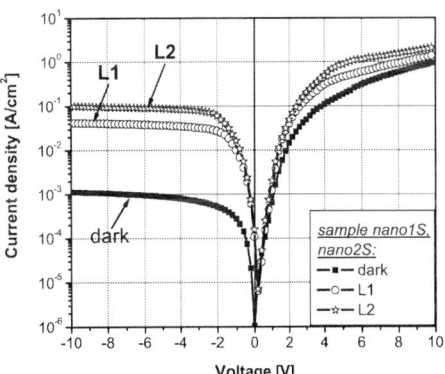

Fig.4 I-V characteristics of the photodiodes with nanoparticles

Fig.6 depicts normalized photocurrent spectra of ZnO/Si photodiodes. The photocurrent spectrum of photodiodes with both types of nanoparticles (samples nano1S - with CdTe nanoparticles; samples nano2S - with CdSe/ZnS nanoparticles) had one noticeable increased sensitivity at around 630 (nm), whereas the spectrum of the photodiodes without embedded nanoparticles (samples S) exhibited more local peaks with maximum at ~815 (nm). It implies that the embedded nanoparticles with the emission wavelength of *570* and *630* (nm) strengthen the absorption around *600* (nm), resulting in the increase of generated photocurrent at this wavelength region. When we compared the photocurrent spectra of the photodiodes with nanoparticles, the photocurrent magnitude corresponding to absorption at around *600* (nm) for the samples nano2S is higher than for the samples nano1S, relatively to the region at around *800* (nm).

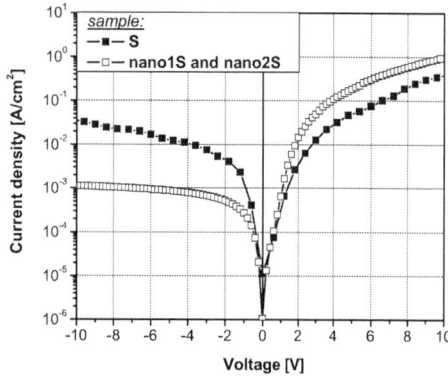

Fig.5 Comparison of dark current densities
of the photodiodes

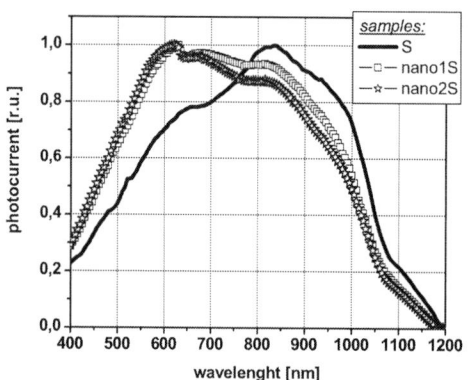

Fig.6 Photocurrent spectra of the photodiodes

4. Conclusions

Embedding of nanoparticles in between Si substrate and ZnO layer considerably influence on the electrical and optical properties of n-ZnO/p-Si photodiodes. Based on the comparison of dark current densities of the photodiodes with and without nanoparticles, it may be concluded that n-ZnO/p-Si photodiodes with embedded nanoparticles demonstrated lower saturated dark leakage current density in comparison to those without nanoparticles and increased light absorption at around *600* (nm).

Acknowledgement

The authors would like to thank T. Kups and L. Spiess from TU Ilmenau for the help with the XRD measurements. This work was supported by projects VEGA 1/0689/09 and MSMT CZ project 1M060310.

References

[1] S. Pillai, M.A.Green, *Solar Energy Materials & SolarCells*, vol. 94, 2010, pp.1481-1486
[2] E. Moulin *et al., Journal of Non-Crystalline Solids*, vol. 354, 2008, pp. 2488-2491
[3] B. Butun *et al., Photonics and Nanostructures - Fundamentals and Applications*, vol. 5, PHOREMOST Special Issue on Advances in Nanophotonics, Oct. 2007, pp. 86-90
[4] E. Moulin *et al., Thin Solid Films*, vol. 516, 2008, pp. 6813-6817
[5] Yu. A. Akimov, W. S. Koh, S. Y. Sian, S. Ren, *Appl. Phys. Lett.*, vol. 96, 2010, pp. 073111
[6] I.-S. Jeong *et al., Appl. Phys. Lett.*, vol. 83, no. 14, 2003, pp. 2946-2948
[7] J. Škriniarová *et al., Journal of Physics: Conference Series*, vol. 100, 2008
[8] M. Cui *et al., Vacuum*, vol. 8, 2007, pp. 899–903

Fabrication of Novel High Frequency and High Breakdown InAlAs-InGaAs pHEMTs

M. Mohamad Isa[1], D. Saguatti[2], G. Verzellesi[2], A. Chini[3], K. W. Ian[1] and M. Missous[1]

[1]M&N Group, School of E&EE, The University of Manchester, Manchester, UK
[2]DISMI, Università di Modena e Reggio Emilia, Reggio Emilia, Italy
[3]DII, Università di Modena e Reggio Emilia, Modena, Italy
E-mail: Muammar.MohamadIsa@postgrad.manchester.ac.uk

This paper presents a Novel low noise, high breakdown InAlAs/InGaAs pseudomorphic High Electron Mobility Transistors (pHEMTs). The improvements in breakdown voltage are brought about by a judicious combination of epitaxial layer design and field plate techniques. No significant degradations of DC and RF characteristics are observed for devices with field plate structures. An outstanding improvement in breakdown voltages of >30% is attained by field plate devices which should allow their usage in efficient high-added power efficiency amplifiers design.

1. Introduction

Amongst all III-V compound semiconductors, the InGaAs-InAlAs materials system has optimum band structure and transport properties making it the device of choice in many optoelectronics, magnetic and electronic applications. Indeed, THz range cut off frequencies have already been reported [1], and this system is road mapped in sub-22nm electronics. However, one of the most relevant issues that limit their applications is the trade-off between frequency performance (quantified by f_T and f_{max}) and low breakdown voltage. We therefore report on novel high frequency and high breakdown delta-doped (δ-doped) InGaAs-InAlAs pHEMT. The fabricated devices have breakdown voltages similar to those obtained by GaAs devices but with all the advantages associated with the InP-based system, such has high mobility and saturation velocity. The fabricated devices will be not only appropriate for low noise applications, but also in high power density, high-efficiency power amplifiers (PAs). Of particular importance to Low Noise Amplifier (LNA) applications, these devices will facilitate the design of receivers with minimal protection circuitry.

In this paper, we report the approaches for development of high breakdown InP pHEMT devices. The devices incorporating field plate (FP) extension from gate toward the drain [2] were processed on the same epitaxial wafers and at the same time as conventional pHEMT devices for comparison purposes. Simulation using Computer Aided Tools (CAD) suggested that variation of field plate extensions and Si_3N_4 thickness would permit higher V_{BR} HEMT devices [3].

2. Material Growth and Device Fabrication

The epitaxial layer growths were performed using an in-house solid-source Molecular Beam Epitaxy (MBE) on a high uniformity multiple wafers RIBER V100H system. From bottom to top, a lattice match InAlAs interfacial layer is grown on top of an InP substrate. A better confinement of quantum well is attained by means of two δ-doped layers. The first δ-doped layer is in-between an un-doped lattice match supply InAlAs layer. The first spacer

978-1-4244-8574-1/10 $26.00 © 2010 IEEE

layer is a lattice matched, un-doped InAlAs of approximately 100 Å thickness. The two Dimensional Electron Gas (2 DEG) is formed in a narrow band gap $In_{0.7}Ga_{0.3}As$ layer. The second supply layer is a thin, highly strained, wide band gap $In_{0.25}Al_{0.75}As$ layer where the second δ-doped layer is located. The exploitation of large Schottky barrier in wide band gap material will not only increased the barrier height (seen by carriers in the 2DEG channel) but also reducing the gate leakage current due to tunnelling and consequently increases the Schottky breakdown voltage [4]. The 200 Å highly doped cap layer is grown at the top and was aimed for very low, alloyed Ohmic contact. The epitaxial layer is summarised in Table 1 and its Hall data at room temperature (RT) and 77 K are shown in Table 2.

Table 1
Epitaxial layer for device under construction

Layers	Material	Thickness
Cap	$In_{0.53}Ga_{0.47}As$	200 Å
Barrier$_2$	$In_{0.25}Al_{0.75}As$	70 Å
δ-doped$_2$	Si	-
Spacer$_2$	$In_{0.25}Ga_{0.75}As$	28 Å
Channel	$In_{0.7}Ga_{0.3}As$	150 Å
Spacer$_1$	$In_{0.52}Al_{0.48}As$	100 Å
δ-doped$_1$	Si	-
Barrier$_1$	$In_{0.52}Al_{0.48}As$	100 Å
Buffer	$In_{0.52}Al_{0.48}As$	4500 Å
Substrate	S.I. InP	-

Table 2
Hall effects data for device under construction

Hall data	Value
Carrier Concentration, n_H at RT / 77 K (x 10^{12} cm^{-2})	8.3 / 5.7
Hall Mobility, μ_H at RT / 77 K (cm^2/V.s)	8212 / 24926

The devices were fabricated by first defining the active areas by means of a non-selective orthophosphoric (H_3PO_4:H_2O_2:H_2O) wet etch to the buffer layer. Then the sidewalls of the active layer were etched with succinic acid for better gate isolation at the later stage of gate deposition. The alloyed Ohmic contacts for Source and Drain were then formed by thermal evaporation and lift-off of AuGe/Au. Subsequently, the 200 Å gate footprints were formed by highly selective succinic acid etching. The 1.5 µm gate electrodes were defined by thermal evaporation and lift-off of Ti/Au metallisation. Later, probing pads were patterned and deposited using thick Ti/Au scheme. Subsequently, 90 nm of Si_3N_4 was deposited in a plasma chamber before etching openings at the end of each terminal for probing and connection to the field plate structures. Finally, the field plate extensions ranging from 0.3 µm to 1.2 µm were formed by thermal evaporation and lift-off of Au. The DC and RF measurements were characterised using HP4142 parameter analyser and HP8510C Vector Network Analyser (VNA). In this work, the non-field plate and field plate devices were fabricated on the same epitaxial samples, eliminating sample-to-sample variations in processing. An illustration of field-plated devices showing the field plate extension, L_{FP} and gate length, L_g are shown in Fig. 1. Note that the field plate and gate electrode are isolated by 90 nm thick of Si_3N_4. In the figure, the field plate extension is illustrated.

3. Results and Discussions

The effective widths of each device under test are 200 µm and 600 µm respectively. Fig. 2 shows the Schottky gate leakage where field plate featured devices are shown having lower leakage current as low as -1 µA/mm, tenth times lower than the conventional pHEMT devices. The same trend was also observed at the device's on-state leakage where the 1.2 µm field plate extensions have the lowest tunnelling leakage current of -1 µA/mm at V_{DS} = -6V. The maximum on-state leakages due to impact ionisation at V_{DS} = 2V for all devices were well below -25 µA/mm as depicted in Fig. 3

978-1-4244-8574-1/10 $26.00 © 2010 IEEE   42

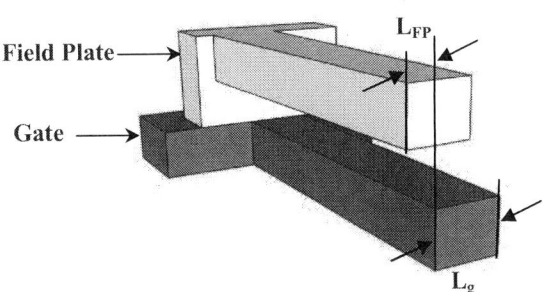

Fig. 1: Field plate terminal and gate electrode. Note for Field Plate extension, L_{FP} dimension.

Before hitting the break down point, all devices show the same output transfer characteristics as illustrated in Fig. 4. This shows that no carriers are lost in the channel even though field plate structures are incorporated. All devices have transconductance of approximately 450 mS/mm.

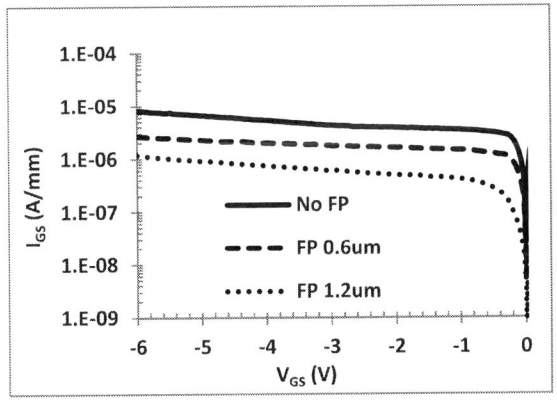

Fig. 2: Schottky diode characteristics for 200 µm devices

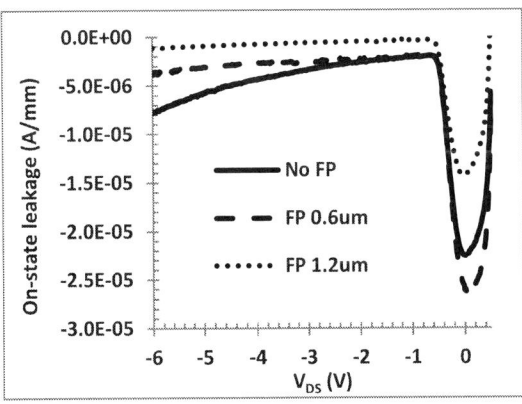

Fig. 3: On-state leakage at $V_{DS} = 2V$ for 200 µm devices

As expected, the V_{BR} has been improved by utilising field plate designs. V_{BR} at I_{DSS} ($V_{GS} = 0V$) and pinch-off ($V_{GS} = -1.2V$) are shown in Fig. 4 and Fig.5 respectively. At $V_{GS} = 0V$, the V_{BR} for $L_{FP} = 1.2$ µm is 5.9V, an improvement of 31% from the conventional gate structures. At pinch off the V_{BR} was improved by 36%. The V_{BR} for conventional and $L_{FP} = 1.2$ µm is 14V and 19V respectively. The device cut-off frequencies biased at $V_{DS} = 1.5$ V and 30% I_{DSS} are shown in Fig.6, where f_T for all devices are ~ 8 GHz. This shows that the field plate structures did not degrade the device's DC and RF performances.

4. Conclusions

High Breakdown InGaAs-InAlAs pHEMTs incorporating improved epitaxial layer and field plate structures have been successfully fabricated and characterised in this work. The newly developed field plate devices have demonstrate more than 30% improvement in V_{BR}, lower Schottky and on-state leakage, in addition to almost similar DC and RF performances.

The low noise and high voltage application will inherent its applications in high voltage, high efficiency Power Amplifiers (PAs) at millimetre-wave band, robust LNAs with streamlined protection circuitry and as well as integrated RF transceiver, optimized for low-noise receivers and high-power PA transmitter.

978-1-4244-8574-1/10 $26.00 © 2010 IEEE

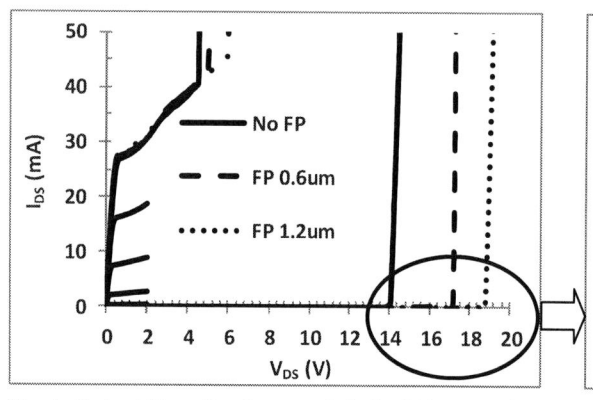

Fig. 4: Output Transfer characteristic for 200 μm devices

Fig. 5: Off-state breakdown for 200 μm devices
($V_{GS} = -1.2V$)

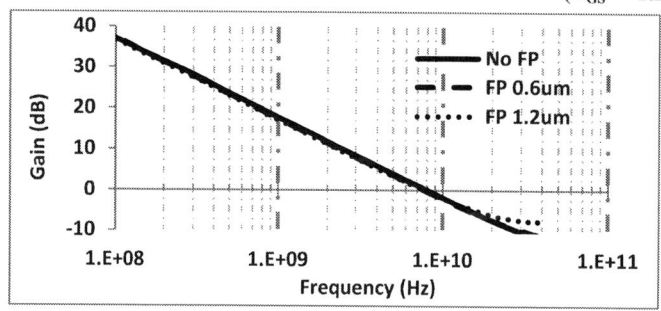

Fig. 6: Cut off Frequency, f_T biased at $V_{DS} = 1.5V$, 30% I_{DSS}

Acknowledgement

This work was supported by both the STFC in the SKADS programme and also partially supported by the *Fondazione Cassa di Risparmio di Modena* (Italy) under the project "Field-plated InGaAs-InAlAs high electron mobility transistors for emerging high-frequency applications", 2009-2011.

References

[1] D. H. Kim, J. A. Del Alamo, "30-nm InAs pseudomorphic HEMTs on an InP substrate with a current-gain cutoff frequency of 628 GHz", *IEEE Electron Device Letters*, Vol 28, Issue 8, pp. 830 -833, 2008.

[2] A. Chini et. al., "Very High Performance GaN HEMT devices by Optimized Buffer and Field Plate Technology", *1st European Microwave Integrated Circuit Conference*, pp. 61-64, Sept.2006.

[3] D. Saguatti, A. Chini, G. Verzellesi,M. Mohamad Isa, K. W. Ian, and M. Missous,"TCAD optimization of field-plated InAlAs-InGaAs HEMTs", *Proc. Of International Conference on Indium Phosphide & Related Materials 2010,* (IPRM 2010), Kagawa, Japan , pp. 1-3, June 2010.

[4] A. Bouloukou, B. Boudjelida, A. Sobih, S. Boulay, J. Sly and M. Missous, "Design of low leakage InGaAs/InAlAs pHEMTs for wide band (300MHz to 2GHz) LNAs", *Proc. of The 7th International Conference on Advanced Semiconductor Devices and Microsystems*, 2008 (ASDAM 2008), Smolenice, Slovakia, pp. 79-82, Oct. 2008.

[5] A. Bouloukou, B. Boudjelida, A. Sobih, S. Boulay, J. Sly and M. Missous, "Very low leakage InGaAs/InAlAs pHEMTs for broad band (300MHz to 2GHz) low-noise applications", *Materials Science in Semiconductor Processing*, Vol. 11, Issues 5-6, pp. 390-393, Oct. 2008.

[6] A. Bouloukou, B. Boudjelida, A. Sobih, S. Boulay, J. Sly and M. Missous, "Novel High-Breakdown InGaAs/InAlAs p HEMTs for Radio Astronomy Applications", *Proc. of the 4th ESA Workshop on Millimetre Wave Technology and Applications* (7th Millimetre Wave International Symposium), Espoo, Finland, Feb. 2006.

[7] A. Bouloukou, B. Boudjelida, A. Sobih, S. Boulay, J. Sly and M. Missous, "Design of low leakage InGaAs/InAlAs pHEMTs for wide band (300MHz to 2GHz) LNAs", *http://www.skads-eu.org/p/memos.php*.

978-1-4244-8574-1/10 $26.00 © 2010 IEEE   44

A Width-Dependent Body-Voltage Model to Obtain Body Resistance in PD SOI MOSFET Technology

Arash Daghighi[1], Azam Asgari-Khoshooie[2]

1- Assistant Professor, Faculty of Engineering, Shahrekord University, Shahrekord, Iran.
E-mail: daghighi-a@eng.sku.ac.ir
2- Graduate Student, Department of Electrical Engineering, Islamic Azad University Najafabad Branch, Isfahan, Iran.

A new width-dependent nonlinear relation to obtain body voltage (V_B) in PD SOI MOSFET is presented. The new model is extracted based on the distributed nature of the body resistance (R_B). Using 3-D simulations of an H-gate 45 nm PD SOI MOSFET, the body voltage variations along the device width were obtained. It was shown that the nonlinear relation approximates the 3-D simulation results. The proposed model was used to obtain the body resistance. Simulation results verified that the proposed model follows the non-linear variations of the body resistance as device width varies. The comparisons of the 3-D device simulation results and proposed body voltage relation show the effectiveness of the model on estimation of the body voltage (V_B).

1. Introduction

SOI MOSFET offers advantages over bulk MOSFET in terms of speed, isolation, density, and performance gain in low power electronics [1]. In PD SOI MOSFET technology, body contacts are used to make a path for the generated holes in the impact ionization process to escape to a contact [2]. The generated holes face the body resistance and a body voltage is build up. The body resistance has a distributed nature [3]. As a result of body voltage variations along the device width, the mobile charges in the neutral body region, hence the body resistance along the device width varies. Therefore, body resistance is bias dependent [4]. In this paper, a new width- dependent body voltage model is proposed. The model is used to obtain the body resistance and the results were compared with the 3-D device simulation findings.

2. 3D Simulation Results

Fig. 1 shows the small-signal equivalent circuit of a body contacted PD SOI MOSFET.
The body node is connected through the body resistance (R_B) to an external voltage (V_{BC}).
The total body current (I_B) is written as:

$$I_B = I_{ii} + I_{gb} + I_{gidl} + I_{gisl} + \frac{dQ_D}{dt} + \frac{dQ_S}{dt} + \frac{dQ_E}{dt} + \frac{dQ_G}{dt} \tag{1}$$

where I_{ii}, I_{gb}, I_{gidl} and I_{gisl} denote the impact ionization, gate tunneling, gate induced drain leakage and gate induced source leakage currents, respectively. The source, drain, gate and BOX leakage currents are shown by dQ_S/dt, dQ_D/dt, dQ_G/dt and dQ_E/dt, respectively. The distributed body resistant model is shown in Fig. 2 where I_{Bi}, is the ith element of total body current I_B, and R_{bshj} is the jth element of the body sheet resistance (i,j=1,...,n). By letting *n* approaches infinity and integrating, the body potential is obtained as:

978-1-4244-8574-1/10 $26.00 © 2010 IEEE

$$V_{(Z)} = V_{ext} + R_{bsh}I_B \frac{z}{W}\left(2 - \frac{z}{W}\right) \qquad (2)$$

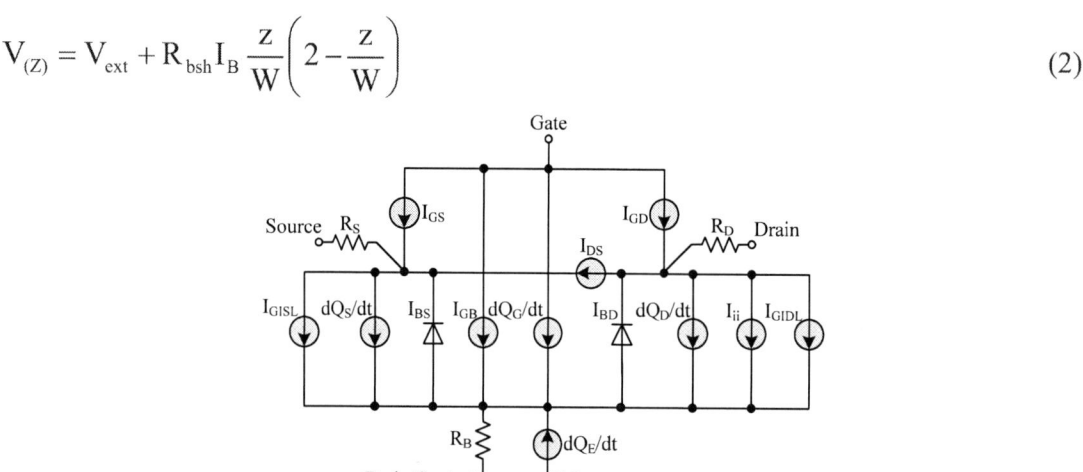

Fig 1: Small-signal equivalent circuit of a PD SOI MOSFET

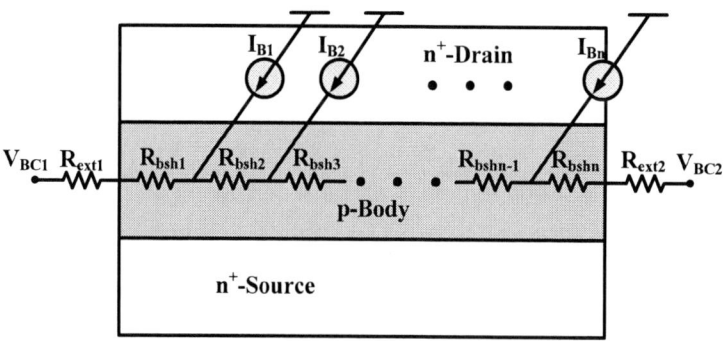

Fig. 2: Distributed body resistance model of the neutral body region, V_{BC} is the external applied voltage

where W denotes the transistor width, z is defined as the equation variable in the Z-axis direction and R_{bsh} is the zero-terminal-voltage body sheet resistance. V_{ext} denotes the external body contact voltage drop and is obtained by:

$$V_{ext} = R_{ext}I_B \qquad (3)$$

where R_{ext} is the external body contact resistance. In deriving Eq. (2), the body is contacted merely at one side of the gate to the ground potential and I_{Bi} (i=1,...,n) was considered independent of the body voltage variations along the device width. The body resistance is defined as [4]:

$$R_B = \frac{W^2}{\mu_B Q_{nbr}} \qquad (4)$$

where μ_B and Q_{nbr} denote the mobility of the majority carriers and the mobile charges in the body, respectively. Using Eq. (2), the distribution of the body voltage along the device width, hence Q_{nbr}, is attained and the body resistance, R_B, can be obtained [4].

In order to verify the body voltage model, an H-gate PD SOI nMOSFET with the channel length of 45 nm was simulated. The device 3-D structure is shown in Fig. 3. As shown, the positive doping relates to donor impurities and acceptor impurities correspond to the negative doping. The drift-diffusion model was used for 3-D simulations [5]. Only one of the contacts was connected to the ground potential (BC1).

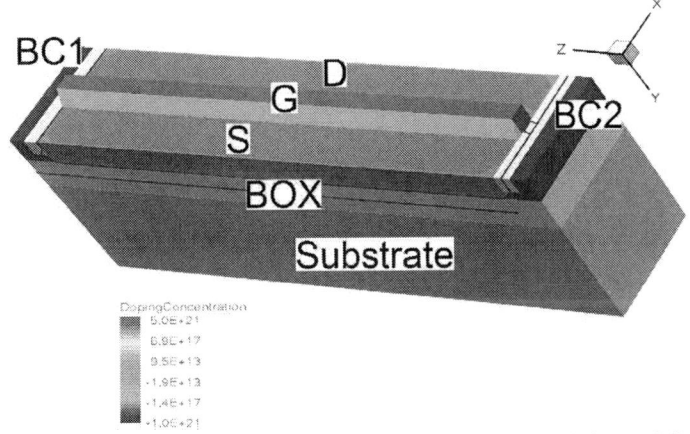

Fig 3: 3-D structure of the simulated 45 nm H-gate PD SOI MOSFET, the body is connected to ground potential at one end and is left floating at the opposite end

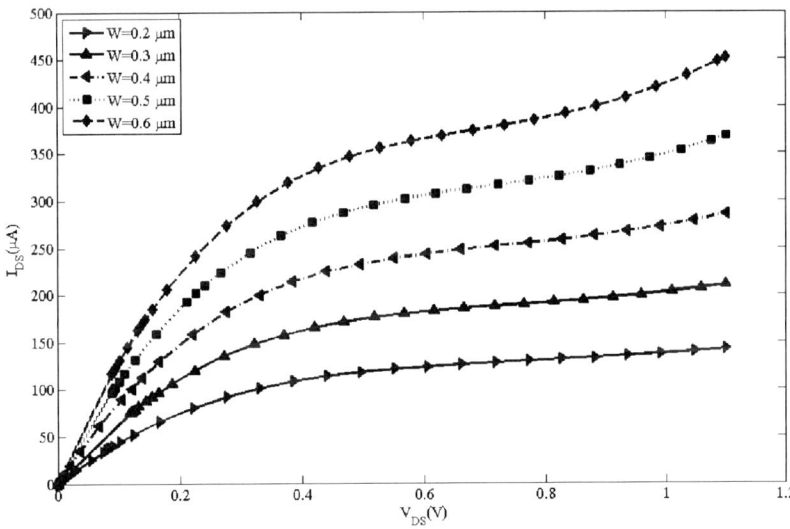

Fig. 4: Drain current vs. drain voltage for various device widths, V_{gs}=1.1 V

Fig. 4 shows drain current vs. drain voltage as device width varies from 0.2 µm to 0.6 µm. A gradual increase in current drive is observed. This is due to the elevated body voltage, where the current increases from the floating body contact at one end (BC2) to the grounded contact at the opposite end (BC1). The comparison of body voltages is made between 3-D simulation results and the body voltage model (Eq. 2). The results are shown in Fig. 5. As can be seen, the model accurately approximates the body voltage as device width increases. In addition, the computed body resistances with the application of proposed model, the 3-D simulation results and the results of the conventional model are depicted in Fig. 6. The conventional model ignores the body potential variation along the device width, hence body resistance shows a linear relation with the device width. However, the proposed model accurately approximates the 3D simulation findings of the body resistance. Therefore, the results confirm the effectiveness of the model to obtain the body resistance (R_B).

978-1-4244-8574-1/10 $26.00 © 2010 IEEE   47

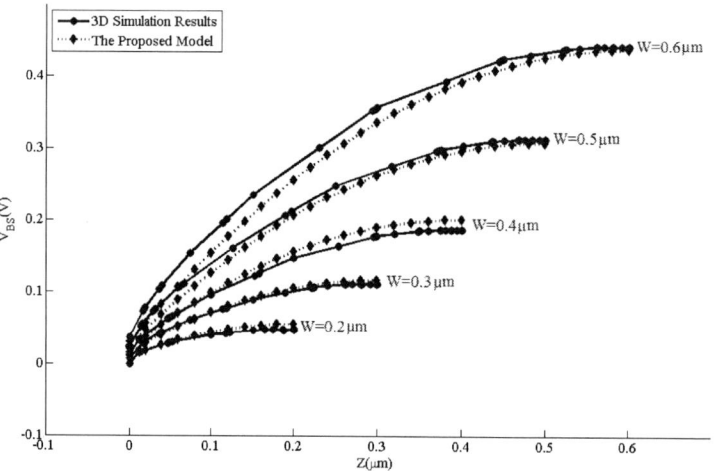

Fig. 5: Body potential variation along the device width for various device widths, $V_{gs}=V_{ds}=1.1$ V

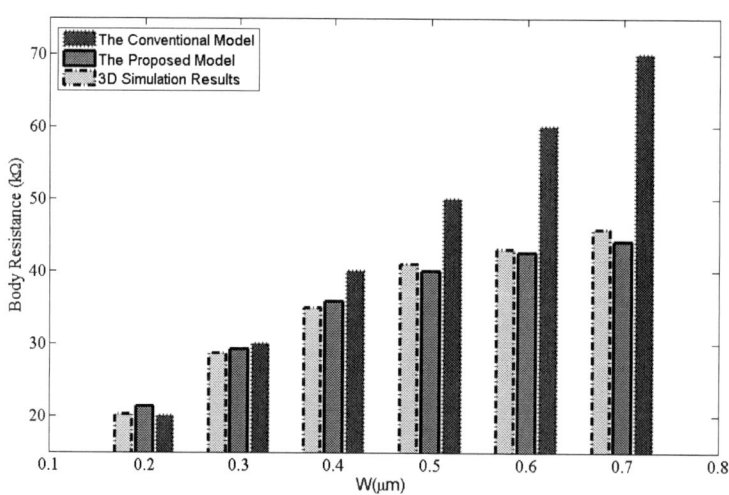

Fig. 6: Total body resistance vs. device width, $V_{gs}=V_{ds}=1.1$ V

3. Conclusion

A new width-dependent body voltage model for body contacted PD SOI MOSFET devices was proposed. The model accurately approximated the body voltage variations along the device width. The model was effectively used to obtain the body resistance.

References

[1] G.G. Shahidi, IBM J. RES. & DEV., Vol. 46, No. 2/3, March/May 2002.

[2] J.P. Colinge, Kluwer Academic Publishers, 3rd edition, 2004.

[3] G.O. Workman, J.G. Fossum, IEEE Trans. on Electron. Devices, Vol. 45, No. 10, pp. 2138-2145, 1998.

[4] G. Gildenblat et al., IEEE Trans Electron Dev., 53(9):1979–93, 2006.

[5] DESSIS Manual. ISE Integrated System Engineering, Version 10.0; 2004.

Reactive ion etching of $Al_xGa_{1-x}N$/GaN heterostructure using Cl_2, BCl_3/Ar gas plasma

W. Oleszkiewicz*, J. Gryglewicz, B. Paszkiewicz, R. Paszkiewicz, A. Szyszka,
M. Ramiączek-Krasowska, A. Stafiniak, , M. Tłaczała

Wroclaw University of Technology, Faculty of Microsystem Electronic
and Photonics, Janiszewskiego 11/17, 50-372 Wrocław
e-mail: Waldemar.Oleszkiewicz@pwr.wroc.pl

The etching processes were carried out in Oxford Instruments Plasmalab80Plus system. Present work is concentrating on discussion of a scope of different $Al_xGa_{1-x}N$/GaN heterostructures in relation to percentage composition of aluminum. The topography of heterostructure surface and slope was controlled using AFM technique.

1. Introduction

Gallium nitride (GaN) and aluminum gallium nitride ($Al_xGa_{1-x}N$) are used in variety of electronic applications, such as high frequency power amplifiers, UV and blue LEDs, detectors and high electron mobility transistors (HEMTs) working in harsh environment [1,3]. Moreover, diversified percentage composition of aluminum in $Al_xGa_{1-x}N$ enables one to design different color LEDs. Unique properties of these materials such as wide band gap, high saturation velocity and high electron mobility makes the both materials good candidates for high power and high frequency applications. However, inertness of those materials to chemical radicals decreases their etch rates significantly. Reactive ion etching seems to be viable technology to remove materials faster than conventional chemical etching. Unfortunately, the diversification in composition of each $Al_xGa_{1-x}N$/GaN heterostructure makes the process of etching difficult to optimize. The process parameters such as RF power, DC bias, pressure, gas flows or temperature of the substrate have to be selected carefully to every heterostructure.

The fabrication of majority of electronic devices is connected with exposing mesa profile from the etched heterostructure. Creating mesa profile in the devices reduces current leakage. Another important process in the fabrication of the devices is gate recess. Thus, the shape of mesa profile has to be adjusted to a particular application. It is especially important to obtain smooth surfaces on the mesa sidewalls under the gate surface. Another issue is to avoid the formation of mesa undercut near the sidewall.

978-1-4244-8574-1/10 $26.00 © 2010 IEEE

Experimental

The test structures of AlGaN/GaN heterostructures, depicted in Fig.1a, for the RIE process were grown on a c-plane sapphire in a vertical flow MOVPE (Metalorganic Vapour Phase Epitaxy) system at atmospheric pressure.

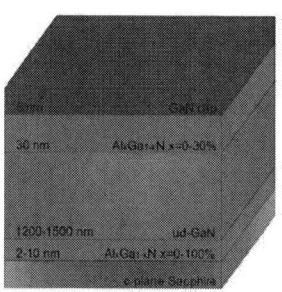

Fig. 1. Scheme of a basic planar $Al_xGa_{1-x}N$/GaN heterostructure

(the figure is not drawn in scale)

The processes of the $Al_xGa_{1-x}N$/GaN heterostructures etching, performed using Oxford Instruments Plasmalab80Plus system in Cl_2/BCl_3 plasma, are strongly dependent on the process parameters. The processes were carried out at substrate temperature of 7°C, RF power 150 W, pressure in the chamber 20 mTorr with constant flow of Ar (5sccm) and Cl_2 (10sccm). The flow of BCl_3 was raging from 2 sccm to 10 sccm. The change of percentage composition of BCl_3 in the process gas affected etching rate, roughness of the heterostructure surface and undercutting of the slope. Two last mentioned effects affect the parameters of the HEMT transistor depicted in Fig 2.a. The proper shape of mesa area is essential for the fabrication of thin submicrometer gate electrodes (Fig 2.b). If the shape is too abrupt, the gate electrodes will be interrupted and the gate bias will not influence the drain current of the transistor. If the slope is too gentle, it will be difficult to align properly the gate between source and drain contact during the lithography process.

a)   b)

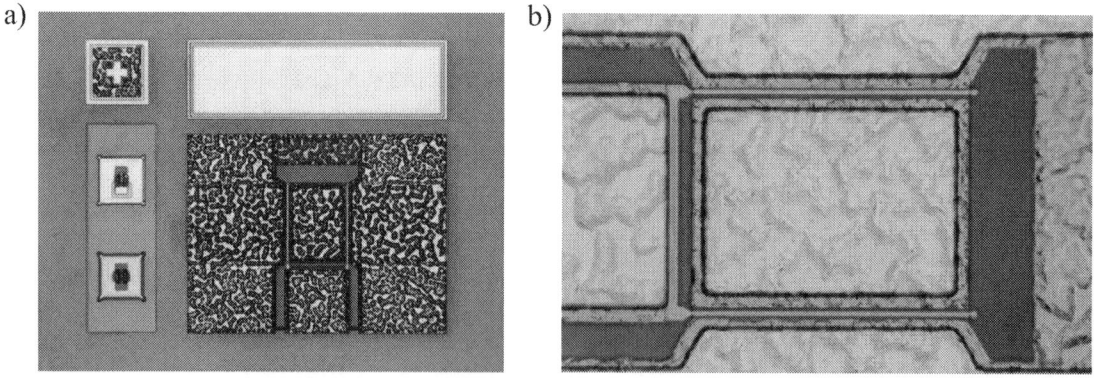

Fig. 2. Picture of the AlGaN/GaN two gate HEMT structure (a), enlarged view

depicting the gate electrodes and the drain contact (b)

The roughness parameters have been examined using AFM technique after performing RIE processes. We observed the change in roughness parameters for unetched (Ra: 5.8 nm, RMS: 7.4 nm) and etched (Ra: 4.1 nm, RMS: 5.1 nm) surface of heterostructures.

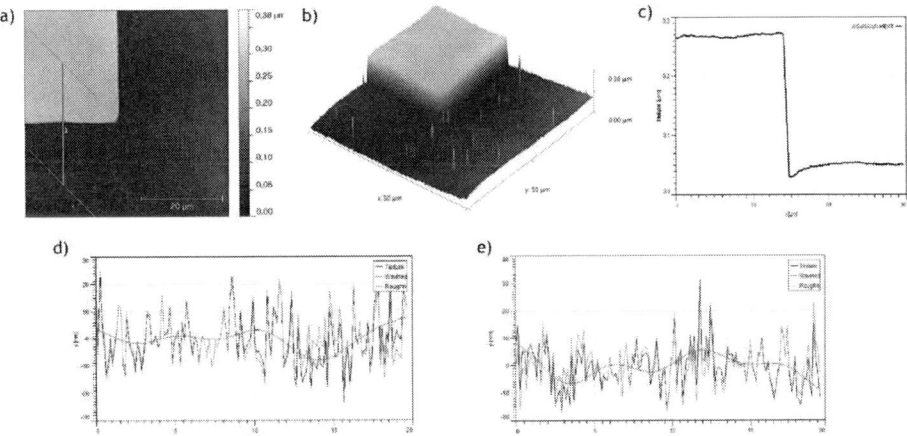

Fig. 3. AFM results of the $Al_{0.3}Ga_{0.7}N$/GaN HEMT heterostrucure after the etching process: topography (a), mesa structure (b), profile of mesa with visible slope undercutting effect (c), roughness of the surface before (d) and after (e) the etching (etching rate: 26 nm/min),

Selectivity of the process parameters allows obtaining smooth surface and reducing the undercutting slope effect. Thus, the BCl_3 and Cl_2 flows influence the process the most. RF power and DC bias are responsible for ion's energy. The higher energy caused by increasing RF power, the faster etch rates can be obtained. However, the fast etch rates are not always advantageous. It would be better to obtain fully controllable process, where the smooth surface is a priority.

The photoluminescence spectra were used to measure the composition of aluminum in $Al_xGa_{1-x}N$ alloy and surface damage after etching process [2]. This measurement was also very useful to indicate the moment, when the $Al_xGa_{1-x}N$ layer was completely removed, during the RIE process.

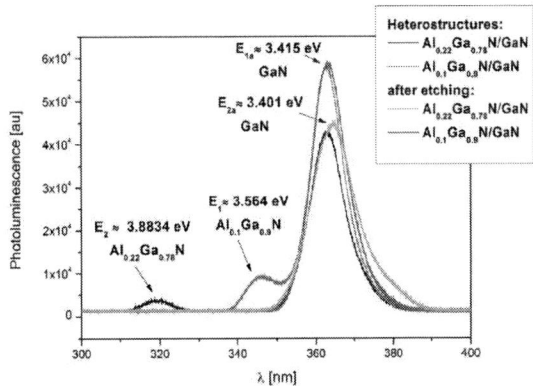

Fig. 3. Photoluminescence spectra of $Al_xGa_{1-x}N$/GaN heterostructures before and after the etching process for varying percentage composition of aluminium.

The employed technique of photoluminescence measurements allowed us to select appropriate etching time for different composition of aluminium in the heterostructures at certain process parameters. Observed shifts of the peaks in the etched heterostructures arise from modification of surface roughness, what results in the change of the optical properties. It also allowed us to evaluate the dependence of the etching rate of the heterostructures on the RIE process parameters.

3. Conclusion

The RIE etching process of the $Al_xGa_{1-x}N/GaN$ heterostructures in chlorine plasma ($Cl_2/BCl_3/Ar$) is stable. Changing the RF power and percentage composition of BCl3 in plasma discharge can effectively control the mesa profile. The used photoluminescence measurement of the unetched and etched heterostructures in the given process conditions, with diverse composition of Al, allowed us to select the appropriate etching time.

Acknowledgement

This work has been supported by the European Union within European Regional Development Fund, through grant Innovative Economy (POIG.01.01.02-00-008/08), Polish Ministry of Science and Higher Education under the grant no. NN 515360436, and by Wroclaw University of Technology statutory grant

References

[1] D. Basak*, T. Nakanishi, S. Sakai, Reactive ion etching of GaN using BCl_3, BCl_3/Ar and BCl_3/ N_2 gas plasmas., *Solid-State Electronics* 44 (2000) 725-728.
[2] J. Gryglewicz*, W. Oleszkiewicz, M. Ramiączek-Krasowska, A. Szyszka, J. Prażmowska, B. Paszkiewicz, R. Paszkiewicz, M. Tłaczała Reactive ion etching of GaN and AlGaN/GaN assisted by Cl_2/BCl_3, *Crystal Research and Technology*, in press.
[3] T.J. Anderson, M.J. Tadjer, M.A. Mastro, J.K. Hite, K.D. Hobart, C.R. Eddy, F.J. Kub, *Journal of Electronic Material*, **39, 5,** 478-481, 2010
[4] W. Oleszkiewicz, J. Gryglewicz, *unpublished.*

Noise in the InAlN/GaN HEMT transistors

K. Rendek, A. Šatka, J. Kováč and D. Donoval

Faculty of Electrical Engineering and Information Technology, Slovak University of
Technology, Bratislava, Slovakia
E-mail: karol.rendek@stuba.sk

*This paper deals with measuring and analyzing input and output low-frequency noise
spectra of Gallium Nitride (GaN) based High Electron Mobility Transistors.
Low-frequency noise spectral densities of drain noise voltage and gate noise current
are shown. The three different measurements of input low–frequency noise of GaN
HEMT transistor were measured for better revealing G-R noise sources in gate
region and their influence to the output noise voltage. The long-term voltage
off-stress was accomplished on the HEMT and impact to the low-frequency noise
characteristics shown.*

1. Introduction

High electron mobility transistors (HEMT) based on AlGaN/GaN and InAlN/GaN technology
has been reported with superior power parameters at very high frequencies unlike FETs or MESFETs.
The heterogeneous structure of HEMT transistors provides improving of the parameters like high
transconductance and a very high cut-off frequency, which are results of the increased carrier density
and mobility in the channel [1]. Unfortunately, advanced HEMT structure brings many issues to
technological process for instance the growth technology of AlGaN and InAlN in terms of reaching
HEMTs with superior quality and reliable structure.

Low-frequency (LF) noise measurement and analysis as factor determining the performance of
various electronic devices is prospective methodology helping to make a progress in HEMTs
technology. LF noise is known to be very sensitive to crystallographic defects, defects introduced
during the growth process or through the device processing technology. In addition,
generation–recombination (G-R) noise in semiconductors indicates the type and quantity of the traps
so that it can be used to evaluate quality of HEMT transistors related to the defects in the device
structure [2].

In this paper, we report on investigation of the input and output low frequency noise properties
of InAlN/GaN HEMT transistors, and the change in InAlN/GaN HEMT transistors noise properties
exposed to the long-term voltage off-stress.

2. Experimental

The HEMT transistor structures were grown on sapphire substrate by MOCVD. The epitaxial
structure consists of 1 μm GaN buffer undoped layer followed by a 1nm AlN barrier and a 10 nm
undoped InAlN barrier layer. The device processing of HEMT transistors started with Ar-based
reactive ion etching for mesa insulation. Optical lithography was used to define Ti/Al/Ni/Au ohmic
contacts and 1 μm long and 100 μm wide Schottky barrier Ni/Au gates. The source-to-drain distance is
~4 μm, and the gate is placed symmetrically between the contacts. Devices under investigation were
not passivated.

The output, input and transfer DC I-V curves of investigated InAlN /GaN HEMTs were
measured using parametric semiconductor analyser Agilent 4155C. The average threshold voltage of
HEMTs as determined from the transfer I-V curves has been $V_{th} \approx -1.95V$.

978-1-4244-8574-1/10 $26.00 © 2010 IEEE

The low-frequency noise measurements at room temperature were accomplished in common source configuration with 470Ω resistor in series with drain and V_{DS} voltage source. The spectrum of the output voltage noise from the drain, amplified by 80dB low-noise preamplifier was measured by signal spectrum analyser SR770 in the 4 Hz-100 kHz range. The output voltage noise measurements were accomplished in the linear and saturation regions of the HEMT transistors. For the input LF current noise measurements, gate current was converted to voltage and preamplified by low-noise current preamplifier SR570. The input noise measurements were accomplished particularly for gate – source, gate – drain and gate – drain plus source in order to find out the difference in the spectral characteristics and reveal contribution of different regions to total input noise of the HEMTs.

In order to determine the evolution of the noise spectra resulting from the device operation, the long-term voltage off-stress was executed in the following steps. First, before starting the voltage off-stress the I-V and noise characteristics were measured. After that the off-stress was started by setting V_{GS} = - 4V and V_{DS} = 50V for 4 hours. Finally after the voltage off-stress the I-V and noise characteristics were measured again. During the drain voltage noise measurement for the different gate voltages was the drain voltage adjusted to the same values.

3.   Results and discussions

The output and input I-V characteristics of one of the InAlN/GaN HEMT transistors before and after the long-term voltage off-stress are shown in Figure 1 (black curve). Interestingly, the output characteristics during the voltage off-stress have changed mainly in the linear region for small V_{DS} whereas in saturation region it changed only slightly and transconductance for $V_{DS} \cong 10$ V remained nearly unchanged. In contrast, the input characteristics reveal effective build-up of the positive charge in the InAlN layer under the gate and negative charge at the InAlN/GaN interface, resulting in a significant shift of the minimum in input I-V curves.

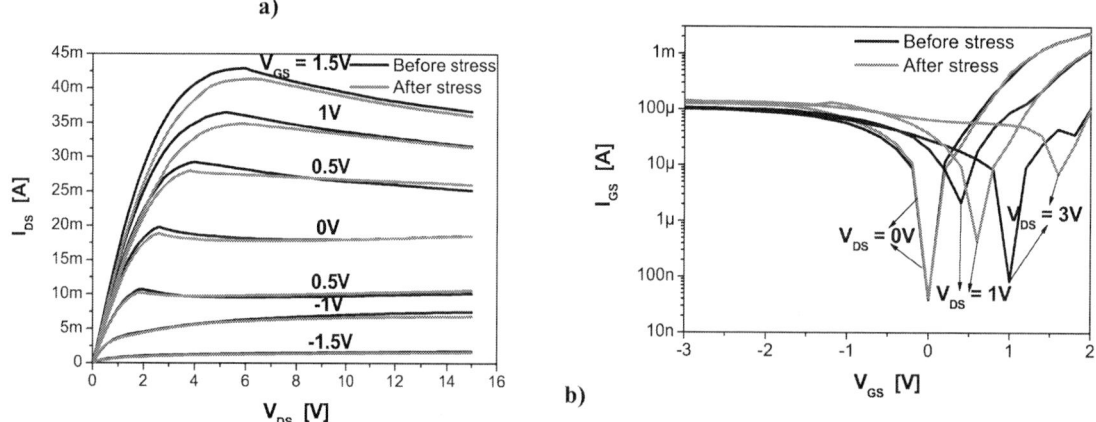

Figure 1. I – V characteristics of GaN HEMT transistor before (black curve) and after (grey curve) the voltage off-stress. a) Output characteristics, b) input characteristics.

The corresponding drain voltage noise spectra of the same transistor were measured for drain voltages V_{DS} from 0 to +4V before the voltage off-stress. The measured spectra shown in Figure 2 are displayed for 0 to 1V V_{DS} range only because the effective value and shape of the noise spectra significantly changes in this V_{DS} range. From the noise spectra measured at V_{DS} below 0.5V, the noise power gradually and smoothly decrease with frequency. If we assume that the total noise is generated by uncorrelated noise sources, contribution of frequency–independent thermal and shot noise sources can be neglected in these spectra.

978-1-4244-8574-1/10 $26.00 © 2010 IEEE   54

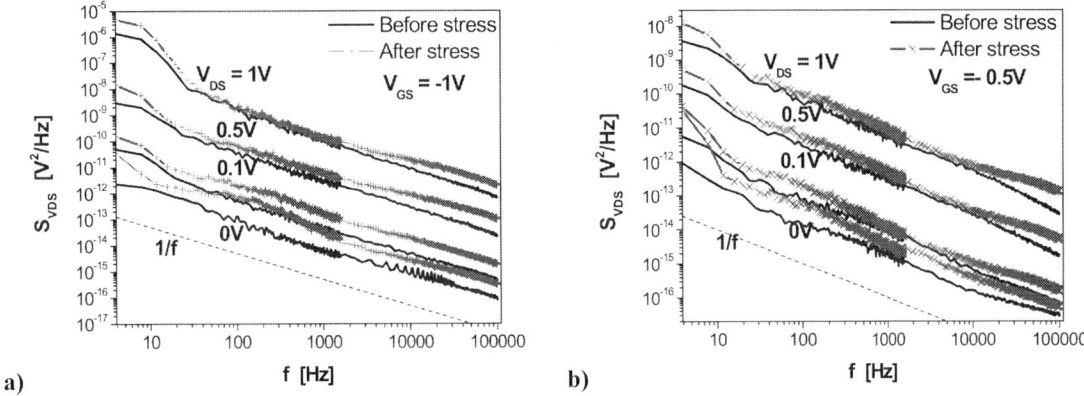

Figure 2. The evolution of the drain voltage noise of GaN HEMT in dependence on the drain voltage V_{DS} at a) $V_{GS} = -1V$ and b) $V_{GS} = -0.5V$, both before (black) and after (grey) the voltage off-stress.

The mean square value of 1/f voltage noise, originated from the conductivity fluctuations in the resistor [3], [4] could be expressed by empirical formula

$$\overline{v_n^2} = K \, r_{DS} \, I^\alpha \frac{1}{f^n} \Delta f \quad \left[V^2 \right] \tag{1}$$

Here, f is the frequency, I is the DC current passing through the device, r_{DS} is dynamical resistance of the transistor channel, $n \cong 1$ for highly conductive devices, α is an empirical constant ($\alpha \approx 1\text{-}2$), K is area and device-dependent constant, and Δf represents the bandwidth of the measurement system. The G-R noise is related to statistical fluctuations in population of charge carriers due to random generation, trapping and release of the carriers in a semiconductor. The mean square value of the voltage G-R noise is described by the formula

$$\overline{v_n^2} = r_{DS} \frac{4\tau \, \overline{\Delta N^2}}{1 + (2\pi f \tau)^2} = r_{DS} \frac{4\overline{I^2}\tau}{\overline{N}(1 + (2\pi f \tau)^2)} \quad \left[V^2 \right] \tag{2}$$

Here, $\overline{\Delta N^2}$ is the average number of traps, $\overline{N}$ is the average number of carriers and τ is a time constant characterising the capture - emission process.

The noise spectra in Figure 2, measured at low V_{DS} before the voltage off-stress show simple 1/f behaviour with $n = 1$, practically without the $1/f^2$ component related to G-R noise sources described by (2). The output voltage noise increases with V_{DS} (and with I_{DS}) according to (1) so it can be concluded that the 1/f noise originated in the channel conductivity fluctuations is the main component of the output voltage noise. The noise behaviour of HEMTs changes for V_{DS} above ≈ 500 mV where G-R noise in the low-frequency region below ~30 Hz becomes dominant over the 1/f noise.

The voltage off-stress remarkably influenced the noise properties of HEMTs, particularly the shape of the spectra (grey curves in Figure 2). The increase of the G-R noise in the low-frequency region below 30 Hz (practically with the same time constant as observed in the spectra before the voltage off-stress) is obvious. Some increase of the G-R noise in a high-frequency part of the spectra is also distinguishable from both the Figure 2. In order to find the origin of the G-R noise in the output voltage noise and to investigate the influence of the gate noise sources to the output noise of the HEMTs, input current noise spectra were measured after the voltage off-stress. Generally, the transistor input current noise should be comprised of two uncorrelated parts, one originated from G-S and the other from G-D regions. On the other hand, some noise contributed from beneath the gate electrode should be correlated [5], [6]. Therefore, three input current noise spectra were measured: $\overline{i_{nGS}^2}$ from G-S region with D electrode kept floating, $\overline{i_{nDS}^2}$ from G-D region with S electrode kept floating and finally $\overline{i_{nGSD}^2}$ from input gate when S and D electrodes are shorted.

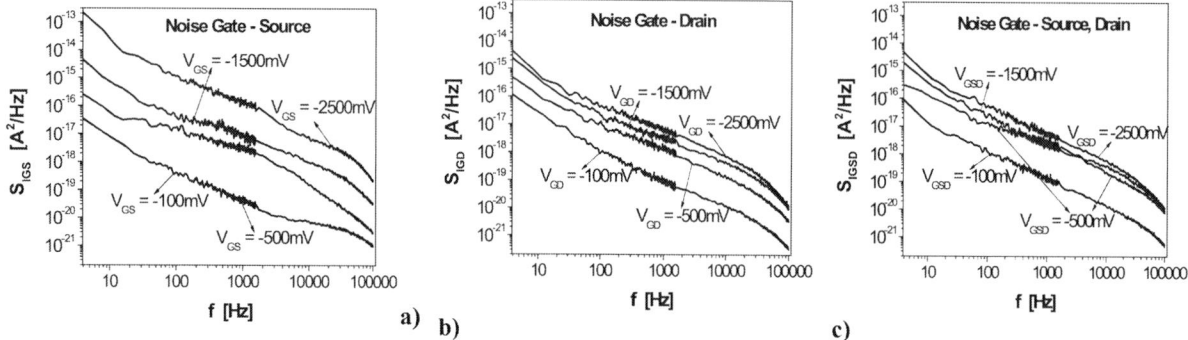

Figure 3. The evolution of the InAlN/GaN HEMT input current noise in dependence on the voltage a) V_{GS}, b) V_{GD}, and c) $V_{G,S+D}$.

The measured input current spectra are compared in Figure 3. The spectra in Figure 3a) consist of superposition of 1/f noise and strong G-R noise. Three different G-R time constants should be determined from these spectra by spectral decomposition. Similarly the noise spectra in Figure 3b) consist of superposition of 1/f and G-R noise, but the shape and relative levels are slightly different which points out significantly higher density of defects/traps at the G-S transistor region (under the gate) than at the G-D region. The total input noise spectra in Figure 3c), which is determined by contribution of both regions weighted by relative conductance of the regions show similar behaviour. The most importantly, the output noise spectra reveal G-R components of time constants similar to input current noise, which results in a similar shape of the spectra. This supports the assumption about the control of the output noise of the HEMTs by the G-R noise sources from under the HEMTs gate and G-D region.

4. Conclusion

The drain voltage noise and gate current noise spectral characteristics of GaN HEMT transistors were investigated. It has been confirmed that 1/f and G-R noise sources are main constituents of the HEMT transistor output noise. Significant injection of the noise from the HEMTs gate region has been found in the output spectral noise characteristics. The output noise spectra measured before and after the voltage off-stress revealed influence of the long-term voltage off-stress to the increase of the transistor output noise. The gate current noise spectra measured at three different configurations revealed influence of the partial transistor regions.

Acknowledgement

The work has been done in Centre of Excellence CENAMOST (VVCE-0049-07) with support of the VEGA project 1/0716/09, NanoNetII ITMS code 26240120018, and the European Community's from Seventh Framework Programme FP7/2008-2011 under grant agreement n°214610, project MORGaN. The authors wish to thank Alcatel-Thales III-V Lab, France for MOCVD growth of structures and Institute of Electrical Engineering, Slovak Academy of Sciences, Slovakia for device processing.

References

[1]   Vasilescu G.: Springer, Verlag Berlin Heidelberg, 1999, 709 pp.; ISBN: 3-540-40741-3
[2]   Mohammadi S. et al.: IEEE Trans. on ED, 47, 677-686, 2000
[3]   Hooge F. N. et al.: Rep. Prog. Phys, 44, 480-532, 1981
[4]   Vandamme L. K. J. et al.: Circuits Devices Syst, 149, 3-12, 2002
[5]   Sozza A. et al.: Microelectronics Reliability, 46, 1725-1730, 2006
[6]   Curutchet A. et al.: Microelectronics Reliability, 43, 1713-1718, 2003

Inter-digitated AlGaN/GaN Schottky diode for monolithic integration

B. Paszkiewicz, R. Paszkiewicz[*], M. Wosko, M. Tlaczala

Wroclaw University of Technology,
Faculty of Microsystem Electronics and Photonics,
Janiszewskiego 11/17, 50-372 Wroclaw

New construction of the AlGaN/GaN/semi-insulating Schottky diode was proposed for operation at gigahertz regime. Based on the performed numerical simulations the planar diode with inter-digitated lay out was elaborated. The test structures of the diode was fabricated in AlGaN/GaN heterostructures grown on a c-plane sapphire by MOVPE technique. The d.c. and high frequency characteristics of the device were measured. The cut-off frequency, f_T, of the diode was 8.2 GHz. The obtained results proved that inter-digitated Schottky diode fabricated in AlGaN/GaN heterostructure grown on semi-insulating substrate is suitable for high frequency operation at gigahertz regime and could be also monolithically integrated with HEMT.

1. Introduction

The AlGaN/GaN material system is attractive for numerous devices application. The relatively well developed research areas are photonic and high-power electronic devices based on nitrides. Nitrides heterostructures are also one of the leading candidates for high frequency application up to THz range (regime). The monolithic integration of AlGaN/GaN HEMT (high Electron Mobility Transistor) and Schottky barrier diodes in AlGaN/GaN heterostructure grown on semi-insulating substrate is important for the next-generation high frequency electronics [1]. But it can be shown that if classical Schottky diode, with axial symmetry (Fig.1a), is fabricated in AlGaN/GaN HEMT heterostructure its cut-off frequency, f_T, can reach only about 1 MHz. It is a result of the fact that in such constructions only the edge part of the Schottky contact influences the device operation. For high frequency operation new constructions of the AlGaN/GaN heterostructures Schottky diode have to be proposed [2].

2. Experimental

Based on the performed numerical simulations at WEMiF WUT the planar inter-digitated structure of AlGaN/GaN/semi-insulating substrate Schottky diodes was elaborated (Fig.1b). The test structures of the diode was fabricated in AlGaN/GaN heterostructures (Fig.1c) grown on a c-plane sapphire in a vertical flow MOVPE (Metalorganic Vapour Phase Epitaxy) system at atmospheric pressure. The 30 nm thick Si-doped ($N_D = 3\times10^{18}$ cm^{-3}) $Al_{0.16}Ga_{0.84}N$ layer was grown on 2 μm thick un-doped high-temperature GaN layer deposited on a low-temperature GaN buffer layer [3]. The deposited AlGaN/GaN heterostructure was Ga-polar and the 2DEG was induced at the AlGaN/GaN interface. The concentration of the 2DEG was determined using capacitance-voltage (C-V) measurement as 3.75×10^{12} cm^{-2} [4]. Mesa structures were fabricated by reactive ion etching technique. The chlorine plasma was applied to obtain the required etching velocity and shape of mesa.

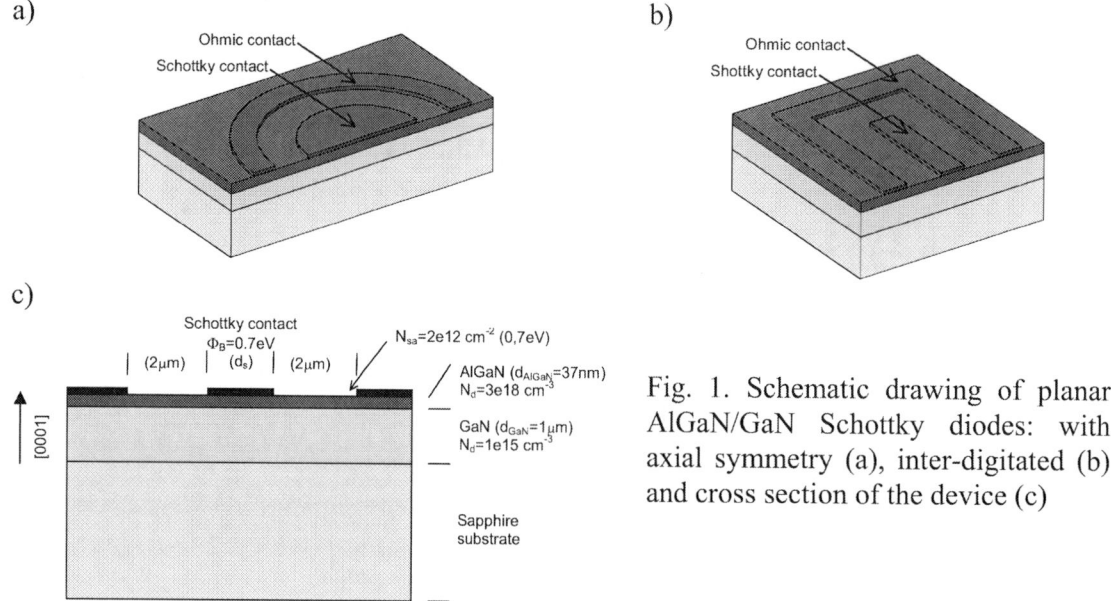

Fig. 1. Schematic drawing of planar AlGaN/GaN Schottky diodes: with axial symmetry (a), inter-digitated (b) and cross section of the device (c)

The ohmic contact was made using Ti/Al/Ni/Au metallization deposited in a single UHV process, formed in the lift-off process, and followed by optimized rapid thermal annealing (RTA) at 830°C for 25 s under a N_2 ambient. The Ru/Au Schottky contacts were evaporated by e-beam technique and formed in the lift-off process. The inter-digitated anode length was 1 μm, and the width was 100 μm.

3. Measurements and discussion

The d.c. and high frequency characteristics of the device were measured. On wafer S parameters measurement of the forward biased diode was performed at frequency range from 10 MHz to 6 GHz for different currents (20 μA÷100 μA) (Fig.2a).

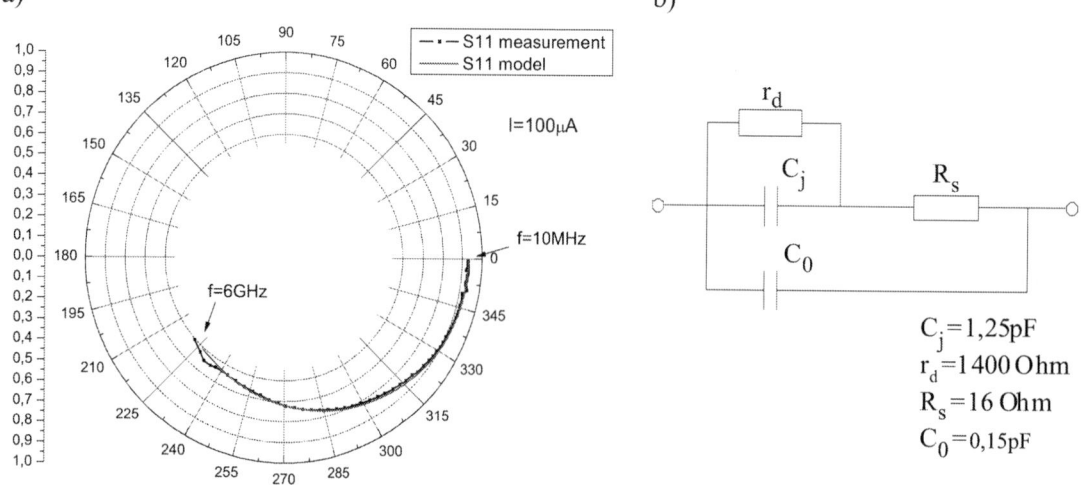

Fig. 2. S parameters (a) and equivalent circuits (b) of the interdigitated AlGaN/GaN Schottky diode (r_d – dynamic resistance, R_s – series resistance, C_j- junction capaciatance, C_0 – parasitic capacitance)

Based on the measured S parameters the values of the equivalent circuit elements of the diode were evaluated (Fig. 2b). The measured diode had the capacitance equal to 1,2 pF, its series resistance was 16 Ω. It means that the cut off frequency of the device was as high as 8,2 GHz.

In Figure 3a d.c. voltage of the diode versus available power of the generator is shown. The measurement was performed at 2,5 GHz. The bias current of the diode was kept constant ($R_L=\infty$). The evaluated sensitivity of the device versus available power of the generator is shown in Figure 3b. The obtained sensitivity of the devices was 0,3 mV/μV.

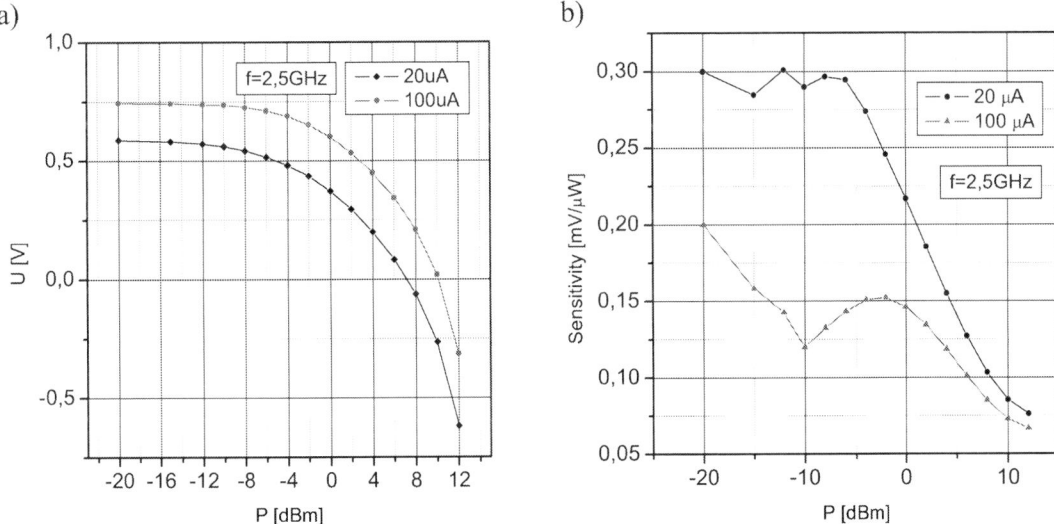

Fig. 3. D.c. voltage of the interdigitated AlGaN/GaN Schottky'ego diode vs. available power of the generator ($R_L=\infty$, f=2,5 GHz) (a), (b) sensitivity of the device versus available power of the generator (b)

4. Conclusion

The obtained results proved that inter-digitated Schottky diode fabricated in AlGaN/GaN heterostructure grown on semi-insulating substrate is suitable for high frequency operation at gigahertz regime and could be monolithically integrated with HEMT. Additionally the fabrication process of both devices is fully compatible. Further optimization of the diode construction and its technology will allow to increase the cut off-frequency.

References

[1] K-Y Wong, W. Chen, K. J. Chen, Integrated Voltage Reference Generator for GaN Smart Power Chip Technology, IEEE Trans. on Elektron Dev., 57, 4, (2010), 95

[2] S. Yoshida, J. Li, N. Ikeda, K. Hataya, AlGaN/GaN field effect Schottky barrier diode (FESBD), phys. stat. sol. (c), 2, 7, (2005), 2602

[3] R. Paszkiewicz, B. Paszkiewicz, J. Kozlowski, T. Piasecki, W. Kosnikowski, M. Tlaczala, Influence of crystallographic structure on electrical characteristics of (Al,Ga)N epitaxial layers grown by MOVPE method, J. Cryst. Growth .248, (2003),487-93

[4] B. Paszkiewicz, Impedance spectroscopy analysis of AlGaN/GaN HFET structures, J. Cryst. Growth, 230 (2001) 590

Acknowledgement

This work has been supported by the European Union within European Regional Development Fund, through grant Innovative Economy (POIG.01.01.02-00-008/08), Polish Ministry of Science and Higher Education under the grant no. NN 515360436, and by Wroclaw University of Technology statutory grant

978-1-4244-8574-1/10 $26.00 © 2010 IEEE

Preparation and properties of ZnO nanomaterials for sensoric applications

T. Brath[1*], D. Búc[1], M. Čaplovičová[2], L. Čaplovič[3], M. Predanocy[1], V. Hrnčiar[4]

[1] Department of Microelectronics, Slovak University of Technology
in Bratislava, Bratislava, Slovakia
[2] Department of Geology of Mineral Deposits, Comenius University, Bratislava, Slovakia
[3] Institute of Material Science, Slovak University of Technology, Trnava, Slovakia
[4] Institute of Technologies and Materials, Slovak University of Technology
in Bratislava, Bratislava, Slovakia
*e-mail: tomas.brath@stuba.sk

ZnO nanorods were grown on SiO_2 substrates using a two-step process. In the first step, the formation of seed layers was performed. In the second step, the growth of nanorods above the seeds was carried out via three different methods. The used methods were magnetron sputtering and two types of hydrothermal methods. The obtained nanomaterial was characterized by scanning electron microscopy and X-ray diffraction. A gas sensor was then prepared using the combination of these techniques and some chosen characteristics of this sensor were measured.

1. Introduction

Recently, zinc oxide has been attracting particular interest because of its remarkable optical and electrical properties. 1D ZnO nanocrystals are considered to be the most promising highly sensing materials of sensors due to the slower electron/hole recombination rate, as well as their high surface-to-volume ratio. Usually, high sensitivity and quick response can be expected for a sensor made from porous materials. In addition, the gases will quickly diffuse into the pores and the surface reactions will take place at a higher rate [1]. Consequently, 1D ZnO nanostructures have been successfully synthesized using various approaches [2]. These methods can be generally classified into two main categories, i.e. vapor-phase and solution phase growth. Vapour-phase processes such as thermal decomposition, chemical vapor deposition and thermal evaporation are favored. Among solution based methods, hydrothermal approach is used for large-scale production of ZnO [2]. In this paper, effects of the growth conditions on the properties of ZnO were investigated. The paper gives a short survey of used methods and the obtained results including a preparation of a gas sensor.

2. Experimental

1.1 Seed layers

Each method of preparation of nanomaterials required the preparation of the so called seed layers. Seed layers were prepared on n-type SiO_2 wafers. These wafers were previously coated with ZnO deposited by RF magnetron sputtering. Consequently the substrates were covered with seed layers. To do this, zinc acetate dehydrate ($Zn(CH_3COO)_2$) was added into ethanol at room temperature to form an ethanol solution of zinc acetate (0.5M). Then the solution was applied dropwise onto the surface of the wafers to create a uniform layer. After evaporating the ethanol the prepared layers were annealed at 380 °C for 3 hours.

1.2 Sputtered ZnO

The first method (*method A*, Tab. 1) utilized to prepare ZnO a nanomaterial was RF magnetron sputtering. The substrates were previously covered with seed layer. Then the substrate with seed layer was deposited via magnetron sputtering using ZnO target. The input magnetron power was 125 W and the process was carried out using argon atmosphere with total pressure of Argon 5×10^{-1} Pa. A 100 nm thick layer of ZnO was deposited.

1.3 High pressure hydrothermal (HPHT) method

This method (*method B*, Tab. 1) uses an autoclave to obtain higher temperature and therefore higher pressure of used solution. Firstly, solutions of two chemicals were prepared. A 0,25 mM solution of deionized water and zinc nitrate hexahydrate ($Zn(NO_3)_2 \cdot 6H_2O$) was prepared. Similarly a 0,25 mM solution of deionized water and hexamethylenetetramine ($C_6H_{12}N_4$) was prepared and then the solutions were mixed together. The prepared sample was immersed in the solution. The autoclave was then hermetically closed and gradually heated to 120 °C. The estimated pressure inside the autoclave according to Wagner and Pruss was approximately 2×10^6 Pa [3]. The temperature was held constant for 7 hours. After 7 hours the autoclave was gradually cooled down to room temperature and the substrate was picked out, rinsed with water, dried and annealed at 380 °C for 1 hour.

1.4 Atmospheric pressure hydrothermal (APHT) method

The third method (*method C*, Tab. 1) uses an atmospheric pressure hydrothermal device. It consists of three main components: a heater, a glass container and a cooler. A 150 ml of the pre-prepared solution mentioned above was put into the glass container. The sample was immersed in the solution. The container was gradually heated to 97 °C. The temperature was then held constant for 7 hours. The evaporating solution condensates on the surface of the cooler and flows back into the container, so the losses of the solution during the process were minimized. After 7 hours the sample was rinsed, dried and annealed for 1 hour at 380 °C.

Tab. 1 Various methods of preparation of the ZnO nanomaterial

Method	1 step	2 step	3 step	ZnO nanomaterial [nm]
A	Seed layer	Annealing 380°C/3 h	MS	30-100 (clusters)
B	Seed layer	Annealing 380°C/3 h	APHT growth	50 (hexagonal nanorods); 100 - 200 (hexagonal nanorods); 40 (flower-like)
C	Seed layer	Annealing 380°C/3 h	HPHT growth	10-30 (nanorods)

1.5 Preparation of a sensor

To confirm the applicability of the previously mentioned processes at a fabrication of sensors, an interdigitally structured sensor was prepared. A platinum interdigital structure was covered with sensing layer via high pressure hydrothermal method (*method B*, Tab. 1).

3. Main Results

Fig. 4 shows a SEM image of the sample prepared by *method A*. We can see the fine dispersed and porous structure in 30 to 100 nm range in diameter. On Fig. 5 SEM image of the sample prepared via atmospheric hydrothermal method is shown. The surface contains

978-1-4244-8574-1/10 $26.00 © 2010 IEEE

nanorods 100-200 nm in diameter. Some of them exhibit a hexagonal structure. This can be understood as an explicit proof that the nanorods grow along the *c*-axis [4].

Fig. 4 SEM image of a sample prepared by magnetron sputtering.

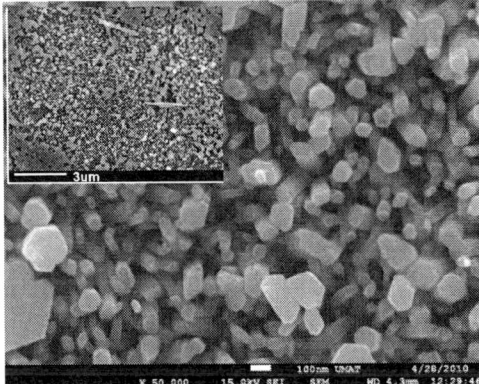

Fig. 5 SEM image of a sample prepared by atmospheric hydrothermal method.

Fig. 6 SEM image of a sample prepared by atmospheric hydrothermal method.

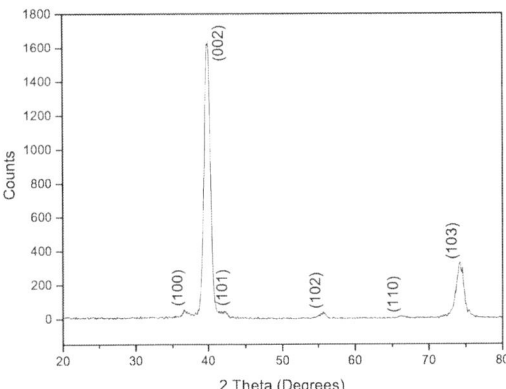

Fig. 7 XRD spectra of the prepared nanomaterial.

SEM image of the sample prepared via high pressure hydrothermal method can be seen on Fig. 6. The surface is homogenously coated with the nanocrystals in hexagonal form. The film consist of nanorods with approximately 10-30 nm in diameter. The nanorods are well vertically aligned and have a high density. It can be seen, that the higher pressure improves the density of the prepared nanomaterial and also allows the preparation of nanorods with smaller diameter. XRD pattern of the grown ZnO layers is shown on Fig. 7. Similar peaks were identified by Zhao et al [5]. It can be seen that the (002) plane has the highest peak intensity suggesting that the layer has a *c*-axis-preferred orientation. Additional peaks along with (100), (101) and (103) planes are also observed.

We measured the resistance vs. time dependence of the sensor as a function of isopropanol $(CH_3)_2CHOH$ concentration. Isopropanol was applied dropwise into the environment (with a 1% concentration) in which the sensor was working (at 120 °C) and the changes of resistance were measured (Fig.8). We found that by adding isopropanol to the air the resistance rose to its maximum and was approximately constant until the contaminated environment was removed, when the resistance dropped back to its original level. The calculated response of the sensor according to equation (1) was approximately 10%.

978-1-4244-8574-1/10 $26.00 © 2010 IEEE   63

$$R = \left(R_g - R_a\right)/R_g \,, \qquad\qquad (1)$$

where R_a and R_g are the resistivity of the sensor exposed to air and isopropanol respectively.

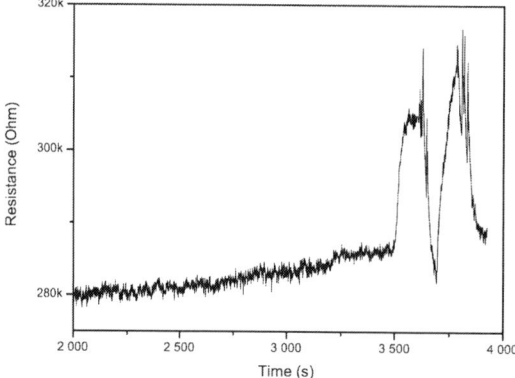

Fig. 8 Sensor response to isopropanol concentration changes

Conclusion

We compared three different methods of preparations of ZnO nanostructures. We had observed some changes in the fine structure as a result of different methods and growth conditions. We confirmed the influence of the pressure in the hydrothermal growth system on the morphology of grown structures. We can explain this effect by the improved kinetics of the reactions in the autoclave. The XRD spectra of the prepared nanorods proved the presence of nanorods grown in a direction perpendicular to the surface of the substrate. The examined surface morphology gives the assumption to use them for different applications in gas and electrochemical MEMS sensors, what was approved by the prepared gas sensor.

Acknowledgements

This paper is the result of the project implementation: Centre for development and application of advanced diagnostic methods in processing of metallic and non-metallic materials, ITMS: 26220120014, supported by ERDF and by the Scientific Grant Agency of the Ministry of Education of the SR and the SAS, No. 1/0337/08, 1/0553/09, 1/0815/08, APVV 0009-07 INTERMATEX.

References

[1] H. Xu, X. Liu, D. Cui, M. Li and M. Jiang. A novel method for improving the performance of ZnO gas sensors. *Sensors and Actuators B*, 114, 301-307, 2006.

[2] D. B. Shouli, C. Liangyuan, L. Dianqing, Y. Wensheng, Y. Pengcheng, L. Zhiyong, C. Aifan, C. C. Liu. *Different morphologies of ZnO nanorods and their sensing property.* Sensors and Actuators B, 146, 129-137, 2010.

[3] W. Wagner and A. Pruss . International Equations for the Saturation Properties of Ordinary Water Substance. Revised According to the International Temperature Scale of 1990. *Journal of Physical and Chemical Reference Data,* 22, 783-787, 1993.

[4] S. N. Bai, H.H. Tsai, T. Y. Tseng. Structural and optical properties of Al-doped ZnO nanowires synthesized by hydrothermal method. *Thin Solid Films,* 516, 155-158, 2007.

[5] J. Zhao, Z. G. Jin, T. Li, X.X. Liu. Nucleation and growth of ZnO nanorods on the ZnO-coated seed surface by solution chemical method. *Journal of the European Ceramic Society,* 26, 2769–2775, 2005.

Influence of thickness on transparency and sheet resistance of ITO thin films

Michał Mazur[*], Danuta Kaczmarek[*], Jarosław Domaradzki[*], Damian Wojcieszak[*], Shigeng Song[**], Frank Placido[**]

[*] Faculty of Microsystem Electronics and Photonics, Wroclaw University of Technology, Janiszewskiego 11/17, 50-372 Wroclaw, Poland
e-mail: michal.mazur@pwr.wroc.pl
[**] Thin Film Centre, University of the West of Scotland, Paisley PA1 2BE, Scotland

Microwave assisted reactive sputtering was applied to obtain homogeneous and high optical quality ITO thin films with thickness of 50, 100, 200 and 280 nm. Electrical properties of deposited ITO thin films were measured using standard four-point probe method together with transmission spectra of ITO thin films in the wavelength range from 330 nm to 880 nm. The figure of merit calculated for all samples has shown meaningful differences in performance of transparent conductive oxides. There was a big difference between 300 nm thick ITO thin films and other samples.

1. Introduction

Indium oxide is a transparent ceramic material, when doped with Tin (usually about 10%) it becomes conducting but stays transparent. This Indium Tin Oxide (ITO) can be applied by vacuum coating to a wide range of glass and plastic materials to make transparent conducting panels for EMC and antistatic applications [1]. ITO thin film is a highly degenerated, widegap semiconductor with a low electrical resistivity and high optical transmission across the visible spectrum [2]. Because of their unique optoelectrical properties, ITO films have many applications, such as in solar cells [3], gas sensor [4], flat panel displays [5]. Conventional ITO films can be reproducibly prepared by various methods including reactive magnetron sputtering [6], electron beam evaporation [7], ion beam assist deposition [8], and pulsed laser deposition (PLD) [9].

In this work ITO thin films prepared by microwave assisted reactive magnetron sputtering process with different thicknesses have been investigated and their electrical and optical properties were determined based on resistivity and optical transmission measurements.

2. Experimental details

ITO thin films were deposited using a magnetron sputtering with a microwave source to improve the plasma ionisation. Microwave assisted reactive sputtering allows to obtain homogeneous and high optical quality ITO thin films. In this project MicroDyn® microwave-assisted reactive magnetron sputtering system was used.

System was equipped with a 120 mm x 372 mm high purity (>99.99%) In-Sn alloy (90:10) target, 10 kW dc power supply and a 3 kW plasma source. The target-to-substrate distance was approximately 122 mm [10]. The substrate holder was rotated with the speed of 1 cycle per second.

Standard microscope slides were placed on a large, rotating drum. Substrates were wiped with isopropanol. The chamber was evacuated using a turbomolecular pump backed by a rotary pump. In the chamber, after reaching the operating pressure, argon and oxygen were introduced and the microwave plasma was turned on for 300 s to pre-clean substrates surfaces from organic residues. The drum was continuously rotating past the microwave position and during the clean step the magnetron was not powered. During rotation of the drum substrates pass magnetron with In-Sn target. Deposited thin film oxidized in plasma enhanced by the microwave outside the sputtered target. Microwave plasma acts as a virtual anode and provides a source of electrons, which reduces arcing. The system was operated with optical monitoring, which allows on controlling of thin film thickness during deposition process. High quality ITO thin films were produced with deposition rate at about 2Å/s.

Four sets of ITO thin films have been prepared with different thickness at. The thickness of ITO thin films varied to observe the dependence of the optical and electrical properties. The thickness of samples was 50, 100, 200 and 280 nm.

During deposition magnetron and microwave power, argon and oxygen flow were kept at the same level. Thickness of ITO thin films was measured during the deposition processes based on optical monitoring of samples. Thickness was fitted to the known n and k value used by this software. After reaching the exact value of thickness the process was stopped.

3. Results

Electrical properties of deposited ITO thin films were determined using standard four-point probe method. The necessary equation for sheet resistance measurements is:

$$R_{sheet} = 4.53 \cdot V/I \ [\Omega/sq], \text{ where: } V - \text{voltage [V], } I - \text{current [A]}$$

The value 4.53 is a correction constant which must be used in sheet resistance calculations. Measurement results are shown in Table 3.1. Results show meaningful differences in sheet resistance between samples with different thickness. Sheet resistance decreases when ITO is thicker. Results of sheet resistance are shown in Fig. 3.1.

The transmittance of the ITO thin films deposited on microscope slide substrates was measured by a Hitachi U-3501 spectrophotometer. Transmittance spectra are also presented in Fig. 3.1. Transmittance is highly dependent on thickness of prepared thin films.

Deposited ITO thin films with different thickness were compared between each other. The figure of merit ϕ_{TC} provides a useful tool for comparing the performance of transparent conductive coatings when their electrical sheet resistance and optical transmission are known. Merit figure should be as high as possible. The figure of merit is defined by [11, 12]:

$$\phi_{TC} = T^{10}/R_{sheet},$$

where:
T – optical transmittance (at 550 nm),
R_{sheet} – sheet resistance.

a)   b)

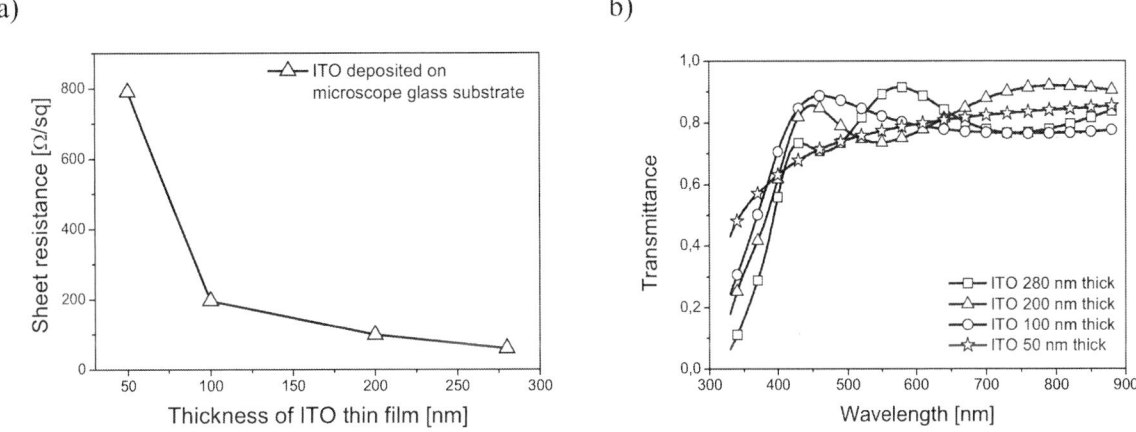

Fig. 3.1. Results of ITO thin films with different thicknesses measurements:
a) sheet resistance, b) transmittance spectra

Results of sheet resistance, average optical transmittance at 550 nm wavelength and merit figure are shown in Table 3.1. Results of merit figure investigation are also shown in Fig 3.2.

Table 3.1. Results of sheet resistance, optical transmittance and merit figure investigation of ITO thin films with various thickness

Parameter:	ITO (50 nm thick)	ITO (100 nm thick)	ITO (200 nm thick)	ITO (280 nm thick)
Sheet resistance [Ω/sq]	790.6	196.62	88.34	59.85
Average transmittance at 550 nm	0.774	0.822	0.804	0.802
Merit figure [1/Ω]	$9.76 \cdot 10^{-5}$	$7.16 \cdot 10^{-4}$	$1.28 \cdot 10^{-3}$	$1.84 \cdot 10^{-3}$

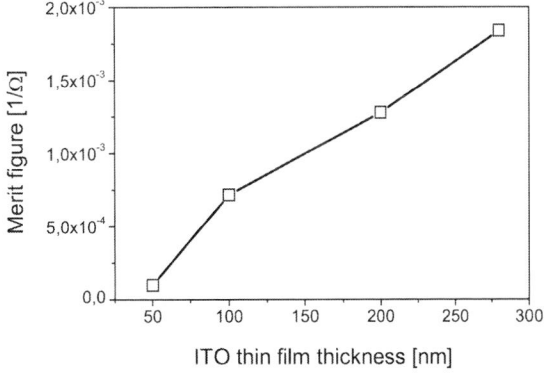

Fig. 3.2. Results of merit figure investigation for ITO thin films with different thickness

The highest merit figure was obtained for 280 nm thick ITO sample due to its high transmittance and low sheet resistance. The lowest merit figure was observed for 50 nm thick ITO sample. It was caused mainly because of very high sheet resistance.

4. Summary

ITO thin films were deposited by microwave-assisted magnetron sputtering process. Samples were deposited on microscope slides. Microwave power, argon and oxygen flow rate and sputtering pressure remained at the same level in case of every sample during the deposition. Electrical investigation has shown big difference in sheet resistance between thin films with different thickness. Samples with thicker ITO film have much lower sheet resistance. Optical properties were obtained using Hitachi spectrophotometer. Transmittance spectra have shown shift of interference peaks which is connected with change of thin films thickness. Performance of ITO thin films was compared for each thickness using figure of merit. This investigation has shown that ITO thin film with thickness of 280 nm has the lowest sheet resistance and simultaneously the highest merit of figure. The smallest figure of merit was obtained for the thinnest ITO thin film.

Acknowledgement

This work was financed from the sources granted by the NCBiR in the years 2008-2010 as a development research project number N R02 0019 04.

References

[1] J. Eite and A.G. Spencer, *Indium Tin Oxide for transparent EMC shielding and Anti-static applications*, Proceedings of EMCUK 2004, Newburry, United Kingdom

[2] J.Y. Lee, J.W. Yang, J.H. Chae, J.H. Park, J.I. Choi, H.J. Park, Daeil Kim, *Dependence of intermediated noble metals on the optical and electrical properties of ITO/metal/ITO multilayers*, Optics Communications 282 (2009) 2362–2366

[3] D. Rached, R. Mostefaoui, Thin Solid Films 516 (2008) 5087

[4] V.S. Vaishnav, P.D. Patel, N.G. Patel, Thin Solid Films 487 (2005) 277

[5] U. Betz, M. Olsson, J. Martly, M. Escola, Surf. Coat. Technol. 200 (2006) 5751

[6] W. Wu, B. Chiou, Thin Solid Films 298 (1997) 221

[7] D. Raoufi, A. Kiasatpour, H. Fallah, A. Rozatian, Appl. Surf. Sci. 253 (2007) 9085

[8] L. Meng, J. Gao, R. Silva, S. Song, Thin Solid Films 516 (2008) 5454

[9] C. Viespe, I. Nicolae, R. Medianu, Thin Solid Films 515 (2007) 8771

[10] L.J. Meng, F. Placido, *Annealing effect on ITO thin films prepared by microwave-enhanced dc reactive magnetron sputtering for telecommunication applications*, Surface and Coatings Technology, vol. 166, 2003, p. 44–50

[11] G. Haacke, *New figure of merit*, Journal of Applied Physics, vol. 47, No. 9, 1976

[12] Y.S. Kim, J.H. Park, D.H. Choi, H.S. Jang, J.H. Lee, H.J. Park, J.I. Choi, D.H. Ju, J.Y. Lee, Daeil Kim, *ITO/Au/ITO multilayer thin films for transparent conducting electrode applications*, Applied Surface Science, vol. 254, 2007, p. 1524–1527

Thermoelectrical properties of TiO$_2$:(Co, Pd) and TiO$_2$:Nb thin films

E. Prociów, M. Mazur, J. Domaradzki, D. Wojcieszak, D. Kaczmarek, T. Gawor

Faculty of Microsystem Electronics and Photonics, Wroclaw University of Technology,
Janiszewskiego 11/17, 50-372 Wroclaw, Poland
e-mail: michal.mazur@pwr.wroc.pl

In this work thermoelectrical properties of TiO$_2$:(Co, Pd) and TiO$_2$:Nb thin films have been described. Thin films were performed by high energy magnetron sputtering method. Sputtering process was carried out from mosaic targets under low pressure of oxygen reactive gas. Electrical and thermoelectrical properties of as deposited and annealed at 800 K thin films were analyzed based on resistivity and thermoelectrical voltage measurements. Results have shown that manufactured thin films had different type of electrical conductivity and good thermoelectrical stability.

1. Introduction

Strict regulations for limiting of exhaust fumes from cars and factories, requirements for environmental monitoring, growing demands for electronic control of production process have stimulated searching for a new generation of sensors [1, 2]. Nowadays industry is searching for new materials for sensors applications, which can work in hard conditions. Materials for sensors applications should keep stable parameters and should be independent for time of work, high temperature etc. One group of materials with good stability is metal oxides. It is reported that oxides have better parameters than classical semiconductors especially from sensors application point of view [3-5].

Nowadays thin films with thermoelectric effects are used in many fields of sensor applications. Temperature difference between two junctions of thermoelectric material generates thermoelectrical voltage, which is proportional to the temperature differences [6]. In case of thin films based on metal oxides it can give thermoelectrical response for changes of temperature or gas atmosphere [7].

In this work thermoelectrical properties of TiO$_2$ thin films doped with Nb (TiO$_2$:Nb) and with Co and Pd (TiO$_2$:(Co, Pd)) have been described. Analysis of electrical and thermoelectrical properties of as-deposited and annealed at 800 K thin films was performed based on resistivity and thermoelectrical voltage measurements.

2. Experimental details

Thin films were manufactured by high energy magnetron sputtering method. Particles were sputtered from mosaic Ti-Co-Pd and Ti-Nb targets under low atmosphere of reactive oxygen gas (0.1 Pa). Deposition process was described in earlier work [8]. All manufactured thin films were deposited on Corning 7059 substrates. For electrical measurements two NiCrSi-Ag ohmic contacts have been deposited on opposite sites of samples thought metallic masks by magnetron sputtering process.

Composition of manufactured thin films was measured by energy dispersive spectroscopy (EDS). EDS results have shown that in TiO$_2$:(Co, Pd) thin film was 15.8 at. % of Co and 6.9

at. % of Pd, while in TiO$_2$:Nb thin film was 4 at. % of Nb-dopant. Thickness of thin films was determined with the aid of Fizeau interferometer equipped with Hg (551 nm) filtered lamp. Results have shown that the thickness of TiO$_2$:(Co, Pd) and TiO$_2$:Nd thin films was 495 nm and 400 nm, respectively.

Measurements of resistivity and thermoelectrical voltage were performed to determined electrical stability of manufactured thin films. Electrical parameters were investigated for as-deposited and annealed at 800 K thin films. Thin films were annealed by 5, 10, 20, 50, 100 and 200 minutes. For electrical measurements Meratronik V533 multimeter was used, while temperature was controlled by Pt100 sensor. Thermoelectrical voltage was investigated with the aid of specialist system, which allows on electrical measurements in temperature range from 0 to about 220°C.

3. Results and discussion

In Fig. 1 characteristics of resistivity in dependence of post-process annealing time of TiO$_2$:(Co, Pd) and TiO$_2$:Nb thin films have been presented. Thin films were annealed at 800 K, while resistivity was measured at room temperature (20°C). Results have shown that according to resistivity criterion both manufactured thin films were semiconducive [9, 10]. Resistivity variations at the begging of annealing process (to about 20 minutes) were caused by process of structure forming, where structural defects were dropped away. This effect was described by authors also in previous paper [4].

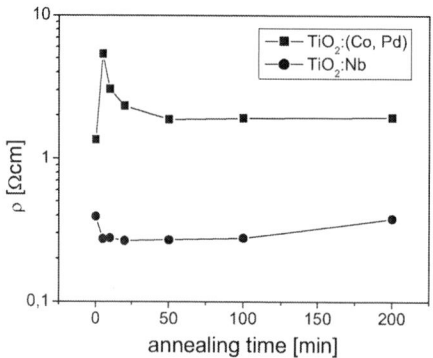

Fig. 1. Characteristics of resistivity in dependence of post-process annealing time of TiO$_2$:(Co, Pd) and TiO$_2$:Nb thin. Thin films were annealed at 800 K, while resistivity was measured at room temperature (20°C).

In Fig. 2 characteristics of thermoelectrical voltage U in differential temperature $\square$T function have been presented. Thermoelectrical voltage of as-deposited and annealed at 800 K thin films was measured from room temperature (20°C) up to 220°C. Thermoelectrical voltage increased after annealing process for both films. In case of TiO$_2$:(Co, Pd) thin film voltage was almost independent from annealing time over annealing by 20 minutes, while for TiO$_2$:Nb thin film thermoelectrical voltage increased with increase of annealing time. From well known formula Seebeck coefficient S for both structures was calculated [11]:

$$S = \lim_{\Delta T \to O} \frac{\Delta U}{\Delta T}$$

978-1-4244-8574-1/10 $26.00 © 2010 IEEE

where: S is Seebeck coefficient, ΔU – difference of thermoelectrical voltage, ΔT – difference of temperature.

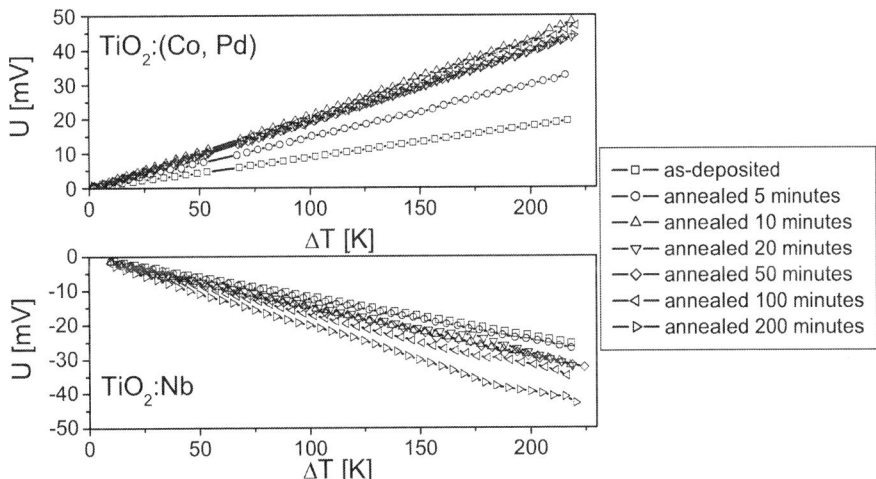

Fig. 2. Characteristics of thermoelectrical voltage vs. temperature difference of as-deposited and annealed at 800 K TiO_2:(Co, Pd) and TiO_2:Nb thin films. Thermoelectrical voltage was measured from room temperature (20°C) up to 220°C.

Seebeck coefficient characteristics have been presented in Fig. 3. Thermoelectrical voltage was measured from room temperature (20°C) up to 220°C. As it can be observed on the plot, module of Seebeck coefficients increased after annealing. The sign of Seebeck coefficient in case of TiO_2:(Co, Pd) thin film was positive, while for TiO_2:Nb thin film it was negative. That testifies about different type of electrical conductivity. TiO_2:Nb thin film had n-type, while TiO_2:(Co, Pd) film had p-type of electrical conductivity.

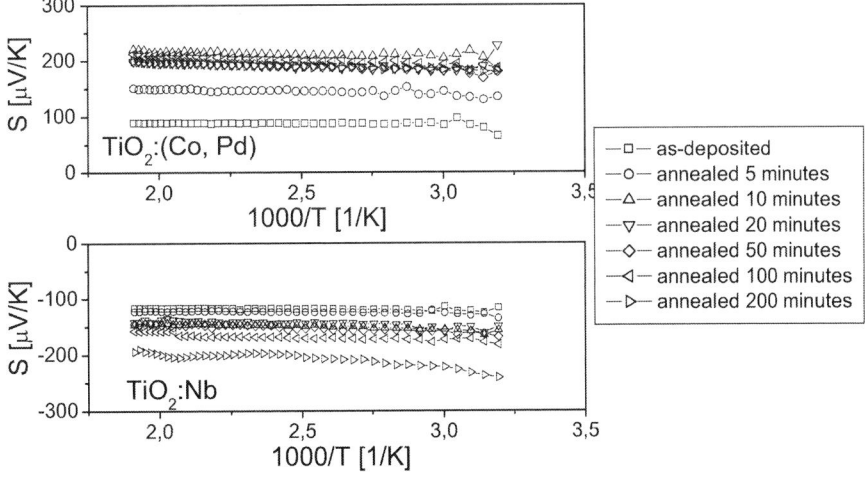

Fig. 3. Seebeck coefficient S vs. temperature difference of as-deposited and annealed at 800 K TiO_2:(Co, Pd) and TiO_2:Nb thin films. Thermoelectrical voltage was measured from room temperature (20°C) up to 220°C.

4. Summary

In this work electrical and thermoelectrical stability of TiO_2:(15.8 at. % Co, 6.9 at. % Pd) and TiO_2:(4 at. % Nb) thin films have been presented. It was confirmed that both thin films had typical resistivity for semiconducive materials. Thermoelectrical measurements have shown that thin film with Nb-dopant had n-type, while film doped with Co and Pd had p-type of electrical conductivity. Moreover, thermoelectrical measurements have shown that in case of both manufactured thin films thermoelectric effect was observed and they were electrically stable even in high temperatures.

Acknowledgement

This work was financed from the sources given by the NCBiR in the years 2009-2012 as a development research project number N N508 1329 37 and from the statute sources given by MNiSW in years 2010 – 2011.

References

[1] McAlfeer J. F.,. Moseley P. T, Bourke P., Norris J. O. W., Stephan R., Tin Dioxide Gas Sensors: Use Of The Seebeck Effect, Sensors and Actuators vol 8 (1985), p. 251 – 257

[2] Ogita M., Higo K., Nakanishi Y., Hatanaka Y., Ga_2O_3 thin film for oxygen sensor at high temperature, Applied Surface Science, vol. 175-776 (2001), p. 721-725

[3] Edelstein A.S., Cammaratra R.C. (Eds.), Nanomaterials: Synthesis Properties and Applications, Institute of Physics Publishing, ISBN 0-7503-0358-1, Bristol 1996,

[4] Prociów E., Łapiński M., Zieliński M., Kaczmarek D., Domaradzki J., Sieradzka K., Mazur M., Gawor T., Badanie właściwości elektrycznych w warunkach starzenia termicznego tlenków metali do zastosowań w czujnikach. XXIX Konferencja Elektroniki i Telekomunikacji Studentów i Młodych Pracowników Nauki, SECON 2010 p. 10,

[5] Fettig R., A view to recent developments in thermoelectric sensors, 15th International Conference on Thermoelectrics, IEEE 1996

[6] Buchner R, Sosna C., Maiwald M., Benecke W., Lang W., A high-temperature thermopile fabrication process for thermal flow sensors, Sensors and Actuators A, vol. 130–131 (2006), p. 262–266

[7] Shin W., Matsumiya M., Izu N., Murayama N., Hydrogen-selective thermoelectric gas sensor, Sensors and Actuators B, vol. 93 (2003), p. 304–308

[8] Domaradzki J., Borkowska A., Kaczmarek D., Prociow E., Transparent oxide semiconductors based on TiO_2 doped with V, Co and Pd elements, Journal of Non-Crystalline Solids, vol. 352 (2006), p. 2324–2327

[9] Schroder D. K., Semiconductor material and device characterization, Third edition; IEEE press, A John Wiley & Sons, Inc., Publication

[10] Seeger K., Semiconductor Physics, Springer-Verlag (1973)

[11] Goto T., Li J.H., Hirai T., Maeda Y., Measurements of the Seebeck Coefficient of Thermoelectric Materials by an AC Method, International Journal of Thermophysics, vol. 18, no.2 (1997), p. 569-577

Effect of Substrate Temperature on Oblique-Angle Sputtered ZnO:Ga Thin Films

I. Novotný[1], D. Kotorová[1], S. Flickyngerová[1], V. Tvarožek[1], L. Spiess[2], P. Schaaf[2], M. Netrvalová[3] and P. Šutta[3]

[1] Department of Microelectronics, Slovak University of Technology, Ilkovičova 3, 812 19 Bratislava, Slovakia

[2] Department of Materials for Electronics, Institute of Materials Engineering and Institute of Micro- and Nanotechnologies, Ilmenau University of Technology, Ilmenau, Germany

[3] Department of Materials & Technology, New technologies - research centre, University of West Bohemia, Univerzitní 8, 306 14 Plzen, Czech Republic

e-mail: ivan.novotny@stuba.sk

A study of the effect of substrate temperature on oblique-angle sputtered galium-doped zinc oxide (GZO) thin films was carried out. Both the oblique-angle sputtering and the substrate temperature lowered the resistivity of GZO thin films down to 4 x 10^{-3} Ωcm together with an increase of their optical transmittance over 90%.

1. Introduction

Transparent conducting oxides based on ZnO are promising for application in thin-film photovoltaic cells and various optoelectronic devices [1]. The "directional" physical vapor deposition techniques, particularly evaporation, are used for preparation of sculptured thin films - a new class of optical materials whose columnar morphology is tailored to elicit desired optical properties - and they are applicable for engineering of the nanoscale morphology (Lakhtakia et al. [2]) as well as in preparation of nanoporous and nanostructured films (Plawsky et al. [3]). The film morphologies are dependent on deposition parameters including the deposition rate, the angular distribution of the incident deposited flux, the film and substrate temperature and the energetic parameters of the surface-substrate interface [4]. We indicated previously (Cerven et al. [5]) the possibility to prepare textured ZnO thin films with inclined c-axis by RF diode sputtering and we deposited transparent conductive ZnO:Ga (GZO) thin films by two oblique-angle sputtering arrangements at room temperature [6].

Our present aim was to investigate the influence of the substrate temperature on the properties of oblique-angle sputtered GZO thin films.

2. Technology and Experimental

Ceramic target, ZnO+2% Ga$_2$O$_3$ of 102.4 mm diameter, was RF diode sputtered in Ar for preparation of GZO thin films (with thicknesses of 0.5 - 1 μm) on Corning glas substrates at temperatures of room and 200°C. Oblique-angle deposition arrangement, „A" was used for preparation of ZnO:Ga thin films (Fig. 1).

The thin film morphology was observed by a SEM Philips XL30 (cross-section) and a X-ray diffractometer AXS Bruker D8 equipped with a position sensitive area detector Histar (crystalic texture). Electrical properties were measured by Van der Pauw method. Optical transmittance was measured by Avantes AvaSpec 2048 spectrometer.

978-1-4244-8574-1/10 $26.00 © 2010 IEEE

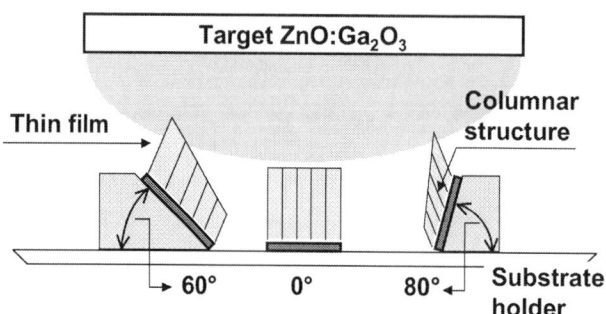

Fig.1. Schematic layout of oblique sputtering arrangement

3. Results and discussion

During thin film growth, the surface diffusion strongly influences the final morphology and texture of the film. Thin films deposited in non-oblique arrangement at the room temperature showed a weak prefered orientation (002) of crystalities (Fig. 2, A0). The elevation of substrate temperature as well as the local self-heating of the film during sputtering has increased the surface diffusion of deposited particles. It caused a strong improvement of both the crystalline texture in [001] direction (Fig. 2, AT0) and the columnar structure perpendicular to substrate (Fig.3, AT0).

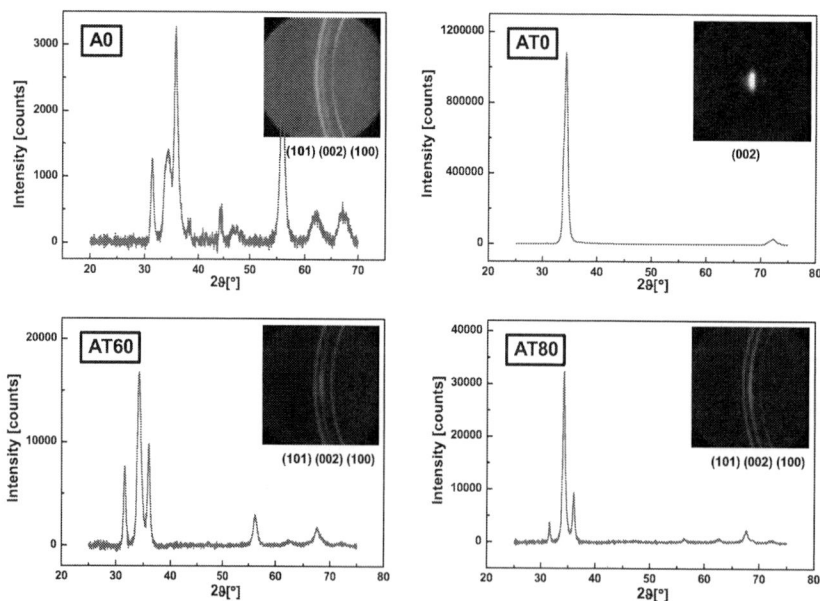

Fig. 2. XRD patterns and azimuthal line profiles of GZO films sputtered at: room substrate temperature/0°-angle (A0), 200°C/0°-angle (AT0), 60°-angle (AT60), 80°-angle (AT80)

An increase of the mean grain size to the value about 100 nm at elevated substrate temperature gave rise to carrier Hall mobility from 1.5 cm^2/Vs (RT) to 3.4 cm^2/Vs (200°C), Fig. 4. Simultaneously carrier density was falling down to value of 2.4 x 10^{19} cm^{-3} what caused only a slight rise of resistivity in the range of 10^{-2} Ωcm.

978-1-4244-8574-1/10 $26.00 © 2010 IEEE   74

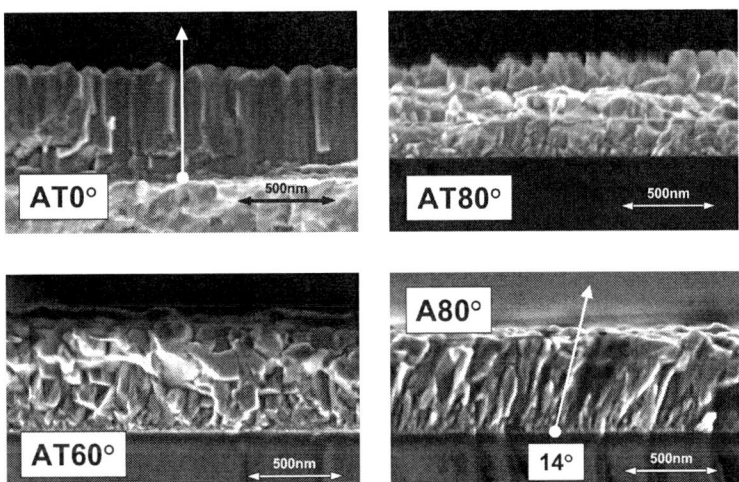

Fig. 3. SEM cross-sections of GZO films sputtered at normal/oblique-angles and substrate temperature 200°C (AT0, AT60, AT80) and room temperature (A80)

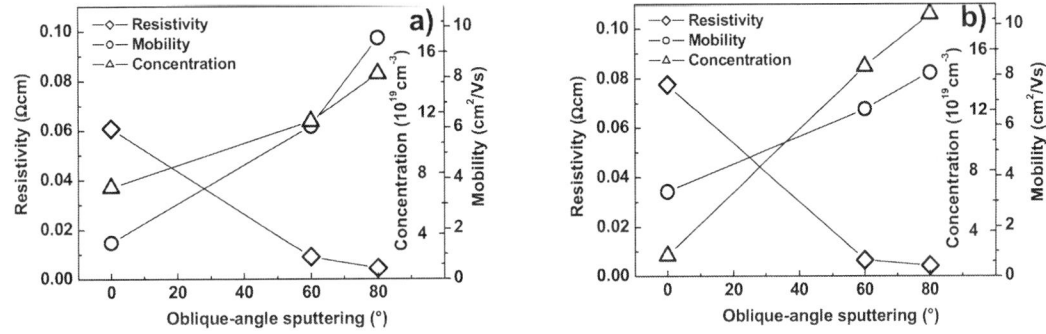

Fig. 4. Electrical properties of GZO films sputtered at different oblique-angles and substrate temperatures: room temperature RT (a), 200 °C (b)

In case of oblique-angle sputtering two main competing mechanisms determine features of the growing film: (i) self-shadowing of impinged atoms by neighbouring atoms, conservation of their parallel momentum / directional surface migration and (ii) random surface diffusion of deposited particles, in contrary. The oblique-angle sputtering itself has improved the GZO thin-film columnar structure which was deviated about 14° from the surface normal (Fig. 3, A80). The consequence of that was an increase of carrier Hall mobility to the maximal value of 9.6 cm^2/Vs suggesting that the grain boundary scattering was relatively constrained (Fig. 4 a). That caused the decrease of resistivity from the range of 10^{-2} Ωcm down to 4 x 10^{-3} Ωcm. Additionally, resistivity was affected by an increase of the carrier concentration up to the range of 10^{20} cm^{-3} which seems to be the limit because the impurity scattering is getting dominant over this value. The oblique-angle sputtering at substrate temperature of 200°C followed the tendency of an improvement of electrical properties mentioned previously (AT80 with the minimal resistivity of 4.32 x 10^{-3} Ωcm, Fig. 4 b) in spite of growing of non-inclined columnar structure with worsen preferred orientation of crystallities (Figs. 2 and 3).

The optical transmittance in the UV spectrum region (blue–shift of the absorption edge) gives the information about the width of optical band-gap (Morkoc and Ozgur [7]). The direct optical band-gap Eg was obtained by plotting and extrapolation of $(\alpha h\nu)2$ vs. hv. It was slightly widened by the oblique-angle sputtering at elevated substrate temperature: 3.36 eV at 200°C in comparison with 3.15 eV at room temperature (Fig. 5a). The oblique-angle sputtering along with an increase of substrate temperature enhanced the transmittance up to 90 % (Fig. 5b).

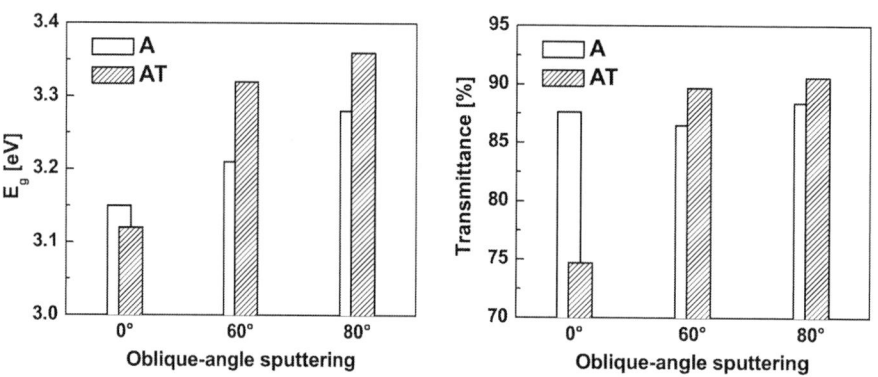

Fig. 5 Influence of different oblique-angles and substrate temperatures on optical properties of sputtered GZO films: room temperature RT (A), 200°C (AT)

4.   Conclusion

Comparison of our results in oblique-angle sputtering confirmed that the surface diffusion determined by the energy of sputtered particles and the substrate/film temperature significantly affected the crystallographic texture and an inclination of the columns. Both technological parameters, the oblique-angle sputtering and the substrate temperature, lowered the resistivity of GZO thin films together with an increase of their optical transmittance. It seems that electro-optical properties of GZO thin films were more affected by the oblique-angle sputtering.

Acknowledgments

Presented work was supported by the MSMT Czech Republic project 1M06031 and SK VEGA project 01/0220/09. We thank to Dr. K. Pfeifer and E. Remdt for the careful realization of SEM pictures.

References

[1] M. Zeman, Advanced Amorphous Silicon Solar Cell Technologies, In: *Thin film solar cells: fabrication, characterization and applications*, page numbers (173-236), John Wiley & Sons, New York, 2007
[2] A. Lakhtakia, M. C. Demirel, M. W. Horn, and J. Xu, *Advances in Solid State Physics*, **46**, 295, 2008.
[3] J. L. Plawsky, J. K. Kim, and E. F. Schubert, *Materials Today*, **12**, 36, 2009.
[4] L. Abelmann, and C. Lodder, *Thin Solid Films*, **305**, 1, 1997.
[5] I. Cerven, T. Lacko, I. Novotny, V. Tvarozek, and M. Harvanka, *Journal of Crystal Growth*, **131**, 546, 1993.
[6] V. Tvarozek, I. Novotny, P. Sutta, M. Netrvalova, S. Flickyngerova, L. Spiess, and P. Schaaf, in *Proceedings of the 27th International Conference on Microelectronics MIEL*, Niš, Serbia, 2010, p. 177.
[7] H. Morkoc, and U. Ozgur, *Zinc Oxide: Fundamentals, Materials and Device Technology*, Wiley-VCH Verlag, Weinheim, 2009.

Nanocrystalline Diamond/amorphous Composite Carbon Films Prepared by PECVD Technology for Photocathode Application

J. Huran[1], N.I. Balalykin[2], G.D. Shirkov[2], P. Boháček[1], A.P. Kobzev[2], A. Valovič[1]

[1]Institute of Electrical Engineering, Slovak Academy of Sciences, Dúbravská cesta 9, Bratislava, 841 04, Slovakia
[2]Joint Institute for Nuclear Research, Joliot-Curie 6, 141980 Dubna, 141980, Russia
e-mail: elekhura@savba.sk

Nanocrystalline diamond/amorphous composite carbon films were deposited by plasma enhanced chemical vapour deposition method. The concentrations of species in the films were determined by RBS (Rutherford backscattering spectrometry) and ERD (elastic recoil detection) methods. The RBS results showed the main concentrations of C in the films. The concentration of hydrogen was approximately 20 at.%. Chemical compositions were analyzed by FTIR spectroscopy. IR results showed the presence of C-H specific bonds. Film was used for photocathode application. The original quantum efficiency of prepared photocathode at energy of FH 15,6 mJ was $1,43x10^{-6}$ %.

1. Introduction

Nanocrystalline diamond, ultrananocrystalline diamond or amorphous carbon embedded NCD (NCD/a-C) films, have advantages of having higher surface flatness, high hardness, high wear resistance, high thermal conductivity, low friction coefficient, high electrical resistance, high optical transparency, high electron emission efficiency and excellent chemical inertness. The properties of deposited films are generally characterized by powerful ex-situ techniques that are commonly available, for instance Scanning Electron Microscopy (SEM), Atomic Force Microscopy (AFM), Transmission Electron Microscopy (TEM), Raman spectroscopy or X-ray Diffraction (XRD). Diamond films have been extensively investigated in field electron emission (FEE) [1], [2]. J.E. Yater et al. showed that grains may impede electron transport in diamond films and argued that the reduction of grain sizes is important for diamond film to be used as a cold cathode electron source [3]. This mechanism is actually similar as recently proposed conduction channel model, in which the grain boundary area can act as electron conduction channel in diamond field electron emission. In the electron conduction model, the diamond grain boundary plays the main role, as grain boundaries of diamond film consist of sp^2 phase. The sp^2- bonded regions are of low electrical resistivity and act as an electron transport path, which facilitates the field electron emission. The plasma was electrically studied by a Langmuir probe in PECVD system [4]. There are several carbon-based photo cathodes, like polycrystalline diamond, hydrogenated amorphous carbon and nanostructured fullerene films. Polycrystalline diamond photocathodes are chemically inert, have a high damage threshold but also a low QE of 10^{-6}. Hydrogenated diamond photocathodes have the highest QE's of $8\cdot10^{-4}$ for 213 nm wavelength, but have low damage threshold and become oxidized after irradiation.

In this study, we investigated properties of nanocrystalline diamond/amorphous composite carbon (NCD/a-C) films prepared by plasma enhanced chemical vapor deposition (PECVD).The properties of films were investigated by RBS, ERD and IR measurement

978-1-4244-8574-1/10 $26.00 © 2010 IEEE

techniques. Property of prepared photocathode was performed by measurement of quantum efficiency.

2. Experiments

The methane was introduced into capacitively coupled plasma reactor through the shower head, which is also an upper electrode with 20 cm diameter. Gas was flown vertically toward the substrate on bottom electrode connected with RF power 150 W and frequency 13.56 MHz. A p-type silicon wafer with resistivity 2-7 Ωcm and (100) orientation was used as the substrate for the carbon films. Prior to deposition, standard cleaning was used to remove impurities from the silicon surface, and the 5% hydrofluoric acid was used to remove the native oxide on the wafer surface. The wafer was then rinsed in deionized water and dried in nitrogen ambient. The flow rate of CH_4 gas was 40 sccm. The deposition temperature was for sample P1 – 400 ^{o}C and P2 – 500 °C. Spectroscopic ellipsometry was used for film thickness measurements and results are: for samples P1 -315 nm and P2 -325 nm. The concentration of species in the carbon films was determined by Rutherford backscattering spectrometry (RBS). Chemical compositions were analyzed by infrared spectroscopy.The hydrogen concentration was determined by the elastic recoil detection (ERD) method. For this purpose the $^4He^+$ ion beam from a Van de Graaff accelerator at JINR Dubna was applied. The energy of E = 2.4 MeV was chosen. The target was tilted at an angle $\alpha = 15°$ with respect to the beam direction and the recoiled protons were measured in forward direction at an angle Θ_1 (30°) by a surface barrier detector. The quantum efficiency (QE) testing of prepared photocathode was performed at JINR. At one side of the cathode test facility vacuum chamber a fused silica window is mounted that transmits UV light. The vacuum condition was 4×10^{-9} mbar. The 15 ns UV laser pulses (quadrupled Nd:YAG laser) are used to illuminate the (NCD/a-C) film as photocathode. Laser spot size 5 mm. During testing, the laser energy was monitored using a calibrated portion of the signal that was picked off from the main beam. To draw the electrons from the cathode a positive voltage was placed on the anode-extractor. This voltage was kept at roughly 5 kV. The photocathode current is measured by using an oscilloscope.

3. Results and discussion

An example of plasma optical emission spectrum generated by a CH_4 glow discharge is show in Fig. 1. By using OES the $CH(X^2\Pi)$ radical number density using actinometry method and plasma composition are determined. Figure 2 show RBS spectra of two samples P1 and P2 with different deposition conditions of the deposited carbon films. The (NCD/a-C) films contained C, H and also other species which were under the detection limit of RBS method. From ERD measurement, it follows that the concentration of hydrogen in thin films depends on the deposition conditions. ERD analyses made on prepared layers show that amount of incorporated H was decreased from 21 at.% to 17 at.% with increasing of deposition temperature. The values were obtained by computer modeling of measured spectra and compared with the results obtained from the Si reference sample implanted with H. In the case of sample P1 the concentration of hydrogen and carbon are 21 and 75 at.% respectively. The concentrations of H and C in sample P2 are 17 and 77 at.%, respectively. The FTIR spectra revealed the main absorption region between 1200-1600 cm^{-1} and 2800-3150 cm^{-1}. The most important result is that the sp^3 hybridization is stronger in the sample deposited at higher temperatures. At lower wave numbers for the low temperature sample the sp^2 CH olefinic related peak is more pronounced compared to the high temperature case. For both samples beside the sp^3 bonds sp^2 and graphite like related peaks can be assigned.

The photocathode current is measured by using an oscilloscope as a voltage on resistance 47 Ω, see Fig.3. QE is defined as ratio of numbers of emitted electrons and injected laser photons. QE in % is expressed with the conventional parameters as QE [%] = 123,8 I/ λ P,

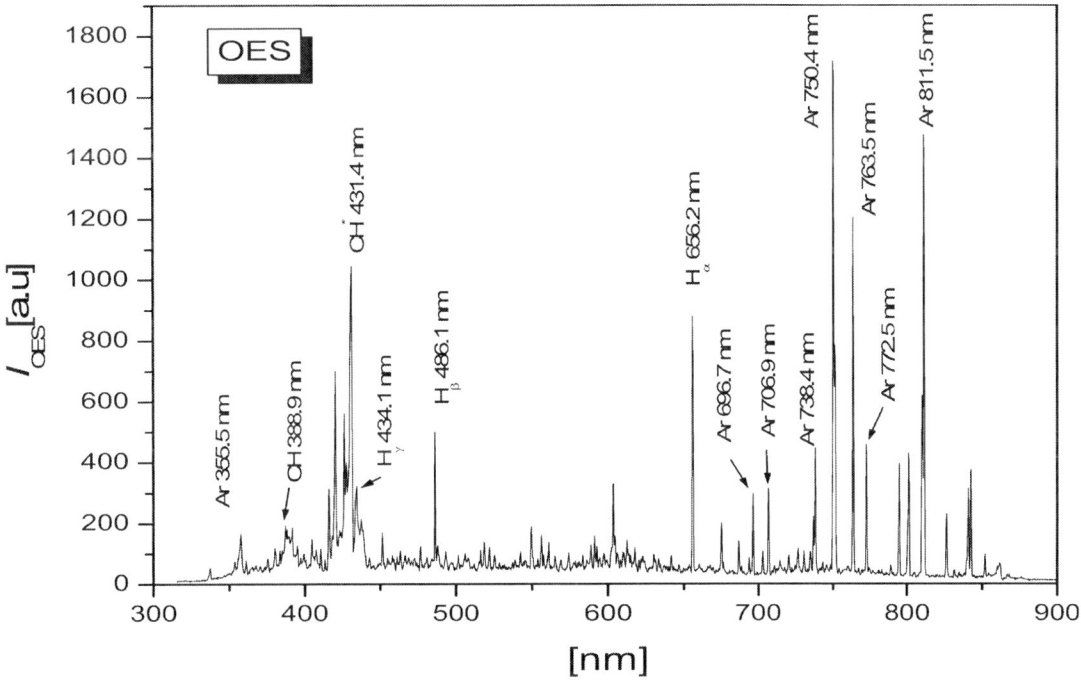

Fig. 1. Optical emission spectrum of CH_4 glow discharge at 10 Pa with small amount of Ar for actinometry method.

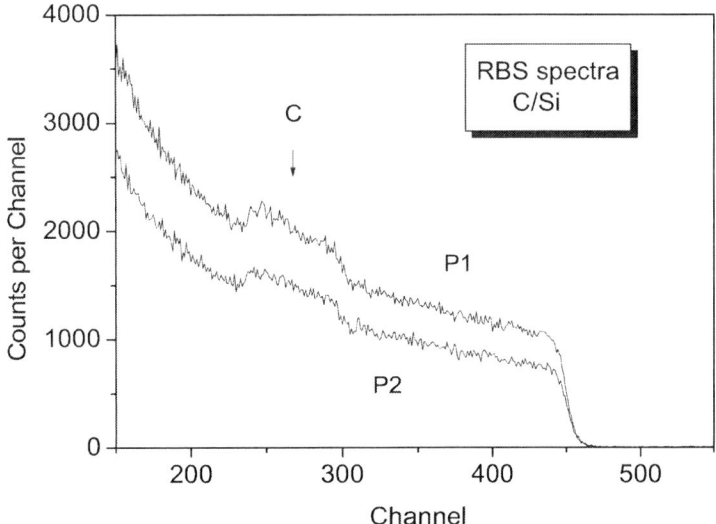

Fig. 2. RBS spectra of (NCD/a-C) films deposited onto a silicon substrate for 2 MeV alpha particles detected at scattering angle of 170°. The spectra are for samples P1 and P2.

where I is emitted current of the electron beam in mA, λ is wave length of the laser light in nm, P is power of the laser light in Watt. In our case: $I = 0,85$ A; $\lambda = 266$ nm; $P = 0,4$ M W. The original quantum efficiency at energy of FH 6 mJ had been 1×10^{-3} %.

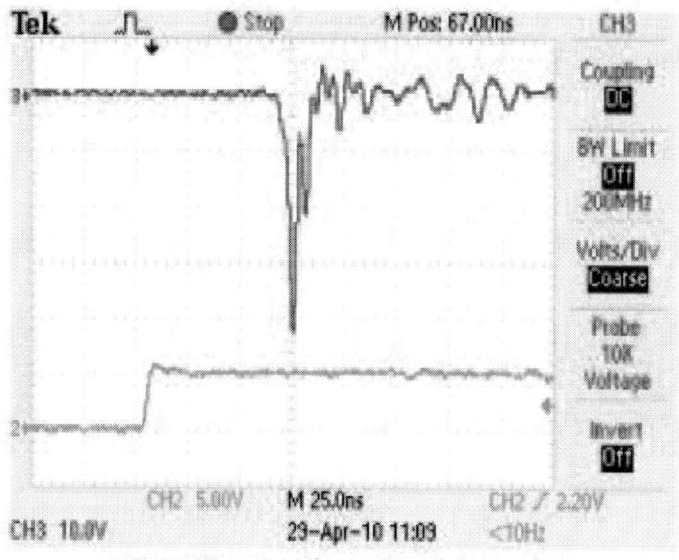

Fig. 3. The (NCD/a-C) photocathode current pulse-red line

3.   Conclusion

The RBS results showed that the concentration of C in the films dependent a little on the deposition temperature. The films contain a small amount of oxygen and nitrogen. The concentration of hydrogen dependent on deposition temperature and increases from 17 to 24 at.% with decreasing of deposition temperature. The results presented above demonstrate that 2.4 MeV $^{4}He^{+}$ ERD analyses may be successfully used to measure the hydrogen concentration. FTIR results showed the presence of C-H specific bonds. The original quantum efficiency of prepared photocathode at energy of FH 6 mJ was 1×10^{-3} %.

Acknowledgement

This research has been supported by the Slovak Research and Development Agency under the contracts APVV-0713-07 and APVV-0459-06 and by the Scientific Grant Agency of the Ministry of Education of the Slovakia and Slovak Academy of Sciences, No. 2/0192/10.

References

[1]   X. Lu, Q. Yang, C. Xiao, and A. Hirose, *Thin Solid Films* **516**, 4217, 2008.
[2]   P.M. Koinkar, P.P. Patil, M.A. More, V.N. Tondare, and D.S. Joag, *Vacuum* **72,** 321, 2004.
[3]   J.E. Yater, A.Shih, J.E. Butler, and P.E. Pehrsson, *J. Appl. Phys.* **93**, 3082, 2003.
[4]   C. Corbella, M.C. Polo, G. Oncins, E. Pascual, J.L. Andújar, and E. Bertran, *Thin Solid Films* **482**, 172, 2005.

N-doped Nanocrystalline Silicon Carbide Films Prepared by PECVD Technology

P. Boháček[1], J. Huran[1], A. Valovič[1], A.P. Kobzev[2], V.N. Shvetsov[2], M. Kučera[1],
L. Malinovský[3]

[1]Institute of Electrical Engineering, Slovak Academy of Sciences, Dúbravská cesta 9,
84104 Bratislava, Slovakia
[2]Joint Institute for Nuclear Research, Joliot-Curie 6, 141980 Dubna, Moscow region, Russia
[3] Institute of Physics, Slovak Academy of Sciences, Dúbravská cesta 9, 84511 Bratislava,
Slovakia
e-mail: pavol.bohacek@savba.sk

This work presents the properties of nanocrystalline SiC(nc-SiC) films prepared by plasma enhanced chemical vapour deposition. A p-type silicon wafer with resistivity 2-7 Ωcm and (100) orientation was used as the substrate for the nc-SiC:H films. The concentration of species in the SiC films was determined by RBS and ERD. Chemical compositions were analyzed by IR spectroscopy. Film morphology was assessed by AFM. The RBS results showed the main concentrations of Si and C in the films. The concentration of hydrogen was approximately 20 at.%. IR results showed the presence of Si-C, Si-O, Si-N, Si-H, N-H, C-H, C-N specific bonds. Results of current-voltage (I-V) measurements before and after samples irradiation by neutrons are presented.

1. Introduction

Thin-film manufacturing methods using gas mixture consisting of two or more gases such as $SiH_4 - CH_4$ is used for the production of thin-film transistors and other electronic devices and industrial products [1]. Silicon carbide has attracted much interest for wide range of applications. With its wide band gap, excellent thermal properties and large bonding energy, silicon carbide films are ideal for optoelectronic blue and ultra-violet wavelength emissions operating at high power levels, high temperatures and caustic environments [2]. In semiconductor nanostructures, quantum confinement effects lead to a modification in the electronic band structure, the vibronic state and the optical emission with respect to the bulk material [3, 4]. SiC has several advantages over other wide-band gap semiconductors at the present time including commercial availability of substrates, known device processing techniques, and the ability to grow a thermal oxide for use as masks in processing, device passivation layers, and gate dielectrics. Furthermore, SiC can also be used as a thin buffer layer for the growth of diamond films on silicon substrates [5]. For example, a-$Si_{1-x}C_x$:H was used as a wide window material to enhance the conversion efficiency of amorphous solar cell. The significance of this material follows from the fact that its electrical and optical properties can be controlled by varying the carbon, silicon and hydrogen composition of the film. PECVD technique offers an attractive opportunity to fabricate amorphous hydrogenated N-doped SiC films at intermediate substrate temperatures and it provides high quality films with good adhesion, good coverage of complicated substrate shapes and high deposition rate [6]. Recently, Si-rich a-SiC_x:H films have attracted new attention in the photovoltaic community, since this material has shown excellent electronic surface passivation of c-Si comparable with thermal SiO_2 and low temperature amorphous silicon nitride (a-SiN_x) passivation [7].

In this contribution the attention has been focused to the properties of silicon carbide films prepared by the plasma enhanced chemical vapour deposition (PECVD) of silane SiH_4 and methane CH_4. The structural properties were investigated by RBS, ERD, IR and PL measurement techniques. Spectroscopic ellipsometry was used for optical characterization of the film. Electrical characterization was made by I-V measurement technique at room temperature before and after neutron irradiation of samples.

2. Experiment

The plasma CVD reactor with parallel plate electrodes was used as reported previously [8]. A p-type silicon wafer with resistivity 2-7 Ωcm and (100) orientation was used as the substrate for the nc-SiC:H films. The flow rates of SiH_4 and CH_4 gases were 10 sccm and 40 sccm, respectively. Small amount of NH_3 (flow 1 sccm for sample B1, 2 sccm-B2 and 4 sccm-B3) was put to gas mixture for N-dopant. The concentration of species in the SiC films was determined by Rutherford backscattering spectrometry (RBS). Chemical compositions were analyzed by infrared spectroscopy. The IR spectra were measured from 4000 to 400 cm^{-1}. The hydrogen concentration was determined by the elastic recoil detection (ERD) method. Film morphology was assessed by AFM. Irradiation of samples by fast neutrons in IREN facility at JINR Dubna was used for radiation hardness investigation. The thickness, refractive index and optical gap were determined by spectroscopic ellipsometry. For this purpose a SpecEl-200 spectroscopic ellipsometer (400 - 900 nm) manufactured by Micropac, software Scout from Wolfgang Theiss and OJL model was used. The transversal diode structures were prepared for electrical characterization of nc-SiC:H films. The circular electrodes of Au (120 nm thick) as a Schottky contacts with two different diameters (0.7 and 1.2 mm) were formed using metal masks on the side with nc-SiC:H film on each sample. The other side of samples was fully covered by Al ohmic contact (260 nm thick). Metal systems were thermally evaporated in a dry high-vacuum system. The electrical properties of nc-SiC:H films were determined by I-V measurement with HP 4140B pA meter/DC voltage source controlled by personal computer at 295 K. Photoluminescence spectra were measured at 6 K and 300 K. The properties of nc-SiC:H films are discussed on the base of the obtained results.

3. Results and discussion

The SiC films obtained had smooth and flat surface. RBS and ERD analysis indicated that the films contain silicon, carbon, nitrogen, hydrogen and small amount of oxygen. Concentration were: Sample B1 (silicon 35 at.%, carbon 40 at.%, hydrogen 20 at.%, nitrogen 5 at.%); sample B2 (35, 35, 22, 7); sample B3 (35, 32, 22, 10); respectively. Spectroscopic ellipsometry analysis indicated that the refractive index decreases with increasing concentration of nitrogen in the films and were for B1 – 2.4; B2 – 2.3; B3 – 2.1; respectively. The measured IR spectrum revealed the main absorption region between 400 and 2000 cm^{-1}. IR results showed the presence of Si-C, Si-O, Si-N, Si-H, N-H, C-H, C-N specific bonds. The main phonon or vibration frequency is related to SiC and have the following characteristics determined from the reflection spectra: center position 795 cm^{-1}, width 178 cm^{-1}; center position 795 cm^{-1}, width 45 cm^{-1}; center position 804 cm^{-1}, width 42 cm^{-1}. The non stressed phonon position of cubic SiC is 796 cm^{-1}. In amorphous material a shift to higher values indicate on recrystallisation or nucleation of small crystallites. Figure 2 shows PL intensities measured at two different temperatures for sample B1. It can be seen that PL intensity is higher at 6 K measurement temperature and PL peak emission wavelength decrease.

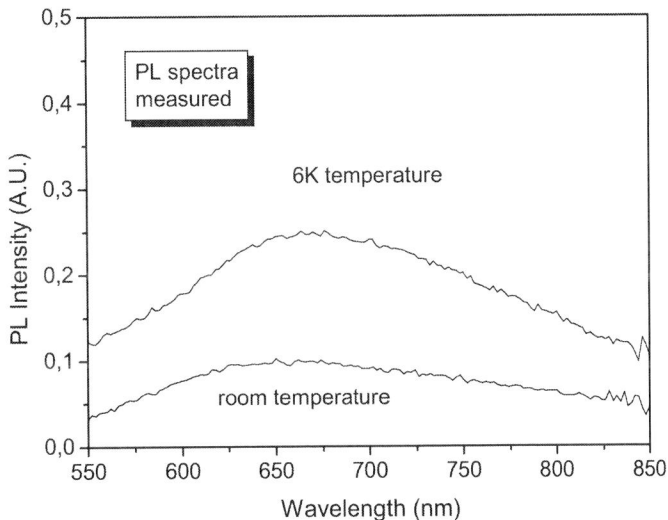

Fig. 2. Photoluminescence spectra measured at different temperatures for sample B1.

From I-V characteristics of prepared diode structures Au/SiC/Si/Al before irradiation by neutrons, depicted in Fig. 3, we can see, that the conductance of the SiC layer prepared at 450 $^{\circ}$C for sample B1 is very small about 1.10^{-12} S for reverse direction. The forward current increases slowly at low bias voltages in a range of $0.05 \div 0.20$ V. At higher voltages, over 1 V, the forward current is limited by the series resistance due to ohmic contact and the bulk resistance of SiC layer. We observed some dispersion in characteristics that is due to the

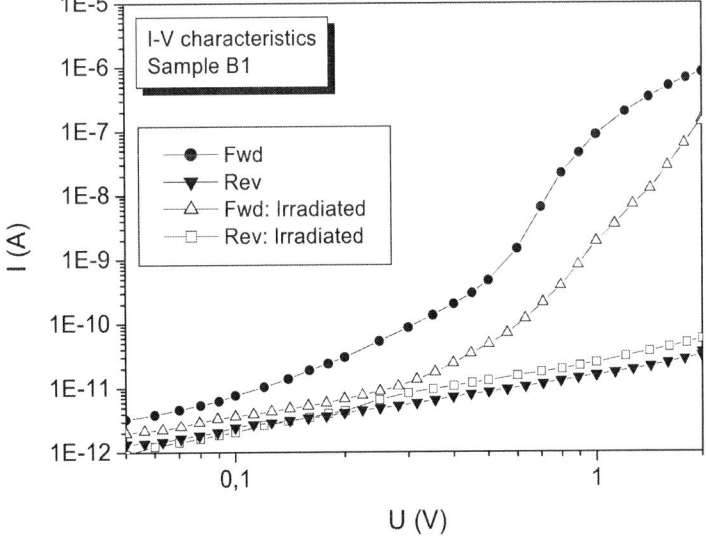

978-1-4244-8574-1/10 $26.00 © 2010 IEEE

Fig. 3. I-V characteristics for diode structure Au/SiC/Si/Al with Au top contact (Φ=1.2 mm) on SiC film and full-area Al back contact, before (full symbols) and after irradiation (open symbols) by fast neutrons with fluence of 5×10^{12} cm^{-2} plotted in a log-linear scale.

inhomogenity of SiC film parameters. In the same figure are shown I-V dependencies for diode structure Au/SiC/Si/Al after irradiation by fast neutrons with fluence of 5×10^{12} cm^{-2}. From measured reversed I-V characteristics we can see small increase in concuctance after irradiation. It can be explained by the degradation of interface between top contact and film surface - increasing density of interface state. Greater decrease of forward current can be explained by change of bulk film properties after neutron irradiation.

4. Conclussion

We have investigated the structural and electrical properties of nc-SiC films prepared by plasma enhanced chemical vapor deposition with different nitrogen concentration. The RBS results showed that the concentrations of Si, C and N in the films are changed by the change of NH$_3$ flow. The concentration of hydrogen was determined by the ERD method and the value is approximately 20 at.%. IR results showed the presence of Si-C, Si-N, Si-H, C-H and Si-O bonds. PL intensity is higher at 6 K measurement temperature and PL peak emission wavelength decrease. From measured I-V characteristics we can see differences in forward and reverse currents for diode structure Au/SiC/Si/Al before and after irradiation by fast neutrons.

Acknowledgement

This research has been supported by the Slovak Research and Development Agency under the contracts APVV-0713-07, SK-UA-0011-09, APVV-0577-07 and by the Scientific Grant Agency of the Ministry of Education of the Slovakia and Slovak Academy of Sciences, No. 2/0192/10.

References

[1] M. Motohashi, K. Ashibu, Y. Hiruta, and K. Homma, *Electronics and Communications in Japan,* **90,** 9, 2007

[2] V. M. Ng, M. Xu, S.Y. Huang, J. D. Long, and S. Xu, *Thin Solid Films,* **506-507,** 283, 2006

[3] G. S. Solomon, M. Pelton, and Y. Yamamoto, *Phys. Rev. Lett.,* **86,** 3903, 2001

[4] A. Kassiba, M. Makowska-Janusik, J. Bouclé, J. F. Bardeau, A. Bulou, and N. Herlin-Boime, *Phys. Rev.,* **B66,** 155317, 2002

[5] E.G. Wang, *Physica B,* **185,** 85, 1993

[6] H. Colder, P. Marie, L. Pichon, and R. Rizk, *Phys. Stat. Sol. (c),* **1,** 269, 2004

[7] M. Vetter, C. Voz, R. Ferre, I. Martin, A. Orpella, J. Puigdollers, J. Andreu, and R. Alcubila, *Thin Solid Films,* **511-512,** 290, 2006

[8] J. Huran, I. Hotovy, J. Pezoltd, N. I. Balalykin, and A. P. Kobzev, *Thin Solid Films,* **515,** 651, 2006

Material optimization of the alignment marks for the EBDW lithography

L. Matay[1], R. Andok[1], V. Barák[1], A. Ritomský[1], A. Konečníková[1], I. Kostič[1],
S. Partel[2], P. Hudek[2]

[1] Institute of Informatics, SAS, Dúbravská cesta 9, SK-845 07 Bratislava, Slovakia.
[2] Vorarlberg University of Applied Sciences, Hochschulstrasse 1, A-6850 Dornbirn, Austria.

We present results of material optimization for the alignment marks used in the Electron-Beam Direct-Write (EBDW) lithography. Such marks have been proposed both for negative (grooves) as well as for positive (elevated) topographies. The primary mask for the alignment mark patterns is done by photolithography and e-beam lithography. The negative topography of the marks was transferred by Deep Reactive Ion Etching (DRIE), while the positive topographies were realized by evaporation of various metals followed by lift-off process. We measured signals at characteristic beam-steps when the e-beam scans across the mark. This was done on two scanning positions. Out of these measurements we obtained the detection signal characteristics for different fiducial alignment marks with positive and negative topographies, respectively.

1. Introduction

Electron-Beam Direct-Write lithography (EBDW) offers a solution for patterning of structures where it is not possible to use the classical optical lithography due to increasing requirements for the resolution and precision of the structures. Processing of the individual layers depends on the quality of the used tool as well as on the requirements and conditions of the technological processes. The determining advantage of the EBDW lithography is that it operates "maskless" [1] and thus allows patterns to be generated from a data file. An important parameter of technological processes, besides the resolution, CD-linearity/uniformity, is the precision of the pattern placement, i.e. the registration and alignment of structures in the particular chip layer. For this reason design and realization of well-detectable alignment marks delivering a high contrast in the detection plays the key role in the high resolution micro- and nano-lithography [2, 3]. The concept of such alignment patterns must be process compatible and at the same time it must satisfy the parameters of the particular targeted technology node.

2. Process flow

As substrates, we used 100 mm p-type (100) standard-cleaned silicon wafers. Spin coating of 1 μm thick high molecular weight PMMA from a solvent at 2000 rpm followed for e-beam lithography (EBL) and of 2.1 μm thick AZ 1518 for optical lithography (OL), respectively. For elevated alignment marks (positive topography – PT), thin metal layers of Ti, Pt, Cr, Ag, Au and Ni with thicknesses from 80 nm to 160 nm were then deposited onto the resist patterned wafers using an electron gun evaporation system (base pressure ~ 10^{-5} Pa) and lifted-off in acetone under ultrasonic agitation. The alignment marks with negative topography (NT) were fabricated using DRIE of the resist patterns into the silicon wafers using fluorine-based chemistry. The etching process parameters (anisotropy, bias, depth) have been strictly related to the used optical signal coding/decoding method of the final alignment marks. In the final step the wafers were spin coated with PMMA resist in order to simulate the real sample condition using the so called fiducial marks. The technological and material parameters of both types of alignment marks with resist thickness of h_R and deposited metal thickness and/or depth of Si etching h_L are presented in Tab. 1.

Sample	Resist	h_R [μm]	Layer	h_L [μm]	Notice
Si-17-01	PMMA	1	Ti/Pt	0.087	EBL, PT
Si-17-02	PMMA	1	Ti/Pt	0.112	EBL, PT
Si-17-03	PMMA	1	Ti	0.077	EBL, PT
Si-15-01	AZ 1518	2.1	-	14	OL, ICP RIE, NT
Si-15-02	AZ 1518	2.1	-	55	OL, ICP RIE, NT
Si-15-03	AZ 1518	2.1	Cr/Ag	0.160	OL, PT
Si-15-04	AZ 1518	2.1	Ti/Au	0.124	OL, PT
Si-15-05	AZ 1518	2.1	Cr	0.108	OL, PT
Si-15-06	AZ 1518	2.1	Ni	0.129	OL, PT

Tab. 1: Technological parameters of used alignment marks: h_R – resist thickness, h_L – deposited metal thickness and/or depth of the grooved mark (Si-trench).

3. Experimental and discussion

The lithography experiments have been carried out on the ZBA 21 variable shaped e-beam pattern generator operating at 20 keV. The used tool was upgraded by *Vistec Electron Beam GmbH*, Jena (Germany) and is currently equipped with the RT11 operation system controlled by a PC with own software tool enabling processing of large amount of exposure data with exposure dose control possibility and easy operation. The RT11 operating system is equipped with manual and automatic e-beam adjustment which is carried out by scanning of the beam across the alignment marks. Two types of topographical patterns for marks, indicated as type I and II, were used. The type I represents a rectangle of 3200 x 1000 μm^2 and the type II contains a rectangle of 3186 x 500 μm^2. Additionally, the mark type II has an inserted "cramp" with the size of the long arm of 3200 μm and two short arms of 460 μm. The line-width of these arms is 4 μm. The first experiments with negative topography (trenches) were published in [4] clearly demonstrating that the output signal from 200 μm deep grooved mark is 10 V higher than for the marks etched only 10 μm deep.

In order to obtain signals from the marks we scanned marks across their edge. In the first step, the pattern generator always checks the position of the marks. The obtained signals from grooved marks showed a very good contrast with a good detectability, however, with an opposite polarity compared to the standard elevated material topographical marks. This came out to be an unbeatable problem despite the fact that the control system offers various scanning possibilities. Based on these restrictions we came to the conclusion that satisfactory result can be achieved, but by choosing a different design of the mark.

We measured various signals at characteristic beam-steps when the pattern generator scans across the mark edge. This was done on two scanning positions (at the beginning and end of a mark). Out of these measurements we obtained the signal characteristics for the samples with both topographies. Fig. 1a shows the topography of the test marks type I and II together with their direction and positions of measurement (1, 2). Fig. 1b shows the signal at the Si/metal edge with indicated scan direction as well as the amplitude of the signal change ΔU. The respective results are given in Tab. 2. A detailed recorded signal for the Si-15-04 sample with Au layer for scanning across the positions 1 and 2 through the type I mark with positive topography is shown in Fig. 2. The amplitude of the measured signal ΔU at the Si/Au boundary is 6 V. Fig. 3 shows the signal from the type II mark (sample Si-15-06) consisting of a thin Ni layer across the positions 1 and 2 for the positive topography. The measured signal amplitude ΔU at the Ni/Si interface is 1.2 V. The combined mark of type II was chosen as a combination of two marks mentioned in the ZBA 21-1 tool manual to improve the overall pattern placement precision as well as for finding out the signal output quality depending on the proportion of the marker dimensions 536/3/4 μm (line/space/line).

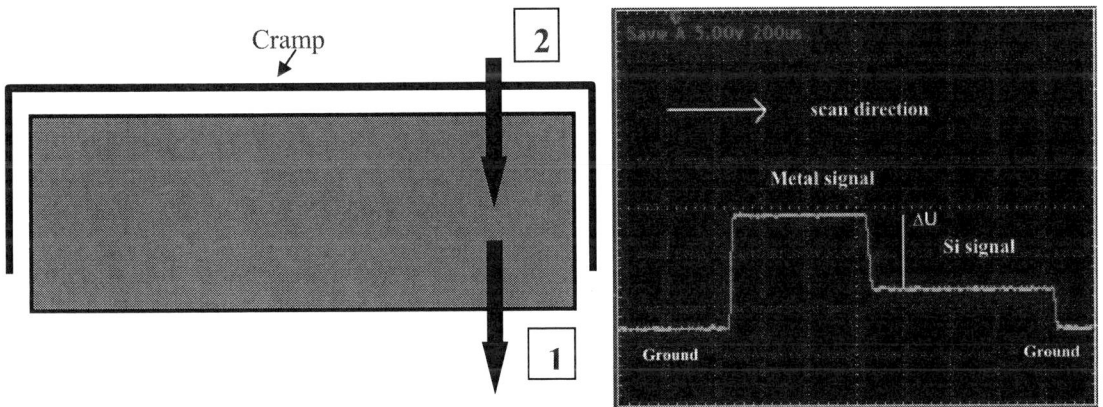

Fig. 1. Left: schematic view of the test mark type II; right: measured signal ΔU from the metal/Si across the measurement position 1.

Sample	Si-17-01	Si-15-04	Si-15-05	Si-15-06
Mark material	PT, type I, Pt	PT, type I, Au	PT, type I, Cr	PT, type II, Ni
ΔU (V)	7	6	3.5	1.2
h_L (μm)	0.087	0.124	0.108	0.129

Tab. 2. Dependency of the signal amplitude ΔU on the applied metallic layer.

The highest signal was obtained for the alignment mark with Au and Pt metallic layers ($\Delta U = 6$ and 7 V); the lowest signals have been measured for alignment marks with Cr and Ni layers ($\Delta U = 3.5$ and 1.2 V). It turns out that the layer thickness for given metals does not play the critical role in the signal amplitude, while the most important parameter is the element atomic number (Au-79, Pt-78, Ni-28 and Cr-24, respectively). From the point of view of applicability of the marks also their geometry is important for the optimum signal evaluation and processing with the emphasis to obtain the best possible global and/or local placement uniformity of the written pattern details on the wafer over all technology layers.

Acknowledgement

This work was supported by the grant of the Slovak Research and Development Agency No. APVV-SK-AT-012-08 and the Austrian OeAW Programm WTZ (Slovak-Austrian Science and Technology Cooperation), Scientific Grant Agency of the Ministry of

Education of Slovak Republic, the Slovak Academy of Sciences No. VEGA- 2/0214/09 and CENTE I No. ITMS – 26240120011.

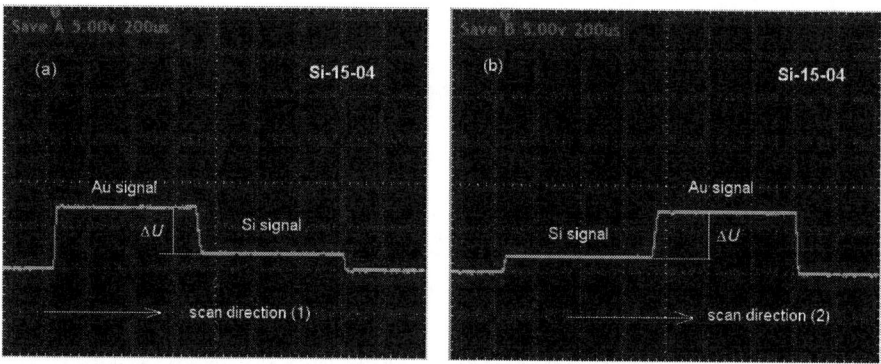

Fig. 2. Signal scanned for the sample Si-15-04 (Au/Si) across *(a)* position 1 and *(b)* position 2 for the marks with a positive topography (mark type I, $\Delta U = 6$ V).

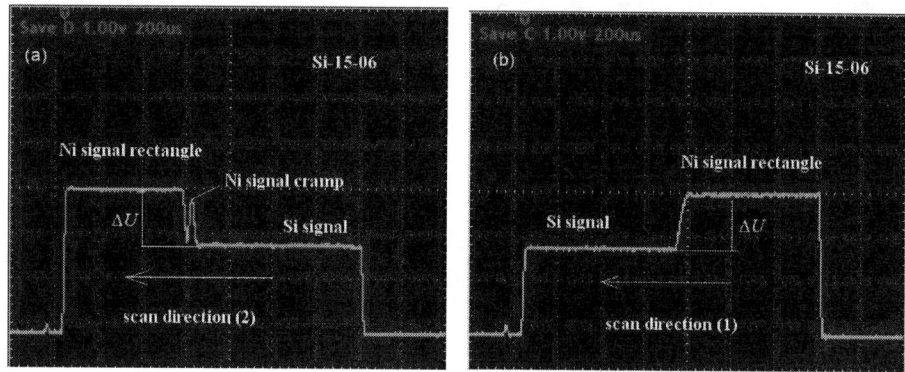

Fig. 3. Signals scanned for the Si-15-06 (Ni/Si) sample across the position *(a)* 2 and *(b)* 1 for the marks with positive topography of max. values of 1.2 V (mark type II).

References

[1]   N. Saitou, *Electron - beam lithography.* In: Y. Nishi, R. Doering, Handbook of Semiconductor Manufacturing Technology, 2000, pp. 571-587.

[2]   U. Weidenmueller, P. Hahmann, L. Pain, M. Jurdit, D. Henry, et al., *Microelectronic Journal,* **78-79**, 16, 2005.

[3]   E. Kratschmer, D. P. Klaus, R. Wiswanathan, M. L. Turnidge, P. L. Reed, B. PcPhail, *J. Vac. Sci. Technol. B*, **27**, 2563, 2009.

[4]   L. Matay, R. Andok, A. Ritomský, V. Barák, I. Kostič, S. Partel, P. Hudek, in *Proceedings of the 16th International Conference on Applied Physics of Condensed Matter (APCOM 2010).* ISBN 978-80-227-3307-6, Malá Lučivná, June 16 – 18, 2010, pp. 252-255.

Patterning of nanometer structures by using direct-write e-beam lithography for the sensor development

P. Ďurina[1], M. Štefečka[1,2], T. Roch[1], J. Noskovič[1], M. Trgala[1], A. Pidík[1], I. Kostič[3], A. Konečníková[3], L. Matay[3], P. Kúš[1] and Plecenik[1]

[1]Department of Experimental Physics, FMFI UK, Mlynská dolina F2, 842 48 Bratislava, Slovakia
[2]BIONT, a.s., Karloveská 63, 842 29 Bratislava, Slovakia
[3]Institute of Informatics, Slovak Academy of Sciences, Dúbravská cesta 9, 845 07 Bratislava, Slovakia
e-mail: durina.pavol@gmail.com and ivan.kostic@savba.sk

In this work, the optimalisation of e-beam parameters and the writing strategy have been performed. Various positive and negative e-beam resists have been evaluated for high resolution e-beam lithography and pattern transfer. Both, lift-off method and ion beam etching have been investigated for the pattern transfer into thin Pt and MoC layer on saphire substrate.

1. Introduction

Direct-write e-beam lithography (DWEBL) is widely used in nanoscale device research due to nano-scale resolution, flexibility in layout design [1], and the possibility of being performed in a commercially built scanning electron microscope (SEM). There are some commercial SEM based nanolithography systems [2]. In this work, SEM VEGA II (*TESCAN*) equiped with pattern generator has been used for the investigation of pattern transfer of nanometer structures on spahire substrates.

Saphire as a substrate for sensor development becomes more attractive in recent years because of properties as an excellent electrical insulator and the cost of substrates. The use of DWEBL for patterning of structures on saphire substrate requires a specific approach in comparison with DWEBL semiconductor materials. The resolution and application are often limited by the surface charging effect.

The motivation of this work was the application of the patterning of sub-100 nm structures for gas sensors development.

2. Experimental

2.1 Lithography

All experiments have been done by using scanning electron microscope TESCAN VEGA II SBH (*TESCAN*) equiped with tungsten heated filament, electron energy in the range of 1-30 keV, full computer control of the optical system, beam blanker, pattern generator and Tescan lithography software [3].

Positive e-beam resists PMMA A2 (*Microchem*) with 950 000 high molecular weight and PMMA ARP 679.04 (*Allrezist*) have been chosen for nanostructure pattern transfer due to its uniform resist coating, high resolution and high contrast for DWEBL exposures, long shelf life, and good adhesion to most substrates. The bilayer PMMA A2/copolymer MMA(8.5)MAA (*Microchem*) has been investigated for lift-off method. The investigation of negative organic resists AZ nLOF 2070 (*MicroChemicals*) and negative resist SU-8

(*Microchem*) under e-beam exposure have been performed. The masking capability of PMMA and SU-8 resists during ion beam etching has been investigated.

The e-gun evaporation process of thin Pt layer has been utilised.

2.2 Etching

In our experiments, we used dry ion etching with Argon ions from source of ions fy KLAN-53M with source fy SEF-53M.

As a mask for etching of Platinum layer (20 nm thick) we used positive resist PMMA-A2 with thickness of layer 100 nm. We etched in ion energy range from 400 to 600 eV with ions current range from 15 to 25 mA. Etching time depends on value of energy and current of argon ions. Etching time range was from 5 to 12 min. On the ground of the results from our experiments, we defined optimal parameters of ion etching for Platinum layer with thickness 20 nm as: Ion energy E= 450 eV, Ion current I= 20 mA and etching time t = 8 min.

As a mask for etching of MoC (molybdenum-carbides) layer with thickness 50 nm we used negative resist SU-8 with thickness of layer 85 nm. We etched MoC layer in ion energy range from 400 to 600 eV with ions current range from 15 to 25 mA. Etching time depends on value of energy and current of argon ions. Time etching range was from 25 to 50 min. On the ground of the results from our experiments, we defined optimal parameters of ion etching for MoC layer with thickeness 50 nm as: Ion energy E= 500 eV, Ion current I= 20 mA and time etching t = 40 min.

3. Results

For the highest resolution PMMA exposure on insulating saphire substrate, the optimised parameters for DWEBL process were found to be: acceleration voltage of 30 keV, beam current of 20-40 pA and single pixel line exposure. Thin metallic Al layer evaporated on PMMA was found as sufficient to reduce the amount of surface charges during DWEBL process (Fig. 1). Detail of the lines in 20 nm thin Pt layer on silicon with the linewidth 80 nm patterned via lift-off and single layer PMMA A2 at 30 keV is shown in Fig. 2.

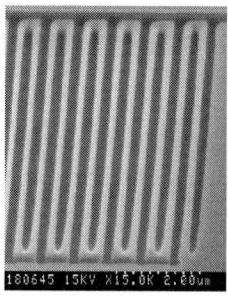

Figure 1 Detail of the meander structure in PMMA 495k resist on Saphire, the thickness 350 nm, e-beam exposure at 30 keV, Line/Space 200 nm.

Figure 2 Detail of the grating, Pt layer on silicon with the thickness 20 nm patterned via lift-off and single layer PMMA A2 at 30 keV. The linewidth 80 nm.

The use of bilayer PMMA A2/copolymer MMA(8.5)MAA (*Microchem*) has been optimised to create an undercut profile suitable for lift-off of metals (Figs. 3,4). Minimal size of structures 80 nm has been achieved in thin PMMA resist layer (Fig. 3), 100 nm linewidth in 30 nm platinum layer on saphire substrate for structures patterned via lift-off (Figs. 4).

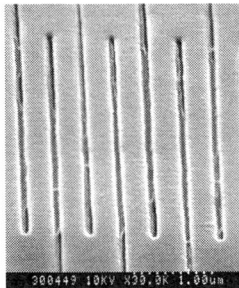

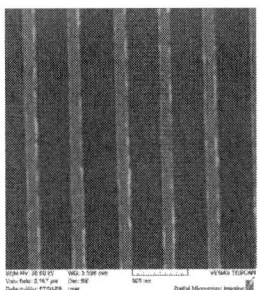

Figure 3 Detail of the grating in bilayer PMMA-A2-resist/copolymer MMA(8.5) MAA on saphire, the thickness 260 nm, the linewidth 80 nm on the top PMMA A2 resist layer. 30 nm Pt layer evaporated on the top of PMMA A2.

Figure 4 Detail of the grating, 30 nm Pt layer on saphire patterned via bilayer PMMA A2 / MMA(8.5)MAA on saphire and lift-off, e-beam exposure at 30 keV. The linewidth 100 nm.

In case of ion beam etching, the smallest achieved dimensions of Platinum structures (grooves) on saphire substrate were 85 nm (Fig. 5) and in the case of MoC structures reproducible linewidth was 100 nm (Fig. 6). The smallest dimension is demonstrated less than 50 nm (Fig. 7) in MoC layer which was achieved after a tentative overetching.

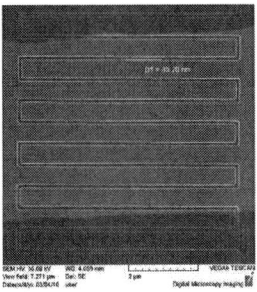

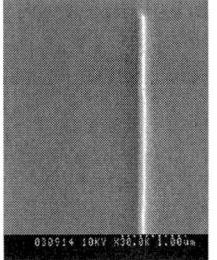

Figure 5 Detail of the meander structure in Pt layer on saphire patterned via ion beam etching, the thickness 20 nm. The masking layer was 80 nm thin PMMA A2 resist. The width of the groove is 85 nm.

Figure 6 Detail of the MoC line on saphire substrate after ion beam etching. The linewidth 100 nm.

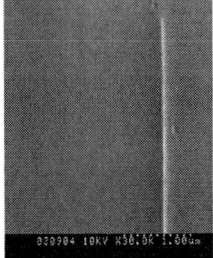

Figure 7 Detail of the MoC line on saphire after ion beam etching. The linewidth <50 nm.

For the patterning of nanostructures on sapphire substrate using lift-off method, bilayer 80 nm thin PMMA A2 950k resist/ 100 nm thin copolymer MMA(8.5)MAA was found more appropriate in comparison to single layer 80 nm thin PMMA A2 950k resist. Bilayer was covered with 20 nm thin Al or Cr layer evaporated on PMMA A2 before the exposure to reduce the amount of surface charges during DWEBL process.

4. Applications

The patterning methods described in this work have been applied in the investigation of the dependence of gas sensors characteristics on the size of structures. The patterned structures in the range of 100 nm and below were applied in the experiments with the gas-sensing properties of titanium dioxide thin layer. Measurement results with gas sensors are published in another paper.

5. Conclusions

The process of e-beam exposure was investigated and the process conditions were optimised to achieve conductive metallic patterns with minimal linewidth below 100 nm on saphire substrates. Comparable results were achieved using lift-off method and ion beam etching for the patterning. The use of SEM based e-beam lithography system VEGA II (*TESCAN*) for nanostructure patterning was demonstrated. The method was used for the patterning of sub-100 nm structures for gas sensors development.

Acknowledgement

This work was supported by the Slovak Scientific Grant Agency under the contracts VVCE-0058-07, APVV-0034-07, APVV-0432-07, and grants VEGA-1/0096/08, VEGA-2/0214/09, and VEGA-1/0162/10.

References

[1] M. A. McCord and M. J. Rooks, *Handbook of Microlithography, Micromachining and Microfabrication*, 1st ed., edited by P. Rai-Choudhury Vol. I, Chap. 2, p. 139–249, IEE, London, 1997.

[2] D. M. Tennant and A. R. Bleier, *Handbook of Nanofabrication*, 1st ed., edited by Gary P. Wiederrecht, Chap. 4, p.121-148, Elsevier, 2010

[3] Tescan VEGA-II Technical Documentation, Brno, Czech Republic

Temperature dependence of the pyroelectric behaviour in GaN/AlGaN

A. Laposa, J. Jakovenko and M. Husak

Department of microelectronics, Czech Technical University,
Technicka 2, 16627 Prague, Czech Republic
e-mail: laposale@fel.cvut.cz, jakovenk@fel.cvut.cz and husak@fel.cvut.cz

So far the dependence of the spontaneous polarization coefficient for GaN and AlN on temperature has been measured to be minimal, which corresponds with expectation that the spontaneous polarization is reduced at the elevated temperatures of interest. There are also no reports on the piezoelectric polarization at higher temperature. This paper is initial study on the influence of temperature related pyroelectric behaviour in GaN/AlGaN. Summarize recent findings and consideration.

1. Introduction

The A^{III}-N semiconductors have been a subject of constant intense investigation in the past two decades as they are suitable in high power and high temperature electronic devices. This generally leads to better understanding of physical effects in III-nitrides structures at a fundamental level. It is well known that the wurtzite A^{III}-N materials, in comparison to traditional semiconductors such as A^{III}-B^V or Si, have a strong polarization effect. The temperature dependence of the spontaneous polarization (dipole moment per unit volume) is defined as pyroelectricity. The pyroelectric effect is in other words a change in electric dipole moment of unit volume of material when temperature is varied over the volume. This phenomenon has to be taken into account in device design process, could be used to study thermal stability, but it is also promising for application in high temperature pyroelectric sensors [1]. The studies of pyroelectric properties of GaN and AlGaN are still at a preliminary stage. Usually, the spontaneous polarization effect plays a very important role in AlGaN/GaN heterostructure-based devices [2]. In principle, in order to understand pyroelectricity in any material, one has to consider the variatious mechanism of the spontaneous polarization (such as ionic, electronic, orientational or surface charge) and study their variation with temperature [3].

2. Polarization in A^{III}-N

One of the contributions to polarization in III-nitrides is due to asymmetry of wurtzite crystal structure (member of the hexagonal crystal system); the nitrides (usually grown in the 0001 direction) have a nonzero polarization parallel to the hexagonal axes in a zero electric field state, known as spontaneous polarization (true intrinsic polarization). It is not feasible to directly measure the spontaneous polarization in bulk materials, since the induced charges at the surfaces are compensated by ambient charges. There is only one report (fourteen years old), on measurement the induced voltage in GaN sample with temperature variations, known as pyroelectric effect [4]. Epitaxial films of GaN and AlGaN layers contribute an additional piezoelectric polarization arising from the lattice mismatches between the substrate and different III-nitrides layers. The lattice misfit issue in the growth of heterostructural epitaxies is as old as the technology itself.

The large values of polarization charge densities make it possible to the formation of a two dimensional electron gas at the AlGaN/GaN heterointerface, in other words depends on spontaneous and piezoelectric polarization field. 2DEG density is very sensitive to any change in stress field of the structure because this built-in strain is responsible for generation of piezoelectric field [5]. Such high sensitivity to stress variation enables AlGaN/GaN heterostructures to cover important segment of pressure sensors.

3.   Pyroelectric and piezoelectric effect

In heterostructures, each interfaces between different materials carriers a charge that equals the divergence in the spontaneous and/or strain-induced piezoelectric polarization across the interface. This charge cannot be compensated by charged states or free carriers [6]. Spontaneous pyroelectric polarization induces strong built-in electric field in pyroelectric-semiconductor granular systems. The magnitude of polarization depends on temperature, and, when temperature changes, in investigating either of these phenomena, the stray charges cannot respond instantaneously and the induced voltage in response to the corresponding dynamic variations can be measured. These charges variations include the variation of the spontaneous polarization (primary pyroelectric effect) and the variations caused by changes in temperature-related strain (secondary pyroelectric effect) [7]. Secondary pyroelectricity arising from piezoelectricity and thermal expansion is sometimes larger than the primary pyroelectric effect. GaN-like group-III nitrides have large internal electric fields of both pyro- and piezoelectric origin.

Pyroelectricity describes a relationship between thermal and electrical variables: a change in temperature ΔT creates a change in the electric polarization P. Polarization is a vector (= first rank tensor) and temperature is a scalar (= zero rank tensor). Therefore the pyroelectric coefficient, defined by $P_i = p_i \Delta T$ is a first rank tensor property. The pyroelectric coefficient is the temperature derivative of the spontaneous polarization [8].

Hence for a free pyroelectric crystal, the total pyroelectric coefficient can be expressed as the sum of the primary pyroelectric coefficient as the first term on the right side of the following equation and the secondary pyroelectric coefficient as the second term on the right side [8, 9]

$$\left(\frac{\partial p}{\partial T}\right)_{\sigma,E} = \left(\frac{\partial p}{\partial T}\right)_{S,E} + \Sigma_i \left(\frac{\partial p}{\partial s_i}\right)_{T,E} \left(\frac{\partial s_i}{\partial T}\right)_{\sigma,E} \quad (1)$$

Where p is the macroscopic dipole moment per mole, σ is the macroscopic stress tensor, and s is the macroscopic strain tensor with elements s_i and E is electric field. Eqn. (1) can be written as

$$p^{\sigma,E} = p^{S,E} + \left(\frac{dp}{d\sigma}\right)_{T,E} \left(\frac{dS}{ds}\right)_{T,E} \left(\frac{ds}{dT}\right)_{\sigma,E} = p^{S,E} + d_{ijk}^{T,E} C_{jklm}^{T,E} \alpha_{jk}^{\sigma,E} \quad (2)$$

Where $p^{\sigma,E}$ is the total pyroelectric coefficient; $p^{S,E}$ is the primary pyroelectric coefficient; d_{ijk}, C_{jklm} and α_{jk} denote respectively the piezoelectric moduli, elastic compliance coefficient and coefficient of thermal expansion (the secondary pyroelectric coefficient is given by product of d_{ijk}, C_{jklm} and α_{jk}). The details of the relationship between the stress, strain, piezoelectric polarization and polarization tensors are in [12]. Due to symmetries of wurtzite structures, only 3 components of the 27 elements of d_{ijk} are independent, a simpler matrix notation, common in the nitride literature [13, 15]

$$P_{SP} = P_{SP,AlN} + P_{SP,GaN}(1-x) \tag{3}$$

$$P_{PZ} = 2 \cdot \frac{a-a_0}{a_0}\left(e_{31} - e_{33} \cdot \frac{C_{13}}{C_{33}}\right) \tag{4}$$

Where a and a_0 are the lattice constants, e_{31} and e_{33} are the piezoelectric coefficients, C_{13} and C_{33} denote the elastic constants.

4. Temperature dependence consideration

In paper [1] was indicated that primary pyroelectric coefficient would be proportional to temperature, later [16] was predicted the T^3 law for pyroelectric coefficient. In a recent paper [2] were calculated dependence of the spontaneous polarization coefficients for GaN and AlN from 0 K to 1000 K with results that the closer temperature to zero, the more the pyroelectric coefficient is strictly proportional to specific heat and at low temperatures becomes proportional to the T^3. In very recent papers on high-temperature dependence modeling were calculated thermal conductivities and electron mobilities and compared to with data from various groups [14, 15]; differences were explained as over the years the electron mobilities increase due to improved quality of the material samples. There is only one report focused on measurement the induced voltage with temperature variations and only for low temperature from 200 K to 300 K and only for GaN sample [4]. The pyroelectric and piezoelectric effects play an important role in AlGaN/GaN based heterostructures [1] and good understanding of the electrical polarization effects at AlGaN/GaN interface is a key to the proper device simulation [12].

To study the effect of this polarization induced interface charge on carrier transport and heterostructure device characteristics, we focused on single $Al_{25}Ga_{75}N$/GaN heterostructure first. In a tensile strained 25 nm thick $Al_{25}Ga_{75}N$ layer on 1,5 µm thick GaN substrate, there is both pyro- and piezoelectrical polarization, while the relaxed 1,5 µm thick GaN layer only contains a pyroelectric moment. Assuming charge neutrality for the whole structure, there is a polarization induced charge density σ at the $Al_{25}Ga_{75}N$/GaN interface and a compensating charge -σ/2 at the surface and at the nucleation layer.

As a very initial result, based on theoretical understanding of the polarization induced charge we calculated the sheet carrier concentration of polarization induced 2DEG for pseudomorphic Ga-face AlGaN/GaN heterostructure for different temperatures varying from 300K to 700K as can be seen in Figure 1. Also we calculated the interface piezoelectric charge at position 1.500 nm of 9.329^{-03} C/m^2 which corresponds to $5.829*10^{12}$ electrons/cm^2 and the interface pyroelectric charge at position 1.500 nm of 10.063^{-03} C/m^2 which corresponds to $6.281*10^{12}$ electrons/cm^2

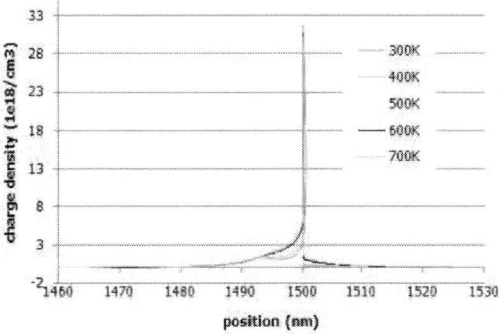

Figure 1: Charge density caused by pyroelectric effect for different temperature

5. Conclusion

So far the dependence of the spontaneous polarization coefficient for GaN and AlN on temperature has been measured to be minimal, which corresponds with expectation that the spontaneous polarization is reduced at the elevated temperatures of interest. There are also no reports on the piezoelectric polarization at higher temperature. This paper is initial study on the influence of temperature related behaviour in GaN/AlGaN. Much of the work in this area remains to be done and we expect improvements of our calculation followed by next step, measured the real structure to observe the pyroelectric effects in the sample and results compared to those simulated. Further theoretical and experimental studies are required to understand the characteristics of the pyroelectric effect of III-nitrides. Precise temperature dependence of the pyroelectric effect in nitrides is not fully known. For the proper device design is crucial investigation of the change of spontaneous polarization of GaN and AlN and their alloys.

Acknowledgement

This research has been supported in the frame of the European Union Project MORGAN (FP7 contract 214610) and partially by the Czech Science Foundation project No. 102/09/1601 "Intelligent Micro and Nano Structures for Microsensor Realized Using Nanotechnologies".

References

[1] M.S. Shur, A. D. Bykhovski and R. Gaska, *MRS Internet J. Nitride Semicond. Res.* **4S1**, G1.6, 1999.

[2] W. Yan, R. Zhang, X. Xiu, Z. Xie, P. Han, R. Jiang, et al., *Appl Phys Lett* **90**, 21, 2007.

[3] M. R. Srinivasan, *Bull. Mater. Sci.* **6**, 2, 1984.

[4] A.D. Bykhovski, V. V. Kaminsky, M. S. Shur, Q. C. Chen and M. A. Khan, *Appl. Phys. Lett.* **69**, 21, 1996.

[5] O. Ambacher, B. Foutz, J. Smart, J. Shealy, N. Weimann, K. Chu, et al, *J Appl Phys* **87**, 2000.

[6] G. Zandler, J. A. Majewski, and P. Vogl, *J. Vac. Sci. Technol. B* **17**, 1617, 1999.

[7] M. S. Shur, R. Gaska and A. Bykhovski, *Solid-State Electronics* **43**, 8, 1999.

[8] J. F. Nye, *Physical Properties of Crystals: Their Representation by Tensors and Matrices*, Oxford University Press, Oxford, 1985.

[9] M. Born, *Rev. Mod. Phys.* **17**, 245, 1945.

[10] B. Szigetti, *Phys. Rev. Lett.* **35**, 1532, 1975.

[11] R. E. Newnham - *Properties of Materials - Anisotropy, Symmetry, Structure*, Oxford University. Press, 2005.

[12] R. F. Tinder, *Tensor Properties of Solids*, Morgan & Claypool Publishers, 2008.

[13] T. Piasecki, W. Kosnikowski, B. Paskiewicz, *Optica Applicata* **35**, 3, 2005.

[14] S. Vitanov, V. Palanovski, S. Maroldt, R. Quay, *Semiconductor Device Research Symposium*, College Park, USA, 2009.

[15] S. Vitanov, V. Palanovski, S. Maroldt, R. Quay, *Solid-State Electronics* **54**, 8, 2010.

[16] M. Born and K. Huang, *Dynamical Theory of Crystal Lattices*, Oxford University Press, New York, 1954.

Off-state stress investigation of InAlN/GaN HFETs with different AlN buffer layer

M. Florovič[1], J. Kováč[1], H. Behmenburg[2], P. Kordoš[1,3], J. Škriniarová[1],
D. Donoval[1], M.Heuken[2]

[1]Department of Microelectronics, Faculty of Electrical Engineering and Information
Technology, Slovak University of Technology, Ilkovičova 3, 812 19 Bratislava, Slovakia
[2]AIXTRON AG, Kaiserstr. 98, 52134 Herzogenrath, Germany
[3]Institute of Electrical Engineering SAS, Dúbravská cesta 9, 841 04 Bratislava, Slovakia
E-mail: martin.florovic@stuba.sk

This study is focused on static electrical properties of the InAlN/GaN HFETs with the AlN buffer layer, which have different mechanical strain due to different growth conditions. Investigated devices were tested under off-state high drain bias of 55 V and their performance before and after the stress is analysed. Significant influence of the AlN buffer on the properties of virgin as well as stressed devices, particularly the gate current, is observed.

1. Introduction

The GaN-based heterostructure is a promising candidate as a field-effect transistor (HFET) device due to its wide band gap, superior carrier saturation velocity, thermal conductivity, and high breakdown field all of which are required for high temperature and high speed applications [1,2]. Most research on HFETs has focused on understanding and improving the performance, structure and fabrication processes. SiC is a reference substrate, mainly due to its high thermal conductivity and a relatively small lattice mismatch to GaN of 3.49% [3]. Several requirements exist for a buffer structure suitable for such devices. Insulating properties as well as low defect densities are necessary to suppress parasitic conduction and to enhance breakdown voltage and carrier mobility, respectively.

Excellent performance of GaN devices can be obtained routinely. However, achieving acceptable reliability and stability under continuous high performance operations is still needed for the commercialization. Reliability in high drain biases and high frequency conditions should be ensured for use in applications of high performance microwave devices and monolithic integrated circuits (MMICs). Improvement in reliability requires a better understanding of the failure features and mechanisms. Lattice constants of AlN and 6H-SiC suggest a compressive in-plane strain of 1.2% of the AlN layer [4]. The slightly smaller in-plane thermal expansion coefficient of 6H-SiC compared to AlN will add about 0.03% of compressive strain during cooling down. In this study $In_{0.14}Al_{0.86}N$/GaN HFETs with different AlN buffer were off-state tested under high drain bias of 55 V and analysed.

2. Experimental

Investigated InAlN/GaN samples were grown on semi-insulating 6H-SiC substrates by metal-organic vapour phase epitaxy (MOVPE). They consist of an AlN buffer layer 270–350 nm thick, followed by a 2.5 μm GaN layer and 7 nm $In_{0.14}Al_{0.86}N$ barrier layer (thereafter designated as InAlN only). The difference in the samples investigated is only in the preparation of the AlN buffer layer, all other layers were grown at identical conditions. The AlN buffer layer was grown by changing the NH_3 flux resulting in different V/III flow

978-1-4244-8574-1/10 $26.00 © 2010 IEEE

ratios (Tab. I.). This results in different residual strain in the AlN layer, as evaluated by micro-Raman spectroscopy and XRD [5]. Obtained strain data are shown in Tab. I. too. The strain decreases continuously with increased V/III flow ratio from 2 GPa tensile strain for the lower flow ratio to −0.8 GPa compressive strain for the highest flow ratio measured by Raman spectroscopy and 1.63GPa to –2.24 GPa measured by XRD as shown in Tab.I.

The InAlN/GaN HFETs were prepared by conventional processing steps. For ohmic contact formation the metallization layer stack of Ti/Al/Ni/Au (12/200/40/100 nm) annealed for 40 s at 825-925 °C (AT) was used (Tab.I). The gate contact with the gate length of 2 μm was formed using Ni/Au (10/300 nm) metallization. Static electrical properties of the fabricated HFETs were measured using Agilent 4155C equipment with wolfram contact probes connected to the expanded contacts. The stress conditions can be described as follows:

i) at first the virgin devices were characterised in details before the stress,

ii) the off-state stress conditions were $V_G = -4$ V (lower than the pinch-off voltage) and $V_{DS} = 55$ V (this value is far below the breakdown voltage) and during the stress in duration of 60 min the gate and drain currents were measured continuously,

iii) the stress was interrupted after 10 min and detailed characterisation was performed,

iv) the stressed devices were characterized after total stress time of 60 min,

v) after the stress the devices were left for 30 min without bias connection and were characterised in details in order to evaluate the stress-recovery conditions.

2.   Results and discussion

The output and transfer $I–V$ characteristics of virgin InAlN/GaN HFETs (i.e. before the stress) yielded different saturation drain current as well as gate leakage current. The saturation drain current for the sample 41 (i.e. with the compressive strained AlN) was 820 mA/mm, while the samples with tensile and lightly strained AlN (Raman measurements) yielded much lower currents, 430–530 mA/mm as summarized in Tab. I. (that the drain saturation I_{DS} and gate currents are before the stress, I_G at $V_{DS} = 55$ V, $V_G = -4$ V).

Tab. I. Properties of investigated InAlN/GaN structures and devices with different AlN buffer

Sample nr.	V/III ratio	Strain Raman [GPa]	Strain XRD [GPa]	I_{DS} at $V_G = 1$ V [mA/mm]	I_G [mA]	R_C [Ω.cm]	AT [°C]
42	240	2.00	1.63	430	0.04	0.59	825
40	1200	0.44	-0.35	530	0.15	0.37	925
43	4700	0.14	-1.80	460	0.015	0.31	850
41	8200	-0.80	-2.24	820	1.9	0.28	900

The gate leakage current of device with tensely stained AlN layer show high forward leakage current (Fig. 1a), while the device with high compressive strain (i.e. with better output and transfer performance) exhibited high reverse leakage current at $V_G= -4$V. Typical transfer characteristics of investigated four device types are shown in Fig.1.b. The devices had nearly identical pinch-off at $V_G \approx -3$ V.

978-1-4244-8574-1/10 $26.00 © 2010 IEEE

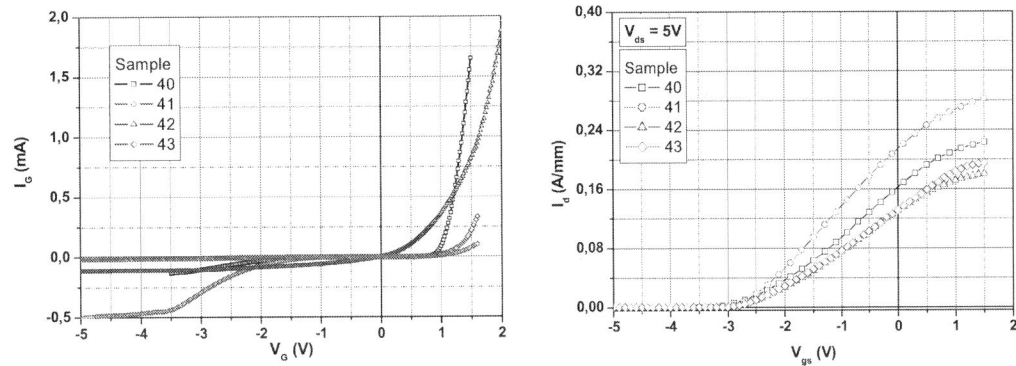

Fig 1. (a) Gate current characteristics and (b) transfer *I–V* characteristics of investigated HFETs with different AlN buffer

The drain and gate currents of the InAlN/GaN HFETs were measured continuously during the off-state stress ($V_{DS} = 55$ V, $V_G = -4$ V), as shown in Fig. 2a.b. The data obtained are nearly identical, which confirms that the drain current measured at the off-state stress conditions consists actually particularly from the gate leakage contribution. The drain/gate current decreased continuously with the stress time and in similar behaviour for all device types investigated. The stress was interrupted after 10 min to perform detailed characterisation of the devices. A partial increase of the drain/gate current was observed after this interruption, which indicates on some recovery effect (see Fig. 2).

During the next stress, i.e. between 10 and 60 min of the total stress time, the drain/gate current follows the previous stress vs time dependences. The devices were characterised in details after the total stress time of 60 min, as well as after 30 min without the stress. The results are shown in Fig. 3a,b. From a comparison of the output characteristics for sample 41 with the high compressive strain (Fig. 3a) it follows that the drain current decreased rapidly during the 10 min stress and remain constant with stress time. However, 30 min after the stress the output characteristic recovered to nearly the same one as before the stress. This is demonstrated on the right side of Fig. 3b, in which the drain currents before the stress as well as during 10 and 60 min of the stress and 30 min after the stress ($V_{DS} = 8$ V and $V_G = 1$ V) are summarized. Qualitatively similar effect was observed on the samples with tensile AlN buffer (42), despite of lower drain currents. For devices with low strain the stress time dependence shows gradually decreasing drain current (42) while minor changes in drain current were observed for device near zero stress (43) measured by Raman spectroscopy.

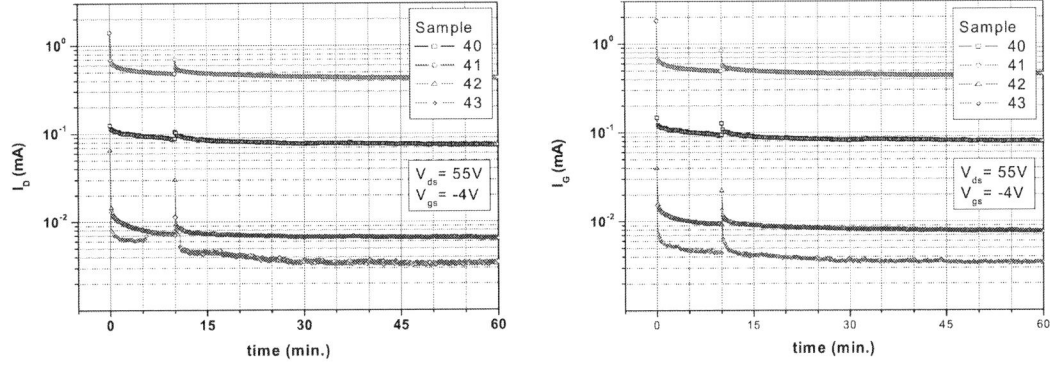

Fig 2. Time dependence of (a) drain current and (b) gate current during off-state stress.

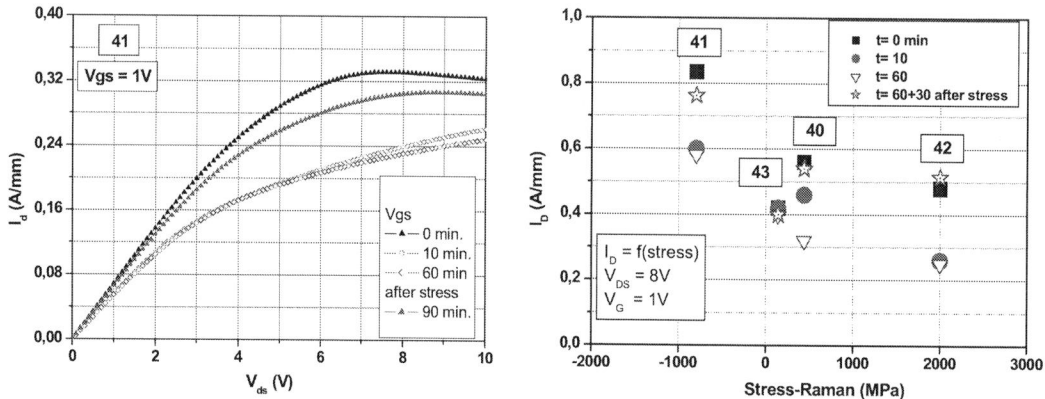

Fig. 3. Output characteristics before, during and after off-state stress for sample with compressive strain (a) and a comparison of drain currents for all samples investigated (b).

3. Conclusions

Static performance of InAlN/GaN HFETs with AlN buffer layer prepared at different conditions were analysed before, during and after the off-state stress. Obtained results can be summarised as follows: i) the best performance was obtained on device with highly compressively strained AlN, while the minor influence of of-stress was observed for near zero stress device ii) a decrease of the drain and gate current, nearly identical for all types of devices in time, was observed during the stress, iii) a recovery process, i.e. an increase of the drain/gate currents was obtained if measured 30 min after the stress, and iv) the variation of obtained characteristics can be attributed to various defect density and annealing temperature of ohmic contacts of individual samples as well. This result shows that during and after the off-state stress some trapping states activation/deactivation occurs. This effect will be studied in details in the next.

Acknowledgement

The research leading to these results was done in the Slovak Center of Excellence CENAMOST (Slovak Research and Development Agency Contract. No. VVCE-0049-07) with funding from the European Community's from Seventh Framework Programme FP7/2008-2011 under grant agreement n°214610, project MORGaN, with support of grants VEGA No. 1/0689/09, 1/0742/08. The authors wish to thank MicroGaN, GmbH, Ulm for HFETs processing.

References

[1] S. M. Sze, *Physics of Semiconductor Devices*, John Wiley & Sons, New York, 1981.
[2] D. Y. Chen, Y. A. Chang, and D. Swenson, *Appl. Phys. Lett.* **68**, 96, 1996.
[3] G. Birch, P. Maple, and F. Oak, in *Proceedings ASDAM'96 Conference*, Smolenice, Slovakia, 2000, p. 3.
[4] Z. J. Reitmeier, S. Einfeldt, R. Davis, X. Zhang, X. Fang, S. Mahajan, Acta Materialia 57 (2009) 4001.
[5] H. Behmenburg, C. Giessen, R. Srnanek, J. Kovac, H. Kalisch, M. Heuken, R. H. Jansen – *IC MOVPE XV*, submitted to Journal of Crystal Growth.

Growth and structural properties of GaAs on Al pseudo-substrates for ultrafast optoelectronics

Z. Sofer[1], D. Sedmidubský[1] and M. Mikulics[2]

[1]Dept. of Inorganic Chemistry, Institute of Chemical Technology, Prague, Technická 5, 166 28 Prague 6, Czech Republic
[2]Institute of Bio- and Nanosystems (IBN-1), Research Centre Jülich, Germany
e -mail: zdenek.sofer@vscht.cz

GaAs is broadly used in modern electronics. On the other hand, application of GaAs-based devices in high power electronics is complicated due to the substantial excess heat generated during device operation. One possibility to remove the excess heat is to employ substrates with high thermal conductivity. In this contribution we present the growth of GaAs layers by MOVPE on aluminum (111) pseudo-substrates designed for an improved heat management in GaAs electronic circuits. Pseudo-substrates for GaAs deposition were prepared by Al evaporation on (100) GaAs substrate and subsequently heat treated. The GaAs layers exhibit polycrystalline character with high preferential orientation in (100) direction. The roughness of the layers was in the range of 10 to 100 nm and the thickness in the range of 500 – 3000 nm. These layers exhibit extremely low carrier lifetime due to the growth-induced defects and are suitable for fabrication of ultrafast MSM photodetectors.

1. Introduction

The heat management in semiconductor devices during device operation at high current densities (transistors) or high laser fluencies (photodetectors) represents a serious issue. Application of substrates with high thermal conductivity can solve the problem with excessive heat removal. However, the heteroepitaxial growth of semiconductors on metallic substrates exhibits a lot of difficulties. Many experimental results have already been published on the growth of semiconductor layers on metallic substrates. The growth of GaAs on Mo substrates employing liquid phase epitaxy was reported in [1], the growth of Si on Ni, W and steel in [2, 3]. The MBE growth of GaN on Mo, Ta and Nb was published in [4, 5, 6] and the growth of GaN on Ag substrates and successful fabrication of a Schottky diode was reported in our previous work in [7]. In this contribution we present the growth of GaAs epitaxial layers on Al/GaAs pseudo-substrates.

2. Experimental part

The Al pseudo-substrates were fabricated by high vacuum evaporation of 7N purity Al on (100) GaAs. The deposition rate of Al was 0.1 nm.s^{-1} and the final thickness of Al layer was 200 nm. To improve the crystal structure of Al layer the temperature of the substrate was 200 °C, which is over the recrystalization temperature of high purity aluminum. Before the deposition of GaAs layer the pseudo-substrates were heated for 60 minutes at 600 °C in nitrogen atmosphere. The subsequent growth was performed at 600 °C in a horizontal reactor using TMGa and AsH$_3$ as gallium and arsenic precursor, respectively. The pressure in the reactor was 20 mbar and the total gas flow was 2900 sccm.min^{-1} at a gas velocity of 0.9 m.s^{-1}.

The arsine partial pressure $p(AsH_3)$ was kept at 124 Pa and the gallium partial pressure $p(TMGa)$ at 1.43 Pa, giving a V/III ratio of 87. To minimize the strain on Al-GaAs interface the cooling rate after deposition was below 1 °C.min^{-1}. The X-ray diffraction was measured in Bragg-Brentano geometry. The AFM measurement was performed in tapping mode. The thickness of the layers was measured by SIMS and RBS. The RBS spectrum was obtained using single charged hydrogen ions accelerated by 850 kV. The SIMS analysis was performed with Perkin Elmer PHI 600 series instrument. A Cs$^+$ primary ion beam with an energy of 3.0 keV and a flux ~18 nA was focused to a diameter of ~30 μm and raster-scanned across a sample area of 400 × 1000 μm under the incidence angle 60 degrees (relative to the sample normal). The photoluminescence was measured at the room temperature by micro-PL setup with 325 nm excitation wavelength from He-Cd laser.

3. Results and discussion

The structure of the fabricated GaAs layers was studied by X-ray diffraction. The X-ray diffraction patterns of Al layers exhibit a (111) orientation with no additional reflections. The improvement of GaAs structure quality was observed with decreasing the GaAs layer thickness. The GaAs growth was highly preferential in (100) direction, however other reflections, (111), (220), (331) and (422), were observed as well. The intensity of additional reflection increased with the layer thickness. The intensity of (400) direction showed more than 200 times higher intensity than the reflections for other crystallographic orientations. The shift of (400) reflection indicates the strain induced by different lattice parameters and thermal expansion coefficients of GaAs and Al. The enhancement of FWHM for (400) refection indicates an increase of dislocation density with layer thickness. The X-ray diffraction of GaAs layers is shown in Fig.1. The thickness of the layers was measured by SIMS and RBS. The results from both methods are in good agreement. The thickness measurement by SIMS is shown on Fig.2. The signal of CsAl$^+$ and CsGa$^+$ clusters was used for the concentration profile measurement. The photoluminescence was measured at the room temperature and the representative PL spectrum is shown in Fig.3. In the measured PL data we observed a red shift of 52 meV induced by a tensile strain in GaAs layer caused by lattice mismatch on GaAs – Al interface. The roughness of the GaAs surface was measured using AFM. We observe that the roughness represented by the root-mean-square (RMS), increases with increasing layer thickness. The RMS measured on 5×5 μm area increased from 13.2 nm in case of 600 nm thick layer to 62.0 nm in case of 2300 nm thick layer. Moreover, formation of rectangular blocks was observed on the surface of the thick layer. This indicates a development of three dimensional growths. The surface of 2300 nm thick GaAs layer is shown on Fig.4.

The lifetime of the photogenerated carriers in GaAs layer was studied using femtosecond time-resolved reflectivity measurements by an all-optical pump/probe system featuring ~ 20 fs temporal resolution. A carrier lifetime of less than 25 fs was obtained for our GaAs material grown on aluminum pseudo substrate at 600 °C. Our experimental pump/probe results also show that the carrier rise- and fall times are in the range from 50 to 100 fs. Pump/probe measurement are shown on Fig.5. We note that such an ultrashort carrier lifetime of only 25 fs on GaAs material is almost 75 % shorter than the values measured for nitrogen implanted GaAs and LT-GaAs materials in our previous study [8]. Next, we have fabricated and tested metal-semiconductor-metal (MSM) photodetectors integrated in coplanar strip lines (CPS) on our GaAs layers grown on aluminum pseudo-substrates. More details to the device fabrication and characterization were presented in our previous study [9]. From electro-optical sampling measurements on these devices we obtained photo-response as short as 300 fs FWHM, showing peak amplitude up to 1.2 V at 10 V bias (see figure 6).

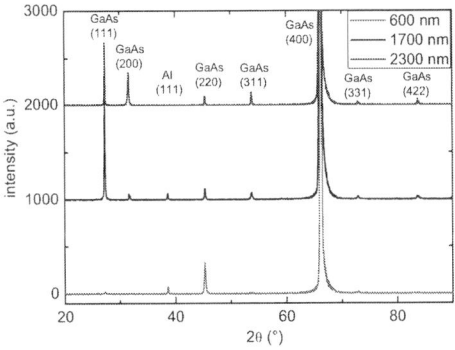

Fig. 1: The x-ray diffraction of GaAs layers on aluminum (111) pseudo-substrate.

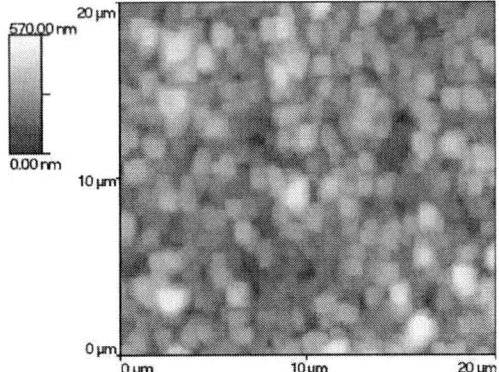

Fig. 2: SIMS measurement of GaAs layer thickness.

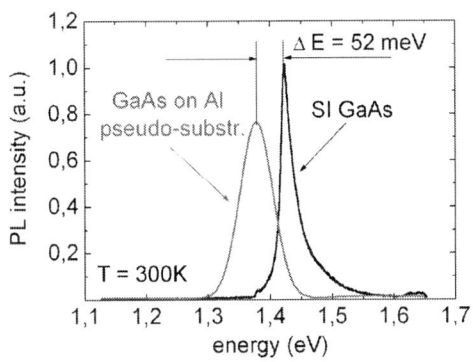

Fig. 3: The PL spectra of SI GaAs and 600 nm thick GaAs layer on Al pseudo-substrate.

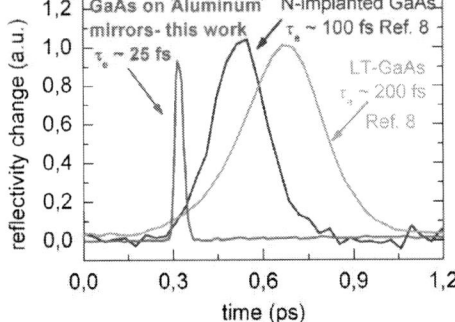

Fig. 4: The surface morphology of 2300 nm thick GaAs layer on Al pseudo-substrate.

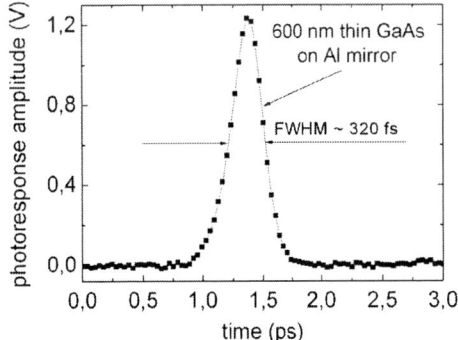

Fig. 5: Carrier lifetime in different GaAs materials.

Fig. 6: Transient photoresponse waveform measured on GaAs/Al MSM photodetector.

4. Conclusion

The growth of GaAs on (111) aluminum pseudo-substrates was performed by MOVPE. The X-ray diffraction measurement showed polycrystalline character of the layers with highly preferential orientation along (100) direction. The increase of the layer thickness led to a degradation of structure quality and surface morphology. The thickness of the layers was measured by RBS and SIMS. The PL measurement performed at room temperature showed red shift which indicates tensile strain in the GaAs layer induced from Al-GaAs interface. The grown layers exhibit an extremely low carrier lifetime due to the high defect concentration. The MSM photodetectors based on the GaAs films exhibit 300 fs FWHM photo-response with peak amplitudes up to 1.2 V with 10 V bias. This material is a promising candidate for highly sensitive ultrafast photodetectors and high-power devices.

Acknowledgments

This work was supported by the Czech Science Foundation (grant N° 104/09/0621 and 203/09/1036) and the Ministry of Education of the Czech Republic (research projects N° MSM6046137302).

References

[1] J. M. Woodall, IBM Technical Disclosure Bulletin 21, 2584 (1978).

[2] G. W. Racette, R. T. Frost, J. Crystal Growth 47, 384 (1979).

[3] D. E. Carlson, Third E.C. Photovoltaic Solar Energy Conference. Dordrecht, Netherlands: Reidel, 294 (1981).

[4] K. Yamada, H. Asahi, H. Tampo, Y. Imanishi, K. Ohnishi, K. Asami, Proceedings of International Workshop on Nitride Semiconductors, Tokyo, Japan: Inst. Pure & Appl. Phys, 556 (2000).

[5] K. Yamada, H. Asahi, H. Tampo, Y. Imanishi, Appl. Phys. Lett. 78, 2849 (2001).

[6] A. V. Andrianov, K. Yamada, H. Tampo, H. Asahi, Semiconductors 36, 878 (2002).

[7] M. Mikulics, M. Kocan, A. Rizzi, P. Javorka, Z. Sofer, J. Stejskal, M. Marso, P. Kordoš, H. Lüth, Appl. Phys. Lett. 87, 212109 (2005).

[8] M. Mikulics, M. Marso, I.C. Mayorga, R. Güsten, S. Stanček, P. Kováč, S. Wu, X. Li, M. Khafizov, R. Sobolewski, E. A. Michael, R. Schieder, M. Wolter, D. Buca, A. Förster, P. Kordoš, H. Lüth, Appl. Phys. Lett. 87, 041106 (2005).

[9] M. Mikulics, R. Adam, Z. Sofer, H. Hardtdegen, S. Stanček, J. Knobbe, M. Kočan, J. Stejskal, D. Sedmidubský, M. Pavlovič, V. Nečas, D. Grützmacher, M. Marso, Semiconductor Science and Technology 25, 075001 (2010).

GaN Power Electronics

Bin Lu, Daniel Piedra and Tomás Palacios

Department of Electrical Engineering and Computer Science,
Massachusetts Institute of Technology,
77 Massachusetts Ave., Bldg. 39-567B, Cambridge, MA 02139, USA
e-mail: tpalacios@mit.edu

Between 5 and 10% of the world's electricity is wasted as dissipated heat in the power electronic circuits needed, for example, in computer power supplies, motor drives or the power inverters of photovoltaic systems. This paper describes how the unique properties of GaN enables a new generation of power transistors has the potential to reduce by at least an order of magnitude the cost, volume and losses of power electronic systems. We will describe three key technologies: Schottky drain contacts and substrate removal to increase the breakdown voltage, and a dual-gate device with superior enhancement-mode characteristics.

1. Introduction

AlGaN/GaN high electron mobility transistors (HEMTs) are excellent candidates for the next generation of power electronics, due to their combination of high electron mobility (μ_e) and high critical electric field (E_c)[1], [2]. As shown in Figure 1, for a given breakdown voltage V_{bk}, normally set by the application, the theoretical specific on-resistance, R_{ON}, of GaN transistors is nearly three orders of magnitude smaller than that of Si transistors and it also surpasses the limit of SiC. These excellent performance enables the use of GaN high voltage transistor in a new generation of power electronic circuits, characterized by at least 10-fold reduction in power losses, volume and cost.

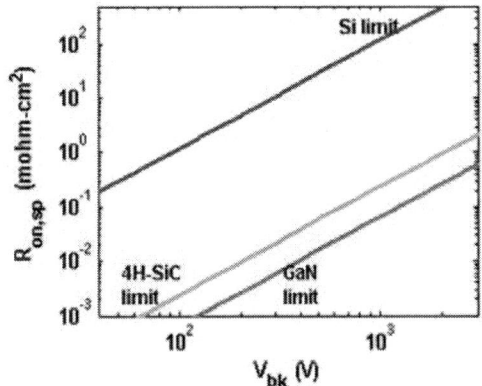

Figure 1: Specific on-resistance as a function of operating voltage for different semiconductor materials.

Most of the reported high-breakdown AlGaN/GaN HEMTs are grown on SiC substrates. However, the limited diameter (up to 4 inch) and high cost of SiC severely hinders the commercialization of GaN-based power electronics in SiC. Intense effort is currently under way to demonstrate the performance GaN transistors on Si substrate, where the low cost and large diameters of the wafers are very attractive from a commercialization point of view.

2. Challenges for GaN-on-Si Power Electronics

In spite of the great potential of GaN on Si high voltage transistors, its application to power electronics is currently limited by three important challenges. On one hand, the breakdown voltage of GaN transistors on Si substrates is lower than when SiC is used as a substrate. To mitigate the effect of the Si substrate, very thick layers of GaN buffer are typically used, which increases the wafer cost and, more importantly, the wafer bow[3]. Wafers with more than 50 μm of bow are very difficult to process using commercial fabrication technologies. In addition, the great majority of GaN transistors are normally-on or depletion mode due to the large charge densities induced by the polarization at the AlGaN/GaN interface. Finally, the leakage current of GaN transistors is still higher that what is required in power electronics (<0.1 μA/mm). In this paper we present three novel technologies to overcome the challenges described above:

1. Schottky-drain metallization
2. Substrate removal
3. Dual gate transisors

3.   Schottky-Drain Technology

To maximize the breakdown voltage of GaN power transistors for a given thickness of the buffer region, it is important to engineer the electric field in the drain access region in a way that it is as uniform as possible. Our group has recently developed a new drain contact technology based on a Schottky metallization that significantly increases the device breakdown voltage[4].

To demonstrate our Schottky drain technology, we used commercially-available GaN/$Al_{0.26}Ga_{0.74}$N/AlN/GaN transistor structures grown by metal organic chemical vapor deposition on Si (111) substrates by Nitronex. The heterostructure has a 20 Å GaN cap layer, a 175 Å $Al_{0.26}Ga_{0.74}$N barrier and a 10 Å AlN interlayer on a 2 μm undoped GaN buffer and transition layer. Standard ohmic contacts were formed by Ti/Al/Ni/Au alloyed for 30 s at 870C in N_2 atmosphere. Unannealed Ti/Au was used as Schottky metallization in the Schottky drain devices. Prior to the Ti/Au deposition, 10 nm recess on the GaN/$Al_{0.26}Ga_{0.74}$N barrier was performed by BCl_3 and Cl_2 plasma with an etch rate of 13~14 Å/min to reduce the series resistance in the Schottky contact. Then 150 nm mesa isolation was achieved by BCl_3 and Cl_2 plasma etching. Finally, Ni/Au/Ni Schottky gates were formed by E-beam evaporation. Both the ohmic drain and Schottky drain devices were fabricated at the same time on the same wafer. All the breakdown voltages were measured with a Tektronix Curve Tracer 576 system. The breakdown voltage is defined as the voltage at which the leakage current reaches 1 mA/mm. The devices were immersed in Fluorinert[TM] FC-770 to prevent surface flash breakdown during measurements.

978-1-4244-8574-1/10 $26.00 © 2010 IEEE   106

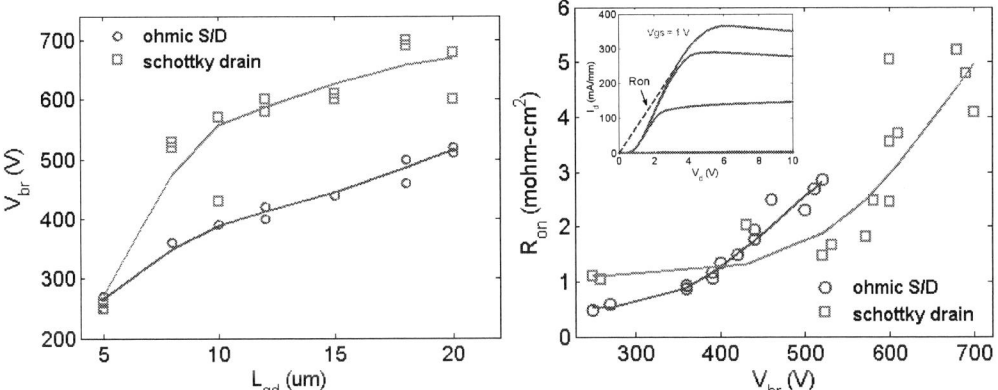

Figure 2: Three terminal breakdown measurement for alloyed ohmic contact and Schottky drain contact.

At least two different mechanisms limit the breakdown voltage in power transistors: buffer/substrate breakdown and gate breakdown. The new Schottky drain devices help improving both of them. We measured the buffer/substrate breakdown voltages of the standard and the Schottky drain devices. In these measurements, a 150 nm deep recess was performed between the contacts to eliminate the 2DEG. The buffer/substrate breakdown of the conventional ohmic drain devices is about 550V. By using Schottky drain contact, the buffer/substrate breakdown voltage is increased above 700 V. This improvement has been associated to the much smoother morphology of the Schottky drain contacts. In conventional ohmic contacts, the high temperature annealing causes metal spikes which we believe increase the electric field, reducing the breakdown voltage.

Three terminal breakdown voltages were also measured on both Schottky drain and ohmic drain HEMTs as shown in Figure 2. Again, the Schottky drain devices had almost 200 V higher breakdown voltage than the standard transistors. This higher performance is obtained without degrading the specific on-resistance, R_{on}. The specific R_{on} resistance was calculated from the I-V curves of the devices when V_{ds}/I_{ds} reaches the lowest value divided by the active area defined as the area between source and drain. The Schottky drain devices have higher breakdown voltage and better V_{br}/R_{on} characteristic at high voltage level.

4. Substrate Removal Technology

For large enough source-to-drain distances, the ultimate breakdown voltage of a GaN power transistor on Si is determined by the distance between the GaN channel and the Si substrate [3]. The Si substrate is typically p-type doped during the grown of the GaN buffer and acts as a highly conductive layer underneath the GaN transistor and the critical breakdown field in the device then becomes vertical instead of horizontal.

978-1-4244-8574-1/10 $26.00 © 2010 IEEE

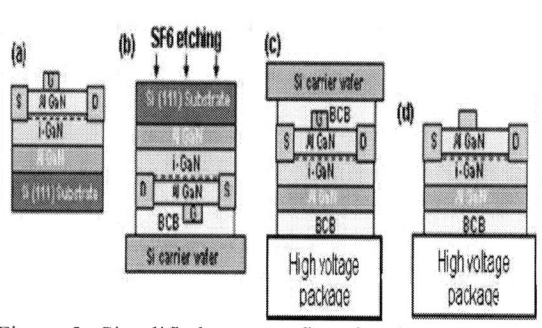

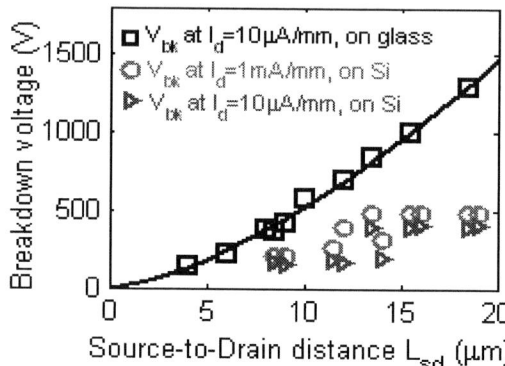

Figure 3. Simplified process flow for the removal of the original substrate in GaN power transistors.

Figure 4. Typical breakdown voltage as a function of source-to-drain distance for AlGaN/GaN structures with and without Si substrate removal technology.

To eliminate the vertical breakdown of the AlGaN/GaN HEMTs on Si, our group has recently demonstrated a new technology based on chemically removing the Si substrate and transferring the AlGaN/GaN HEMTs to a high voltage insulating substrate (glass in our first demonstration) through wafer bonding (Figure 3) [5]. This new device shows a x3-4 fold increase in the maximum breakdown voltage for a given gate-to-drain spacing. For example, device with L_{gd} = 18 μm shows breakdown of 1370 V and on-resistance of 4.3 mΩ•cm^2 with very low leakage current (< 10 μA/mm), much higher than the ~500 V of breakdown obtained in the same device before removing the Si substrate. Figure 4 shows the two terminal buffer breakdown voltage as a function of source-to-drain spacing (L_{sd}). More than 1450 V breakdown and an on-resistance of 5.3 mΩ•cm^2 is achieved on devices with L_{gd} = 20 μm, which is beyond our power supply maximum output voltage.

5. Dual Gate Technology

In spite of the great potential of AlGaN/GaN HEMTs for power electronics applications, its use is severely limited by most of the devices being depletion-mode (D-mode). Enhancement-mode (E-mode) AlGaN/GaN HEMTs are highly desirable for power electronics as they can greatly simplify circuit designs and improve system reliability. Several approaches have been reported in the past to fabricate normally-off GaN HEMTs, including gate recess, AlGaN/GaN/AlN/GaN heterojunctions, fluorine plasma treatment, and p-type AlGaN gate, among others. Many of these methods, although successful in achieving E-mode operation, compromise the on-current, specific-on resistance and the threshold voltage. We have recently developed a new dual-gate AlGaN/GaN E-mode HEMT with a threshold voltage of 2.5 V, maximum drain current of 430 mA/mm, and breakdown voltage of 643 V at zero gate-to-source voltage [6].

978-1-4244-8574-1/10 $26.00 © 2010 IEEE

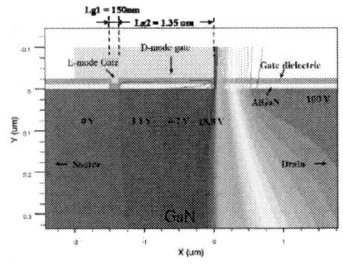

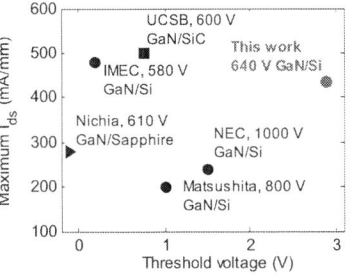

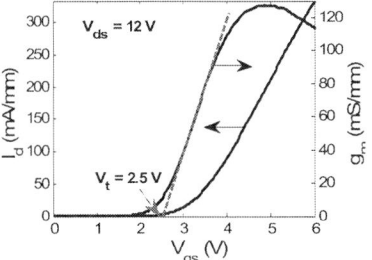

Figure 5: Simulation of the electrostatic potential in AlGaN/GaN HEMTs fabricated with a dual gate technology.

Figure 6: Trade-off between maximum ON current and threshold voltage in different devices reported in the literature.

Figure 7: Transfer characteristic and threshold voltage of a dual-gate GaN power transistor.

The device reported in this paper has the same wafer structure and fabrication technology that it was reported in section 3 of this paper. The only difference is the use of a new dual gate structure, where a very short gate controls the E-mode behavior while a longer gate (D-mode) supports most of the electric field in pinch-off as shown in Figure 5. The first ~ 150-nm-long gate was patterned with electron beam lithography and the AlGaN barrier was then fully recessed with low damage BCl_3/Cl_2 plasma etching. 14 nm Al_2O_3 gate dielectric was then deposited by atomic layer deposition. Finally, a 2-μm-long Ni/Au/Ni gate was deposited overlapping with the first gate-recess region. The 2 μm gate was shifted ~ 1 μm towards the drain side to support the drain voltage. The transistor has a gate-to-source spacing (L_{gs}) of 1.5 μm and a gate-to-drain spacing (L_{gd}) of 18 μm.

The new dual gate device shows a maximum drain current of 430 mA/mm (Figure 6), with a specific on-resistance of 4.1 mΩ.cm2. The double gate structure allows this device to combine a low on-resistance with a large threshold voltage of 2.5 V, extrapolated from the g_m-V_{gs} transfer characteristic curve (Figure 7). The breakdown voltage, measured at $V_{gs} = 0$ V, was 643 V with gate leakage less than our equipment sensitivity of 100 nA/mm. The combination of large threshold and breakdown voltage, with the low on-resistance and leakage current makes this new E-mode device a very attractive option for the next generation of power electronics circuits.

6. Conclusion

In conclusion, GaN-on-Silicon HEMTs offer the potential to revolutionize power electronics by enabling important energy savings and new flexibility for advanced power circuits. In this paper, we have presented three new technologies to overcome some of the main challenges of these devices. First, by using a Schottky drain contact the buffer breakdown can be significantly increased. Second, the removal of the Si substrate allows the fabrication of GaN HEMTs with only 2 μm of GaN buffer thickness and more than 1500 V breakdown. And third, a dual-gate technology enables the combination of low on-resistance and normally-off behavior in the same device. By using these technologies and others currently under development, GaN power electronics will quickly become one of the main markets for GaN devices.

978-1-4244-8574-1/10 $26.00 © 2010 IEEE

Acknowledgement

The authors would like the thank the MIT Energy Initiative, the DOE GIGA project, the ARPA-E ADEPT program and the MARCO IFC program for partially supporting the work described in this paper.

References

[1] U. K. Mishra, S. Likun, T. E. Kazior, and Y. F. Wu, "GaN-based RF power devices and amplifiers," *Proceedings of the IEEE*, vol. 96, no. 2, pp. 287–305, 2008.

[2] N. Ikeda et al., "GaN Power Transistors on Si Substrates for Switching Applications," *Proceedings of the IEEE*, vol. 98, no. 7, pp. 1151-1161, 2010.

[3] S. L. Selvaraj, T. Suzue, and T. Egawa, "Breakdown Enhancement of AlGaN/GAN HEMTs on 4-in Silicon by Improving the GaN Quality on Thick Buffer Layers," *Electron Device Letters, IEEE*, vol. 30, no. 6, pp. 587-589, 2010.

[4] B. Lu, E. Piner, and T. Palacios, "Schottky-Drain Technology for AlGaN/GaN High-Electron Mobility Transistors," *Electron Device Letters, IEEE*, vol. 31, no. 4, pp. 302-304, Apr. 2010.

[5] B. Lu and T. Palacios, "High Breakdown (>1500 V) AlGaN/GaN HEMTs by Substrate-Transfer Technology," *Electron Device Letters, IEEE*, vol. 31, no. 9, pp. 951-953, 2010.

[6] B. Lu, O. I. Saadat, and T. Palacios, "High-Performance Integrated Dual-Gate AlGaN/GaN Enhancement-Mode Transistor," *Electron Device Letters, IEEE*, vol. 31, no. 9, pp. 990-992, 2010.

AlGaN/GaN HEMT on Si (111) substrate for millimeter microwave power applications

S. Bouzid[1,2], V. Hoel[1], N. Defrance[1], H. Maher[2], F. Lecourt[1],
M. Renvoise[2], D. Smith[2] and J.C. De Jaeger[1]

[1]IEMN (Institut d'Electronique, de Microelectronique et de Nanotechnologie),
UMR CNRS 8520 – Lille University, Avenue Poincare 59652, Villeneuve d'Ascq, France
[2]OMMIC, 22 avenue Descartes 94450, Limeil-Brévannes, France
e-mail: virginie.hoel@iemn.univ-lille1.fr and h.maher@ommic.com

This paper reports the capability of AlGaN/GaN HEMTs on Si (111) substrates for microwave power applications above 30GHz. A current gain cut-off frequency f_t=90GHz and a maximum power gain cut-off frequency f_{max}=135GHz are obtained for a 80nm gate-length transistor. These results, associated with low lag effects, demonstrate the capability of these transistors for high performance, cost effective, MMIC fabrication on a Si substrate for high frequency microwave power applications.

1. Introduction

AlGaN/GaN High Electron Mobility Transistors (HEMTs) are being considered as prime candidates for the fabrication of solid-state microwave power amplifiers working above 30GHz. The transistor capabilities are due to the unique material properties associating high electron velocity, high breakdown voltage and high 2D electron gas channel charge density suitable for high temperature applications. Most of the development work so far has been done on epilayers grown on a SiC substrate due to the high thermal conductivity of SiC [1-2], but there is increasing interest and work performed on epilayers grown on highly resistive Si (111) substrates [3] for low cost applications [4-5]. Recently high cut-off frequencies have been demonstrated on both epi-materials [6-7]. To increase the current gain cutoff frequency f_t and the maximum power gain cutoff frequency f_{max}, several issues must be overcome including the development of short gate length device technology and an epi-material including a low barrier layer thickness. Specific steps must be developed to be able to fabricate gate lengths down to 70nm and to reduce the short channel effects and parasitic resistances. This paper reports on the technological process developed to fabricate short gate length transistors on Si (111) substrate for MMICs fabrication. A 80nm gate length transistor exhibits a cutoff frequency f_t=90GHz and f_{max}=135GHz, associated with low lag effects, showing the capabilities of AlGaN/GaN HEMTs on Si substrate for high performance, low cost, microwave power applications above 30GHz.

2. Epitaxy and transistor fabrication

The AlGaN/GaN transistor epitaxy was grown on a high-resistivity Si(111) substrate by metal-organic chemical vapor deposition (MOCVD). The device structure consists of a 12.5 nm $Al_{0.26}Ga_{0.74}N$ barrier and a 2 nm cap layer of GaN. Material characteristics were determined from Van Der Pauw and TLM measurements before and after passivation (Table1). It can be seen that the carrier mobility is higher than 2050 cm²/V.s. Furthermore, the sheet carrier density increases after passivation linked to a drop of the sheet resistance due to an additional tensile constraint.

Transistor fabrication starts with Ti/Al/Ni/Au (12/200/40/100nm) ohmic metallization followed by a rapid thermal annealing (RTA) at 850 °C for 30 s under a nitrogen atmosphere. The device isolation is obtained by multiple ion implantations based on different energies and doses. Contact resistance and specific contact resistivity, measured by the Transmission Line Model (TLM) on different patterns, are 0.38 Ω.mm and 3×10^{-6} Ω.cm^2 respectively remaining quite similar before and after passivation. The T-shaped gate based on Ni/Au (40/300 nm) metallization with a 80 nm footprint is defined by electron-beam lithography using a tri-layer resist stack (PMMA/Copolymer/PMMA). The drain to source spacing is 2.65µm and the gate to source distance is 1.15µm. The gate shape is observed using a Focus Ion Beam (FIB), as shown in Fig. 1. Finally, devices are passivated with SiN/SiO$_2$ (50/100 nm) deposited by plasma-enhanced chemical vapor deposition (PECVD) at 340 °C.

	Hall Measurements			TLM
	μ (cm^2/V.s)	N_s 10^{12} (cm^{-2})	$R_\square$ (Ω)	$R_\square$ (Ω)
Before passivation	2050	7.7	395.8	368
After passivation	2122	9.5	306.5	294

Table 1: Material characteristics

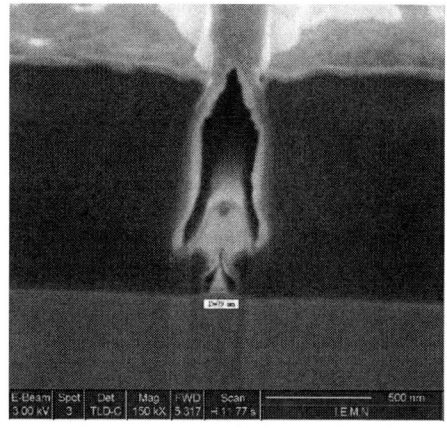

Figure 1: Cross-sectional 80 nm T-shaped gate FIB image before lift-off

Figure 2: Forward and reverse Schottky characteristics

3. DC characterization

All DC and small signal measurements were carried out on 2×50 µm x 0.080 µm HEMT devices using microwave probes.

Figure 2 shows the forward and reverse characteristics of the Schottky contact after passivation. At V_{GS}=-15V, the reverse leakage current is 10µA/mm and the barrier height is 0.69 eV associated to an ideality factor η=3.4. From the DC transistor characterization, a maximum drain current density of 600 mA/mm is obtained at V_{GS} = 0 V with a pinch-off voltage of -2.8 V (Fig.3). The knee voltage is less than 3 V indicating excellent ohmic contact fabrication. The transfer characteristics at $V_{DS} = 6$ V are also presented in Fig.3. The peak extrinsic transconductance is 280 mS/mm at $V_{GS} = -1.7$ V. The extraction of the intrinsic transconductance g_{mi} is performed from the knowledge of the static source resistance $R_s = 8.54\ \Omega$ deduced from the contact resistance and the sheet resistance. An intrinsic transconductance $g_{mi} = 366$ mS/mm is obtained.

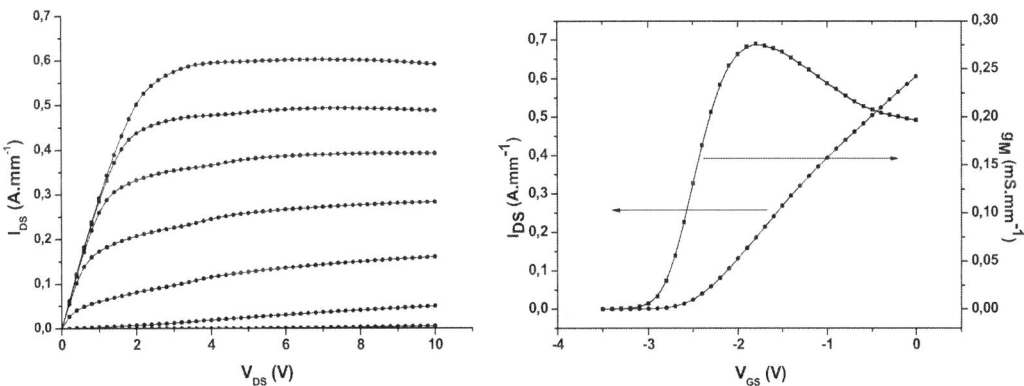

Figure 3: DC characteristics of the 80 nm gate AlGaN/GaN HEMT. (left) I_{DS} (V_{DS}) characteristics ($V_{GS} = -3$ to 0 V, step +0.5 V). (right) Transfer characteristics at $V_{DS} = 6$ V.

4. RF and pulse characteristics

The scattering parameters were measured from 0.25 to 50 GHz using an Agilent Technologies N5245A vector network analyzer and line-reflect-reflect-match (LRRM) calibration. Pad capacitances were de-embedded using dedicated on-wafer calibration structures. Fig. 4 shows the current gain modulus ($|H_{21}|$) and the Mason's maximum unilateral gain (U) derived from S-parameters measurement versus the frequency. At $V_{DS} = 6$ V and $V_{GS} = -1.7$ V, an extrinsic current gain cutoff frequency (f_t) of 90 GHz and a maximum power gain cutoff frequency (f_{max}) of 135 GHz are obtained. These values are very good for AlGaN/GaN HEMT on high resistivity silicon with an 80 nm gate length. Moreover, a good f_{max}/f_t ratio value of 135/90 is observed.

Pulse measurements were carried out in order to study the trap response to an applied electrical field. It relies on the exclusive use of cold quiescent bias points, and short pulse length (500ns) permitting to mitigate current drop due to thermal effects. Fig. 5 shows the DC pulsed characteristics wherein all quiescent bias points are chosen to reveal the gate and drain lag effects. The pulse $I_{DS}(V_{DS})$ characteristics determined at the quiescent point $V_{DS0} = 0$ V, $V_{GS0} = -5$ V (beyond pinch-off voltage) are compared to the $I_{DS}(V_{DS})$ reference determined at the quiescent point $V_{DS0} = 0$ V, $V_{GS0} = 0$ V in order to analyze the gate lag effect. On the same figure, the pulse $I_{DS}(V_{DS})$ characteristics determined at $V_{DS0} = 15$ V, $V_{GS0} = -5$ V are presented in order to show the drain lag effect. Even with high electrical stress upon the gate (about twice the pinch-off voltage), the gate lag remains low with a current drop of only 11 % at $V_{DS} = 6$ V. Regarding the drain lag contribution, it has been evaluated to about 19 % at the same V_{DS}.

These results are very promising and demonstrate the possibility of using AlGaN/GaN HEMTs on Si (111) substrate for microwave power applications above 30 GHz.

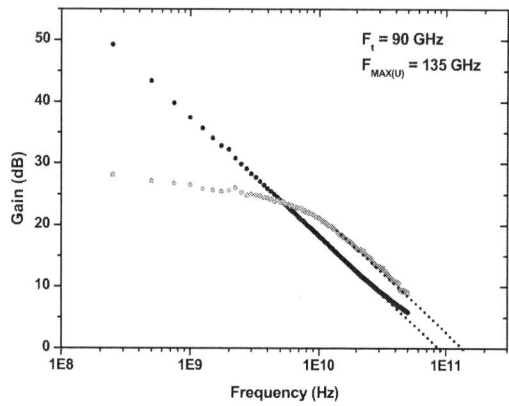

Figure 4: Microwave characteristics of T-shaped 80 nm gate length AlGaN/GaN HEMT at V_{DS} = 6 V and V_{GS} = -1.7 V. Extrapolation at -20dB/dec of $|H_{21}|$ and U yield f_t = GHz and f_{max} = GHz respectively.

Figure 5: DC pulsed $I_{DS}(V_{DS})$ characteristics of T-shaped 80 nm gate length AlGaN/GaN HEMT at three different quiescent bias points. (V_{GS0} = 0 V, V_{DS0} = 0 V), (V_{GS0} = -5 V, V_{DS0} = 0 V) and (V_{GS0} = -5 V, V_{DS0} = 15 V) for V_{GS} from -3 to 0 V (step 1 V).

4.   Conclusion

80 nm gate length AlGaN/GaN HEMTs on epitaxy grown on Si (111) were fabricated. Devices exhibit very interesting performance with cut-off frequencies of f_t = 90GHz and f_{max} = 135GHz. These results, associated with the low lag effects, demonstrate that with the advances in material quality and processing techniques, AlGaN/GaN devices on silicon substrate are showing very promising potential for microwave power in Ka-band. In the future, the device performance is to be further improved by reducing the drain to source spacing.

References

[1]   T. Palacios, A. Chakraborty, S. Rajan, C. Poblenz, S. Keller, et al, *IEEE Electron Device Letters*, **26**, 11, 2005, p. 781.

[2]   D. Kim, V. Kumar, J. Lee, M. Yan, A. M. Dabiran et al, *IEEE Electron Device Letters*, **30**, 9, 2009, p.913

[3]   K. Cheng, M. Leys, S. Degroote, J. Derluyn, B. Sijmus et al, *Japanese Journal of Apllied Physics,* **47**, 3, 2008, p.1553.

[4]   D. Ducatteau, A. Minko, V. Hoel, E. Morvan, E. Delos, et al, *IEEE Electron Device Letters*, **27**, 1, 2006, p.7

[5]   J.W. Johnson, E.L. Piner, A. Vescan, R. Therrien, P. Rajagopal, et al, *IEEE Electron Device Letters*, **25**, 7, 2004, p.459

[6]   J. W. Chung, W.E. Hoke, E.M. Chumbes, and T. Palacios, *IEEE Electron Device Letters*, **31**, 3, 2010, p.195.

[7]   S. Tirelli, D. Marti, H. Sun, A. R. Alt, H. Benedickter et al, *IEEE Electron Device Letters*, **31**, 4, 2010, p.296.

Characterisation of electrical properties of AlGaN/GaN Schottky diode at very high temperature

A. Chvála, D. Donoval, R. Šramatý, J. Marek, J. Kováč, P. Kordoš and J. Škriniarová

Department of Microelectronics, Slovak University of Technology in Bratislava,
3 Ilkovicova 3, 812 19 Bratislava, Slovakia
e-mail: ales.chvala@stuba.sk

Recent progress in GaN based high electron mobility transistors (HEMTs) has revealed them to be strong candidates for future high power devices at high frequency operation. In order to extract and utilize the favorable GaN material properties, however, there are still a lot of areas to be investigated. Among them the most important is to develop new processes, structure design and characterization techniques. Determination of the effective Schottky barrier height ϕ_b on GaN and related compound semiconductors with higher precision is important for further analysis of new combinations of metals and semiconductors and better understanding of physical behaviour at the interface. In this paper we present the modified method of evaluation of the selected parameters on the AlGaN/GaN heterostructure from the I-V measurement in a wide temperature range.

2. Introduction

There are many methods of determining of the Schottky barrier height ϕ_b available involving *I-V* measurements, *C-V* measurements and photoelectron measurements [1][2]. For their simplicity the *I-V* measurements are most popular and commonly used. The method most frequently used in practice presumes pure thermionic emission over the barrier:

$$I_1 = I_{te}\left[\exp\left(\frac{qV}{kT}\right) - 1\right]. \tag{1}$$

As a refinement of this method the series resistance *Rs* and the ideality factor *n* were introduced to include the contributions of other current-transport mechanisms. Then:

$$I = I_{te}\exp\left[\frac{q(V - IR_S)}{nkT}\right]\left\{1 - \exp\left[\frac{q(V - IR_S)}{kT}\right]\right\}, \tag{2}$$

where $I_{te} = AA^{**}T^2\exp\left(\frac{-\phi_b}{kT}\right)$ \hfill (3)

is referred to as a saturation current. *A* is the diode area and *A*** is the modified Richardson constant (for GaN $A^{**} = 3 \times 10^5$ Am^{-2}K^{-2} [3]). If the ideality factor *n* deviates significantly from 1 the other mechanisms of current flow through the Schottky structure play more important role and the evaluation of Schottky barrier height using Eqn. (2) can lead to incorrect non-physical extracted results.

The GaN material is the wide band gap semiconductor where a quantum-mechanical process can not be neglected. The *I-V* characteristics exhibit a considerable presence of other current mechanisms like thermionic emission (TE), tunnelling (TU), generation-recombination (GR) and leakage currents (RL) through a leakage resistance R_L. The total current can be expressed as a sum of all components [4]:

$$I = I_{te}\left[\exp\left(\frac{q(V - IR_S)}{kT}\right) - 1\right] + I_{gr}\left[\exp\left(\frac{q(V - IR_S)}{2kT}\right) - 1\right] +$$

$$+ I_t\left[\exp\left(\frac{q(V - IR_S)}{E_0}\right) - 1\right] + \frac{V - IR_S}{R_L}. \tag{4}$$

where I_{te}, I_t, I_{gr} are the thermionic emission, tunnelling and generation-recombination saturation currents, E_0 is the characteristic constant for the tunneling. The pure thermionic emission current can be easy separated from the total current and then the Schottky barrier height ϕ_b can be evaluated with higher precision.

2. Experiment

The Schottky diode under investigation was prepared by Ni/Au metallization on a top of the AlGaN/GaN heterostructure (Schottky contact). The ohmic contact was created by Ti/Al/Ni/Au metallization. The Schottky diode has a square geometry with 100 μm edge. The forward *I-V* characteristics measured in the wide temperature range are shown in Fig. 1a. The region of the straight line is commonly used to determine the Schottky barrier height ϕ_b and the ideality factor *n* using of the standard method described by Eqn. (2). A strong increase of the apparent barrier height and simultaneously a decrease of the ideality factor with increased temperature are obtained (Fig. 1b). The strong increase of the barrier height with increased temperature cannot be explained theoretically. The high value of the ideality factor (2 - 8) indicates that the other mechanisms of the current transport (tunnelling, generation-recombination and leakage) are becoming dominant, particularly for small forward biases.

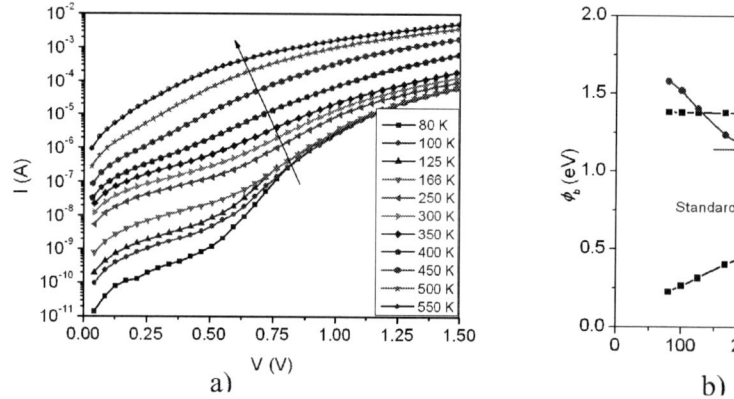
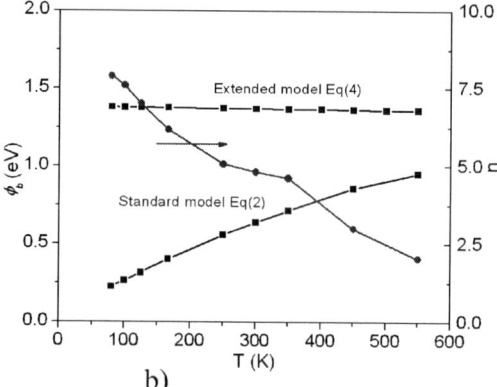

Fig. 1: a) *I-V* characteristics of Ni/Au - AlGaN Schottky structure in a wide temperature range. b) Temperature dependence of a) the Schottky barrier height ϕ_b and the ideality factor *n* extracted by simple (Eqn. (2)) and extended method (Eqn. (4)).

Using the extended model described by Eqn. (4) all four current parts were extracted from the *I-V* curves at various temperatures. For the low current range the leakage current RL is a dominant effect, additionally the tunnelling current TU plays a main role. The thermionic emission current TE and the serial resistance are dominant for the high current density range. The generation-recombination current GR is small and can be neglected. Fig. 2 shows a comparison of the measured and computed *I-V* curve. The computed curve is the sum of all calculated current parts and good agreement is clearly seen (fitting error less then 15 % for all temperatures). From the extracted pure thermionic emission current the effective Schottky barrier height ϕ_b was calculated and the result shows physically reasonable values (ϕ_b is slightly decreasing with temperature proportionally to energy band-gap) (Fig. 1b).

Thus, the termionic emission current depends by square law on temperature (Eqn. (3)), while the tunnelling component which dominates for the diodes investigated is less dependent on temperature. This fact supports a necessity of high-temperature measurement of Schottky diodes with relatively high tunnelling contribution, as it is the case of GaN-based diodes in general. The termionic emission component approaches the total current with increased temperature is remarkable (comparison for temperature 166 K and 550 K in Fig. 2). This explains that an extraction of the pure thermionic emission current from the total current is necessary in order to evaluate correct Schottky barrier height.

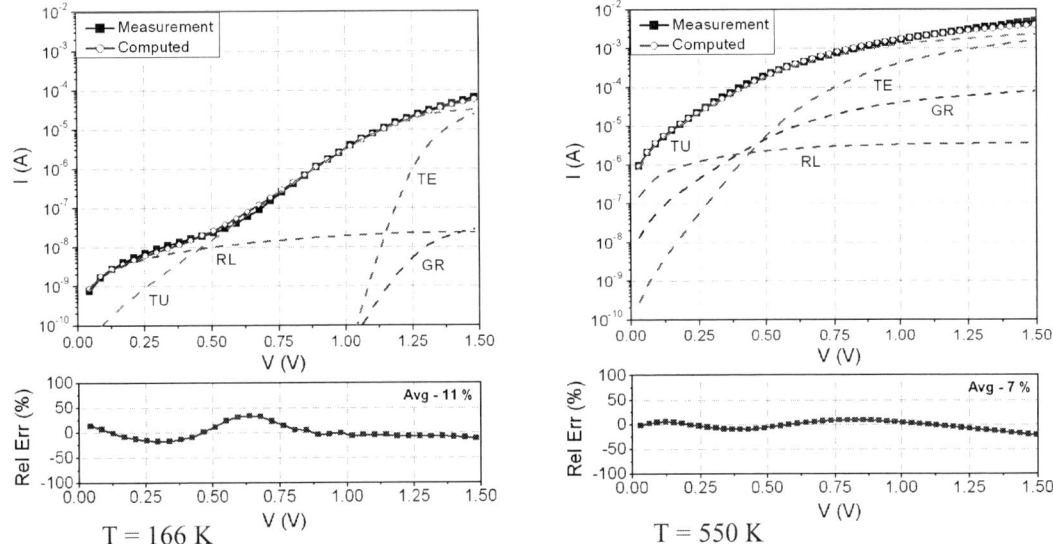

Fig. 2: Fitting results of the forward *I-V* characteristic of Schottky diode at 166 K and 550 K.

Temperature dependence of the thermionic emission saturation current I_{te} on the inverse temperature is given in Fig. 3a. The value of $\phi_b = 1.38$ eV is consistent with ϕ_b values determined from *I-V* measurements using extended model and Richardson constant $A^{**} = 3.82 \times 10^5$ Am^{-2}K^{-2} extracted from this activation energy plot corresponds very well to most quoted values for the Ni/Au - AlGaN Schottky structures.

The tunnelling current is the dominant one, as it follows from performed analysis of the current transport in the Schottky diodes investigated. According to the dislocation model of the tunnelling mechanism in Schottky diodes the tunnelling saturation current $I_{t,0K}$ at zero absolute temperature can be expressed as:

$$I_{t,0K} = A.q.v_D \mathrm{N}_{dis} \exp\left(-\frac{\phi_b}{E_{0,0K}}\right) \tag{5}$$

where v_D is the Debye frequency and N_{dis} is the dislocation density. Thus, characterization of the tunnelling current as a function of temperature allows us to evaluate the dislocation density in the device investigated. The tunnelling saturation current I_t and the tunnelling energy E_0 as a function of temperature are shown in Fig. 3b. From these data we can obtain $I_{t,0K} = 2 \times 10^{-13}$ A and $E_{0,0K} = 66$ meV by extrapolation to 0 K. The dislocation density calculated using Eqn. (5) is about 1×10^6 cm^2, considering $\phi_b = 1.38$ eV and $v_D = 1.68 \times 10^{13}$s^{-1} which is in agreement with the value reported before in [5]. From the tunnelling energy evaluated AlGaN doping is in the upper part of the 10^{19} cm^{-3} range, which is in agreement with the residual doping of the AlGaN layer expected to be around 5×10^{18} cm^{-3}.

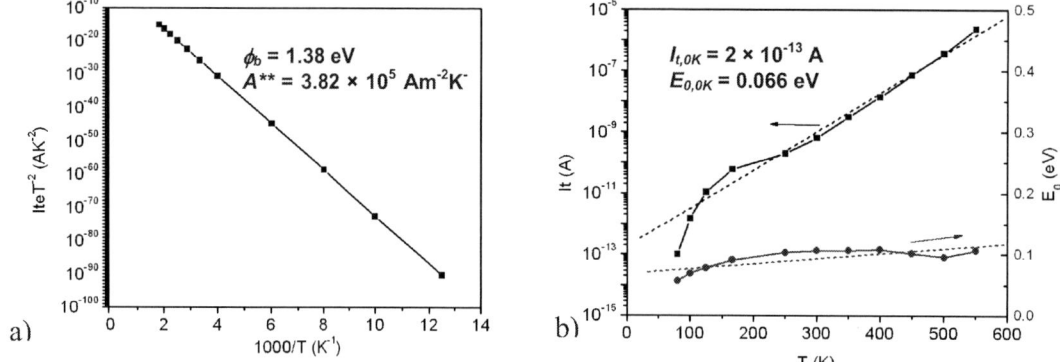

Fig. 3: a) Activation energy plot of the thermionic emission current extracted by Eq.(4). b) Tunnelling saturation current and tunnelling energy as a function of temperature for Schottky diode.

3. Conclusions

The modified approach of evaluation of the Ni/Au-AlGaN/GaN Schottky barrier height ϕ_b from the forward I-V characteristics is presented. This method allows us to assess the effect of particular mechanisms (tunnelling, generation-recombination and leakage currents) on the total current and to determine the thermionic emission current more exactly at various temperature. From the plot of I_{te}/T^2 vs $1/T$ the reasonable Schottky barrier height and modified Richardson constant A^{**} were calculated. Analysis shows that the tunnelling current is dominant for the Schottky diode. The dislocation density of about 1×10^6 cm^2 is evaluated for AlGaN/GaN-based diodes assuming dislocation model of the tunnelling current. Very good agreement between the experiment and theory confirms the validity of our modified approach.

Acknowledgement

This work has been done in Centre of Excellence NanoNET, ITMS code 26240120010 and Center of Excelence CENAMOST (Slovak Research and Development Agency Contract No. VVCE-0049-07) with support by European Project MORGAN No. FP7 NMP IP 214610, projects APVV LPP-0195-09 and VEGA 1/0742/08.

References

[1] Z. Luo et al.: Photoelectric response of Schottky barrier in La0.7Ca0.3MnO3 /Nb:SrTiO3 heterojunctions, Applied Physics Letters 92, 182501, (2008).

[2] N. Miura et al.: Thermal annealing effects on Ni/Au based Schottky contacts on n-GaN and AlGaN/GaN with insertion of high work function metal, Solid-State Electronics 48, pp. 689–695, (2004)

[3] T. Sawada et al.: Properties of GaN and AlGaN Schottky contacts revealed from I–V–T and C–V–T measurements, Applied Surface Science 216, pp. 192–197, (2003).

[4] D. Donoval et al.: Analysis of measurement on PtSi-Si Schottky structures in a wide temperature range, Solid-State Electron 34, pp. 1365-1373, (1991).

[5] E. Arslan, et al.: Dislocation-governed current-transport mechanism in (Ni/Au)-AlGaN/AlN/GaN heterostructur, Journal of Applied Physics 105, 023705, (2009).

On the Identification of Trap Location in AlGaN/GaN HEMTs during Electrical Stress

M. Ťapajna,[1] R. J. T. Simms,[1] Y. Pei,[2] U. K. Mishra,[2] and M. Kuball[1]

[1] Center for Device Thermography and Reliability, University of Bristol, BS8 1TL, UK
[2] Department of ECE, University of Santa Barbara California, Santa Barbara CA 93106, USA
e-mail: milan.tapajna@bristol.ac.uk

Location and properties of traps generated in AlGaN/GaN high electron mobility transistors submitted to electrical stress was studied using an integrated electrical and optical methodology. A spatial and spectral electroluminescence study reveals traps generated during both OFF- and ON-state stress to be located in the gate and access region close to the drain side of the gate edge, while UV-light assisted detrapping analysis in conjunction with photoluminescence illustrate these dominant traps located mostly within an AlGaN subsurface layer.

1. Introduction

In order to utilize the full potential of AlGaN/GaN high electron mobility transistors (HEMTs), it is necessary to address reliability issues, *i.e.* the degradation of performance during HEMT operation [1,2]. However, detailed knowledge of the degradation is still lacking, mostly due to difficulties in identifying the nature and location of the electronic traps generated during stressing. In the present paper, an UV light-assisted trapping analysis in conjunction with an electroluminescence (EL) and photoluminescence (PL) study is employed to identify the location of traps generated in AlGaN/GaN HEMTs submitted to ON- and OFF-state stress. Our results indicate these traps to be located mostly in the AlGaN subsurface layer of the HEMT access region close to the drain side of the gate edge.

2. Experimental Details

$Al_{0.26}Ga_{0.74}N$/GaN HEMTs used in this study were grown on SiC substrates using MOCVD, with standard Ti/Al/Ni/Au ohmic contacts, Ni/Au gate contact (gate length and width of 1 and 150 μm, respectively), and a 160-nm-thick SiN_x passivation layer. The HEMTs were subjected to ON- and OFF-state stress with V_{gs}=0 and -5 V, respectively, and V_{ds}=30 V for 40 hours at room base plate temperature (T_b) and at 100 °C. Detrapping analysis similar to [1] was employed to determine relative trap densities and activation energies. The key addition to this detrapping analysis employed in this work is that the detrapping characteristics were measured after UV light exposure using spectrally filtered UV light ranging from $\hbar\omega$=3.3 to 4.2 eV [3]. The HEMT is first illuminated by UV light (Fig. 1) and changes in the trap occupation are measured by stepping the device into the linear regime after applying a filling pulse and monitoring the drain current (I_d). EL was used to determine the hot electron temperature (T_e) distribution across the device, where T_e is extracted from the high energy tail of the EL spectrum similar to [4,5]. PL measurements were carried out at room temperature using an excitation wavelength of 244 nm, before and after stressing the HEMT. DC and pulsed (V_{gs} pulsed with 100 ns long pulses, V_{ds} DC biased) transistor characteristics were measured. Gate-lag was calculated as a difference between DC and pulsed I_d at given V_{ds}.

3. Results and Discussion

Figures 2(a) and (b) show output characteristics and transconductance (g_m) as a function of V_{gs} measured before and after ON- and OFF-state stress. A decrease of I_d after OFF-state stress is mostly due to a positive threshold voltage shift (V_{th}) of 0.46 V, while after ON-state stress, a negative V_{th} shift of -0.2 V was observed. Such V_{th} shift after OFF-state stress has previously been ascribed to a generation of negatively charged traps underneath the gate as a consequence of reverse biased source-gate diode [2], while during ON-state stress, thermally induced gate contact degradation may explain the V_{th} shift. Both stress conditions result in the drain voltage knee walkout (Fig. 2(a)), a decrease in g_m for lower V_{gs} which is, however, almost unchanged at higher V_{gs} (Fig. 2(b)), and an increase in gate-lag, being larger for OFF-state stress in agreement with the larger g_m decrease.

Fig. 3(a) compares a spatial linescan of T_e across the active area of HEMTs before and after ON- and OFF-state stress. The most pronounced changes in T_e profiles takes place in the gate and access region at the drain side of the gate after both stress conditions, with the largest decrease in T_e at the drain side of the gate edge. This is consistent with trap generation taking place in these device regions. Traps generated in the HEMT access region close to the drain side of the gate can mitigate electric field that results in lowering of electron temperature. However, from these results, it is not possible to distinguish whether traps generation takes place in AlGaN barrier or GaN buffer layer.

To determine in which device layer the generated traps are located, PL measurements were carried out. Fig. 3(b) shows PL spectra taken on the HEMT before and after OFF-state stress. AlGaN band-edge PL decreased by about factor of 4 after OFF-state stress. This would be consistent with the generation of non-radiative electronic traps inside the AlGaN layer. In order to prove this hypothesis, I_d detrapping analyses were carried out on HEMTs stressed in ON- and OFF-state stress conditions (T_b=25 °C). First, we measured I_d detrapping characteristics without UV light exposure. Three traps denoted as Tp1, Tp2, Tp3 were observed in agreement with our previous study [2]. Fig. 4(a) shows detrapping analysis together with corresponding I_d transients (inset). OFF-state stress results in an increase of all three trap amplitudes, while ON-state stress induced an increase mostly of the Tp1 amplitude only. The trap amplitude is related to the trap density. In the following, we focus on the

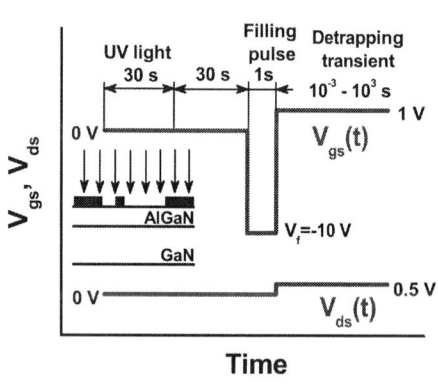

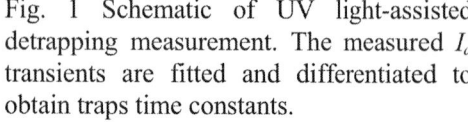

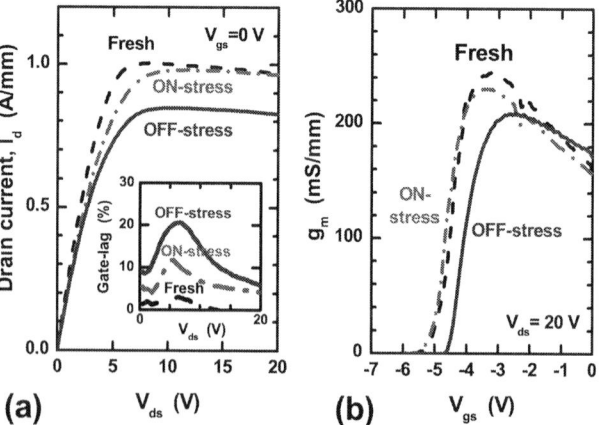

Fig. 1 Schematic of UV light-assisted detrapping measurement. The measured I_d transients are fitted and differentiated to obtain traps time constants.

Fig. 2 (a) DC output, g_m-V_{gs} characteristics, and gate-lag measurements (inset) of AlGaN/GaN HEMT before and after ON-/OFF-state stress. Gate-lag was determined as $(I_d^{DC}-I_d^{pulse})/I_d^{DC}$.

978-1-4244-8574-1/10 $26.00 © 2010 IEEE

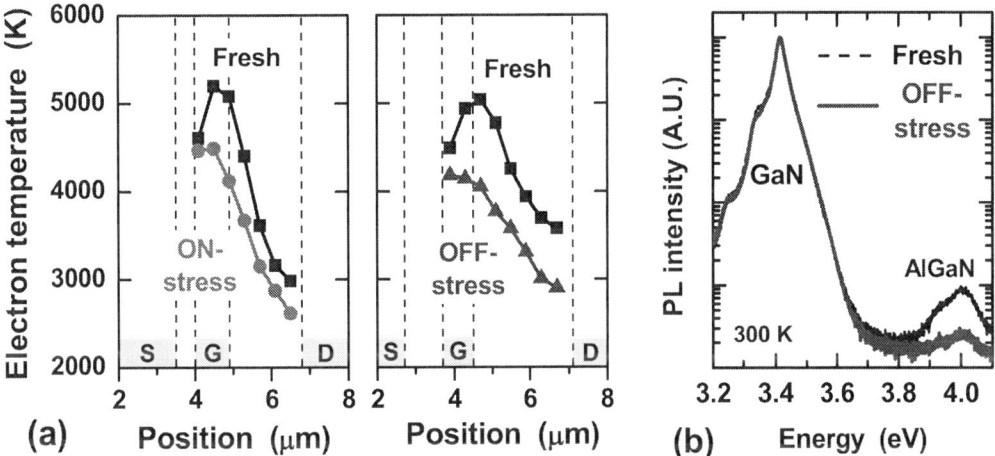

Fig. 3 (a) Electron temperature profiles across HEMT active region measured before and after ON- (left panel) and OFF-state (middle panel) stress. (b) PL spectra obtained at gate-drain spacing of a transistor before and after OFF-state stress and normalized to GaN band-edge luminescence peak. The PL measurements were performed at a device stressed at T_b=100 °C.

dominant trap, Tp1, as it is mostly affected by both stresses. I_d detrapping analysis was performed at different T_b to determine trap activation energies. An activation energy for Tp1 time constant of 0.45 eV before and after ON-state stress was obtained and it increased to 0.65 eV after OFF-state stress [2]. Next, UV light-assisted I_d detrapping analysis was carried out on stressed HEMTs. Fig. 4 shows Tp1 trap amplitude as a function of incident photon energy. UV light has a distinct influence on the Tp1 amplitude at photon energies 3.6 and 3.4 eV for devices stressed in ON- and OFF-state stress condition, respectively. As this corresponds to 0.5 and 0.7 eV below AlGaN conduction band edge, *i.e.* consistent to their activation energies, this illustrates Tp1 traps are predominantly located in the AlGaN device layer. This is also consistent with the result of the PL measurement. As illustrated in the T_e profiles (Fig. 3(a)) the largest changes in the device, during both stresses, take place in the access region close to the drain side of the gate, being larger for OFF-state stress, and this is the spatial location of these traps inside the AlGaN layer. In addition, there is a feature 0.3 eV below AlGaN conduction band edge, whose origin is not clear at present.

The influence of the UV light exposure on the Tp1 amplitude can be understood as UV light-induced trap population (transition of electrons excited from the AlGaN or GaN valence band into the trap level, inset of Fig. 4(b)). During subsequent transient measurements, electrons are emitted from Tp1 into AlGaN conduction band and extracted by the gate for example via surface leakage. Traps involved in the I_d detrapping transients are assumed to be also filled via surface leakage and tunneling. It is therefore likely that Tp1 traps are located in AlGaN subsurface within the tunneling distance (~3 nm). Generation of dominant traps located in AlGaN subsurface layer at the drain side of the gate that become negligible towards the drain contact is also consistent with gate-lag increase, knee voltage walkout, and g_m decrease for low V_{gs} only observed (Fig. 2) as demonstrated in [6]. Moreover, it would be in agreement with our previous results on a linear dependence of the EL intensity on V_{ds} before and after OFF-state stress indicating a negligible change of scattering mechanism of electrons in the transistor channel [2]. A possible mechanism of the generation of such traps could be diffusion related processes [7].

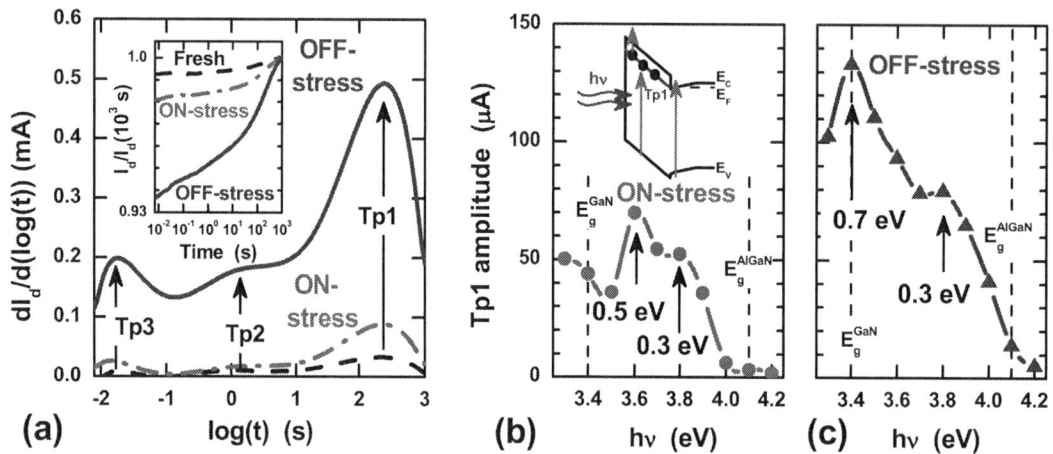

Fig. 4 (a) Derivative of $I_d(\log(t))$ before (dashed line) and one-day after ON- (dash-dotted line) and OFF-state stress (solid line) measured without UV light exposure. Inset of (a) shows corresponding normalized I_d transients. (b-c) Tp1 amplitude measured by detrapping analysis with UV light exposure as a function of UV photon energy on HEMTs submitted to ON- (b) and OFF-state (c) stress. Inset of (b) depicts a population of trap level in AlGaN by UV light.

4. Summary

Early stage degradation of AlGaN/GaN HEMTs upon ON- and OFF-sate stress was investigated. Both stress conditions lead to trap generation in the transistor access region close to the drain side of the gate edge. UV light-assisted trapping analysis and photoluminescence illustrate these traps to be located in the AlGaN subsurface layer of the AlGaN/GaN HEMT.

Acknowledgement

Funding from Office of Naval Research and ONR Global (N00014-08-1-1091) through the DRIFT program (monitored by Dr. Paul Maki) is gratefully acknowledged. The authors would like to thank Dr. D. Wolverson (University of Bath) for providing us access to a deep-UV-PL system and for support to the measurements, and N. Killat (University of Bristol).

References

[1] J. Joh and J. del Alamo, *IEDM Tech. Dig.*, 2008, p. 1.

[2] M. Ťapajna, R. J. T. Simms, Y. Pei, U. K. Mishra, and M. Kuball, *IEEE Electron Dev. Lett.* **31**, 662, 2010.

[3] M. Ťapajna, R. J. T. Simms, M. Faqir, Y. Pei, U. K. Mishra, and M. Kuball, in *Proc. of IEEE Int. Reliab. Phys. Sym.*, 2010, p. 152.

[4] N. Shigekawa, K. Shiojima, and T. Suemitsu, *J. Appl. Phys.*, **92**, 531, 2002.

[5] J. W. Pomeroy, M. Kuball, M. J. Uren, K. P. Hilton, R. S. Balmer, and T. Martin, *Appl. Phys. Lett.*, **88**, 023507, 2006.

[6] A. Chini, V. Di Lecce, M. Esposto, G. Meneghesso, E. Zanoni, *IEEE Electron Dev. Lett.* **30**, 1021, 2009.

[7] M. Ťapajna, U.K. Mishra, and M. Kuball, *Appl. Phys. Lett.*, **97**, 023503, 2010.

Study of temperature distribution in the channels of AlGaN/GaN HEMT devices by μ- Raman characterization techniques

J. Kováč jr. [a,b], S. K. Jha [b], E. V. Jelenković [c], O. Kutsay [b], M. Pejović [d], C. Surya [c]
J. A. Zapien [b], I. Bello [b], R. Srnánek [a], J. Kováč [a], S. Flickyngerová [a,b]

a) Faculty of Electrical Engineering and Information Technology, Slovak Technical University, Ilkovičova 3, 81219 Bratislava

b) Department of Physics and Materials Science and Center of Super-Diamond and Advanced Films, City University of Hong Kong, 83 Tat Chee Avenue, Kowloon, Hong Kong, PR China

c) Department of Electronic and Information Engineering, The Hong Kong Polytechnic University, Hung Hom, Kowloon, Hong Kong, PR China

d) Faculty of Electronic Engineering, Nis University, P.O. Box 73, 18000 Niš, Serbia

e-mail: jaroslav_kovac@stuba.sk

AlGaN/GaN and InAlN/GaN high electron mobility transistors (HEMTs) were fabricated and their electrical and thermal properties were examined. The influence of irradiation to direct current (DC) and low-frequency noise properties of these devices have already been investigated and published. However, the thermal processes in the active layers of these devices can also induce significant changes in the performance of devices. Since thermal processes are directly associated with temperature distribution, temperature mapping in the devices has been suggested using a μ-Raman technique. Considering that the phonon frequencies are sensitive to the sample temperature, the shift of first-order Raman scattering can conveniently be used to directly measure the device temperature with a high spatial resolution. In this context, we report time resolved and sample mapping Raman spectra inside the HEMTs channel as measures of temperatures.

1. Introduction

The epitaxial layers of AlGaN/GaN high electron mobility transistors (HEMTs) are composed of a carbon doped GaN layer, ~20 nm thick undoped AlGaN barrier layer with Al content of 30 at%. Some of the devices were irradiated to cumulative doses up to 10^7 rad. The influence of this irradiation to direct current (DC) and low-frequency noise properties of these devices were investigated and published elsewhere [1]. The characteristics show deteriorating device performance upon the gamma exposure. Moreover, the thermal conditions of the active layer in these devices can cause significant changes in the performance of devices. Recently, the μ-Raman technique was considered as a very suitable tool for temperature mapping around the whole active regions of these devices [2, 3]. Considering that the phonon frequencies are sensitive to the sample temperature, first-order Raman scattering alteration can conveniently be used for direct measurement of the device temperature with a high spatial resolution. By acquiring of this information together with photoluminescence mapping measurements, we are able to understand the device behavior better. As a result we can identify heating sources and implement changes in the device architecture to prevent regional overheating of the devices. The μ-Raman mapping measurements of unirradiated AlGaN/GaN and InAlN/GaN undoped HEMT structures of open gate TLM devices grown on sapphire substrates (with similar structures) were performed by a Renishaw InVia Raman microscope with a 50-× long working distance objective, while the time-resolved μ-Raman measurements were acquired by a Jobin Yvon LabRam microscope with 80-× long working distance objective. Both the instruments used 180° backscattering collection geometry and 633 nm excitation He-Ne laser lines. Since at given conditions, calibration measurements were impossible, the temperature was calculated based on the knowledge of the linear Raman shift of ~0.02 cm⁻¹/1K in the range of 270-700 K for single crystal GaN [4] and the presumption of an equivalent Raman shift for the measured samples.

2. Experimental

We have used knowledge reported in previous works with similar topic [2, 3, 4] to investigate the energy change of the E_2 phonon mode upon the variable temperature of GaN using a μ-Raman technique. The method assisted to acquire a time continuance and lateral distribution of temperature inside the HEMT channel. A typical Raman spectra of AlGaN and InAlN based HEMT structure acquired at room temperature shows A_1, E_2 peaks at 736.8 cm⁻¹, 570.6 cm⁻¹ and 733.5 cm⁻¹, 568.4 cm⁻¹ (Fig. 1), respectively. These values were then been used as references for the determination of the temperature inside the channel employing and according to [4] the Raman temperature shift of 0.02 cm⁻¹/1K in the temperature range from 270 to 700K [4]. To obtain better understanding of this measurement technique and to prove its reliability we have measured the Raman spectra inside 4 μm and 8 μm channels between two TLM contacts instead of measuring inside the real HEMT channel with a gate between.

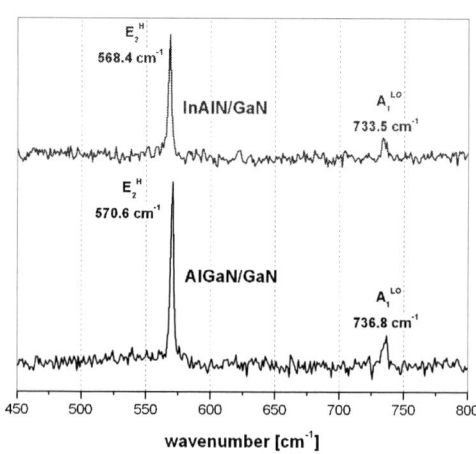

Fig. 1: Typical Raman spectra of AlGaN/GaN and InAlN/GaN HEMT structure inside the channel at room temperature

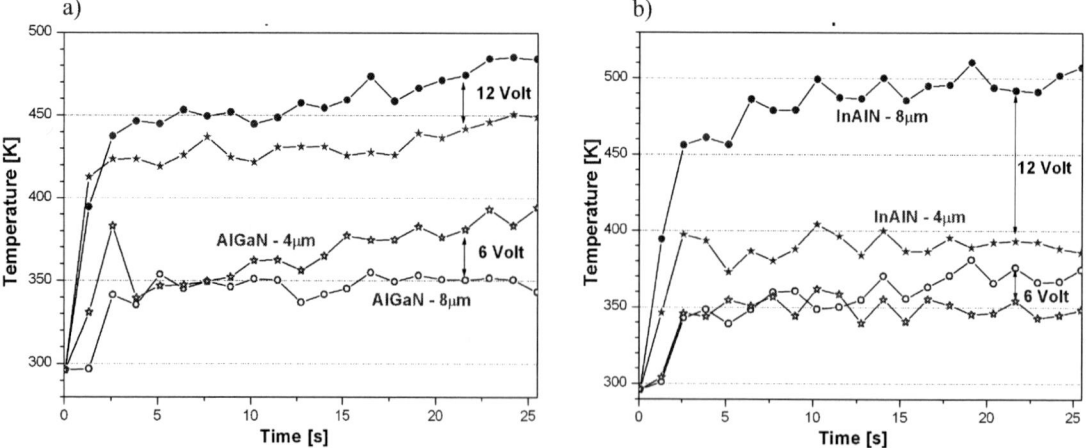

Fig. 2: Time resolved temperature change in **a)** AlGaN/GaN **b)** InAlN/GaN channel by different bias voltage loads

The Raman spectra in time resolved measurements have been taken each 1.25 s after applying a bias voltage to a 4 or 8 μm channel. Adequate power for applied voltage varied from 0.4 W for 6 V to 1.14 W for 12 V. The results infer the stable temperature inside the channels after the 10 s biasing (Fig. 2). The temperature in the 4 μm channel was lower due to better heat sink from the centre of the channel to the metal contact. The aberrance, in Figure 2a, where the temperature for 6V bias in the 8 μm channel is lower than that in the 4 μm channel is likely caused by fluctuations of ambient temperature.

The contact layers were fabricated by standard device processing on top of the HEMT structure and coupled with the Ag wires (Fig. 3). Because the active spot size of the laser is ~1 μm with using the 50-× long working distance objective, mapping with aid of Raman spectra was carried out on an 8 μm channel to obtain greater spatial resolution in aquired images. During the measurements, the samples were loaded with power varying from 0 to 0.4 W.

The resulting maps of the E_2 phonon mode Raman peak maxima infer rising the temperature with increasing of the bias voltage. Thus at the higher bias voltage the power dissipation inside the channel is greater (Fig. 4). The 1.5 cm^{-1} difference of the Raman shifts (from 568.5 to 567 cm^{-1}) is equivalent to the temperature change of 75 K in average. The map of the channel with no bias voltage applied obviously indicates that the dispersion of the peak maxima is low across the channel (Fig. 4a)) due to uniform temperature. As the temperature inside the channel raises (at higher bias voltage) the peak maxima close to the left contact shows slightly lower temperature than on the right hand side. This discrepancy is more likely caused by either the better heat coupling of the left contact or different contact resistance of the contacts.

Furthermore the images illustrate that μ-Raman technique may be used for primary testing the HEMT structures or monitoring the temperature distribution inside the channels. However, the drawback is that the spatial

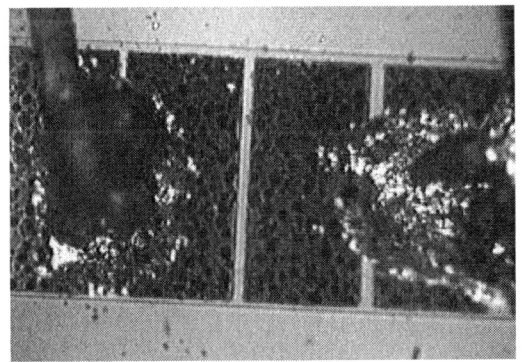

Fig. 3: Image of 8 μm channel with contacts on AlGaN/GaN channel of TLM structure

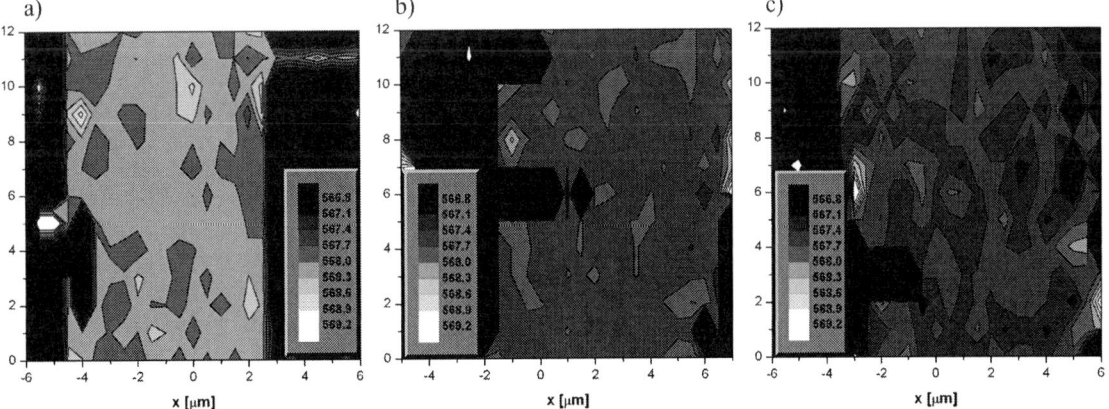

Fig. 4: Maps of Raman scattering E_2^H phonon mode maxima in InAlN/GaN TLM channel at different bias voltages, a) 0V, b) 4V (187 mW) and c) 8V (347 mW), show different temperature distributions inside the channel.

resolution of μ-Raman is comparable with the channel width. This shortcoming can potentially be solved with using an NSOM tip to induce localized sample excitation. By this technique a spatial resolution superior to that of the conventional Raman can be achieved inside the small HEMT channel, in addition to the corresponding topography image.

3. Conclusion

We have measured the mapping and time resolved Raman spectra inside a HEMT structure channel formed by two contacts with the channel widths of 4 and 8 μm to reveal temperature distribution. The results show, that the temperature inside the channels may be considered as stable after the 10 s biasing. It has also been found that at the same bias, the temperature inside the 8 μm-channel is higher than that in the 4 μm channel due to better heat sink from the centre of the channel.

The resulting maps of E_2 phonon mode Raman peak maxima show rise of the temperature with increasing the bias voltage and thus higher power dissipation inside the channel.

The information about the power dissipation inside the HEMT channel is very useful because these transistors transfer a large amount of energy and therefore they operate at high temperature which causes the degradation of its structure inside the channel. This non-contact technique can be used to examine the power dissipation in various types of structures and contacts to identify the heating sources and suppress them by moderation of the device architectures and thus prolonging the lifetime of these devices.

The drawback of low spatial resolution of μ-Raman, being comparable with the HEMT channel width, is suggested to solve with using an NSOM tip and scan-controlled localized sample excitation. The suggested technique will provide a superior spatial resolution inside the small HEMT channel and possibility of acquisition of topographical images.

Acknowledgement

The work has been done in City University of Hong Kong and COSDAF with generous support by General Research Fund (CityU 103208) of Research Grants Council of Hong Kong and the Centre of Excellence CENAMOST (VVCE-0049-07) with support of the VEGA project 1/0689/09 and FP7/2008-2011, n°214610, project MORGaN. The authors wish to thank Alcatel-Thales III-V Lab, France for MOCVD growth of InAlN/GaN structure and Institute of Electrical Engineering, Slovak Academy of Sciences, Slovakia for device processing.

References

[1] S. Jha, E. V. Jelenković, M. M. Pejović, G.S. Ristić, M. Pejović, K. Y. Tong, C. Surya, I. Bello, W. J. Zhang: Stability of submicron AlGaN/GaN HEMT devices irradiated by gamma rays, *Microelectronic Engineering* **86**, p. 37–40 (2009)

[2] D. S. Green, B. Vembu, D. Hepper, S. R. Gibb, D. Jin, R. Vetury, J. B. Shealy, L. T. Beechem, S. Graham: GaN HEMT thermal behavior and implications for reliability testing and analysis, *phys. stat. sol. (c)* **5**, 6, p. 2026–2029 (2008)

[3] J. Kim J.A. Freitas Jr., J. Mittereder, R. Fitch, B.S. Kang, S.J. Pearton, F. Ren: Effective temperature measurements of AlGaN/GaN-based HEMT under various load lines using micro-Raman technique, *Solid-State Electronics* **50**, p. 408–411 (2006)

[4] M. S. Liu, L. A. Bursill, S. Prawer, K. W. Nugent, Y. Z. Tong, G. Y. Zhang: Temperature dependence of Raman scattering in single crystal GaN films, *Applied Physics Letters* **74**, 21, p. 3125-3127 (1999)

Modelling and optimisation of a sapphire/GaN-based diaphragm structure for pressure sensing in harsh environments

M. J. Edwards[1,2], S. Vittoz[3], R. Amen[4], L. Rufer[3], P. Johander[4], C. R. Bowen[1] and D. W. E. Allsopp[2]

[1]Department of Mechanical Engineering, University of Bath, UK,
[2]Department of Electronics and Electrical Engineering, University of Bath, UK,
[3]TIMA Laboratory (UJF, CNRS, G-INP) Grenoble, France,
[4]IVF-Swerea, Mölndal, Sweden.

GaN is a potential sensor material for harsh environments due to its piezoelectric and mechanical properties. In this paper an 8mm diameter sensor structure is proposed based on a GaN / AlGaN / sapphire HEMT wafer. The discs will be glass-bonded to an alumina package, creating a 'drumskin' type sensor that is sensitive to pressure changes. The electro-mechanical behaviour of the sensor is studied in an attempt to optimise the design of a pressure sensor (HEMT position and sapphire thickness) for operation in the range of 10 – 50 bar (5 MPa) and above 300°C.

1. Introduction

Silicon diaphragm pressure sensors have been used since the discovery of piezoresistivity in silicon by Smith in 1954 [1, 2]. Although there are a number of examples of silicon-based 'drumskin' or circular diaphragm pressure sensors [1, 2], little work has examined GaN/sapphire based diaphragms. One of the advantages of using a GaN/sapphire structure, rather than silicon, is that flexural rigidity will be increased due to both GaN and sapphire having a higher Young's modulus than silicon. The reduction in deflection and the superior mechanical properties of sapphire compared to silicon [4] offers the possibility to develop a sensor for operation at higher pressures than Si-based diaphragm devices. Patents were filed in the early 1990's where sapphire diaphragms were used in conjunction with silicon piezoresistive sensors, taking advantage of the piezoresistive properties of silicon and the superior structural properties of sapphire [3, 4]. There are potential benefits in replacing the silicon with GaN to create a GaN/sapphire based device. While the piezoresistivity of GaN is small [5], it is a chemically inert material that is stable up to 1000°C and has large piezoelectric coefficients. GaN is also routinely grown on single-crystal sapphire substrates for applications such as LED's and high electron mobility transistors (HEMTs).

This paper therefore proposes using a GaN/AlGaN/sapphire based device and using HEMTs as the sensing element. This is a different approach to other diaphragm devices where the piezoresistive properties of the material are dominant. By using HEMTs as sensing elements along the diaphragm, the piezoelectric and semiconducting properties of GaN could be harnessed into a pressure sensor for extreme environments.

2. Modelling

The proposed device was modelled using ANSYS V11.0 finite element (FE) software [2] since it has suitable 3D coupled-field elements (SOLID 98) that combines the structural, thermal and piezoelectric behaviour of the device. Figure 1 shows a schematic of the drumskin device; without an outer alumina package that is bonded to the device by the glass frit. The model assumes that the alumina casing is rigid and constraints have been applied to take this into account, figure 1. This allows improved mesh density around the drumskin. For simplicity, it is assumed that the structure is perfectly elastic since GaN and sapphire are linearly elastic, brittle materials.

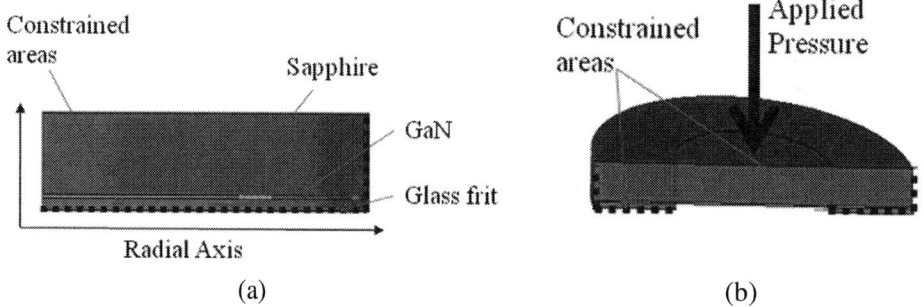

(a)   (b)

Figure 1: Schematic of the GaN/Sapphire drumskin pressure sensor. Sapphire, GaN and the glass frit are purple, turquoise and red respectively. (a) side view (b) cross-sectional view. Diameter is 8mm.

The model consists of an 8 mm diameter GaN/sapphire disk, where the GaN layer is fixed to a glass frit that is constrained in all dimensions at the base. The glass frit radially covers the outer 2 mm of the device, as shown in figure 1(b), resulting in an effective diaphragm diameter of 4 mm. Computational constraints have limited the GaN layer thickness to 15 μm, approximately three times greater than the likely layer thickness. In the case of Figure 1, the thicknesses for the sapphire substrate and glass frit were 350 and 50 μm respectively. The room temperature material properties used for GaN, sapphire and glass frit are contained in the appendix; although higher temperature operation is anticipated.

3. Results and Discussion

3.1 Piezoelectric polarisation and optimisation of positions for the HEMT sensors
Initially, the drumskin behaviour was modelled at 10 - 50 bar (1-5MPa) pressures and the resultant electric-field perpendicular to the radial direction formed by the piezoelectric polarisation of GaN was calculated. It was assumed that the sensor was operating at room temperature and internal stresses in the GaN/sapphire wafer caused by lattice and thermal mismatch were neglected. The results, shown in figure 2, follow a set pattern and this is useful for calibration purposes. For all pressure models, there is a point approximately 1.5 mm from the centre of the drumskin where there is no piezoelectric charge, point (i) in Figure 2. This is analogous to the point of zero strain in piezoresistive diaphragm sensors and is present since the piezoelectric polarisation is directly caused by the stress applied to the drumskin. From this result, HEMTs should ideally be placed at regions of maximum electric field which correspond to the centre of the drumskin; although HEMTs can be positioned anywhere on the drumskin to compare theory with experiment.

978-1-4244-8574-1/10 $26.00 © 2010 IEEE   128

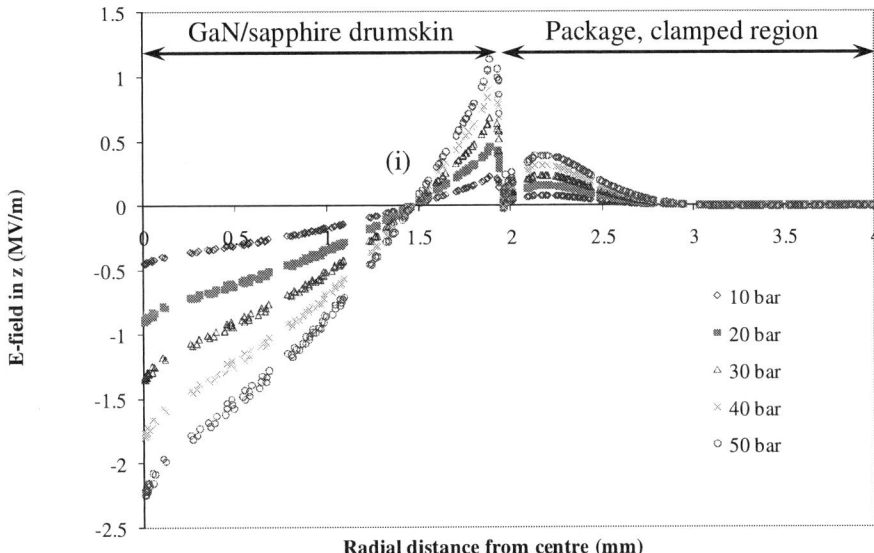

Figure 2: Piezoelectric E-field from drumskin sensor at room temperature and 10 - 50 bar pressures.

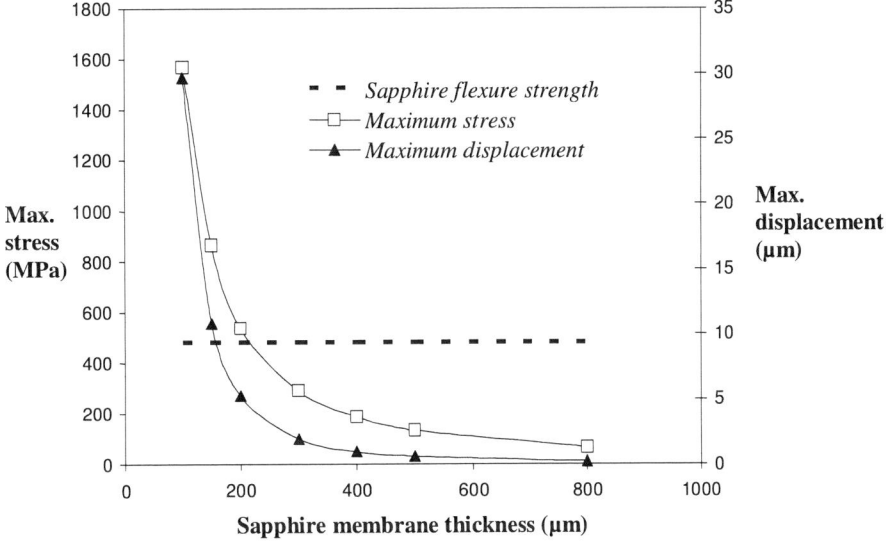

Figure 3. Maximum displacement and maximum stress against the sapphire membrane thickness. The sapphire flexure strength is indicated at 450 MPa.

3.2 Optimisation of sapphire membrane thickness

By studying the maximum displacement of the membrane under a fixed pressure, the mechanical response of the device can be optimized to provide high sensitivity. The maximum displacement of the membrane is found at the center of the membrane and the simplest way to increase the deflection, and sensitivity, of the sensor is to reduce the sapphire layer thickness. However, as the sapphire thickness is reduced the stress in the structure is accordingly increased. Therefore, a compromise must be found between maximum displacement and maximum stress in the membrane. To examine to this issue, a study of the mechanical response of the pressure sensing structure under a fixed pressure of 50 atm (50.66 bar) has been undertaken, knowing that the flexural strength of sapphire is 450 MPa. Figure 3 shows that a good compromise between sensitivity and stress level can be achieved with a membrane thickness between 250-350µm. A further increase in the membrane thickness reduces the displacement unnecessarily because the non-failure stress condition is fulfilled.

4. Conclusions

The paper has presented an overview of the design of a GaN / AlGaN / sapphire drumskin sensor. The combination of GaN and sapphire (250-350µm) offers the opportunity to develop diaphragm based sensors that operate of elevated temperature (>300°C) and pressure (~50 bar). Future efforts aim to manufacture prototype structures for experimental testing and these will be compared with model predictions.

Acknowledgements: The research leading to these results has received funding from the European Community's Seventh Framework Programme FP7/2007-2011 under grant agreement n°214610, project MORGaN. This publication reflects only the author's views and that the Community is not liable for any use that may be made of the information contained therein.

Appendix: Material Properties used in FE Model

Rel. permittivity and 'e_{ij}' piezo coefficients (C/m^2) of GaN			
ε_x	10.4	e_{31}	0.33
ε_y	10.4	e_{15}	-0.33
ε_z	9.5	e_{33}	0.66

Young's Modulus (GPa) and Poisson's Ratio of Glass Frit	
E	55
ν	0.3

Stiffness coefficients (GPa) of GaN and Sapphire		
	GaN	Sapphire
c_{11}	390	495
c_{12}	145	171
c_{13}	106	130
c_{14}	-	20
c_{33}	390	486
c_{44}	105	130
c_{66}	122.5	162

References:

1. Eaton, W.P. and J.H. Smith, *Micromachined pressure sensors: review and recent developments.* Smart Materials and Structures, 1997. **6**(5): p. 530.
2. Yasukawa, A., et al., *Design considerations for silicon circular diaphragm pressure sensors.* Japanese Journal of Applied Physics, 1982. **21**: p. 1049.
3. Sahagen, A.N., *Pressure sensing transducer employing piezoresistive elements on sapphire.* 1991, Pat. No. 4994781.
4. Sahagen, A.N., *Piezoresistive pressure transducer.* 1991, Pub. No. WO/1991/017418
5. Strittmatter, R., *Development of micro-electromechanical systems in GaN.* Dissertation (Ph.D.), California Institute of Technology, 2003.

HEMT-SAW Structures for Chemical Gas Sensors in Harsh Environment

I. Rýger[1], T. Lalinský[1], G. Vanko[1], M. Tomáška[2], I. Kostič[3], Š. Haščík[1], M. Vallo[1]

[1]Institute of Electrical Engineering of the Slovak Academy of Sciences, Bratislava, Slovakia ,
[2]Slovak University of Technology, Faculty of Electrical Engineering and Information
Technology, Department of Microelectronics, Bratislava, Slovakia, [3]Institute of Informatics,
Slovak Academy of Sciences, Bratislava, Slovakia

om1air@gmail.com

A growing thirst for highly sensitive and sufficiently selective sensors for extremely harsh conditions can be seen. This fact excludes the use of conventional sensing devices and gives a space for Surface Acoustic Wave sensors with monolithically-integrated electronics. We have chosen the AlGaN/GaN material as a suitable material due to its excellent chemical inertness and stability of piezoelectric parameters. In this paper we test the possible HEMT transistor/SAW transducer monolithic integration and propose the design of an oscillator based on SAW.

1. Introduction

This work present a new approach to design and fabrication of AlGaN/GaN based SAW-HEMT structures to be applied for chemical gas sensors operating in harsh environment. The proposed solution takes advantage of excellent physical, chemical and mechanical properties of III-Nitride (III-N) compounds, especially gallium nitride (GaN) based heterostructures [1]. A direct on-chip integrated compatibility in the process technology of surface acoustic wave (SAW) structure and high electron mobility transistor (HEMT) was demonstrated. It enabled to integrate two different principles of gas sensing based on both SAW and HEMT with specific chemical absorbing layers (nano-structured TiO_2 and Pt layers) considered as gate electrodes. The keystone principle of gas sensing with SAW device is the fact, if chemical sensing layer binds molecule, the physical properties of this compound are different from initial state. That can cause the phase velocity or dumping change which can be easily measured. Moreover, integrated HEMT sensing device can also serve as a selective thermal heater. It can increase the operating temperature of chemical absorbing layer placed on the gate area of HEMT, provided that thermal isolation conditions are fulfilled [2].

2. Technology process description

An undoped AlGaN/GaN heterostructure grown by metal-organic chemical vapor-phase deposition (MOCVD) on (0001) sapphire substrate was used to define interdigital transducers (IDTs) of SAW structure. A long gate length HEMT device with the chemical absorbing layer is integrated in space between IDTs (Fig. 1).

The direct interaction of forcing electrostatic field with free carriers of 2DEG prevents the acousto-electric transduction in interdigital transducers (IDTs) due to 2DEG layer electrostatic shielding effect [3]. Therefore, SAW signals were excited directly on GaN layer after selective etching of AlGaN barrier layer in the area under IDTs.

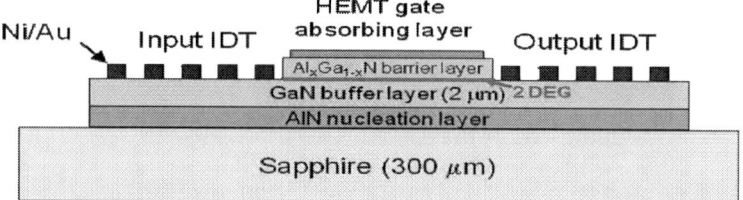

Figure 1- Cross-section through AlGaN/GaN based HEMT-SAW structure

Ni/Au e-beam evaporation and lift-off were carried out subsequently to form the Schottky fingers of the IDTs. Both the width and the spacing between the fingers were designed to be 1 μm. Consequently, the SAW wavelength λ is equal to 4 μm. A real view of fabricated SAW-HEMT structure and fingers of IDT are shown in Fig. 2.

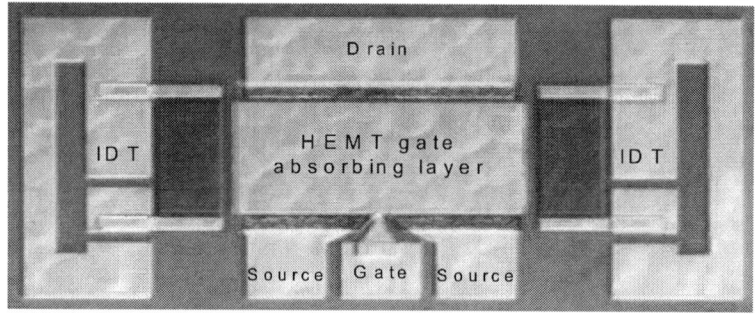

Figure 2- SEM image of fabricated HEMT-SAW structure

3. *Fabricated samples testing*

For basic structure functionality testing, the common I-V/C-V characterization was used. An Agilent Technologies, E8363B network analyzer with 50-Ohm terminal coplanar microprobes was used to measure S- parameters of HEMT-SAW structure. The influence of MESA-isolation and the presence of the HEMT on SAW propagation were analyzed. The obtained amplitude characteristic of SAW sensor structures is shown in Fig. 3. Also we tested the function of the HEMT by measuring the DC characteristics. The measured output characteristics are demonstrated in Fig. 4. The analyses revealed the convenience of the HEMT-SAW structure designed for chemical gas sensor applications

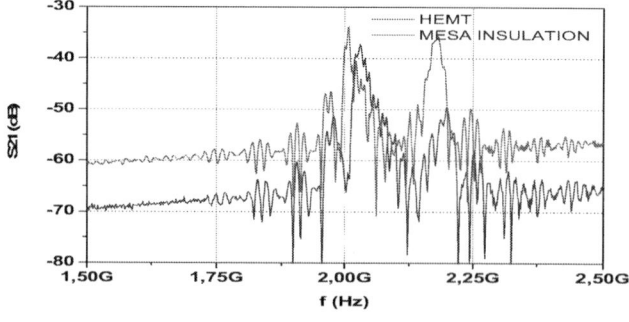

Figure 3- HEMT SAW structure amplitude characteristics.

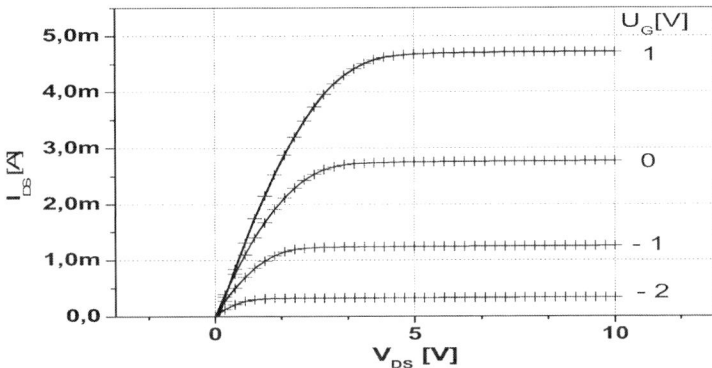

Figure 4- HEMT heating transistor output characteristics

4. *SAW oscillator design*

For testing purpose, simple oscillator using test SAW device in feedback loop was designed. In comparison to circuit proposed in article [4], we chose the discrete solution using silicon-germanium heterobipolar transistors BFP620 with high unity-gain frequency. In the future, heterobipolar transistors can be replaced by HEMT transistors fabricated on chip and, thus the whole oscillator can be monolithically integrated.

The appropriate SAW device working sample was chosen and characterized. Data obtained from measurement were imported into simulation program HSPICE as the two-port scattering matrix parameters in frequency domain. Consequently, three-stage common-emitter amplifier was designed. This amplifier has open loop gain approx. 55dB at 1.7GHz. In this frequency range, the impedance matching appears to be crucial. Out of resonance, SAW filter exhibits strongly capacitive input/output impedances, what results to high return loss. Therefore, impedance matching network was designed. This network is formed as a low-pass filter, to suppress the influence of higher-order harmonics, and thus, the possible effect of "mode hoping". This effect is clearly illustrated in [4]. Matching networks was designed using Smith-chart utility of program MIDE. Provided that the operational frequency is 1.7 GHz, the matching circuit was created from lumped elements. These matching networks were optimized in HSPICE and frequency characteristics have been tested using small-signal AC analysis. Buffer amplifier including pi-attenuator suppresses the influence of load connected to oscillator. Oscillator circuit diagram is displayed in Fig. 5.

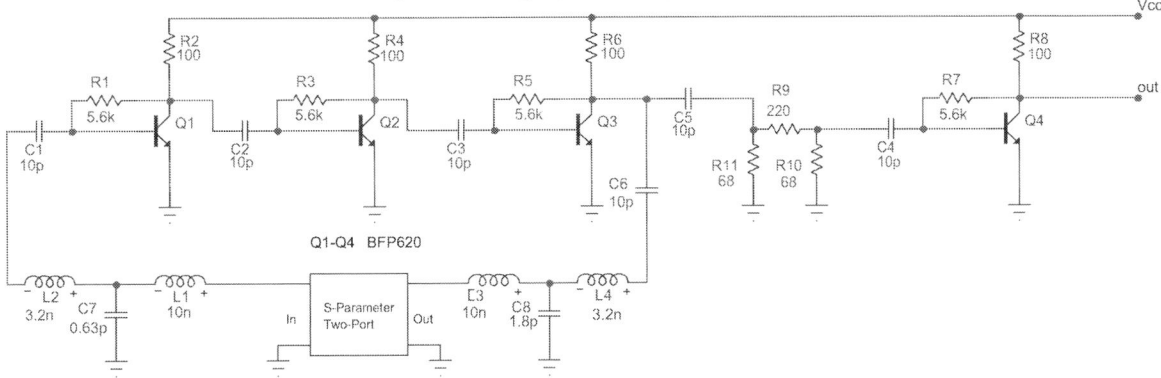

Figure 5- SAW oscillator schematic diagram

From simulation results, we can see that oscillator does not work in linear regime, provided that it does not contain the amplitude stabilization circuit. Three-stage amplifier small-signal gain exceeds the pass band attenuation in SAW filter and becomes saturated. For frequency measurements this drawback is not crucial, but it may lead to above-mentioned mode hoping in worst case, provided that distorted signal at amplifier output contains higher-order harmonics, at which amplitude and phase condition for oscillations can be met.

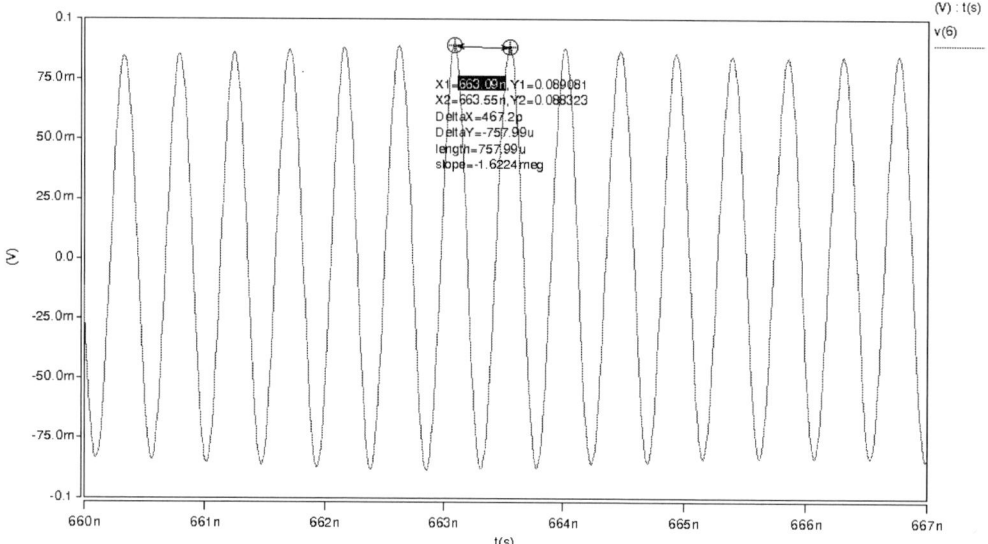

Figure 6, Transient analysis, output of SAW filter connected to base of Q1

In our case this should be suppressed using reflective matching circuit with resonant frequency equal to base band working frequency. At SAW filter output, clean sine signal is present due to band pass filter operation (Fig. 6).

Acknowledgement

This work has been supported in part by the Slovak Research and Development Agency under contracts APVV-0655-07, APVV-VVCE-0049-07, Slovak-French project SK-FR-0041-09 and VEGA project 2/0163/09.

References

[1] V. Cimalla, J. Pezoldt and O. Ambacher, Group III nitride and SiC based MEMS and NEMS: materials properties, technology and applications, J. Phys. D: Appl. Phys. 40 (2007) 6386-6434.

[2] Lalinský, T., Vanko, G., Jakovenko, J., Kutiš, V., Ivanova, M., Haščík, Š., Murín, J., Husák, M. and Kostič, I., AlGaN/GaN HEMT based micro-hotplate for high temperature gas sensors. In: MNE09. The 35th Inter. Conf. on Micro & Nano Engineering., Ghent (2009).

[3] T. Lalinský, I. Rýger, L. Rufer, G. Vanko, Š. Haščík, Ž. Mozolová, J. Škriniarová, M. Tomáška, I. Kostič, A. Vincze, Vacuum 84 (2009) 231-4.

[4] Schmitt, R. F. et al.: Rapid design of SAW oscillator electronics for sensor applications, Sensors and Actuators B76 (2001) 80-85

Investigation of Deep Energy Levels in Heterostructures based on GaN by DLTS

Ľ. Stuchlíková[1], J. Šebok[1], J. Rybár[1], M. Petrus[1], M. Nemec[1], L. Harmatha[1], J. Benkovská[1], J. Kováč[1], J. Škriniarová[1], T. Lalinský[3], R. Paskiewicz[2], M. Tlaczala[2]

[1]Department of Microelectronics, Faculty of Electrical Engineering and Information Technology, Slovak University of Technology, Ilkovičova 3, 812 19 Bratislava, Slovakia
[2]Faculty of Microsystem Electronics and Photonics, Wroclaw University of Technology, Janiszewskiego Street 11/17, 50-372 Wroclaw, Poland
[3]Institute of Electrical Engineering SAS, Dúbravská cesta 9, 841 04
e-mail: lubica.stuchlikova@stuba.sk, jan.sebok@stuba.sk, jakub.rybar@stuba.sk

In this paper we report our results of DLTS investigations of deep-level defects in Schottky-gate AlGaN/GaN LP-MOVPE structures grown on sapphire substrate. The exact location of heterostructure's interface below the surface (20 nm) was determined from the concentration profile to depletion region width dependence. The free charge carrier density was calculated ($n_{2D} = 4.75 \div 5.09 \times 10^{16}$ m^{-2}). Four deep energy levels have been identified from selected DLTFS spectra (activation energies: $E1 = E_C - 0.545$ eV, $E2 = E_C - 0.599$ eV, $E3 = E_C - 0.642$ eV, and $E4 = E_C - 1,118$ eV).

1. Introduction

Gallium nitride is considered by many to be the next important semiconductor material after silicon. As a brilliant light emitter capable to operate at high temperatures, it is a leading candidate to be the key material for the next generation of high frequency and high power transistors [1]. GaN-based layers are usually epitaxially grown on sapphire substrates. The mismatch in the lattice constants of the substrates and the GaN-based layers and also self-compensation tendencies of wide band-gap semiconductors result in a number of deep-level defects, especially dislocations, which influence the reliability and performance of GaN-based devices [2]. The emission and capture processes are complicated by presence of 2DEG in AlGaN/GaN heterostructures.

In this paper we report our results of DLTS investigations of deep-level defects in Schottky-gate AlGaN/GaN LP-MOVPE structures grown on sapphire substrate.

2. Experiment

The Schottky diodes were prepared on AlGaN/GaN material structure using conventional HFET processing steps [3]. This structure was grown by LP-MOVPE method on sapphire (Al$_2$O$_3$) substrate. It consisted of 2 μm thin GaN buffer layer followed by 20 nm thin AlGaN barrier layer and coated with 5 nm thin GaN layer. The AlGaN layer mentioned before was non-homogeneous doped: the outer regions were intrinsic AlGaN and the inner region was Si-doped. In our experiment we have used the sample A (H1441/III_AX) with Al$_{0.19}$Ga$_{0.81}$N. The gate was formed by RTP annealing during 35 s at temperature 400°C performed at SAS. The Schottky diodes were processed simultaneously with HFET devices. Ni/Au Schottky contacts were prepared by standard Nb/Ti/Al/Ni/Au ohmic metallization process. The investigated sample A had a barrier contact area of 100×100 μm.

Measurements were performed using measurement system BIO-RAD DL 8000 DLTS in laboratory Semitest on Department of Microelectronics. This measurement system uses digital DLTS - DLTFS (Deep Level Transient Fourier Spectroscopy) method. The general idea of DLTFS is as follows: N measuring values are sampled from capacitance transient, and the discrete Fourier coefficients c_n^D are formed by numerical Fourier transformation.

In opposite to the conventional DLTS method, DLTFS method measures the complete transient as a $C(t)$ array and transfers the data into a computer system. Using a Fourier transformation and the direct evaluation the time constant and the transient amplitude can be evaluated for every transient measured at any temperature. The obtained DLTS spectra were evaluated by Direct analysis DLTFS and Tempscan maximum analysis.

3.   Results and discussion

Fig. 1a shows characteristic C-V curves measured on the sample A in dark environment and under illumination at different temperatures. The arrows indicate the shift of measured C-V curves with increasing temperature. The area set in close vicinity of voltage -1.2 V (cross point) is in principle temperature and illumination independent.

To evaluate the measured C-V curves correctly, we have used following parameters of GaN semiconductor: the relative permittivity $\varepsilon \sim 8.9$, the band gap width $E_g(0) = 3.47$ eV, and the intrinsic concentration $n_i = 3.22\times10^{-4}$ m^{-3} (300 K) [4]. After applying reverse bias voltage, the capacitance is slightly decreasing with increasing reverse voltage from 0 V. This effect is related to the shift of wave function of electrons located in heterostructure's interface and also depends on Schottky contact interface [5].

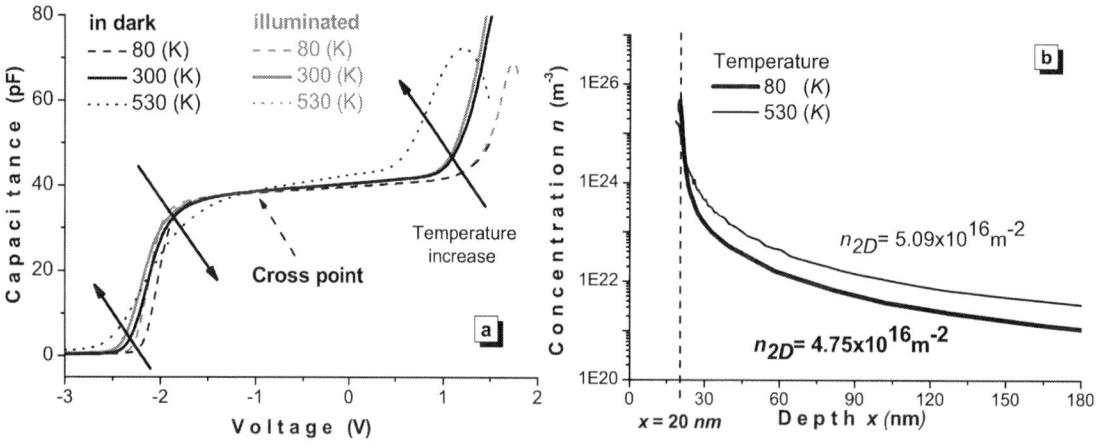

Fig.1: a) CV curves measured on sample A at different temperatures in dark environment and under illumination; b) calculated concentration profile of free charge carriers in sample A at temperatures 80 K and 530 K.

The contribution of capacitance originating in the cap layer of semiconductor (GaN 5 nm) to the final result is very minor, because of its thickness. The sudden fall of capacitance in the range of reverse voltage from -1.5 to -2.5 V is caused by heterostructure's interface depletion. The exact location of heterostructure's interface below the surface (20 nm) was determined from the concentration profile to depletion region width dependence (Fig. 1 b) [6]. The position of heterostructure's interface is set by the maximum of concentration profile.

The free charge carrier density ($n_{2D} = 4.75 \div 5.09 \times 10^{16}\,\mathrm{m^{-2}}$) calculated as an integral of concentration profile corresponds with the value $n_{2D} = 6 \times 10^{16}\,\mathrm{m^{-2}}$ given by engineers (sample producers).

We have produced large set of DLTFS spectra measured on both types of samples. All measured DLTFS spectra exhibit a strong deviation from an exponential dependence.
This can be caused by the presence of several mutually influencing deep energy levels and capture and emission processes in heterostructure's interface.

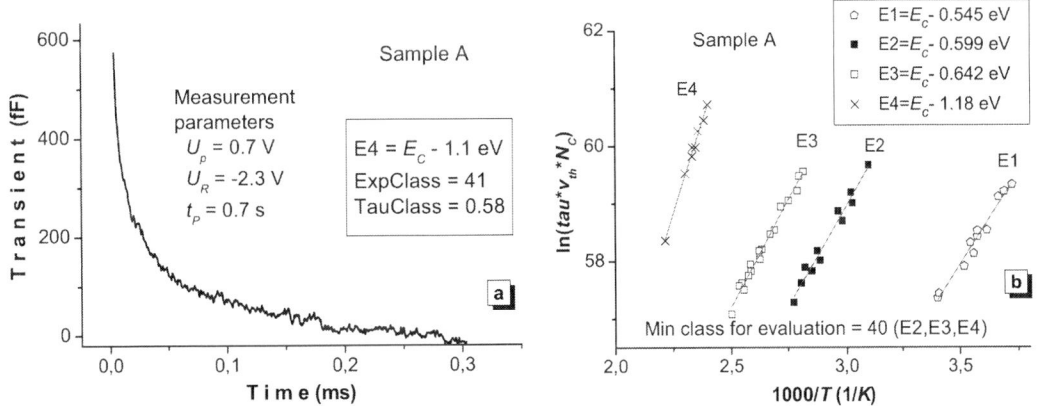

Fig. 2. a) Measured capacitance transient at 500 K on structure A .b) Arrhenius plot for levels E2, E3, E4 gained only by Direct analysis DLTFS with min class for evaluation = 40.

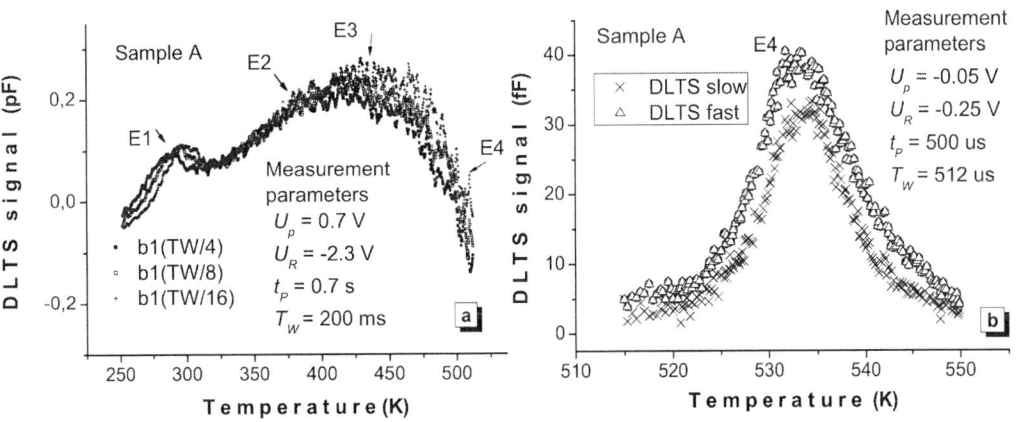

Fig. 3. a), b) Measured DLTFS spectra on structure A at different measurement conditions.

Fig. 2a shows typical measured capacitance transient at 500K, U_p=0,7V, U_R=-2,3V, and t_p=0,7s. Three measured DLTS spectra on structure A at measurement conditions mentioned above are displayed on Fig. 3a. The peaks on Fig. 3b correspond with deep energy level E4 determined from transient on Fig. 2a.

The four deep energy levels have been identified from selected DLTFS spectra (activation energies and capture cross sections: $E1=E_C$-0.545 eV and $\sigma_1 = 2.8 \times 10^{-16}$ cm^2 for E1; $E2=E_C$-0.599 eV and $\sigma_2 = 2.9 \times 10^{-17}$ cm^2 for E2; $E3=E_C$-0.642 eV and $\sigma_3 = 1.8 \times 10^{-17}$ cm^2

for E3; $E4=E_C-1{,}118$ eV and $\sigma_4=1.4\times10^{-13}$ cm^2 for E4) (Fig. 2b). in a Schottky-gate AlGaN/GaN. The parameters of deep energy level E1 was identified by Direct analysis DLTFS (Min class for evaluation=50), Direct analysis DLTFS with multilevel evaluation and Tempscan maximum analysis. In all cases values of determined parameters (activation energies and capture cross sections) were in good equality. The parameters of levels E2 – E4 were identified by Direct analysis DLTFS with multilevel evaluation (Min class for evaluation=40).

It was suggested that the level E1 has been reported in the literature and assigned to the nitrogen antisite point defect NGa (N atom located on a Ga site in the crystal lattice) [7], deep levels E2 – E4 refer to a gallium vacancy–oxygen complex (V_{Ga}–O_N) (complex of missing gallium atom and an oxygen atom placed on one of the nearest nitrogen sites) [8] or dislocations (atoms in the crystal structure are out of their conventional positions) [9].

The behaviour of these defects in our sample was not understood completely which means that more detailed investigation is still needed.

Although we have former experience with the diagnostics of GaN semiconductor using measurement system BIO-RAD DL 8000 DLTS, the presence of 2DEG in heterostructure's interface in examined structure makes the situation more difficult. All gain data are still under investigation and the GaN samples will be analyzed in detail in order to get even more relevant results.

Acknowledgement

This work was performed under the Center of Excellence CENAMOST (VVCE-0049-07) project and supported by the Slovak Scientific Grant Agency VEGA (Contracts 1/0742/08, 1/0507/09 and 1/0689/09), and APVV SK-PL-0017-09.

References

[1]   L. Yarris, *Science beat [online]*, [2009-05-10]. Accessible on www:, <http://www.lbl.gov/Science-Articles/Archive/MSD-gallium-nitride-nanotube.html>.

[2]   T. Tsarova et al., *Deep-Level Defects in MBE-Grown GaN-Based Laser Structure*, Acta Physica Polonica A, Vol. 112, No. 2, p. 331, 2007.

[3]   P. Kordoš et al., *Appl. Phys. Lett.* 92, 152113, 2008.

[4]   *GaN – Gallium Nitride, [online]*, [2009-03-10]. Accessible on www: <http://www.ioffe.rssi.ru/SVA/NSM/Semicond/GaN/bandstr.html#Basic>.

[5]   F. Schubert, *Delta-doping of semiconductors*, Cambridge University Press, ISBN 0521482887, p. 226, 1996.

[6]   H. Kroemer et al., *Appl. Phys. Lett.*, vol. 36, p.295, 1980.

[7]   H.K. Cho, C.S. Kim, C.-H. Hong, *J. Appl. Phys.* **94**, 1485, 2003.

[8]   T. Ito et al., *Journal of Crystal Growth*, vol. 310, pp. 4896-4899, 2008.

[9]   A.Y. Polyakov, et al., *Electrical properties and deep traps spectra in undoped M-plane GaN films prepared by standard MOCVD and byselective lateral overgrowth*, in Journal of Crystal Growth, 2009.

978-1-4244-8574-1/10 $26.00 © 2010 IEEE

Molecular dynamics and Electrical Simulation of a Novel GaN/4H-SiC Hetero-structure Optically Triggered Vertical NPN Device

Srikanta Bose, and Sudip K. Mazumder

Tel.: +1-312-996-6548; E-mail address: sribose@ece.uic.edu

Laboratory for Energy and Switching Electronics System, Department of Electrical and Computer Engineering, University of Illinois at Chicago, 851 South Morgan Street, Science and Engineering Office, Room No. 1020, Chicago, IL: 60607-7053

In this paper, an atomistic molecular dynamics followed by an electrical simulation study is made to study the turn-on and turn-off characteristics, gain, and breakdown voltage of an optically triggered GaN/4H-SiC hetero-structure vertical NPN device with 1 nm AlN as the buffer layer and the results are compared with the optically triggered all-4H-SiC NPN device. The optically triggered GaN/4H-SiC hetero-structure provides better switching characteristics, however, the breakdown strength of all-4H-SiC NPN device shows better results than GaN/4H-SiC device because of high critical field strength of 4H-SiC material.

1. Introduction

The semiconductor materials, GaN and 4H-SiC, have very good electrical and thermal properties [1-3] which lead to their potential applications in high-power electronics. In this work, our focus is on vertical conduction of optically triggered GaN/4H-SiC NPN device with 1 nm AlN, as the buffer material to avoid lattice mismatch between GaN and 4H-SiC. The reason to study optically triggering is to avoid any electro-magnetic-interference (EMI) and GaN has good optical absorption capacity. To investigate whether AlN as the buffer material should not affect the vertical conduction, as AlN being the high bandgap material among GaN and 4H-SiC, an atomistic simulation is made for the heavily doped p-(Ga-face)GaN/n-doped (Si-face)4H-SiC hetero-material system with one-layer of (Al-face)AlN, as the interface material, using molecular dynamics simulation approach [4] in DMol^3 first-principle atomistic simulator [5] module of Material Studio 5.0 [6] on TeraGrid Super Clusters of NCSA (National Center for Supercomputing Applications at University of Illinois Urbana-Champaign, US), Intel 64 Cluster Abe. Further, an electrical simulation study is carried out and the results are compared with optically triggered all-4H-SiC vertical NPN device. By all-4H-SiC NPN device, we mean the device dimensions remain the same as GaN/4H-SiC structure, but there is no AlN buffer layer and GaN material is replaced by 4H-SiC material. ATLAS [7] device simulation software package has been used for the above electrical simulation study.

2. Simulation approach

2.1. Atomistic simulation

For performing atomistic simulation, the supercell approach [8] is adopted. The total number of atoms in the cell is kept sixty and the atoms in (Si-face)4H-SiC are constrained whereas Ga, Al, and N atoms relaxed. The major constraints set in the DMol^3 first-principle atomistic simulator are, Ensemble: NVT; DFT exchange-correlation: LDA/PWC; Thermostat: Simple Nose-Hoover; Temperature: 800 K (This value of temperature is considered in view of experimental settings [9, 10]); Simulation time: 0.5 ps; Core treatment: All-electron with Harris approximation; k-point set: Medium. The results obtained from molecular dynamics simulation are described in results and discussion section.

978-1-4244-8574-1/10 $26.00 © 2010 IEEE

2.2. Electrical simulation

The structure of the device under study is depicted in Fig. 1. The optical pulse (350 nm wave length) of intensity of 15 W/cm^2 is allowed to incident on the *p*-base region of the device with an optical window of 15 μm. In principle, when the light falls on the base of the device, the extra carriers are generated either through band-to-band transitions in case of direct bandgap like GaN or through transitions via forbidden gap levels like indirect bandgap 4H-SiC, resulting in an increase of conductivity [11]. The results obtained are provided in Table 1 and are explained in results and discussion section.

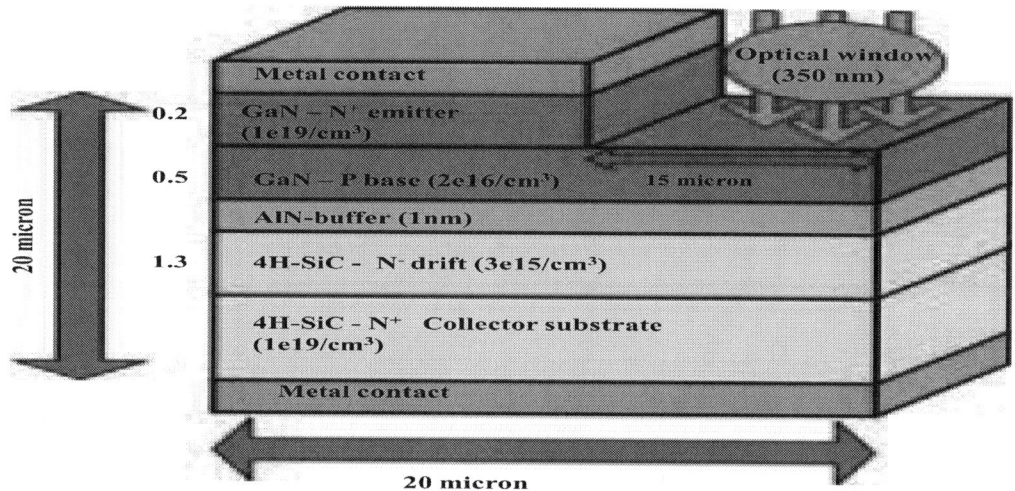

Fig. 1. Schematic of Optically Triggered Vertical NPN GaN/4H-SiC Hetero-structure Device with 1 nm AlN as the buffer material.

Table 1

NPN Vertical Device Structure	Breakdown voltage (V)	Forward drop (V)	Leakage current (A/μm)	Rise time (μs)	Fall time (μs)	Gain
GaN/AlN/4H-SiC	1950	4.0	1×10^{-10}	0.04	0.2	190.0
All-4H-SiC	2000	3.8	4×10^{-11}	0.552	1.569	345.661

3. Results and discussion

Fig. 2 shows the total density of states (DOS) of various cases of AlN interfaced hetero-epitaxial systems, studied through atomistic simulation. The inset shows the DOS of heavily *p*-doped (Ga-face)GaN over *n*-doped (Si-face)4H-SiC, with one-layer of (Al-face)AlN as the interface material. We replace the Ga-site with one Magnesium(Mg) atom for *p*-doped GaN and two Mg atoms for heavily *p*-doped GaN and similarly, C-site of 4H-SiC is replaced with one Nitrogen(N) atom for *n*-

978-1-4244-8574-1/10 $26.00 © 2010 IEEE   140

doped 4H-SiC and two N atoms for heavily *n*-doped (Si-face)4H-SiC. The energy unit has been converted from Hartree to ElectronVolt (1 Ha = 27.2 eV) while reporting the DOS value.

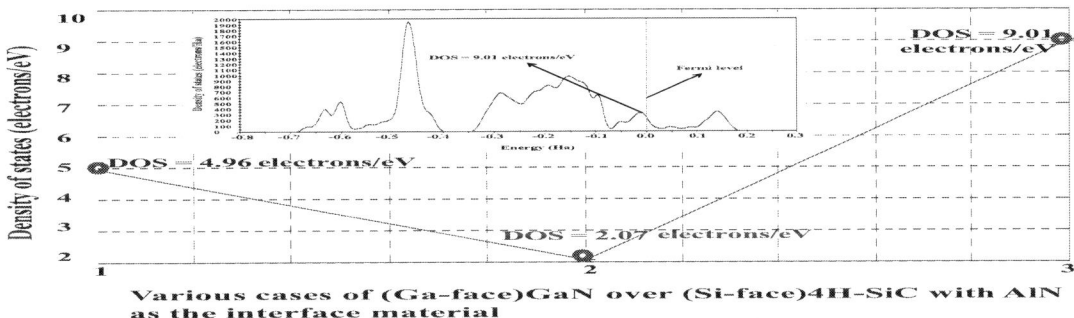

Fig. 2. The total density of states (DOS) of various cases of GaN over (Si-face)4H-SiC hetero-epitaxial systems with one-layer of (Al-face)AlN as the interface material, studied through molecular dynamics simulation approach [1] *p*-doped GaN/*n*-doped 4H-SiC with (Al-face)AlN as the interface material, 2) *p*-doped GaN/heavily *n*-doped 4H-SiC with (Al-face)AlN as the interface material, and 3) Heavily *p*-doped GaN/*n*-doped 4H-SiC with (Al-face)AlN as the interface material]. (Inset)The total density of states (DOS) of heavily *p*-doped GaN over *n*-doped (Si-face)4H-SiC hetero-epitaxial material system, with one-layer of (Al-face)AlN as the interface material.

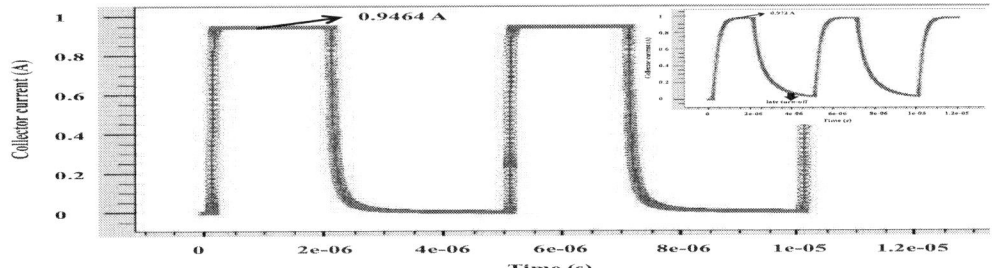

Fig. 3. Collector current variation with time for GaN/4H-SiC NPN device with 1 nm AlN as the buffer material. (Inset) Collector current variation with time for all-4H-SiC NPN device.

A comparatively higher DOS is seen at the Fermi level for heavily *p*-doped GaN over *n*-doped (Si-face)4H-SiC hetero-epitaxial system, with one-layer of (Al-face)AlN as the interface material. The element Mg has valence electrons in $3s^2 3p^0 3d^0$. That means the p and d-orbitals are vacant which means, these are holes (or minority carrier density) ready to be occupied by electrons. The maximum no. of electrons that can be accommodated in p and d- orbitals are 6 and 10, respectively. So, the minority carrier density is quite high in case of Mg-*dopant*. The element N has valence electrons in $2s^2 2p^3$, which means there are 3 unpaired electrons available out of which 2 will go to Si so that it can satisfy the Octet. So, the element N is left with 1 electron which will act as free electron i.e., the majority carrier density is quite low in case of N-*dopant*. The element Ga has valence electrons in $4s^2 4p^1 4d^0 4f^0$. The element Al has valence electrons in $3s^2 3p^1$. Since no free electrons are available in the one-layer of (Al-face)AlN material, it does not affect the DOS of the hetero-epitaxial system and prevents the one freely available electron of N-*dopant* of 4H-SiC to be shared either by Mg-*dopant* or Ga of GaN. From Table 1, we see the breakdown voltage and leakage current of all-4H-SiC NPN device is far better than the vertical GaN/4H-SiC device implying the fact that 4H-SiC material has high breakdown field strength. The gain in GaN/4H-SiC NPN device is less than that of all-4H-SiC device. This is because, though the collector current is higher in GaN/4H-SiC device than all-4H-SiC device, the photo-generation is also higher in GaN/4H-SiC leading to more photo-injected base current, and hence the ratio gets reduced. Fig. 3 shows the transient plots, which displays the collector current variation with time. We can seen that the turn-on time in case of GaN/4H-SiC NPN device is much faster owing to the fact that GaN has much short carrier life time and hence the recombination rate is much faster. The turn-off delay is less in the GaN/4H-SiC NPN device than that of the all-4H-SiC NPN device structure. In case of all-4H-SiC NPN device, incident light will get penetrated from base to drift to collector region as because, the energy of the photon is comparatively higher than its bandgap and hence, even though it has poor absorption coefficient, photo carriers will be generated not only in base but also in the drift and collector regions though the amount will be small whereas in case of GaN/4H-SiC NPN device, the photogeneration is confined to base only. Thus, during the turn-off, due to lower recombination rate, there still exists some extra carriers due to photogeneration in case of all-4H-SiC NPN device, contributing to the total current density and hence, the delay (which we can see from the inset of Fig. 3).

4. Conclusion

The present simulation work outlines that by using 1 nm of AlN as the buffer material between GaN and 4H-SiC, the conductivity is not getting affected drastically which we observed from the atomistic simulation in the form of DOS and also from the electrical simulation of optically triggered vertical GaN/4H-SiC NPN device, as well. One more key aspect of the simulation study shows the fast switching of GaN/4H-SiC NPN device than vertical all-4H-SiC NPN device, the reason of which can be related to the shorter carrier life time in GaN material. The breakdown strength of all-4H-SiC NPN device shows better results than GaN/4H-SiC device because of high critical field strength of 4H-SiC material.

Acknowledgement

We are thankful to National Science Foundation, US (Award No: 0823983) for necessary financial support.

References

[1] http://cst-www.nrl.navy.mil/

[2] http://www.ioffe.ru/SVA/NSM/Semicond

[3] *Group IV elements, IV-IV and III-V compounds. Part a - lattice properties*, Springer-Verlag, 2006. [Also available at http://www.springermaterials.com/navigation/navigation.do?m=l_2_132697_Group+IV+Elements%2C+IV-IV+and+III-V+Compounds.+Part+a+-+Lattice+Properties]

[4] http://en.wikipedia.org/wiki/Molecular_dynamics

[5] B. Delley, *J Chem. Phys.* **92,** 508, 1990.

[6] http://accelrys.com/

[7] http://www.silvaco.com

[8] M. Jarvis, I. White, R. Godby, and M. Payne, *Phys. Rev. B* **56,** 14972, 1997.

[9] Y. Nakano,J. Suda,and T. Kimoto, Phys. Stat. Sol. (c) **2,** 2208, 2005.

[10] A. Brown, M. Lusurdo,T. Kim, M. Giangregorio,S. Choi, M. Morse, P. Wu, P. Capezzuto, and G. Bruno, *Cryst. Res. Technol.* **40,** 997, 2005.

[11] S. Sze, *Physics of Semiconductor Devices*, John Wiley & Sons, New York, p. 654, 1969.

Analysis of structure geometry and interface charge on electrical characteristics of InAlN/GaN HEMTs

Juraj Marek, Daniel Donoval, Jaroslav Kováč, Marian Molnár, Aleš Chvála and Peter Kordos

Department of Microelectronics, Slovak University of Technology,
Ilkovičova 3, 81219 Bratislava, Slovak republic
e-mail: juraj.marek@stuba.sk

In this paper the results obtained from simulations and measurements on InAlN/GaN HEMTs are presented. The HEMT material structure was modelled by Synopsys TCAD tools and electrical characteristics of the device were simulated by DESSIS. Several effects of the geometry and concentration of interface charges on the electrical characteristics are studied. The interface and surface charges as well as deep level traps were taken into account for the numerical simulations. An influence of variation of several structure parameters on the transfer and output characteristics was studied. The results can contribute to better understanding and further calibration of electro physical models used for InAlN/GaN heterostructures.

1. Introduction

GaN-based devices have demonstrated a large potential for applications in high-power and high frequency electronics. In recent years, an alternative approach wherein the Al-GaN layer is replaced by an AlInN barrier has been implemented for improving the HEMT performance after the original proposal of Kuzmík. [1] The advantage of using an AlInN barrier is to adjust the composition of the alloys to obtain a lattice or polarization matched heterostructure and to reduce the defect density in the barrier layer.

2. Material and device structure

The InAlN/GaN material structure under study is the typical one used for HEMT preparation and consists of a GaN buffer layer, followed by a 1 nm AlN spacer layer and a 9 nm thick InAlN barrier layer, grown on SiC substrate (Fig. 1). The indium content in the InAlN barrier

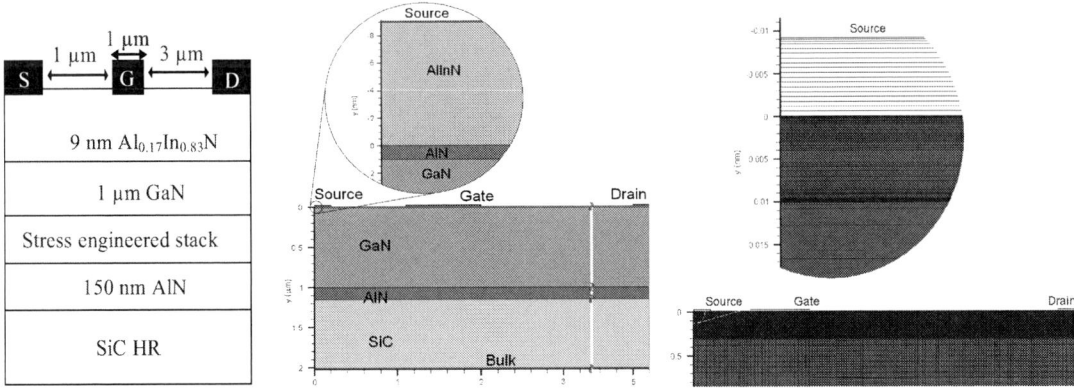

Fig. 1: left - Schematic view of InAlN/AlN/GaN HEMT structure. Gate to drain distance is 3 μm. middle - Model of HEMT structure prepared in MDRAW tool. right – Generated mesh for electrical simulations with 25000 vertexes.

is kept between 17% and 18.8%, close to perfect lattice matching to GaN or in slight compression. By doing so, structural degradation due to strain relaxation may only occur when growing the AlN interlayer. Conventional field-effect transistor fabrication steps were used for the device preparation. The processing started with Ar-based reactive ion etching for a mesa isolation. After that ohmic contacts were prepared by evaporation of Ti/Al/Ni/Au metal stack and subsequent rapid thermal annealing. Finally, Ni/Au gate contacts were patterned by electron-beam lithography. Study of Schotky diodes was carried out in our previous work [2].

3.   Numerical simulations

The model of HEMT structure was prepared using structure editor MDRAW tool from Synopys TCAD package [3]. All dimensions of our first HEMT model were set up to match dimensions of real devices. This structure was used for the physical models calibration by comparison with measured characteristics and was prepared using technological templates of Synopsys for TCAD tools. Several other structures were designed to study an influence of geometry on transistor transfer and output characteristics. Very important step between preparation of structure design and simulations of electrical characteristics is a proper mesh generation. For structures with quantum wells (QW) very dense mesh is typical. In a QW region mesh spacing must be at least 10 – 100 times smaller than QW width to ensure a good convergence of numerical electro-physical solver (Fig.1). Electro-physical behavior of structure was studied using simulation tool DESSIS [4]. In the simulation we employ both the drift-diffusion (DD) and hydrodynamic (HD) models. Since the HEMT's operation is characterized by very high and inhomogeneous fields, the high-field effects on the electron transport must be taken into account. In the case of the DD model high-field effects are taken into account only via field-dependent mobility, while in the case of the HD model the hot-electrons effects are incorporated using the electron temperature approximation. Due to numerical problems to attain convergence with optional transferred electron effect model which takes to account a negative differential mobility, the field and the electron temperature dependences of the mobility are described by Canali model [3]. The static I-V characteristics obtained in the HD model were very close to those obtained from the DD model what is in good correlation with work of A. Brannick [4]. If deep level traps and interface charges are incorporated into the simulation, the HD model must be used which is crucial as well for non isothermal, AC coupled and for transient simulations.

4.   Experimental results and discussion

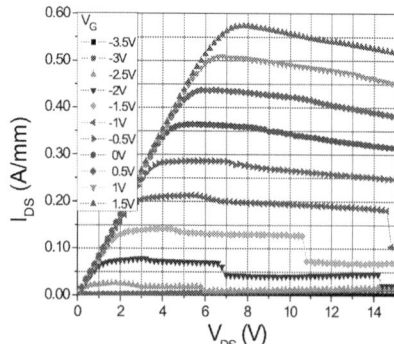
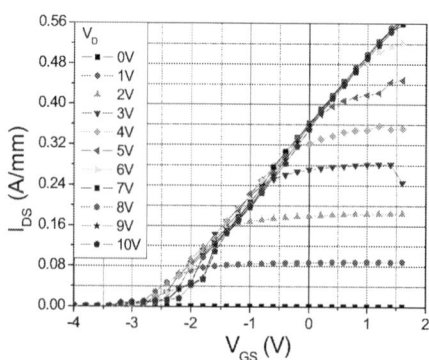

Fig.2: Measured output (left) and transfer (right) characteristics for gate voltages $V_G = -3.5 \sim 1.5$ V and drain voltages $V_D = 0 \sim 10$ V as parameters.

978-1-4244-8574-1/10 $26.00 © 2010 IEEE

Measured output and transfer characteristics of InAlN/AlN/GaN HEMT are shown on Fig.2. The evaluated Schotky barrier height is ~1.46 eV at room temperature [2]. Maximum drain current for $V_{GS} = 1$ V is $I_D = 0.51$ A/mm. This current is lower than published results [5,6] probably due to absence of a passivation layer on the top of structure and due to large gate length (1 μm). Measured pinch off voltage is $V_P = -2.5$ V. For the first simulations basic model of HEMT without additional vacuum layer shown in Fig. 1 was used. In this model no layer on top of InAlN layer was defined, therefore surface of structure was not taken as interface that leads to absence of surface potential. However, it provides easier convergence of simulations. This model was used to study an influence of the structure geometry on electrical characteristics and obtained results were used for the electro-physical model calibration. First of all an influence of the InAlN layer thickness was studied (Fig. 3). The thickness was varied from 7 nm to 25 nm. One can see that with increased thickness the pinch-off voltage is decreasing. This is in accordance with the theory. From these simulations we found that the variation of InAlN thickness around 10 nm has effect only on V_P and not on the slope of transfer characteristics, they are only shifted.

Further, an influence of the gate length on the transfer characteristics was studied. Results are depicted in Fig.3. One can clearly see that the gate length has an influence on the slope of transfer characteristics but the pinch-off voltage stays constant. Decrease of the current with increasing the gate length can be due to a dramatic reduction of the electron mobility under the gate caused by transversal electric field of the gate contact. For longer gate contacts, the carrier mobility is decreased in longer area; therefore the channel resistivity is higher.

An influence of the defect states on the bottom GaN/AlN interface was studied as well. For the negative charges no impact on the transfer characteristics was observed, however, an influence of the positive charge has increased with decreasing the GaN layer thickness. For these simulations the energy of deep level traps in GaN was set to 0.152 eV below the conduction band with the concentration of $N_t = 1e15$ cm^{-3}. With decreasing the GaN layer thickness the leakage current strongly increased. This is due to formation of the parasitic conduction channel at the GaN/AlN interface. This channel provides an additional drain to source current in both on- and off-states. An influence of the polarization charge concentration, defined at the interface of heterostructure, on the transistor characteristics was studied as well. Amount of this charge strongly affects the transfer characteristics, pinch-off voltage and sheet concentration of electrons in the 2DEG. Results of the simulation for the

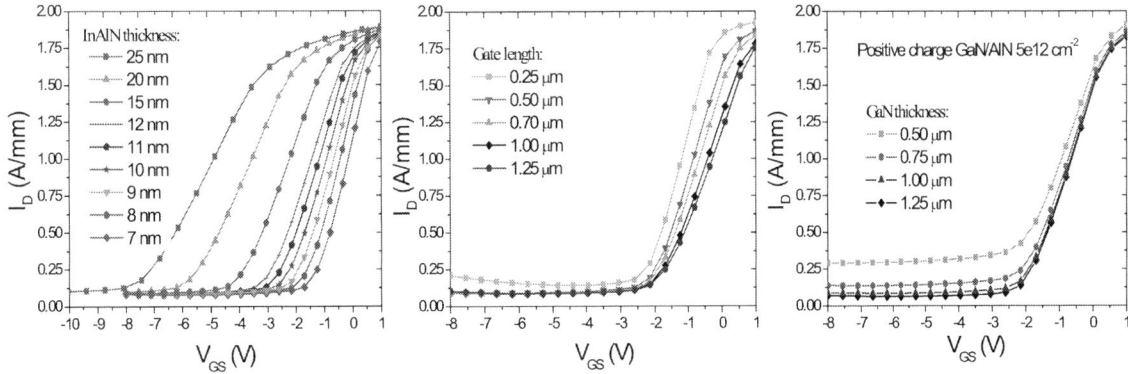

Fig. 3 (left) Simulated transfer characteristics for different InAlN layer thicknesses $d_{InAlN} = 7 \sim 25$ nm and $V_{DS} = 5$V. (right) Simulated transfer characteristics for different gate lengths $L_G = 0.25 \sim 1.25$ μm and $V_{DS} = 5$V.

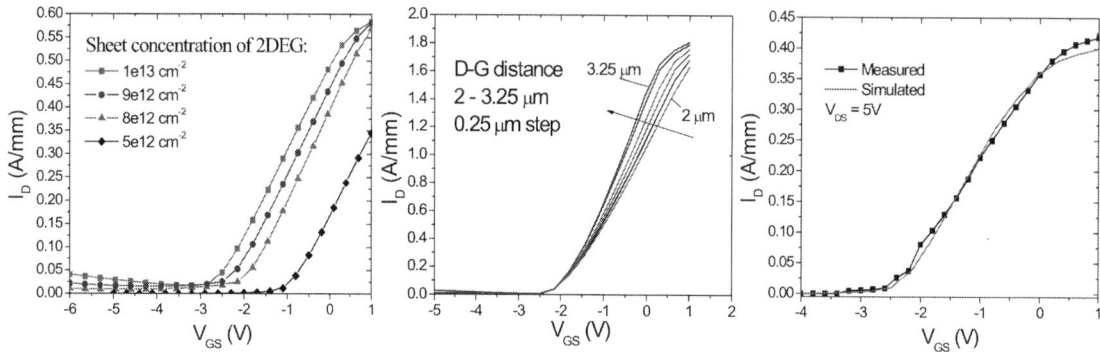

Fig. 4: Influence of polarization charge on transfer characteristic for V_{DS} = 5V. (left) Increase of transfer characteristics slope with increase of G-D distance and calibrated model characteristics vs. measurement.

fixed charge from 5e12 to 1e13 cm^{-2} are shown in Fig. 4 (left). We found that the sheet concentration of fixed charge N = 9e12 cm^{-2} must be used to achieve value of pinch-off voltage V_P = -2.5 V. Finally, an influence of the D-G distance was studied. Results are depicted in the middle of Fig. 4. One can see that with decreasing of G-D distance the slope of transfer characteristics is decreasing. This can be due to the high electric field that degrades the electron mobility in the area between the drain and gate contacts. Samples used for measurements were without passivation, therefore additional vacuum layer was added to surface of structure. After this we obtained very good match between the measured and simulated characteristics (Fig. 4).

5. Conclusion

Using the DESSIS simulation tool for modelling of InAlN/GaN HEMT, we have shown that simulations can provide accurate results also for new materials by implementing the calibrated models. However a lot of work has to be done on better calibration of tunnelling models through InAlN and AlN layers and in area of breakdown regime of HEMTs.

Acknowledgement

This work has been done in Center of Excelence CENAMOST (Slovak Research and Development Agency Contract No. VVCE-0049-07 with support of grant VEGA 1/0742/08. Samples were prepared by AIXTRON AG, Herzogenrath, Germany.

References

[1]   KuzmikJ. *Semicond. Sci. Technol. 17, 540–544 2002*
[2]   Donoval D., et al., Appl. Phys. Lett. **96**, 223501, 2010
[3]   DESSIS - ISE, User manual, ver. 10.0, ISE Zurich, 2004
[4]   Brannick, A. et al., Physica status solidi, **4**, 2007, 651-5
[5]   Medjdoub, et al., in IEDM Tech. Dig., 2006, 927–4
[6]   Medjdoub, F. et al., Open Elect. Electron. Eng. J., **2**, 2008, 1–7

GaN for THz Sources

M. Marso

Faculty of Science, Technology and Communication, University of Luxembourg,
6, rue Richard Coudenhove-Kalergi, L-1359 Luxembourg
e-mail: Michel.marso@uni.lu

The unique electrical and thermal properties of GaN are used to improve two different approaches to generate THz radiation. One method is heterodyne photomixing, where two laser beams with slightly different wavelengths illuminate an ultrafast photodetector. The electrical and mainly the thermal limits of the conventionally used LT-GaAs restrict the THz output power generated by this method up to now. In a second approach, ultrafast transistors, e.g. hetero field effect transistors, are applied in high power high frequency oscillator circuits that act as input for a frequency multiplier chain. In this approach we investigate the utilization of GaN based transistors. Devices in this material system are usually used for high power applications at moderate frequencies, but the very high electron saturation velocity of GaN allows the application above 100 GHz as well.

1. Introduction

The terahertz frequency band, the region between 300 GHz and 3 THz (Fig. 1), has for a long time been known as the "terra incognita" of the electromagnetic spectrum. On the one side, the frequencies are too high for an electronic approach. On the other side, the photon energy is too low for photonic methods. However, terahertz radiation has tremendous potential for various applications such as radio astronomy, terahertz imaging, high-resolution spectroscopy, medicine, security ("body scanner") and defence [1-4].

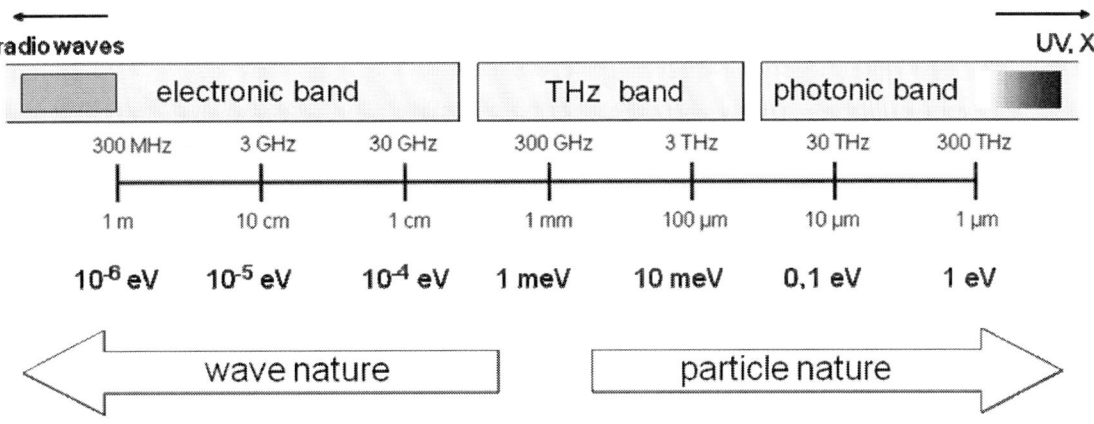

Figure 1: Electromagnetic spectrum, from radio waves to X-rays

During the last years, different methods have been investigated as sources of terahertz radiation. Continuous-wave terahertz sources can be realized by different methods, all with specific strengths and weaknesses (Fig. 2). Backward wave tubes, e.g., are very expensive and huge and have large energy consumption, while far infrared lasers need to be cooled down to work at these (for photonic devices) low frequencies.

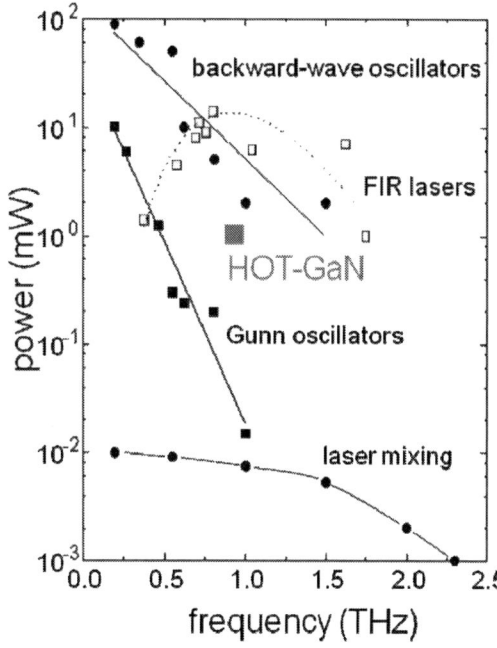

Figure 2: Output power of different continuous-wave THZ sources

Gunn oscillators:
+ reliable and compact technique
- low power for f > 1 THz, expensive

Backward-wave oscillators
+ good tunable
- expensive

F I R lasers
+ high power at > 1 THz
- only discrete frequencies possible
- cooling needed

Laser mixing in LT-GaAs [5-7]
+ up to 3.8 THz, tuneable
- low efficiency

Heterodyne photomixing is a compact and inexpensive approach to generate continuous electromagnetic radiation in the terahertz range, with tuneable frequency [7]. The method uses two lasers with slightly different wavelengths that illuminate an ultrafast photoconductor. The interference of both laser beams creates an oscillation of the illumination intensity in the terahertz range, namely with the difference of both laser frequencies. GaAs grown at low temperatures (LT-GaAs) is the conventionally used photoconductor material [8]. One drawback is the relatively low THz power in the nW to µW range [5-7]. The aim of our work is the improvement of the output power by replacing the LT-GaAs by other semiconductors. For this purpose we investigate GaN that is rather known as basic material for blue LEDs and lasers [9], but it has also remarkable electrical and thermal properties.

A more conventional electronic approach to generate THz radiation consists of the fabrication of an oscillator circuit based on ultrafast transistors, e.g. Hetero Field Effect Transistors (HFET or HEMT) based on InP [10]. These circuits can be designed up to about 100 GHz oscillation frequency [11]. The THz region is achieved by frequency multipliers, e.g. realized by very small-sized Schottky diodes. However, each multiplier stage considerable reduces the output power. In this field we investigate GaN based transistor devices to profit from the much better power performance of this material, compared to classical semiconductors.

2.   Photomixing with GaAs:N

The photoconductor is the key element of the heterodyne photomixer. It converts the laser light into an electrical current. Because both laser beams have slightly different wavelengths, the illumination intensity varies with the difference of both laser frequencies. This results in an AC component of the photocurrent that is converted into electromagnetic radiation by an integrated antenna. Fig. 3 depicts a photomixer chip with finger-shaped photoconductor, dipole antenna and bias lines with integrated RF filter.

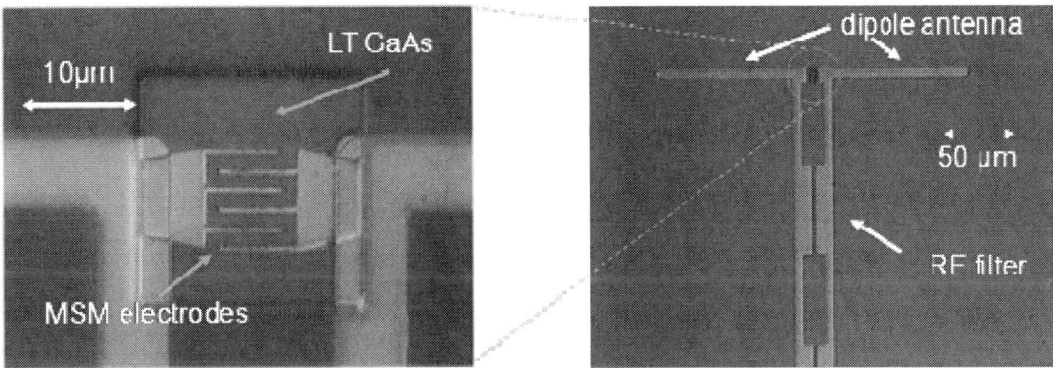

Figure 3: Example of a photomixer chip with dipole antenna, designed for 460 GHz [12]. The photoconductor area is 8x8 μm^2.

The highest frequency that can be generated depends on the speed of the photoconductor. For a well-designed device it is defined by the life time of the photogenerated carriers. The current amplitude and thus the power of the generated electromagnetic radiation depend on the responsivity and the applied bias voltage. The conventionally used material of choice for the photoconductor is GaAs, fabricated by molecular-beam epitaxy at a relatively low temperature of about 200 - 300°C [8]. This LT-GaAs has many defects that act as carrier traps. It shows a subpicosecond photocarrier trapping time (Fig. 4, left) and acceptably high carrier mobility [13, 14]. LT-GaAs photodetectors allow photocurrent pulses in the sub-ps range [15]. The output power of an LT-GaAs photomixer depends on the illumination density and on the applied bias voltage; it is limited by the thermal damage threshold of the material of about 1 mW/μm^2). Best published values are in the μW-range for a frequency of 1 THz [5-7].

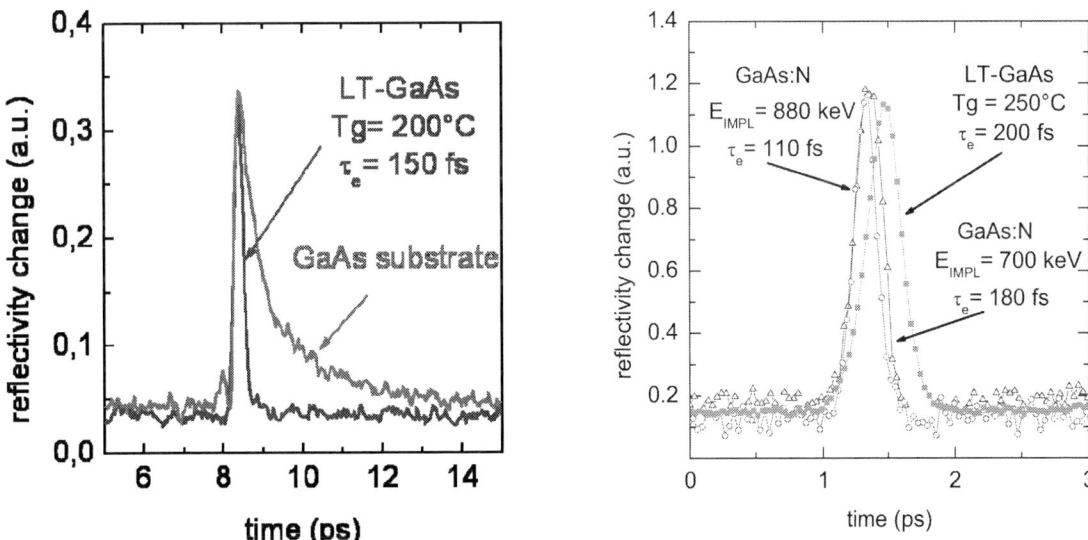

Figure 4: Life time of photogenerated carriers in different material systems, determined by optical reflectivity change measured using femtosecond pump/probe spectroscopy.

$\tau_e = 1/e$ - decay time. [16, 17]

One attempt to increase the output power of the photomixer is the exchange of the LT-GaAs by nitrogen-implanted GaAs (GaAs:N). Implantation of nitrogen ions creates carrier defects similar as in LT-GaAs, but with different carrier trap properties. The appropriate choice of doping dose, implantation energy and annealing temperature gives a material with photocarrier trapping times comparable and even lower than for LT-GaAs (Fig. 4, right). Photomixer circuits based on this material show a three-times higher output power than similar devices based on LT-GaAs (Fig, 5) [17, 18]. Also the dependence on the optical input power shows an improved performance, compared to the conventional LT-GaAs based systems. We attribute this superior behavior of the implanted material to the different, as compared to LT-GaAs, physical origin of the implantation defects that act as carrier traps.

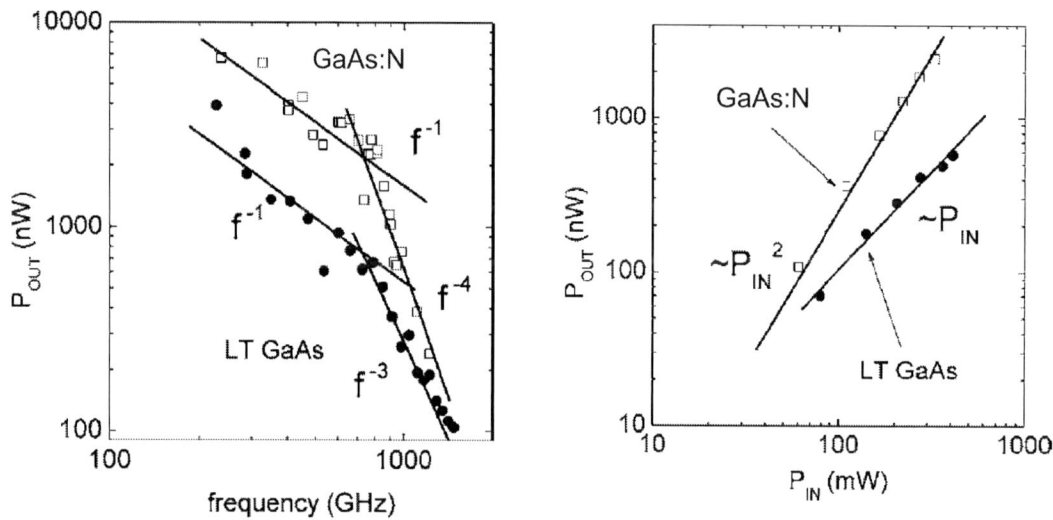

Figure 5: Comparison of the terahertz output power of traveling-wave photomixers fabricated on LT GaAs and on nitrogen implanted GaAs [18].
Left: optical input power: 400mW, bias voltage: 15 V
Right: frequency: 850 GHz, bias voltage: 15 V

3. Terahertz generation by frequency multiplying

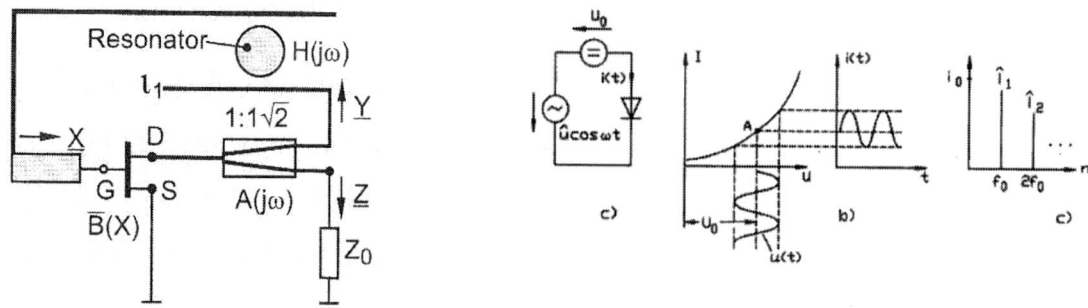

Figure 6: Schematic diagram of the components for the electronic generation of THz radiation.

978-1-4244-8574-1/10 $26.00 © 2010 IEEE   150

Circuit of a transistor-based oscillator (left).

Frequency multiplication by a nonlinear device characteristic (right). [19]

Because of the lack of amplifier devices with cutoff frequencies above 1 THz the electronic generation of THz radiation must be performed by the combination of an oscillator with a lower frequency (typically below 100 GHz) and one or several frequency multiplier stages (Fig. 6). The output power of the oscillator must be high to compensate the power loss in the frequency multipliers. Any dipole with nonlinear characteristic can be used as multiplier, e.g. a Schottky diode [20] or a heterostructure quantum well device [21]. A hybrid solution with an external microwave source and cascaded multiplier modules have already been commercialized [22]. The key for a low-cost and compact THz source, however, lies in the monolithic integration of oscillator and frequency multipliers. InP-based devices are up to now commonly used for ultra high frequency applications [10].

3.1 The project "HOT-GaN"

GaN with its very high electron saturation velocity (Fig. 7, left), high breakdown voltage (Fig. 7, right) and elevated operation temperature shows the highest power density and combined frequency-power performance of all commonly used semiconductors (Fig.7, right). While GaN-based transistors are widely used for high power applications at frequencies in the lower GHz range [23], the material system has also shown its potential for high frequency integrated circuits, e.g. as monolithically integrated power amplifier at 76 GHz with an output power of 12 dBm [24].

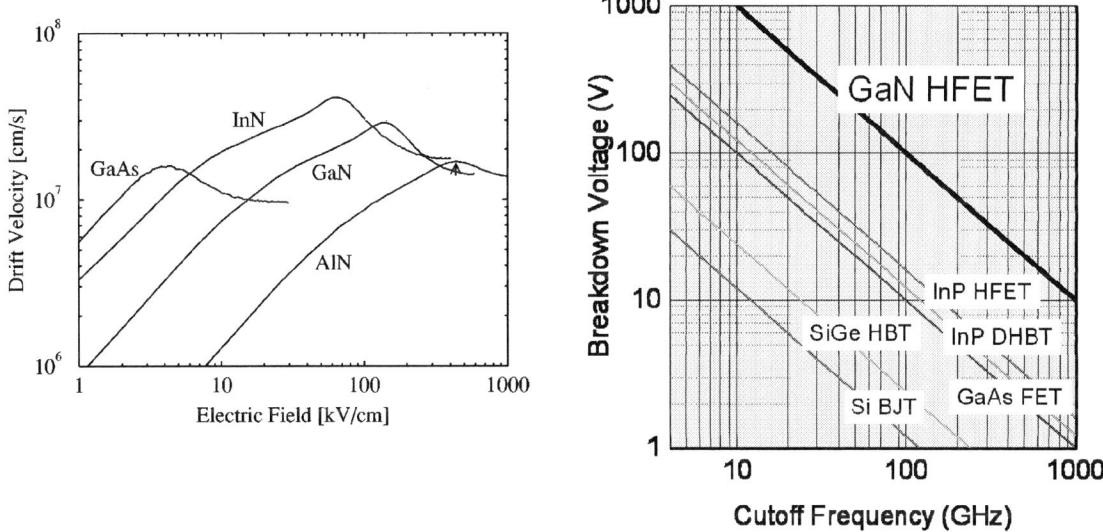

Figure 7: Comparison of electron velocity (left) and device performance
(Johnson's limit, right) for different semiconductor materials and devices. [25, 26]

The aim of our project "HOT-GaN" (HFET Oscillator for Terahertz Generation based on GaN) is the development of a monolithically integrated source for THz radiation (Fig. 8). The oscillator needs AlGaN/GaN HFET devices with f_{max}- values above the operation frequency of 100 GHz. In a first step we have developed a device with a gate length of 90 nm that exhibits a cutoff frequency of 100 GHz, without any gate recess or passivation [27] (Fig. 9).

In a second step the f_{max}-value will be increased to the required value (above 150 GHz) by application of a T-gate process and recessed gate. Published values show that this challenging high frequency performance is achievable [28, 29].

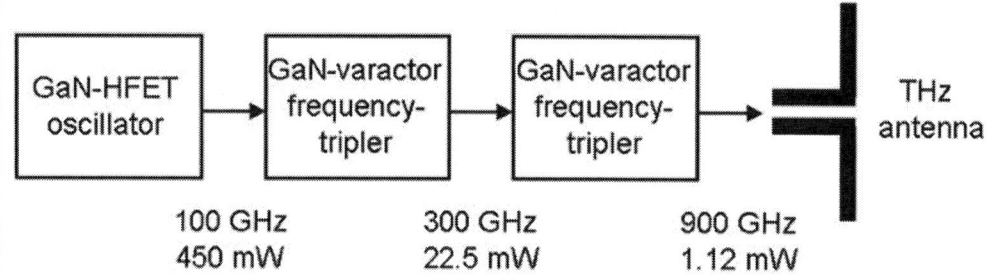

Figure 8: "HOT-GaN" concept: all components are integrated on one chip.

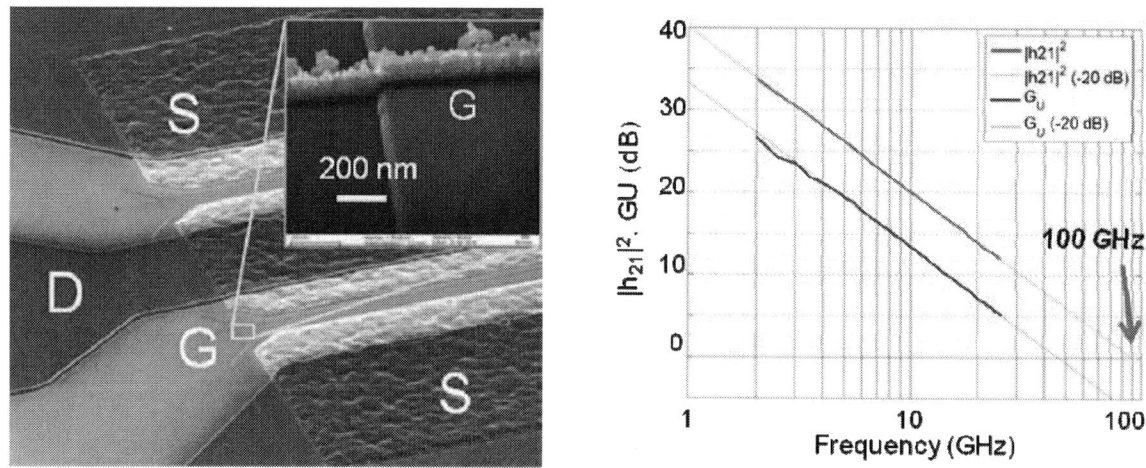

Figure 9: SEM picture and RF performance of an AlGaN/GaN HFET with 90 nm gate length.

The frequency multipliers will be fabricated by GaN MSM diodes. For this purpose we will investigate GaN nanowires that allow the design of diodes with very small areas needed for ultralow capacitances [b5].

The intended performance of the HOT-GaN elements (Fig. 8) will allow the realization of a THz source that is very attractive compared to the existing sources, considering frequency, power, costs and compactness (Fig. 2).

4. LT-GaN for photomixing

Photomixers based on GaN with its high electric breakdown field, high operation temperature and large thermal conductivity could be operated with much higher optical input power and bias voltage than GaAs-based circuits. These excellent properties of GaN will allow sources with much higher THz power, provided that GaN can be grown with ultrafast carrier traps. Up to now, GaN is always grown with a very high defect density (compared to classical semiconductors) because of the lattice mismatch between GaN and substrate. While these defects reduce the quality of GaN-based transistors, they can be used for fast photodetectors, especially when the material is grown at low temperature (LT-GaN) [31].

The layer structure (100nm AlN /150nm GaN, n = 10^{17} cm^{-3}) was grown by plasma induced molecular beam epitaxy on 6H i-SiC substrate at 650°C. The photodetector with an active area of 15x15 µm^2 was fabricated on the upper LT GaN layer. Interdigitated MSM structures with finger width and spacing of 1 µm and 2 µm, respectively, consist of standard Ni/Au Schottky contacts [32]. The time-dependent performance of the prepared LT GaN MSM photodetector was investigated by a standard pump-probe time resolved experiment using a femtosecond sapphire laser with frequency doubling (360 nm). The pulse energy used in the measurements was 68 pJ per 100 fs pulse. Figure 10 shows the response of the photodetector with a full width at half maximum value of 0.9 ps. The frequency response, calculated by Fourier transform, yields a bandwidth of 410 GHz. Comparison of these results with typical LT-GaAs values (0.58 ps and 550 GHz for FWHM and bandwidth, respectively [15]) prove the suitability of this novel material for a photomixer with both high output frequency and power.

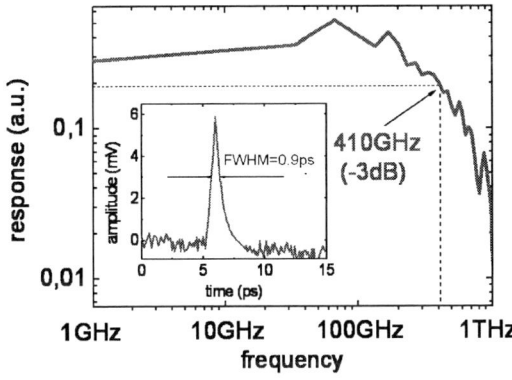

Figure 10:
Time-resolved response of a LT-GaN photodetector (inset) and frequency response (calculated by Fourier transform of the time-resolved response) (2V bias voltage)

5. Conclusions

We have demonstrated the potential of GaN for the improvement of THz sources. The GaAs:N material system has yielded a threefold output power of heterodyne photomixers, compared to the conventional LT-GaAs. A photodetector based on low-temperature grown GaN has shown a sub-ps photoresponse. In combination with the superior electrical and thermal performance of GaN, this result will allow the realization of photomixing systems in the THz range with a large increase of the output power, compared with the existing material systems.

Because GaN based transistor devices and circuits allow a very high power density AND can operate at very high frequencies, this material system is the ideal candidate for the realization of compact monolithically integrated THz sources based on electronic oscillator circuits with integrated frequency multipliers.

Acknowledgement

It is a great pleasure to acknowledge the contributions of Martin Mikulics, Peter Kordoš, Benjamin Strang and Alfred Fox to this work.

978-1-4244-8574-1/10 $26.00 © 2010 IEEE

References

[1] P. Mukherjee and B. Gupta, Int. J. Infrared. Milli. Waves 29, 1091 (2008)

[2] A. Cheville and D. Grischkowsky, Appl. Phys. Lett. 67, 1960 (1995)

[3] B. B. Hu and M. C. Nuss, Opt. Lett. 20, 1716 (1995)

[4] M. Nagel, P. Haring Bolivar, M. Brucherseifer, H. Kurz, A. Bosserhoff and R. Büttner, Appl. Phys. Lett. 80, 154 (2002)

[5] S. Verghese, K. A. McIntosh, and E. R. Brown, Appl. Phys. Lett. 71, 2743 (1997)

[6] T.-F. Kao et al. , Appl. Phys. Lett. 88, 093501 (2006)

[7] E. R. Brown, K. A. McIntosh, K. B. Nichols, and C. L. Dennis, Appl. Phys. Lett. 66, 285 (1995)

[8] D.C. Look, Thin Solid Films, 231, 61 (1993)

[9] S. Nakamura, Science 281, 956 (1998)

[10] D. Pukala et al, IEEE Microw.Wireless Compon. Lett. 18, 61 (2008)

[11] S. Kim et al., Proc. Int. Conf. Indium Phosphide Rel. Mat. 2004, 20

[12] M. Mikulics et al., Proc. ASDAM, Smolenice 2002, 129

[13] E. R. Brown, F. W. Smith, and K. A. McIntosh, J. Appl. Phys. 73, 1480 (1993)

[14] E. Peytavit et al., Appl. Phys. Lett. 81, 1174 (2002)

[15] P. Kordos, A. Förster, M. Marso, and F. Rüders, Electron. Lett. 34, 119 (1998)

[16] M. Mikulics, PhD thesis, RWTH Aachen 2004

[17] M. Mikulics et al., Appl. Phys. Lett. 87, 041106 (2005)

[18] M. Mikulics et al., Appl. Phys. Lett. 89, 071103 (2006)

[19] Edgar Voges, Hochfrequenztechnik, Hüthig 2004

[20] H. Xu, G. S. Schoenthal, J. L. Hesler, T.W. Crowe, and Robert M. Weikle II, IEEE Trans. Microw. Theory Tech. 55, 648 (2007)

[21] V. Duez, X. Mélique, O. Vanbésien, P. Mounaix, F. Mollot and D. Lippens, Electron. Lett. 34, 1860 (1998)

[22] Virginia Diodes Inc., Charlottesville, VA ; www.virginiadiodes.com

[23] J. H. Leach and H. Morkoç, Proc. IEEE 98, 1127 (2010)

[24] S. Yoshida et al., IEEE MTT-S International Microwave Symposium Digest 2009, 665

[25] B. E. Foutz S. K. O'Leary M. S. Shur, J. Appl. Phys. 85, 7727 (1999)

[26] S. Nakajima et al., GaN HEMT: Revolution in High Power Microwave Applications, International Workshop on Nitride Semiconductors, 22. – 27. 10. 2006, Kyoto, Japan

[27] B. Strang et al., to be published

[28] M. Higashiwaki, T. Matsui, and T. Mimura, IEEE Electron Device Lett. 27, 16 (2006)

[29] J. W. Chung, W. E. Hoke, E. M. Chumbes, and T. Palacios, IEEE Electron Device Lett. 31, 195 (2010)

[30] T. Richter, H. Lüth, R. Meijers, R. Calarco, and M. Marso, Nano Letters 8, 3056 (2008)

[31] M. Mikulics et al., Appl. Phys. Lett. 86, 211110 (2005)

[32] M. Mikulics et al., to be published

Preparation and properties of AlGaN/GaN MOS-HFETs with atomic layer deposited Al₂O₃ as gate oxide

R. Stoklas[1], D. Gregušová[1], M. Blaho[1], P. Kordoš[1], M.Tajima[2], and T. Hashizume[2]

[1]Institute of Electrical Engineering, Slovak Academy of Sciences,
SK-84104 Bratislava, Slovakia
[2]Research Center for Integrated Quantum Electronics (RCIQE),
Hokkaido University, Sapporo 060-8628, Japan
e-mail: roman.stoklas@savba.sk

Atomic layer deposition (ALD) technique at 300 °C was used to prepare an Al_2O_3 dielectric layer, to form MOS-HFETs. The static (output and transfer) and dynamic (Capacitance-Voltage) characteristics were used for evaluation of investigated devices. From the static characteristic, an increase of the saturation drain current (up to 35%) and extrinsic transconductance (up to 10%) of the MOS-HFETs were observed. Higher n_S on the MOS-structure, evaluated from the C-V measurement, can be responsible for these effects. The gate leakage current was also reduced about four orders of magnitude in comparison to the HFET.

1. Introduction

AlGaN/GaN heterostructure field effect transistors (HFETs), in which the polarization-induced charge at the AlGaN/GaN interface allows for very high sheet electron densities n_S ($\sim 1 \times 10^{13}$ cm^{-2}) of 2-dimensional electron gas (2DEG) with no intentional doping, are characterized by very high carrier mobility in the channel μ (~ 1800 cm^2/Vs), very high breakdown fields E_C ($\sim 2.10^7$ V/cm), unusual robustness and thermal capability of structures [1]. Thanks to the properties the HFETs based on the AlGaN/GaN heterostructure find wide range of applications in high frequency and high power electronics due to large breakdown voltage, large saturation electron velocity, and can operate in the high-temperature applications. HFETs have already been fabricated on SiC substrates with a f_T of 181 GHz, and a f_{MAX} of 186 GHz at a gate length of 30 nm [2], an output power density of 41.4 W/mm and a power-added efficiency (PAE) of 60% at 4 GHz [3]. The main obstacle to progress has been, and continues to be, reducing the trap densities in the bulk and surface of the materials and solving the problem of significant dispersion between DC and RF characteristics [4]. To address the gate lag effect, various solutions have been proposed to passivate the AlGaN surface. The next main problem of the AlGaN/GaN HFETs is a high gate leakage current due to a high density of defects in the structure. AlGaN/GaN metal-oxide-semiconductor HFETs (MOS-HFETs) using gate oxides such as Ga_2O_3 (Gd_2O_3), SiO_2, MgO, Sc_2O_3, AlN and Si_3N_4 prepared at different deposition techniques (MOCVD, ALD, Sputtering and oxidation, etc.) offer lower gate leakage currents and larger voltage swings than conventional HFETs [5 - 7]. The DC/RF disperssion is also reduced. However, the question of a suitable gate and passivation dielectrics and suitable deposition techniques are still open. The ALD technique has many advantages, such as low deposition temperature and good thickness controllability. Recently, improved transport properties and high microwave-noise performance of the ALD Al_2O_3/AlGaN/GaN MOS-HFETs have been reported [8].

Aluminum oxide (Al_2O_3) is an attractive material for use in the gate dielectric layer of MOSHFETs. Al_2O_3 possesses a large band gap (9 eV), high dielectric constant (~ 9), high breakdown field (10^7 V/cm), thermal stability (amorphous up to at least 1000 °C), and chemical stability [9]. Furthermore, Al_2O_3 is a native oxide of AlGaN.

In this paper, the static properties of AlGaN/GaN metal-oxide-semiconductor HFETs (MOS-HFETs) with 10-nm-thick Al_2O_3 gate oxide, prepared by ALD, and unpassivated HFETs are compared. The static characterization of the MOSHFETs yielded higher drain current as well as a higher transconductance than those of the HFETs. This can be explained due to a higher effective velocity of the channel electrons and a reduced series resistance in the MOSHFET devices.

2.   Experiment

The layer structure was grown on high-resistive SiC substrate by MOCVD. It consisted of a 3-μm-thick GaN undoped followed by a 30-nm-thick undoped $Al_{0.25}Ga_{0.75}N$ barrier and a 3-nm-thick GaN cap layer. The device processing consisted of conventional HFET fabrication steps: (1) Mesa isolation achieved with argon sputtering; (2) evaporation a Ti/Al/Ni/Au multiplayer for ohmic contact; (3) rapid thermal annealing of the multiplayer at 850 °C for 30 s in an N_2 ambient to prepare ohmic contacts. As to the MOS-HFETs, a 10-nm-thick Al_2O_3 gate oxide was deposited by ALD at 300°C using trimethylaluminum and water vapor as precursors and N_2 as a carrier gas. The gate oxide was deposited before Schottky metallization on the source-drain access region, i.e. it served also as a passivation layer. The gate oxide thickness was controlled by elipsometry measurement. The dielectric constant of ~9.5 was evaluated. The Schottky gate metalization consisted of a double Ni/Au layer patterned by optical lithography. The devices had a gate length of 2 and 2.5 μm and a gate width of 50 (one finger) and 100 μm (two fingers). Also processed along with the HFETs and MOSHFETs, there were large-gate "fat"-FET structures with a gate length of 100 μm and van der Pauw patterns with an active area of 0.3×0.3 mm^2. The devices were characterized by the static (output and transfer) and dynamic (Capacitance-Voltage) measurements. Simultaneously processed Hall-effect structures and fat-HFETs were used to characterize the carrier transport properties of used layer structure.

3.   Results and Discussion

The electrical characterization (output, transfer and C-V) of prepared devices was realized at room temperature. Using these measurements the functionality of the devices and the quality of the layer structure can be analyzed. The static output and transfer characteristics were measured at first, and were used for evaluation of parameters as the saturation drain current I_{DS}, extrinsic transconductance g_m and leakage current I_{leak}. The DC output and transfer characteristics (Fig. 1) yielded higher drain current (up to 35% higher saturated drain current at $V_G = 1$ V) and extrinsic transconductances (up to 10% of peak values at $V_{DS} = 6$ V) than those for the HFETs ($I_{DS} = 370$ mA/mm and $g_m = 90$ mS/mm, respectively). The sheet carrier concentration and drift velocity increase in the MOS-HFETs are responsible for these effects, as we have reported recently [10].

978-1-4244-8574-1/10 $26.00 © 2010 IEEE   156

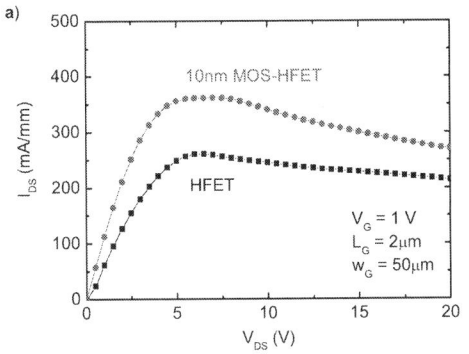 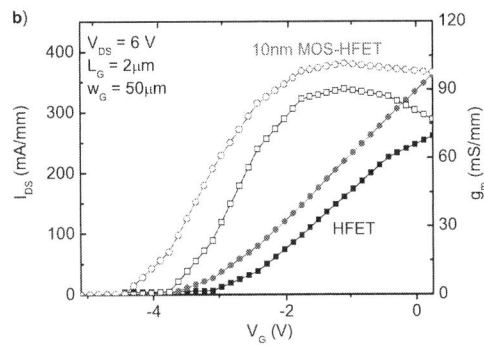

Fig. 1 Comparison of the output characteristics (a) and extrinsic transconductances (b) of AlGaN/GaN HFET and MOS-HFETs with 10-nm-thick Al_2O_3 gate oxide, at $V_G = 1$ V and $V_{DS} = 6$ V respectively.

Typical reverse-bias *I–V* characteristics of the HFET and MOS-HFETs are shown in Fig. 2. The gate leakage current of the MOS-HFETs was found to be 4 orders of magnitude lower ($\sim 10^{-7}$ A/mm at -15 V) than that of the HFET, as shown in Fig. 2. A lower leakage current indicates an efficient reduction of traps in the AlGaN/GaN heterostructure because of the Al_2O_3 gate oxide.

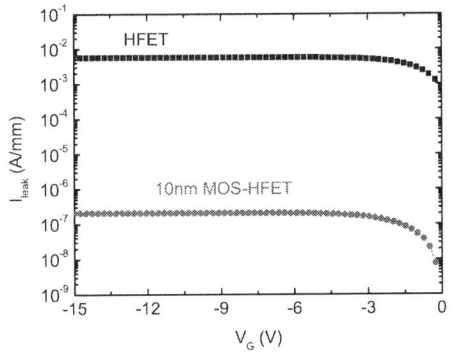

Fig. 2 Gate leakage current on AlGaN/GaN HFET and MOS-HFETs ($L_G = 2$ μm).

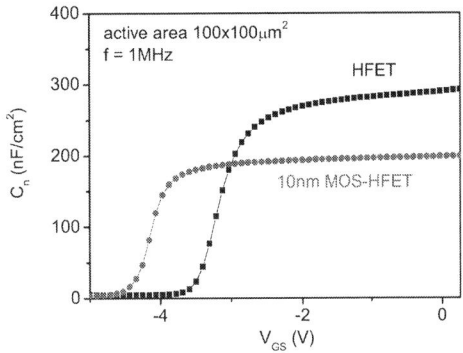

Fig. 3 Typical C−V characteristics of AlGaN/GaN fat-HFETs without and with 10-nm-thick Al_2O_3 gate oxide.

And finally, the capacitance-voltage (*C-V*) measurements in the frequency range from 1 kHz to 1 MHz, by mean of Agilent 4284A LCR-meter, were used to characterize the layer structure used. An active area of the investigated devices was 100×100 μm^2. Fig. 3 shows typical *C-V* dependences for HFETs and MOS-HFETs at 1 MHz frequency. The AlGaN thickness of 30-nm-thick was evaluated from the capacitance $C_0 = 297$ nF/cm^2 on HFETs. Average capacitances for the MOS-HFETs with different contact areas, $C_0 = 199$ nF/cm^2 for 10-nm-thick Al_2O_3, yielded the dielectric constant of Al_2O_3 $\varepsilon = 9.5$. This is in very good agreement with published data for ALD Al_2O_3 layer [9]. And together the sharp capacitance transition from depletion to 2DEG accumulation indicates high quality of prepared gate dielectric. The Capacitance-Voltage (*C-V*) measurement on the fat-HFETs was performed to evaluate the sheet carrier density n_S and threshold voltage V_{th}. Values of $n_S = 6 \times 10^{12}$ and

7.25×10^{12} cm^{-2} and V_{th} = -3.7 and -4.6 V were obtained for the HFET and MOS-HFET, respectively. Higher n$_s$ on the MOS-structure indicate the passivation effect [11].

Additional experiments concerning properties of the ALD Al$_2$O$_3$ layer, as the pulsed and conductance-frequency measurements, are in progress.

4. Conclusions

In summary, the influence of the 10-nm-thick Al$_2$O$_3$ dielectric layer deposited by ALD at 300 °C was analyzed in details. The static measurement yielded an increase of the drain current I_{DS} (up to 35%) and extrinsic transconductance g_m (up to 10%) of the MOS-HFETs in comparison to the HFETs were observed. The gate leakage current was lowered by about four orders of magnitude, which is a sign of an efficient reduction of traps in the AlGaN/GaN heterostructure because of the Al$_2$O$_3$ gate oxide. A higher sheet carrier density n_S on the MOS-structure was evaluated from the *C-V* measurement, which indicates the passivation effect.

Acknowledgement

The authors would like to acknowledge and Mr. P. Eliáš for proof-reading the manuscript. This work reported here was supported by the Slovak Research and Development Agency APVV (Grant No. LPP- 0162-09), the Centre of Excellence CENAMOST (Grant No. VVCE-0049-07) and the Slovak Scientific Grant Agency VEGA (Contract Nos. 2/0098/09).

References

[1] P. Kordoš, G. Heidelberger, J. Bernát, A. Fox, and H. Lüth, *Appl. Phys. Lett.* **87**, 1, 2005.

[2] M. Higashiwaki, T. Mimura, and T. Matsui, *Jpn J Appl Phys*, **45**, 111, 2006.

[3] Y. F. Wu, M. Moore, A. Saxler, T. Wisleder, and P. Parikh, in *Proceedings of the 64th Device Research Conference*, New York, USA, 2006. p. 151.

[4] L. Semra, A. Teliaa, and A. Soltanib, *Surf. Interface Anal.*, **42**, 799, 2010

[5] R. Stoklas, D. Gregušová, J. Novák, A. Vescan, and P. Kordoš, *Appl. Phys. Lett.* **93**, 124103, 2008.

[6] D. Gregušová, R. Stoklas, Ch. Mizue, Y. Hori, J. Novák, T. Hashizume, and P. Kordoš, *J Appl. Phys.* **107**, 106104, 2010.

[7] R. Stoklas, Š. Gaži, D. Gregušová, J. Novák, and P. Kordoš, *Phys. Stat. Sol. (c)* **5**, 1935, 2008.

[8] Z. H. Liu, G. I. Ng, S. Arulkumaran, Y. K. T. Maung, K. L. Teo, S. C. Foo, and V. Sahmuganathan, *Appl. Phys. Lett.* **95**, 223501, 2009.

[9] Y. Y. Zheng, H. Yue, F. Qian, Z. J. Cheng, M. X. Hua, and N. J. Yu, *Sci. China (E) Tech. Sci.* **59**, 2762, 2009.

[10] P. Kordoš, D. Gregušová, R. Stoklas, K. Čičo, and J. Novák, *Appl. Phys. Lett.* **90**, 123513, 2007.

[11] D. Gregušová, R. Stoklas, M. Eickelkamp, A. Fox, J. Novák, A. Vescan, D. Grützmacher, and P. Kordoš, *Semicond. Sci. Technol.* **24**, 075014, 2009.

Comparison of AlGaN/GaN HFETs and MOSHFETs in prospect of oscillator design

A. Fox[1], M. Mikulics[1], B. Strang[1], M. Marso[1,2], D. Grützmacher[1], and P. Kordoš[3,4]

[1]Institute of Bio- and Nanosystems (IBN-1), Research Centre Jülich, Germany
[2]Faculty of Science, Technology and Communication, University of Luxembourg
[3]Department of Microelectronics, Slovak Technical University, SK-81219 Bratislava,
[4]Institute of Electrical Engineering, Slovak Academy of Sciences, SK-84104 Bratislava,
e -mail: A.Fox@fz-juelich.de

The AlGaN/GaN heterostructure without oxide (HFET) and with additional aluminium oxide underneath the gate (MOSHFET) were investigated with regards to RF performance. Both cutoff frequency and maximum frequency of oscillation were measured and compared by means of small signal analyses and equivalent circuit parameter extraction. The maximum oscillation frequency f_{max} is an important figure of merit because it is a defining factor for oscillator design. The cutoff frequency f_t of the HFET with gate length of 300 nm resulted in 28 GHz and showed an increase up to 39 GHz for the MOSHFET device. The maximum frequency of oscillation f_{max} showed contrary behavior. The HFET showed an f_{max} of 120 GHz, while the MOSHFET revealed a reducedd RF performance of 78 GHz. RF simulations based on measured S-parameters indicated an increased gate drain capacitance C_{gd} for devices with a dielectric layer underneath the gate metallization.

1. Introduction

AlGaN/GaN heterostructure field-effect transistors (HFETs) are primary devices for microwave power sources. Another perspective application of these devices is a millimeter-wave oscillator [1]. Oscillation is possible through a wide range of frequencies up to the f_{max} of the devices. Optimization of the cutoff frequency f_t as well as the maximum oscillation frequency f_{max} is a crucial point to achieve high RF performance for both applications. In this work we investigate AlGaN/GaN HFETs and also AlGaN/GaN metal-oxide-semiconductor HFETs (MOSHFETs) which are under systematic study for a short time [2]. Both the gate resistance R_g and gate-drain capacitance C_{gd} are determining a time constant $\tau = R_g \cdot C_{gd}$ which influences the frequency of oscillation f_{max}.

2. Device properties

The layer structure used for device preparation consisted of 3 µm undoped GaN followed by a 30 nm undoped $Al_{0.25}Ga_{0.75}N$ barrier and a 3 nm GaN cap layers, grown by MOCVD on a SiC substrate. Conventional HFET fabrication steps were used for the device fabrication [3]. In the case of the MOSHFETs an Al_2O_3 layer was prepared by room-temperature oxidation of a sputtered Al layer deposited between the source and drain contacts before the gate deposition. The Al_2O_3 thickness was 4.2 nm, evaluated from the reflectivity measurements. Devices with 0.3, 0.5 and 0.7 µm gate length and 2×50 and 2×100 µm gate width, respectively, patterned by electron-beam lithography, were prepared.

978-1-4244-8574-1/10 $26.00 © 2010 IEEE

3. I/V characterization

Static characterization of the prepared devices yielded data typical for AlGaN/GaN transistors. The HFET devices exhibited a saturation drain current of 650 mA/mm at $V_{GS} = 1$ V, and an extrinsic transconductance of 175 mS/mm at $V_{DS} = 6$ V (Fig. 1), Lg=300nm. The MOSHFETs showed slightly increased drain current (750 mA/mm at $V_{GS} = 1$ V) and nearly identical transconductance. In case of the MOSHFET the oxide layer causes an dielectric layer between the gate metallization and the semiconductor. The threshold voltage was about −3.5 V and −5.5 V for the HFETs and MOSHFETs, respectively.

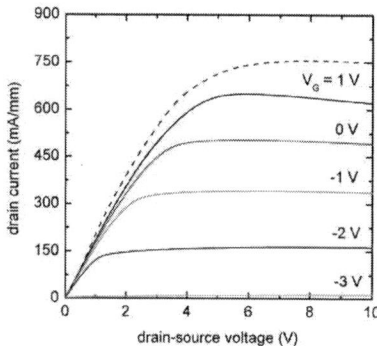

Fig. 1. Static I–V characteristics of AlGaN/GaN HFET (full lines). I–V characteristic at $V_G = 1$ V of AlGaN/GaN MOSHFET is shown for comparison (dashed line).

4. Small signal characterization

The influence of the oxide layer on device performance is demonstrated in the results of parameter extraction from the measured small signal S parameters. The S parameters were measured with the Vector Network analyzer system 8510XF in the frequency range from 1 to 51 GHz. The drain voltage was investigated in the range of 0 to 20V for all devices. The gate voltage for the HFET was varied from 1V to −5V and for the MOSHFET V_{gs} was varied from 1V to −10V to consider the higher negative threshold voltage that can also be seen in Fig 5 by the declining of gate-source capacitance C_{gs} vs. gate-source voltage V_{GS}. The influence of the dielectric layer on the small signal performance is investigated by comparing unilateral gain Gu and current gain H_{21} vs. frequency for both the MOSHFET and the HFET, as shown in Fig 2. The cutoff frequency f_t is defined as the frequency for zero gain value of H_{21}. The cutoff frequency f_t increased from 24 GHz for the HFET up to 39 GHz for the MOSHFET. An important figure of merit is the maximum frequency of oscillation f_{max} which is the frequency at which the unilateral gain Gu drops to unity. In contrast to f_t, f_{max} increases from 78 GHz for MOSHFET to 120GHz for HFET with 300nm gate length. Due to different threshold voltages the best cutoff frequencies were achieved with different gate source voltages at −2.5V and −4.5V for HFET and MOSHFET respectively at a fixed drain source voltage of 20V. This result shows an influence of the parasitic capacitances on f_{max} [4]. The maximum frequency of oscillation f_{max} decreases for all types of devices with increasing gate length, as shown in Fig 3.

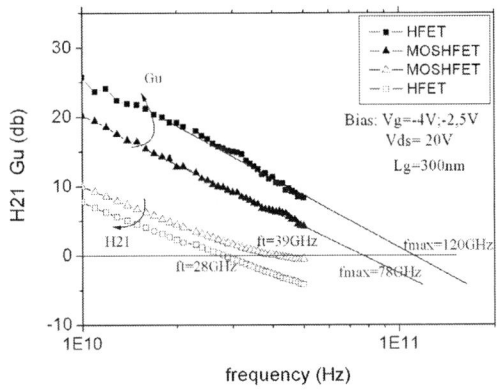
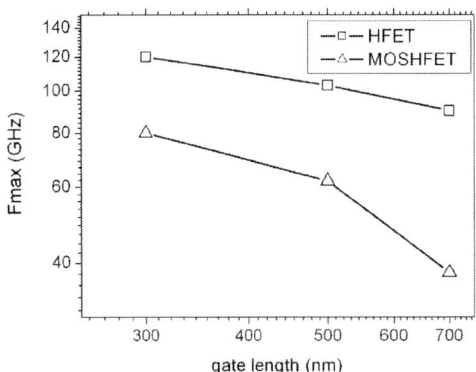

Fig 2: current gain H$_{21,}$ unilateral Gu vs. frequency for HFET, and MOSHFET with f_t and f_{max} @ unity

Fig 3: max frequency of oscillation vs. gate length l$_g$ for HFET, and MOSHFET.

5. Modelling

Device modelling based on the measured S-parameter was done to evaluate the extrinsic and intrinsic parameters, e.g. C$_{gd}$, g$_m$ and C$_{gs,}$ of the device model by means of the extraction software TOPAS based on the 15 element equivalent circuit diagram. The extracted elements confirm the behavior of HFET and MOSHFET described in former works [4]. The enlarged distance from gate to the channel due to the oxide underneath the gate reduces the transconductance g$_m$ as well as the gate-source capacitance C$_{gs}$ of the MOSHFET. Figs 5 and 6 show the dependence of both equivalent circuit elements as function of V$_{gs}$ and V$_{ds}$ respectively. V$_{gs}$ the value for the highest cutoff frequency, it is -2.5 V for the HFET and -4.5 V for the MOSHFET, respectively.The cutoff frequency depends on the ratio of g$_m$ to C$_{gs}$, shown in Fig. 7. The calculated values of f_t by the ratio gm/C$_{gs}$ is in agreement with the measured increase of the cutoff frequency f_t for the MOSHFET compared to the HFET, depicted in Fig 2. Obviously, the reduction of C$_{gs}$ overcompensates the decrease of g$_m$. This phenomena is due to the influence of the oxide layer on scattering, decribed in [4].The gate-drain capacitance, *C$_{gd}$*, is setup by the space charge region between the gate and drain, similar to *C$_{gs}$*, but is typically an order of magnitude smaller. Fig 4 shows C$_{gd}$ of HFET and MOSHFET as function of the drain voltage V$_{ds}$. Obviously the dielectric layer increases this capacitance. This explains the higher f_{max} of the HFET devices despite the lower cutoff frequency: the gate-drain capacitance influences the maximum oscillation frequency as shown in the formula:

$$f_{max} = \frac{f_t}{2\sqrt{R_1 f_t \tau}} \qquad \tau = 2 pi R_g C_{gd} \qquad R_1 = (R_G + R_i + R_s) / R_{DS}$$

Thus to maximize f_{max} the cutoff frequency f_t has to be maximized and C$_{gd}$ as well as R$_g$ must be minimized. This gives a further guide to choose material and thickness of the dielectric layer of the MOSHFET.

6. Conclusion

The AlGaN/GaN material is an excellent candidate for RF-transistor applications. In our investigations devices with oxide underneath the gate showed a good RF performance

978-1-4244-8574-1/10 $26.00 © 2010 IEEE   161

concerning power applications because of good current gain . In contradiction, devices without oxide showed a better frequency of oscillation and thus better RF performance regarding oscillator applications. In general, f_{max} is a better figure of merit for RF transistors because it depends not only on f_t but also on the parasitics of the transistor as well.

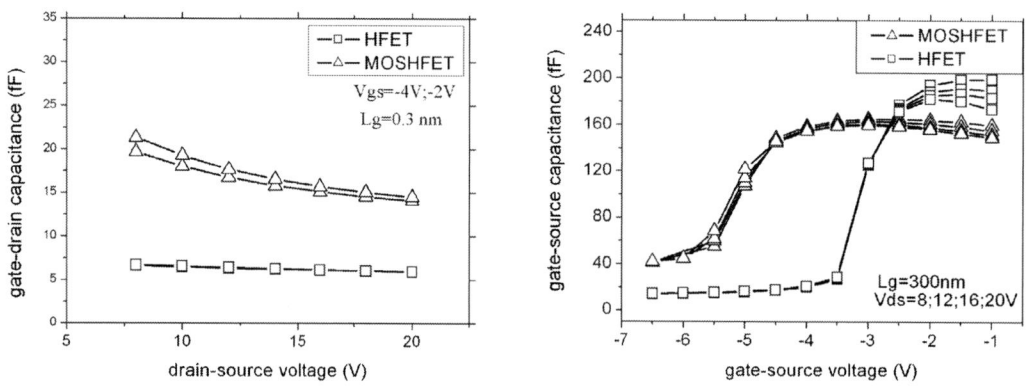

Fig 4: gate drain capacitance C_{gd} vs. drain source voltage V_{ds}

Fig5: gate source capacitance C_{gs} vs gate source voltage V_{gs}.

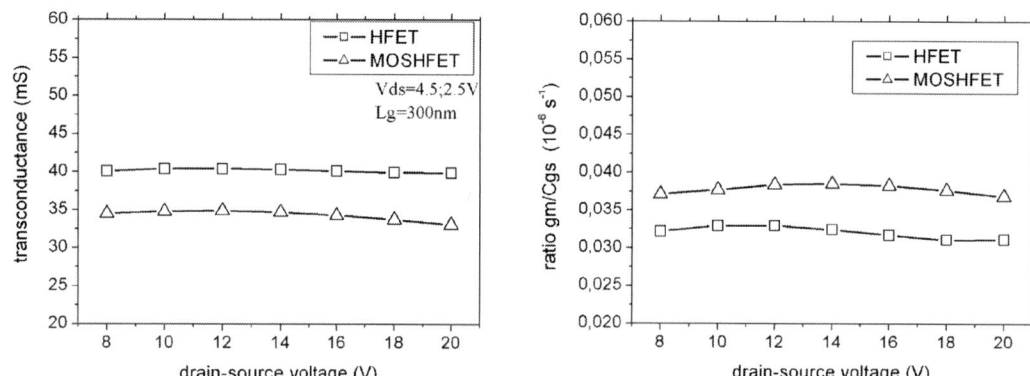

Fig 6: intrinsic transconductance g_m vs. drain source voltage Vds for HFET and MOSHFET.

Fig 7: calculated ratio of intrinsic gm/Cgs for HFET and MOSHFET

References

[1] V. S. Kaper, V. Tilak, H. Kim, A. V. Vertiatchikh, R. M. Thompson,T. R. Prunty, L. F Eastman, and J R.Shealy,J IEEE J. Solid-State Circuits,vol. 8,no. 9, pp. 1457–1461, Sep. 2003

[2] Gregušová D., Stoklas R., Eickelkamp M., Fox A., Novák J., Grützmacher D., and Kordoš P., Semicond. Sci. Technol. 24, 075014 (2009).

[3] Kordoš P., Gregušová D., Stoklas R., Gaži Š., and Novák J., Solid-St. Electron. 52, 973 (2008).

[4] M. Marso, G. Heidelberger, K. M. Indlekofer, J. Bernát, A. Fox, P. Kordoš, and H. Lüth, IEEE Transactions on Electron Devices 53, 1517 (2006).

[5] S.M. Sze, Physics of Semiconductor Devices, John Wiley & Sons, second edition.

Role of the gate-to-drain distance in the performance of the normally-off InAlN/GaN HEMTs.

J. Kuzmik[1,4], Ostermaier[1], G. Pozzovivo[1], B. Basnar[1], W. Schrenk[1], J.-F. Carlin[2], M. Gonschorek[2], E. Feltin[2], N. Grandjean[2], Y. Douvry[3], Ch. Gaquière[3], J.-C. De Jaeger[3], G. Strasser[1], D. Pogany[1], and E. Gornik[1]

[1]Institute for Solid State Electronics Vienna University of Technology, Floragasse 7, A-1040 Vienna, Austria
[2]Institute of Condensed Matter Physics, EPFL Lausanne, CH-1015 Lausanne, Switzerland
[3]IEMN, av. Poincaré, BP 69, 59652 Villeneuve d'Ascq, France
[4]Institute of Electrical Engineering SAS, Dubravska cesta 9, 841 04 Bratislava, Slovakia
e-mail: jan.kuzmik@tuwien.ac.at

We correlate dc maximal drain current I_{DSmax}, pulsed output characteristics, rf small-signal and breakdown performance of normally-off InAlN/GaN HEMTs with varied gate-to-drain distance d_{GD}. It is found that parasitic lag effects which are related to the possible surface states are not appearing at longer d_{GD}. On the other hand compromise need to be found between the improved gate performance and impaired I_{DSmax} and f_T as the d_{GD} is increased. The leakage current of the 0.5 µm-long gate may be reduced by up-to three orders of magnitude, down to µA/mm at -20 V bias if d_{GD} increases from 3 to 15 µm. On the other hand I_{DSmax} and f_T drop by about one third of the original value (from about 0.6 A/mm and 34 GHz down to 0.4 A/mm and 21 GHz, respectively) if d_{GD} changes.

1. Introduction

Normally-off III-N HEMTs may become a key element for future high-power transducers with potential applications in e.g. automotive industry. Other novel applications may include logic ICs suitable for a harsh environment, and safe *rf* transistors. To obtain a maximal switching efficiency of III-N HEMTs several design and technological problems need to be solved. It is desirable to have maximum I_{DSmax} / I_{off} ratio, together with a high off state breakdown V_{broff} and a high threshold voltage V_T. V_{broff} may be increased and the gate leakage current may be decreased by increasing the gate-to-drain distance d_{GD} [1]. On the other hand the same design may lead to profound impact of the surface states, seen as so called current collapse [2]. Similarly, by increasing d_{GD} we may expect decreased I_{DSmax} and f_T. In this work we investigate the effect of the varied d_{GD} on the performance of the normally-off InAlN/GaN HEMTs in detail. Our concept of n^{++} GaN/InAlN/AlN/GaN HEMT uses a self-aligned gate recess, where highly doped GaN cap is supposed to screen the parasitic effects of the surface traps [3].

2. Experiments

6 nm GaN: Si (2×10^{20} cm^{-3}) /1 nm lattice-matched InAlN/ 1 nm AlN/ 2 µm GaN was grown on sapphire by MOCVD [4]. The MESA insulation was performed by reactive ion etching (RIE) in Ar gas. Ohmic contacts are formed by electron-beam evaporation of Ti/Al/Ni/Au and rapid thermal annealing at 800 °C for 30 seconds. The source-to-drain

978-1-4244-8574-1/10 $26.00 © 2010 IEEE

distance d_{SD} was varied from 4 to 16 μm, the source-to-gate distance d_{SG} was fixed at 1 μm. HEMTs with a gate length L_G = 0.5 and 0.25 μm and 2 × 25 μm gate width were patterned by e-beam into PMMA resist. Simple one-step lithography was used for the gate recess selective dry etching and the lift-off. $SiCl_4/SF_6$ was used to perform a damage-free selective RIE of GaN over InAlN in the regime of the inductively coupled plasma [5]. The selectivity of the etching was 15, self-induced bias was kept below 100 V to limit the plasma-induced damage. After the RIE process the sample was loaded into the evaporator to deposit 15nm Ni/150nm Au thick gate stack.

The hard breakdown is obtained by a simple increase of V_{DS} by measuring HEMT output characteristic under the pinch-off conditions. To evaluate the drain current collapse effect we use Keithley 4200-SCF/F by pulsing the gate alone from V_{GS} = -3 V while dc drain bias is sequentially increased so that the last quiescent point was V_{GS} = -3 V / V_{DS} = 20 V. The single pulse duration was 200 ns with the off-state period of 100 μs. Finally the cut-off frequency performance were established from small signal s-parameters for different L_G.

3. Device performance

HEMT with the gate length 0.25 μm and d_{GD} = 3 μm reaches I_{DSmax} = 0.8 A/mm, a transconductance 440 mS/mm, V_T ~ 0.4 V and f_T ~ 50 GHz (not shown). Figure 1 shows the impact of increased d_{GD} on the HEMT output current (at V_{GS} = 2 V) and on the level of the current collapse. The main expected impact of the varied drain access resistance (represented by the gate-to-drain access region) on the HEMT output characteristics can be a shift of the knee voltage for the current saturation, and a decreased current density in the linear

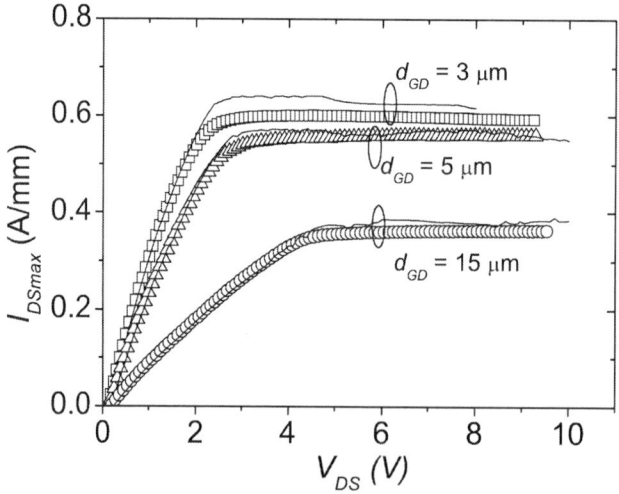

Figure 1 Pulsed (opened symbols) and dc (full line) characteristics of n^{++} GaN/InAlN/AlN/GaN HEMTs with L_G = 0.5 μm for different d_{GD} = 3, 5 and 15 μm. Gate is pulsed from V_{GS} = - 3 V to 2 V, drain is dc biased up-to 10 V. Duration of the pulse is 200 ns.

part of the characteristics as d_{GD} increases. This, indeed can be observed in Fig. 1, however I_{DSmax} drops also, by about 30 % by increasing d_{GD} from 3 to 15 μm. Thus it seems that significant change in the drain access resistance influences the current transport also in the intrinsic part of the HEMT channel. On the other hand screening of the surface traps by highly doped GaN cap remains effective regardless of the d_{GD} value as no significant difference between *dc* and pulsed characteristics are seen for any d_{GD}. Proportionally to I_{DSmax} reduction we have observed a drop in f_T, see Fig. 2 for $L_G = 0.5$ μm. f_T changes from about 34 GHz for $d_{GD} = 3$ μm to about 21 GHz for $d_{GD} = 15$ μm.

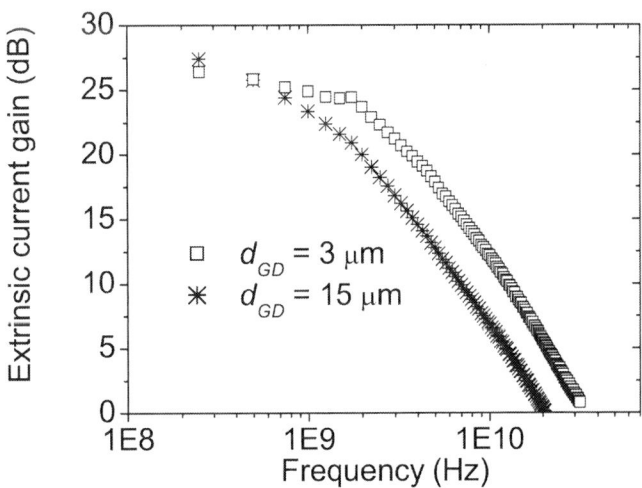

Figure 2 Frequency dependence of gain for n^{++} GaN/InAlN/AlN/GaN HEMTs with $d_{GD} = 3$ μm and 15 μm.

In Figure 3 we show that the two-terminal gate current leakage can be effectively decreased by increasing d_{GD}, by about three orders of magnitude at bias -20 V. On the other hand the two-terminal breakdown voltage was improved by only about 20 V by increasing d_{GD} from 3 to 15 μm. Similarly, three-terminal hard breakdown V_{broff} has increased from about 38 V to only about 50 V (not shown). Highly-resistive GaN buffer will be implemented in our next experiments to obtain more profound V_{broff} increase.

.

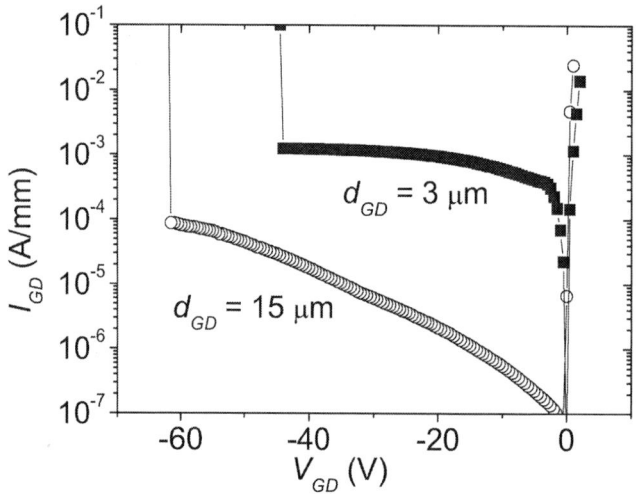

Figure 3 Two-terminal gate-to-drain breakdown behaviour of HEMTs with L_g = 0.5 µm and d_{GD} = 3 µm and 15 µm.

4. Conclusions

Gate-to-drain distance in the normally-off n^{++} GaN/InAlN/AlN/GaN HEMTs is a tuning parameter for the optimal device performance. The distance extension leads to the better gate contact performance without compromise in the negligible current lag effect. On the other hand, drain current and cut-off frequency are being reduced.

Acknowledgement

This work was supported by MORGAN European project FP7 NMP IP 214610 and by Gesellschaft für Mikro und Nanoelektronik GME.

References

[1] J. Kuzmik, G. Pozzovivo, J.-F. Carlin, M. Gonschorek, E. Feltin, N. Grandjean, et al, *phys. stat. sol. (c),* **6**, S925, 2009.

[2] J. Kuzmik, J.-F. Carlin, M. Gonschorek, A. Kostopoulos, G. Konstantinidis, G. Pozzovivo, et al, *phys. stat. sol. (a)* **204**, 2019, 2007.

[3] J. Kuzmik, C. Ostermaier, G. Pozzovivo, B. Basnar, W. Schrenk, J.-F. Carlin, et al, *IEEE Trans. On Electron Dev.* **57**, 2144, 2010.

[4] M. Gonschorek, J.-F. Carlin, E. Feltin, M. A. Py, and N. Grandjean, *Appl. Phys. Lett.,* **89**, 062106, 2006.

[5] C. Ostermaier, G. Pozzovivo, B. Basnar, W. Schrenk, J.-F. Carlin, M. Gonschorek, N. Grandjean, et al, submitted.

Influence of interface states on *C-V* characteristics of AlGaN/GaN heterostructures

J. Osvald

Institute of Electrical Engineering, Slovak Academy of Sciences,
Dúbravská cesta 9, 841 04 Bratislava, Slovakia
e-mail: elekosva@savba.sk

We tried to assess the influence of interface states between AlGaN barrier and GaN buffer layer and interface states on C-V characteristics of the AlGaN/GaN structure. Interface donor states were modelled with discrete peak distribution in energy with certain energy with respect to the conduction band minimum and also by continuous distribution of interface traps in energy. We found that interface states shift C-V curves to more negative voltages and change the plateau of C-V curves as well as do deep levels situated in AlGaN layer and the deep levels in GaN change the slope of capacitance decrease after the two-dimensional electron gas depletion.

1. Introduction

AlGaN/GaN heterostructures are intensively studied in recent years as a basic tool for high frequency, high power and high temperature active semiconductor devices. Its physics and technology is developed to relatively high level. But, certainly, there are still open questions that deserve a solution. One of them is the influence of surface and interface states and deep levels in the heterostructure on its electrical characteristics. Surface states are localized at the surface of AlGaN barrier layer surface and interface states are at AlGaN/GaN interface. Their influence on electrical parameters is prevalently studied by *I-V* and *C-V* measurements. It was shown, *e. g.*, that a current collapse in AlGaN/GaN high electron mobility transistors (HEMT) is more pronounced for higher deep acceptor density in the buffer layer because the trapping effects have greater influence. DLTS measurement has shown that the traps are mostly located at the interface [1]. The influence of interface states on the *C-V* behaviour of AlGaN/GaN MIS heterostructures was studied by Miczek *et al.* [2,3]. They found parallel shift of *C-V* curves as a consequence of interface states increase, instead of expected change of the curves slope as it is obvious in MOS structures. This effect was explained by the different position of the Fermi level at the interface.

We tried to explore the effect of surface and interface states and also the effect of deep levels on *C-V* curves of heterostructures by simulation of electrical transport in such structures.

2. Theory

For finding out the influence of interface states and deep levels we used drift-diffusion approximation. We have solved simultaneously Poisson and drift-diffusion equations. In order to study how interface states influence the *C-V* curve we insert additional charge in the AlGaN layer depending on the quasi Fermi level position. From this reason another loop was necessary to calculate the potential and charge carrier concentration in the situation where the fixed charge of ionised doping atoms is not constant and depends on external potential. The same procedure was then used for assessment of the influence of the interface states between

AlGaN and GaN layers on the *C-V* curves. Finally, very often discussed deep levels placed in the buffer GaN layer and their influence was also studied.

We assumed in our calculations, that the interface states and deep levels were donor-like, it means that they are neutral when occupied and positive when empty. We simulated interface donor states with discrete peak distribution in certain energy with respect to the conduction band minimum and also by continuous distribution of interface traps in energy.

In the case of deep levels we placed them energetically in 0.9 eV under the conduction band minimum. The energy 0.9 eV is the deep level, which has been ascribed to point defects – nitrogen interstitials or gallium interstitials [4,5].

Simulated structures consist were 25 nm thick AlGaN layer and 75 nm GaN layers. GaN layer was simulated in this thickness because of the limited number of mesh points for obtaining reasonable calculation times. The doping concentration of AlGaN barrier layer and GaN channel layer were assumed to be 1×10^{18} cm^{-3} and 1×10^{17} cm^{-3}, respectively. The sheet concentration of two-dimensional electron gas (2DEG) was assumed to be 8×10^{12} cm^{-2}.

3. Results and discussion

The population of the states is determined by electron quasi Fermi level position since the structure is not in equilibrium. It is especially true for forward bias. By the above-mentioned procedure we obtained *C-V* curves corresponding to low frequency *C-V* curves, because we assumed that all free charges are able to respond not only to polarization bias but also to the measuring signal whatever its frequency is.

For both cases, discrete energy level and continuous energy spectrum of interface states, we found that the existence of the states between AlGaN and GaN layers shifts *C-V* curves toward more negative voltages (Figs.1 and 2). This shift increases with increasing interface states concentration.

The simulation shows also another interesting feature at lower negative and positive voltages. The capacitance value has not a plateau, but having reached its maximum slowly decreases and then increases again at high forward voltages. Similar capacitance peak before the capacitance decrease with increasing reverse bias was observed by Kordoš *et al.* [6] and Irokawa *et al.* [7], but in their case the capacitance plateau was still present in very low negative or even positive voltage. This peak was not observed in our simulation for the

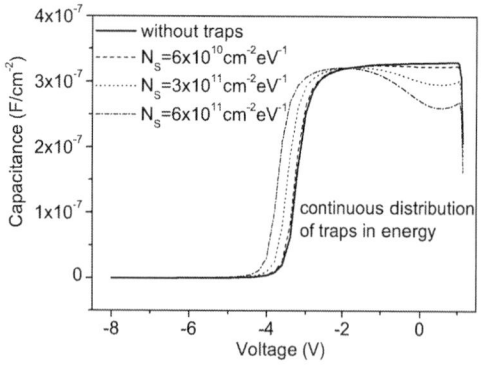 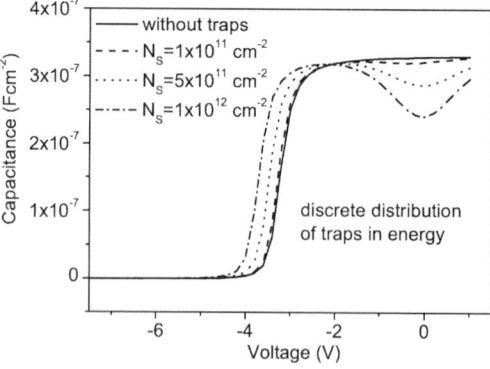

Fig. 1. Continuous distribution of traps   Fig. 2. Discrete distribution of structures

structures without interface states. The only difference between the published experimental results and our simulations is the fact that we have not obtained the capacitance plateau. The mentioned behaviour of *C-V* curves of experimental structures was not observed for measurement frequency 1 MHz, which is normally used for *C-V* characterization but for lower frequencies.

Our explanation of this lower capacitance is the following. If we have 2DEG at the interface, it may react directly to the potential change by its space redistribution and generation and recombination. The response of interface charge reservoir connected with interface states whose concentration depends on the electron quasipotential is not so large in magnitude and that is the reason for capacitance decrease in the voltage region where we could have expected almost constant capacitance connected with 2DEG.

We have also studied the influence of the possible deep levels situated separately in the AlGaN and the GaN layers. Since the shallow doping concentration for AlGaN was higher than the one for GaN also the concentration of the deep levels had to be higher for AlGaN to make a visible effect on this higher charge background. It is seen that the deep levels in AlGaN cause a shift of the *C-V* curve to the left which may be confused with higher 2DEG concentration or higher Schottky barrier height [8]. The slope of the capacitance decrease remains the same as for the situation without the deep levels.

Rather different is the case when the deep levels are situated in GaN channel layer. In Fig. 3 we see that in this case the slope of the capacitance decrease changes and is more expressed in the lower part of the curve. This dependence of the *C-V* on the deep levels concentration is again similar as was found for the *C-V* curves dependence on GaN doping concentration [8] in spite of the fact that in this case the concentration of free charge from the deep levels depends on the external voltage. For lower negative voltage there is again predicted the capacitance valley that could be connected with the low frequency model of the simulation and that is the reason that it is not observed by higher measurement frequencies in experiment.

4.   Conclusion

We have studied the influence of interface states between AlGaN and GaN layers and influence of the deep levels on the shape and behaviour of the *C-V* characteristics of such

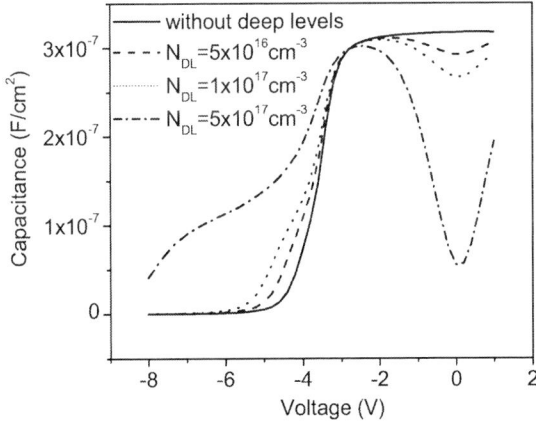

Fig. 3. Deep levels in GaN layer

structures. It was shown that the presence of interface states shifts the C-V curves to more negative voltages. The shift is large for large interface states concentration. For less negative or even positive voltages there appears the valley in the capacitance, which is more pronounced for discrete energy distribution of interface traps.

In principle similar behaviour with the shift of the C-V curves to more negative voltages has been found when we assumed the existence of the deep levels situated in the AlGaN barrier layer. If the deep levels are situated in GaN channel layer the slope of the capacitance decrease is lower than for the structures without the deep levels presence. In both cases the capacitance valley is also observed in less negative voltages.

Acknowledgement

The authors are thankful the financial support received during the development of this work from Slovak Grant Agency for Science under Contract No. 2/0163/09, Agency for Research and Development APVV-0655-07, and The Research-Educational Centre of Excellence VVCE-0049-07.

References

[1] W. Ckickahoui *et al. phys. stat. sol.* (c) **7**, 92, 2010.

[2] M. Miczek, Ch. Mizue, T. Hashizume, and B. Adamowicz, *J. Appl. Phys.* **103**, 104510, 2010.

[3] M. Miczek, B. Adamowicz, Ch. Mizue, and T. Hashizume, *Jap. J. Appl. Phys.* **48**, 04C092, 2009.

[4] F. D. Auret, S. a. Goodman, F. K. Koschnick, J.-M. Spaeth, B. Beaumont, and P. Gibault **73**, 3475, 1998.

[5] A. Y. Polyakov, N. B. Smirnov, A. V. Govorkov, A. V. Markov, S. J. Pearton, N. G. Kolin, D. I. Merkurisov, V. M. Boiko, C-R. Lee, and L.-H. Lee, *J. Vac. Sci. Technol. B* **25**, 436, 2007.

[6] P. Kordoš, D. Gregušová, R. Stoklas, Š. Gaži, and J. Novák, *Solid-St. Electron.* **52**, 973 2008.

[7] Y. Irokawa, N. Matsuki, M. Sumiya, Y. Sakuma, T, Sekiguchi, T. Chikyo, Y. Sumida, and Y. Nakano, *phys. stat. sol. (c)* **7**, 1928, 2010.

[8] J. Osvald, *J. Appl. Phys.* **106**, 013708, 2009.

Effects of soft-UV irradiation on organic thin film transistors with different gate dielectrics.

N. Wrachien[1], A. Cester[1], G. Meneghesso[1], J. Kovac[2], J. Jakabovic[2], D. Donoval[2]

[1] Department Information Engineering, University of Padova,
Via Gradenigo 6\B, 35131, Padova - Italy
e-mail: wrachien@dei.unipd.it

[2] Department of Microelectronics, Faculty of Electrical Engineering and Information
Technology, Slovak University of Technology
Ilkovičova 3, 812 19 Bratislava, Slovakia

We subjected organic thin film transistors with different gate dielectrics to soft-UV irradiation. Irradiation reduces the transconductance on all devices, regardless the gate dielectric employed. However, UV irradiation differently impacts on the drain current: it decreases on devices with hexamethyl-disilazane treated gate dielectric, mainly due to transconductance degradation. Conversely, negative charge trapping dominates over the transconductance degradation on devices with silicon nanoparticles leading to a drain current increase.

1. Introduction

Organic thin-film-transistors (OTFT) are attracting much attention due to their recent achievements in terms of performances [1]. One of the most interesting applications of OTFT is its employment as pixel driver for LCD or AMOLED displays. Because of the intrinsic needs of transparent substrates, the OTFTs of such displays could be subjected to direct sunlight, when they are used outdoor. This is just one of the possible cases in which OTFTs could experience near-UV irradiation. In fact, the sunlight has a non-negligible near-UV component, which might reach intensities (at the sea level) in the order of mW/cm^2 [2,3].

This work aims to investigate the effects of soft-UV irradiation on organic TFTs.

2. Experimental and Devices

We analyzed two sets of top-contact bottom-gate pentacene-based OTFTs, which differ in the gate dielectric stack (Fig. 1). Both devices feature a 40-nm SiO$_2$ layer, thermally grown over a heavily doped n-type silicon substrate. The SiO$_2$ of the first set of devices was treated with hexamethyl-disilazane (HMDS hereafter). A monolayer of 5 nm silicon nanoparticles (SiNPs hereafter) (Meliorum Technologies, Rochester, NY) stabilized with sodium n-dodecylbenzenesulfonic acid was deposited over the SiO$_2$ of the second set of devices.

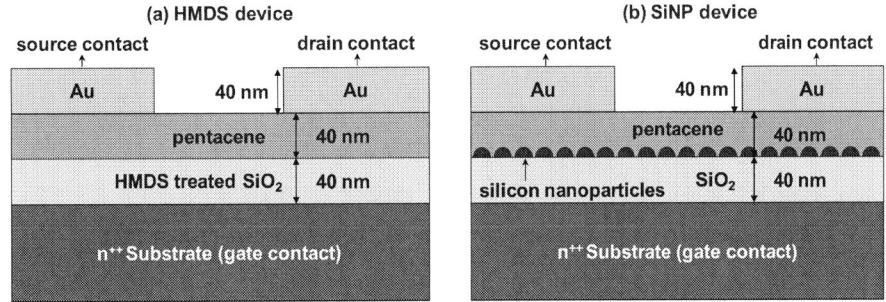

Fig. 1. Cross-section of the devices with HMDS-treated SiO$_2$ (a) and with silicon nanoparticles (b).

After the initial characterization, the devices were subjected to 11 cycles of 2 steps:

1) Irradiation step with grounded terminals. The irradiation time is T_{irr}.
2) Device characterization (I_D-V_{GS}) in dark.

After each cycle, the irradiation time T_{irr} was doubled, with a 4s initial T_{irr} value. We used two irradiation peak wavelengths: 400nm and 385nm. The irradiation intensity was set to ~9 mW/cm^2 for both wavelengths. Measurements were performed in air. Since organic devices may degrade in air, we also repeated the same experiment on reference devices, without irradiation. In this case, the step 1 consisted only of an "idle" step in which the devices were kept with grounded terminals in dark for T_{irr} time. In this way, we can distinguish between the effects of intrinsic drift and UV irradiation on the electrical characteristics.

3. Results and Discussions

In Fig. 2, we plot the I_D-V_{GS} curves measured in the initial characterization and after each irradiation step for HMDS (Fig. 2a) and SiNP (Fig. 2b) at different irradiation levels (dashed lines). For comparison we show in the same figure the effect of the electrical measurement, without irradiation (solid line). Such electrical characteristics still feature some variations, due to either degradation or charge trapping. However, these modifications are much smaller than those observed during irradiation (dashed lines).

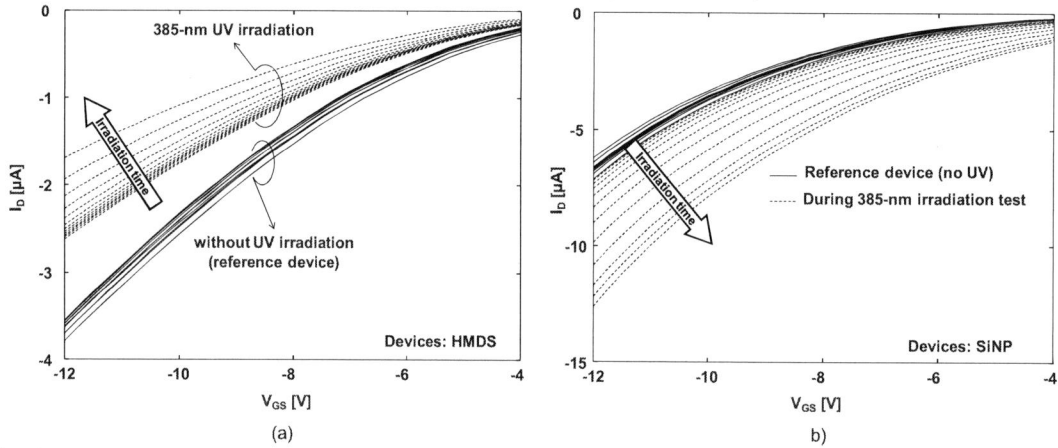

(a)

b)

Fig. 2. Drain current vs. gate voltage characteristics measured at different intervals with (dashed lines) and without (solid lines) irradiation. Fig.a refers to HMDS devices, and Fig.b refers to SiNP devices.

In Figs. 3-5 we plot:

1) The V* variation (ΔV^*), where V* is defined as the V_{GS} required to achieve a fixed drain current (4.5µA for SiNP devices and 1µA for HMDS).
2) The drain current measured at V_{DS}=-8V and V_{GS}=-12V ($I_D^{(M)}$).
3) The transconductance (g_m) at 1µA and 4.5µA for HMDS and SiNP, respectively.

$I_D^{(M)}$ (Fig. 3) and g_m (Fig. 4) are plotted normalized to their initial values.

Without irradiation, the g_m variations are smaller than 1% on HMDS and 2% on SiNP devices, respectively. $I_D^{(M)}$ features a larger variation on SiNP, because of the threshold voltage shift due to the intrinsic charge trapping induced by the measurement itself. At the end of the 11st irradiation cycle, the shift on V* are about 10mV and -500mV on HMDS and SiNP devices, respectively. Conversely, much larger modifications are observed on both kinds of devices when they are subjected to UV-irradiation, with stronger effects at the shorter wavelength. Table 1 summarizes the variations. Noticeably, SiNP and HMDS both feature a substantial g_m drop when irradiated, but opposite $I_D^{(M)}$ and V* variations.

978-1-4244-8574-1/10 $26.00 © 2010 IEEE

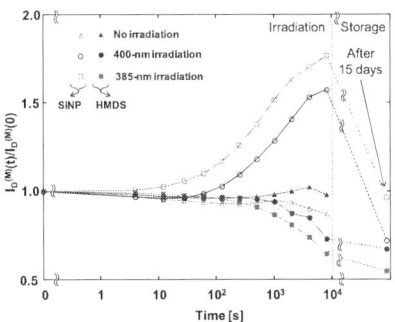

Fig. 3. Evolution of the drain currents evaluated at V_{GS}=-12V and V_{DS}=-8V, normalized to their initial value, for the different devices and irradiation conditions.

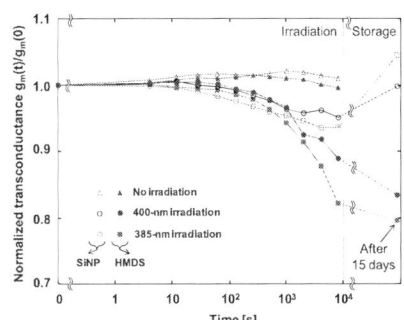

Fig. 4 Evolution of the transconductance evaluated at I_D=1µA (HMDS) and I_D=4.5µA (SiNP) normalized to their initial value, for the different devices and irradiation conditions.

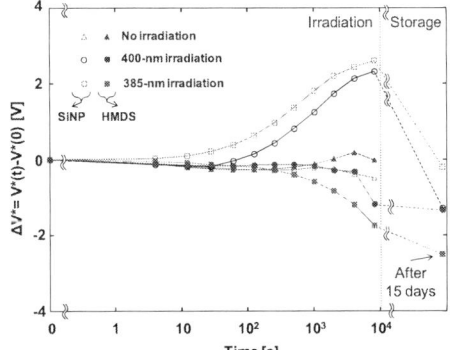

Fig. 5. Variation of V^* for the different devices and irradiation conditions.

TABLE I
VARIATIONS OF THE DEVICE PARAMETERS AT
THE END OF THE IRRADIATION

DEVICE	$I_D^{(M)}$ [%]	g_m [%]	ΔV^* [V]
SiNP – dark	-13	1.0	-0.5
SiNP – irr. @400nm	57	-5.1	2.3
SiNP – irr. @385nm	76	-6.4	2.6
HMDS – dark	-2.3	-0.5	0.01
HMDS – irr. @400nm	-23	-11	-1.2
HMDS – irr. @385nm	-35	-18	-1.7

To assess if irradiation induced any permanent degradation, we also measured the devices 15 days after the last irradiation step. During this time, the devices were stored under vacuum. The last point of each curve of Figs. 3-5 is the value calculated from these measurements. Noticeably, SiNP devices featured a recovery on all the calculated parameters, but HMDS devices exhibited either a negligible variation or even a further degradation.

The very different curves of Figs. 3-5 suggest that the effects on the bulk of the pentacene (if any) are overwhelmed by the variations related to the different gate dielectric of SiNP and HMDS devices. We believe that the main causes of the variations are the negative charge trapping for SiNP devices and photo-induced chemical reactions for HMDS devices.

In fact, the trapped charge produces a flatband voltage shift, inducing the V^* and $I_D^{(M)}$ variation. Charges trapped near the interface might also act as scattering defects, reducing the mobility. Moreover, unstable negative trapped charge neutralization during I-V measurement induces a stretch-out of the I-V curves. As soon as all those charges are removed, the g_m is restored because the lack of mobility degradation and I-V stretch-out. Hence, trapped charge neutralization on SiNP might explain the recovery of not only $I_D^{(M)}$ and V^*, but also g_m.

HMDS devices show no recovery of the electrical characteristics, suggesting that either the charge is much more stably trapped or that UV induces some sort of degradation of the HMDS-treated SiO_2/pentacene interface (reacting either with pentacene, water, or oxygen). In [5] it has been observed that UV-induced direct decomposition of pentacene is unlikely to occur (in vacuum, without other reactants) at least at photon energies up to 25 eV. Indeed, if pentacene degradation had been the main responsible of the g_m drop, we would have observed a strong and stable g_m variation on SiNP devices too. Still, some works [5,6] showed that, in

978-1-4244-8574-1/10 $26.00 © 2010 IEEE   173

presence of oxygen, UV irradiation could enhance the reaction with pentacene. This would induce a strong degradation of the electrical characteristics in both kinds of devices. However, in [6] the authors supposed that the pentacene morphology strongly impacts the combined effect of UV and oxygen, because some crystal structures of pentacene films do not allow for oxygen diffusion. This also might explain the different degradation and recovery kinetics of HMDS and SiNP devices. Incidentally the SiNP OTFTs feature a much better crystalline quality with respect to HMDS OTFTs, as confirmed by AFM measurements [7].

Shorter wavelengths enhance photogeneration of electron-hole pairs or they might induce the onset or enhancement of some photochemical reactions.

As a last remark, the observed strong variations of the electrical characteristics might mark the device failure in few hours. Still, several factors must be taken into account. First, we used an optical power intensity that is 2-3 times the optical power of the UV sunlight spectrum in a clear day. Furthermore, our devices are not encapsulated, hence there is a continuous flux of oxygen and water, which can react and degrade the OTFT performances. A good encapsulation would reduce both the incident optical power, and the flux of reactive substances, strongly enhancing the OTFT lifetime well beyond the hour range.

4. Conclusions

Soft-UV irradiation induces noticeable modifications of the electrical characteristics on our OTFTs, whose entity and stability over time depend on the particular dielectric interface. In particular, the devices with silicon nanoparticles showed large variations mostly due to unstable trapped charge, which anneals in a time as short as 15 days. The devices with HMDS treated SiO_2 showed strong transconductance degradation, which does not recover at least within 15 days. These results highlight that encapsulation and UV filtering are required for these kinds of devices, for a reliable operation over time under sunlight conditions.

Acknowledgement

This work was done in Center of Excellence CENAMOST (Slovak Research and Development Agency Contract. No. VVCE-0049-07) with support of projects APVV-0290-06 and VEGA-1/0689/09.

This work was partially supported by Progetto di Ateneo 2009 – Università di Padova, Italy (Project Number CPDA083941).

References

[1]   C.-C. Kuo; M.M. Payne, J.E. Anthony, and J.E. Jackson, "TES anthradithiophene solution-processed OTFTs with 1 cm^2/V-s mobility," IEEE International Electron Devices Meeting, 2004. IEDM Technical Digest. , pp. 373-376, 13-15 Dec. 2004.

[2]   A. V. Parisi and J. Turner, "Variations in the short wavelength cut-off of the solar UV spectra", Photochem. Photobiol. Sci., 2006, 5, 331–335.

[3]   "Standard Tables for Reference Solar Spectral Irradiances: Direct Normal and Hemispherical on 37° Tilted Surface", ASTM G173-03 2008.

[4]   Tse Nga Ng, M.L. Chabinyc, R.A. Street, and A. Salleo, "Bias Stress Effects in Organic Thin Film Transistors," proc. of 45th annual IEEE International Reliability physics symposium, pp.243-247, 15-19 April 2007.

[5]   A. Vollmer et al., Surface Science 600 (2006), 4004–4007.

[6]   Ashok Maliakal et al., Chem. Mater. 2004, 16, 4980-4986.

[7]   J.Jakabovic, J. Kovac, R. Srnanek, M. Weis, M. Sokolsky, A. Hinderhofer, K. Broch, F. Schreiber, D. Donoval, and J. Cirak, "Pentacene-gate dielectric interface modification with silicon nanoparticles fort OTFTs", IVC-18, Beijing, 25-27 August, 2010.

978-1-4244-8574-1/10 $26.00 © 2010 IEEE   174

Design, preparation and properties of spin-LED structures based on InMnAs

P. Telek[1], S. Hasenöhrl[1], J. Šoltýs[1], I. Vávra[1], M. Držík[2], J. Novák[1]

[1]Institute of Electrical Engineering, Slovak Academy of Sciences, Bratislava, Slovakia
[2]International Lacer Center, Bratislava, Slovakia
Corresponding author: peter.telek@gmail.com

Great advances in the development of III-V diluted magnetic semiconductors materials (DMS) allow for the incorporation of ferromagnetic epitaxial layers into advanced structures. In this contribution, we report on the growth of GaAs/InMnAs layers by metalorganic vapour phase epitaxy (MOVPE) over an AlGaAs/GaAs MQW light-emitting diode structure. In particular, results of electrical and structural characterization of structures are presented. We prepared a single-phase ferromagnetic $In_{1-x}Mn_xAs$ ternary with x close to 0.075 on (100) GaAs substrates using MOVPE. The material exhibited room temperature ferromagnetic behavior with a Curie temperature close to 330K. In addition, all InMnAs ternary samples showed p-type conductivity. The ferromagnetic material was incorporated into four different AlGaAs/GaAs LED structure.

1.   Introduction

Single-phase ferromagnetic $In_{1-x}Mn_xAs$ with x close to 0.1 was recently prepared using MOVPE. The material exhibited room temperature ferromagnetic behavior with a Curie temperature close to 330K, which was independent of the Mn concentration. The surprisingly high T_C was attributed to 'the formation of atomic clusters' [1, 2]. InMnAs is a dilute ferromagnetic semiconductor that is compatible with GaAs/AlGaAs LED technology, and has a Curie temperature T_C above room temperature. Therefore, it is a candidate for a spin aligner material in spin LEDs, as was discussed in recent publications [3, 4, 5].

The spin injection from ferromagnetic material into GaAs is studied by detailed analysis of circular polarization of light from a light emitting diode [5, 6]. A ternary compound material prepared under the condition of a low MnAs molar fraction ($x<0.1$) had a zincblende crystal structure. Above this molar fraction value, a two phase system (i.e. combination of zincblende and hexagonal structure) is usually achieved. To integrate an InMnAs epitaxial layer into a GaAs based LED structure, it is necessary to find an acceptable compromise between two contradictory demands. One is the necessity to incorporate the highest possible MnAs content with the aim to obtain good magnetic properties. The second one is to prevent the creation of hexagonal MnAs clusters inside the InMnAs matrix. The presence of such clusters substantially damages the crystallographic quality of an epitaxial layer with a drastic impact on the LED properties. Considering the above mentioned problems, we decided to prepare InMnAs with $x \sim 0.075$ layer and incorporate it into an AlGaAs/GaAs MQW LED structure.

The LED structure was grown by metalorganic vapor phase epitaxy (MOVPE) on n-type (100)-oriented GaAs substrates with a 400 nm thick n-type GaAs buffer layer. The active region consisted of three 9 nm thick undoped GaAs wells separated by a 24 nm thick AlGaAs barrier sandwiched between two 40 nm thick undoped AlGaAs spacer layers. A p-doped AlGaAs layer followed by a p^+ cap GaAs layer was grown on top of the MQW structure.

2. Structure design and experimental

Epitaxial structures under study were prepared in an AIXTRON AIX 200 low-pressure thermally heated horizontal reactor. Trimethylgallium (TMGa) and trimethylindium (TMIn) were used as the group III precursors, and arsine (AsH₃) as the group V source. As the source of manganese, the bis-(methylcyclopentadienyl) manganese ((MeCp)₂Mn) was used. Palladium diffused hydrogen with a total flow of 6.5 slpm was used as a carrier gas. A set of $In_{1-x}Mn_xAs$ samples with various compositions was prepared with the to aim to study their properties in relation to the increase of the Mn content. In this procedure, the partial pressure of (MeCp)₂Mn was kept constant at a value of 3.59×10^{-2} mbar and the partial pressure TMIn as decreased. The Mn/In ratio in the gas phase increased from 0 to a maximum value of 0,453. This maximal gas phase ratio led to the growth of $In_{1-x}Mn_xAs$ with $x=0.075$. The growth temperature was $Tg=500$ °C.

To integrate an InMnAs epitaxial layer into a GaAs-based LED structure, it is necessary to carefully tune two parameters of the InMnAs layer: Mn content and thickness. An efficient spin aligner would require a thick InMnAs layer with as much Mn as possible. However, the Mn content has to be limited to 12% because it is necessary to prevent the formation of hexagonal clusters in the InMnAs matrix [1]. The lattice mismatch between the GaAs-based heterostructure and the InMnAs layer was very high (up to 7%). The incorporation of such a strained layer into the LED structure led to the formation of misfit dislocations and to a decrease in the quality of the LED structure. The problem can be partially remedied by the insertion of an appropriate spacer layer between the spin aligner layer and recombination region. To cast more light on the problem, we prepared four LED structures with a variable position of the InMnAs layer (see Fig. 1): The reference structure without InMnAs is designated as (A); (B) 300 nm thick InMnAs layer was incorporated directly at the interface between the AlGaAs spacer and p-AlGaAs layer; (C) A 100 nm thick InMnAs layer with $x = 0.075$ was incorporated inside the p-AlGaAs layer. The distance from the interface to the undoped p-type AlGaAs layer was 100 nm; (D) The fourth structure was similar to the third one, only the value of x was higher ($x = 0.10$).

The device processing consisted of conventional LED fabrications steps: (1) preparation of metallic system for ohmic contacts (Au+Zn) on the p-side and (Au+Sn) on the n-side (substrate), (2) annealing of ohmic contacts at 450°C for 180 s in a H₂+N₂ ambient, (3) double reverse process (using negative resist) by dry chemical etching, (4) chip separation (with the chip size of 450x450 μm) and bonding with 21 μm Au wire.

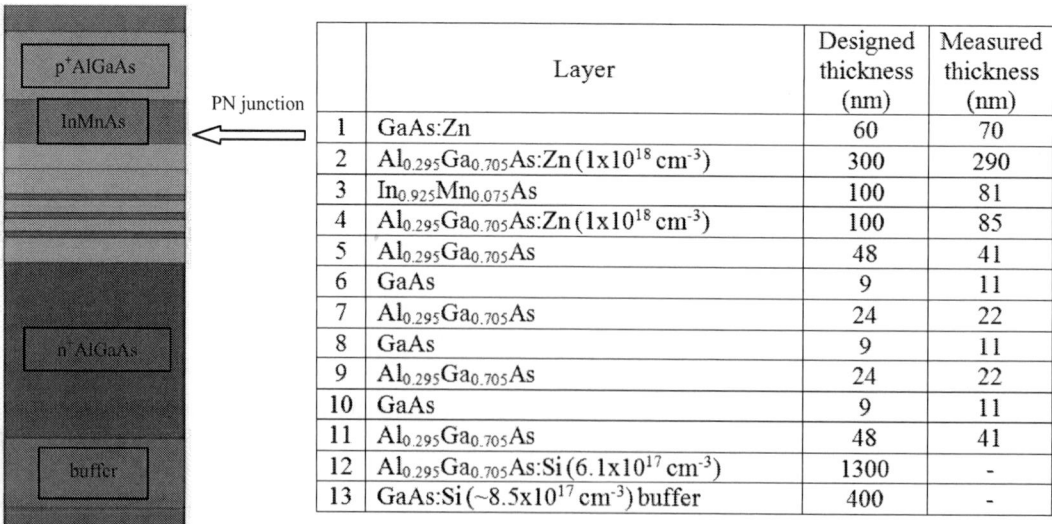

	Layer	Designed thickness (nm)	Measured thickness (nm)
1	GaAs:Zn	60	70
2	$Al_{0.295}Ga_{0.705}As$:Zn $(1 \times 10^{18}$ cm$^{-3})$	300	290
3	$In_{0.925}Mn_{0.075}As$	100	81
4	$Al_{0.295}Ga_{0.705}As$:Zn $(1 \times 10^{18}$ cm$^{-3})$	100	85
5	$Al_{0.295}Ga_{0.705}As$	48	41
6	GaAs	9	11
7	$Al_{0.295}Ga_{0.705}As$	24	22
8	GaAs	9	11
9	$Al_{0.295}Ga_{0.705}As$	24	22
10	GaAs	9	11
11	$Al_{0.295}Ga_{0.705}As$	48	41
12	$Al_{0.295}Ga_{0.705}As$:Si $(6.1 \times 10^{17}$ cm$^{-3})$	1300	-
13	GaAs:Si $(\sim 8.5 \times 10^{17}$ cm$^{-3})$ buffer	400	-

Fig.1 Schmatic cross-sectional view of the LED (structures C, D) with description of layers

3. Results and discussion

All epitaxial layers showed p-type conductivity with Hall hole concentrations between 1 and 4. 10^{18} cm^{-3}. The T_C of the layers was estimated by SQUID measurements as T_C=343 K [7]. The resistivity was in a range of $10^{-2}\Omega$cm, which predicts a low serial resistance after incorporation into the LED structure. We found a strong influence of the increasing Mn content on the surface roughness of the InMnAs epitaxial layers. This InMnAs property was transformed into the surface roughness of LED structure also. The surface of the sample without an InMnAs layer is extremely flat, and we are observe atomic growth steps on it. The rms value is lower than 0.1 nm. On the other hand, the surface morphology of samples MO1065, MO1094, MO1095 (structures B, C, D) follows roughness of the InMnAs layer.

Transmission electron microscopy revealed the presence of strain and dislocation loops in the active parts of LEDs based on structure B (see Fig. 2a). Fig.2b shows that the thinner InMnAs layers led to a substantial improvement in the crystallographic quality. Altough many defects were still present at the bottom interface, no dislocation loop penetrated to the active region and quantum wells. Transmission electron microscopy proved the good dividing lines in the active region of the LEDs (C and D). Improved crystallographic quality of structures C and D led to a substantial improvement in the electrical and electroluminescent

Fig.2 TEM cross-sectional view of the active region of structure B (left) and C,D (right) in the [110] direction

properties of the LED structures. Fig.3 compares reverse I-V characteristics of the four LED structures studied. It was found that the reverse current of reference LED structure A was nearly two orders of magnitude lower than those of structures C and D. The reverse current (measured at V_R=-5V) was about 20 nA; the reverse current at -5V for structures C and D was 1 and 6 μA, respectively. The reverse current in LED structure B, with a very thick InMnAs layer deposited directly at the PN junction, was five orders of magnitude higher than that of the reference structure. We suppose that crystallographic defects were responsible for such an increase in the reverse current.

All samples exhibited room temperature electroluminescence maxima of varied intensity at an energy of 1.467 eV, in accordance with the design of the quantum wells. No influence of the InMnAs composition on the energy of emitted photons was observed. The reference LEDs (A) exhibited $1,5 \times 10^{-4}$ W of output power at I_F = 50 mA. The (B) structure exhibited no measureable level of output power, but the adjusted (C and D) structures showed the output power at level of 6×10^{-5} W at 50 mA forward current.

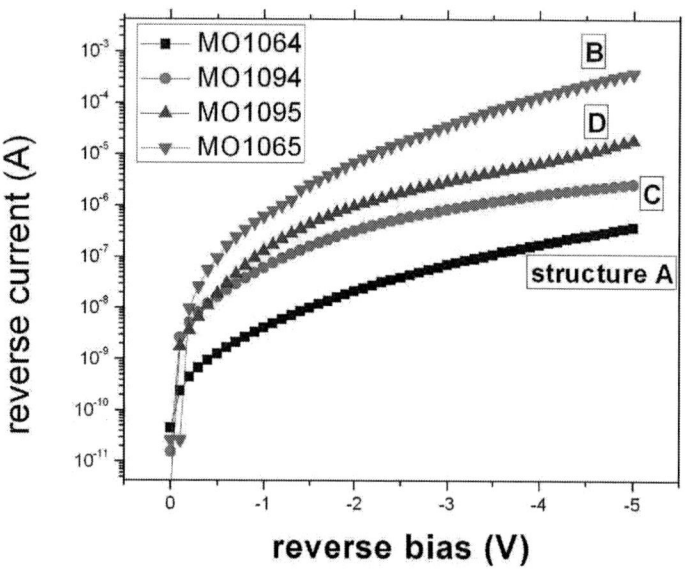

Structure C was used for the preliminary polarization measurement at room temperature. The degree of polarization close to 3% was estimated at magnetic fields between 0.5 and 0.7 T, and a driving current of 50 mA. A detailed description of those experiment will be published elsewhere.

Fig.3 Reverse I-V characteristics of the LED structures

4. CONCLUSIONS

The influence of the incorporation of a thin InMnAs epitaxial layer on the properties of an MQW LED structure was studied by means of TEM, AFM, room temperature photoluminescence, and van der Pauw method. The results obtained showed that a 100 nm thick InMnAs layer was successfully incorporated into the LED structure. A substantial influence on the surface morphology and I-V characteristics of LED structures was observed. The spin-polarization was detected by measuring, the circular polarization electroluminescent light from the top of the LED sturcure where external magnetization was applied.

Acknowledgements

This research was partially supported by VEGA Agency projects No. 2/0081/09 and 2/0007/09 and by APVV project VVCE-0049-07 CENAMOST. Liquid nitrogen for the experiments was sponsored by the U.S. Steel, Košice.

References

[1] A.J. Blattner and B.W. Wessels: Applied Surf. Sci. **221** (2004) 155
[2] J. Novák, J. Šoltýs, P. Eliáš, S. Hasenöhrl and I. Vávra, Study of the growth and structural properties of InMnAs dots grown on high-index surfaces by MOVPE, Materials Science in Semiconductor Processing (in press)
[3] W. Löffler, D. Tröndle, J. Fallert, E. Tsisishvili, H. Kalt, D. Litvinov, D. Gerthsen, J. Lupaca-Schomber, T. Pasow, B. Daniel, J. Kvietkova and M. Hetterich: phys. stat. sol. (c) **3** (2006) 2406
[4] S.V. Zaitsev, V.D. Kulakovskii, M.V. Dorokhin, Yu.A. Danilov, P.B. Demina, M.V. Sapozhnikov, O.V. Vikhrova and B.N. Zvonkov: Physica E **41** (2009) 652
[5] W.Van Roy, P. Van Dorpe, J. De Boeck, G. Borghs: Materials Sci. and Engineering **B126** (2006) 155-163
[6] M. Ramsteiner, Y.H. Hao, A. Kawaharazuka, H.J. Kastner, R. Hey, L. Daweritz, H.T. Grahn and K.H. Ploog: Electrical spin injection from ferromagnetic MnAs metal layers into GaAs, Phys. Rew. **B66** (2002) 081304(R)
[7] J. Novák, I. Vávra, Z. Križanová, S. Hasenöhrl, J. Šoltýs, M. Reiffers and P. Štrichovanec: Applied Surf. Sci. **256** (2010) 5672

STUDY OF ZnO FILMS GROWN WITH DIFFERENT DOPANTS - PHYSICAL PROPERTIES AND THEIR COMPARISON

[1]L. Prušáková, [1]M. Netrvalová, [1]P. Šutta

University of West Bohemia, New Technology Research Centre
Univerzitni 8, 306 14 Plzen, Czech Republic
e-mail: lprusak@ntc.zcu.cz

Transparent and electrically conductive (TCO) thin films of ZnO:Al, ZnO:Ga and ZnO:Sc, used in solar cells as well as optoelectronic devices, have been successfully deposited by rf magnetron sputter deposition using ZnO(98%) / X_2O_3(2%) ceramic target, X ∈{Al, Ga, Sc}, in the inert atmosphere of argon. In this contribution we focused on the changes in physical properties and their comparison in dependence on doping element. The XRD analyses, four probe measurements and UV-VIS spectroscopy were applied to investigate the structure (texture, lattice stress, grain size), electrical (resistivity) and optical (transmittance, optical band gap) properties of TCOs on the glass substrates.

1. Introduction

Transparent conducting oxides (TCOs) have long been a subject of various investigations due to its unique physical properties and applications in commercial devices. Among TCOs, zinc oxide (ZnO) is one of the most promising materials for the fabrication of the next generation of optoelectronic devices in the UV region and optical or display devices. It exhibits numerous characteristics that may enable its efficient utilization in many novel devices such as sensors [1], surface acoustic devices, transparent electrodes and solar cells. Zinc oxide has a wide band gap of 3.37 eV [2]. As-grown, nominally undoped, ZnO usually demonstrates n-type conductivity due to the presence of either Zn interstitials or O vacancies [3]. Unfortunately the as-grown films are not suitable for device applications as the resistivity is too high and the reoxidation of the Zn rich films at ambient temperatures removes the source of conductivity. Undoped ZnO thin films are not stable due to changes in the surface conductance under oxygen chemisorptions and adsorptions. ZnO can be doped with a wide variety of ions to meet the demands of several application fields. Typical dopant elements (F, B, Al, Ga, In, Sn, Sc, etc.) have been used to produce conducting films of ZnO. The ZnO doping is achieved by replacing Zn^{2+} atoms with atoms of elements of higher valence such as In^{3+}, Al^{3+}, Sn^{4+}, etc.

These types of oxide materials have been prepared with many techniques such as sputtering [4], MOCVD, vapour transport, pulsed laser deposition, spray pyrolysis. Compared to other deposition methods, ZnO films obtained by sputtering can be deposited on large areas at low substrate temperatures and exhibit good adhesion to the substrate.

In this work, optical and electrical characteristic of ZnO films with Al, Ga and Sc dopants have been investigated to reveal the effect of their incorporation and to make a comparison between them. We study the effect of different doping elements on the crystalline structure, optical transmittance and electrical resistivity of doped ZnO thin films prepared by rf magnetron sputtering for use in photovoltaic applications, specially as front transparent contact and back reflector.

2. Experimental details

The ZnO:X films under study, where $X \in \{Al. Ga. Sc\}$, were prepared by 13.56 MHz radio frequency (rf) magnetron sputtering using BOC Edwards TF 600 coating system. Thin films with thickness of 350 nm were sputtered on Corning Eagle 2000 glass substrates at two temperatures (T_S = room temperature (RT) and 100°C) using the high purity ZnO ceramic target with 2% X_2O_3. The sputtering rf power was kept at 400W, argon working gas at constant pressure of 1 Pa was used. The target to substrate distance was kept at 150 mm.

The structural properties of the films were studied by X-ray diffraction (XRD) using an automatic Philips X-ray powder diffractometer X´pert PRO equipped by a fast semiconductor detector Pixcel. Copper Kα radiation (λ = 0.154 nm) was used. The XRD patterns were recorded using Bragg-Brentano geometry and the diffraction angle 2θ varied from 25 to 75 degrees. Evaluation of real crystalline structure (biaxial lattice stress $<\sigma_1 + \sigma_2>$, average micro-strain $<\varepsilon>$ and average crystallite size $<D>$) was performed.

For evaluating the optical properties of the films the transmittance spectra in the range of 190 to 1100 nm wavelength were recorded using the SPECORD 210 spectrophotometer. Integrated transmittance was calculated in the range of 400 to 1100 nm wavelength. The electrical resistivity of the films was obtained by four point probe measurements and their thickness was measured by the single profile method using the nanoindentor XP.

3. Results and discussion

The XRD measurements indicate that all investigated ZnO films have polycrystalline structure and a strong preferred orientation of crystallites in the [001] direction along the c-axis perpendicular to the substrate surface. The XRD revealed the presence of a compressive stress in all ZnO films. which is characteristic for films prepared by sputtering techniques. Fig. 1 shows the diffraction patterns of the Al, Ga and Sc doped ZnO films deposited at room temperature (RT) and 100°C.

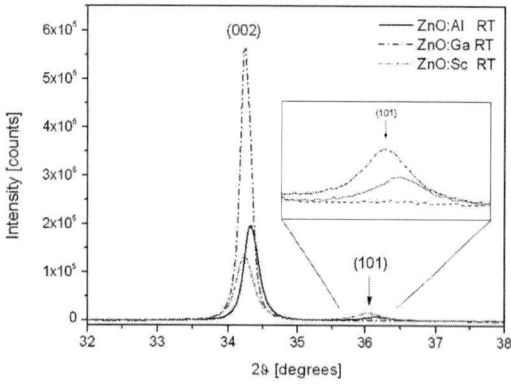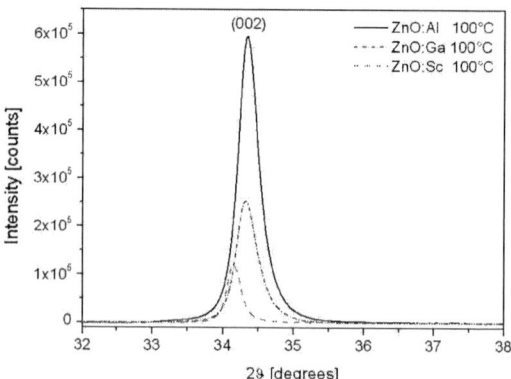

Fig. 1a: Diffractions patterns of Al, Ga and Sc doped ZnO films deposited without heating of the substrate.

Fig. 1b: Diffraction patterns of Al, Ga and Sc doped ZnO films deposited with substrate heated at 100°C.

The position, height, integrated intensity and full-width in half maximum are the main four parameters that characterize the diffraction lines. We used a procedure utilizing the integral breadth of a diffraction line for determining the average size of crystallites $<D>$ and micro-strains $<\varepsilon>$. For determination of the biaxial lattice stress $<\sigma_1 + \sigma_2>$ the shift of the (002) line was evaluated (Tab. 1).

Tab. 1: Summary of structural, optical and electrical parameters of ZnO films.

Dopping element	Deposition temperature	Resistivity	T	Band-gap	$<D>$	$<\sigma_1+\sigma_2>$	$<\varepsilon>*10^3$
$X \in \{Al.Ga.Sc\}$	[°C]	[Ω.cm]	[%]	[eV]	[nm]	[GPa]	[-]
ZnO:Al	RT	0.394	88.8	3.04	85	-1.174	1.91
	100	0.065	88.9	3.03	80	-1.057	2.13
ZnO:Ga	RT	3.05	91.38	3.01	91	-2.747	1.25
	100	0.036	88.6	3.02	106	-1.235	3.39
ZnO:Sc	RT	31.32	89.83	3.04	51	-2.074	2.32
	100	0.524	90.63	3.04	84	-2.855	1.81

There are two dependencies valuable for all doping elements: a) Decrease of resistivity with decreasing of the biaxial lattice stress; b) Decrease of transmittance with decreasing of the biaxial lattice stress (Fig. 2).

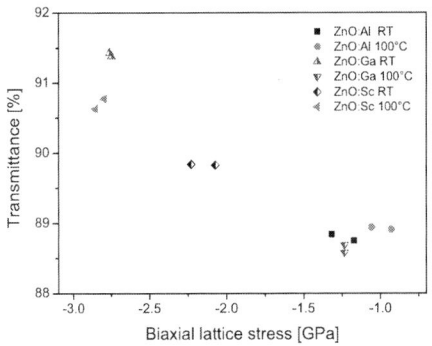

Fig. 2: With decreasing compressive biaxial stress in the ZnO structure the transmittance of the films decreases.

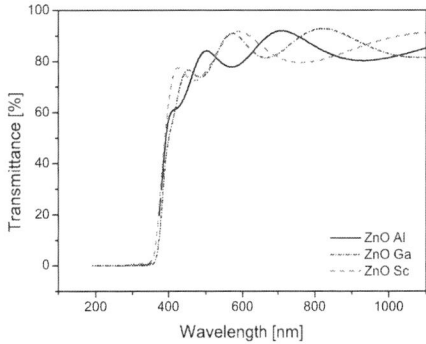

Fig. 3: Transmittance spectra of Al, Ga and Sc doped ZnO films which all exhibit integral transmittance higher than 88%.

In case of Al doping the 100°C is not as sufficient temperature to influence the size of crystallites $<D>$ as well as the biaxial lattice stress $< \sigma_1 + \sigma_2>$ which slightly decreased resulting in decrease of the resistivity. On the other hand ZnO doped with Ga demonstrated considerable shift of $< \sigma_1 + \sigma_2>$ which decreased twice after substrate heating comparing the RT and 100°C temperature during the deposition. It caused decrease of the transmittance and on the other hand drop-off the resistivity in two orders of magnitude. In case of Sc doping the $<\sigma_1 + \sigma_2>$ increased after substrate heating. It influenced the transmittance which slightly increased. Despite increase of biaxial lattice stress the resistivity of the films dropped down in two orders of magnitude from 31 to 0.5 Ω.cm due to a combination of mainly two effects: increase of the size of crystallites $<D>$ and higher concentration of free carriers.

All of the investigated ZnO films exhibit more than 88% transmittance in the visible and near infrared range (Fig. 3) and the optical band-gap of 3 eV (Tab. 1) which is sufficient for application in opto-electronic devices.

3.   Conclusions

Transparent and conductive ZnO films doped with aluminium, gallium and scandium were deposited by rf magnetron sputtering onto glass substrate of RT and 100°C. The effect of different dopants and the substrate temperature on the structural and opto-electrical properties of ZnO films was investigated. Higher temperature during the deposition caused that more of the doping atoms get into substitute position of ZnO lattice and acount of free carriers increased. For all Al, Ga and Sc doped ZnO films the resistivity decreased with heating of the substrate. Despite, the resistivity of the ZnO:Sc films is weak in comparison with other used dopants. It could be explain with first principle calculations which indicated that Sc ions form larger atomic volume in the film despite equal ionic radius of Sc and Zn atoms. This could induce non zero magnetic moments which can influence the scattering of the charged particles of the film [5].

The transmittance have changed in accordance with the dependence on the biaxial lattice stress $<\sigma_1 + \sigma_2>$ described above (Fig. 3). From performed experiments it can not be exactly concluded which of these dopants is more or less suitable for use in photovoltaic applications.

Acknowledgement

This work was supported by the project of Ministry of Education, Sports and Youth of the Czech Republic No. 1M06031.

References

[1]   S. C. Minne, S. R. Manallis, and C. F. Quante: „Paralel atomic force microscopy using cantilevers with integrated piezoresistive sensors and integrated piezoelectric actuators," *Applied Physics Letters*, vol. 67, no. 26, pp. 3918-3920, 1995.

[2]   H. Morkoc, U. Ozgur: Zinc Oxide – Fundamentals, Materials and Device Technology, Wiley-VCH Verlag, Weinheim, Germany 2009, ISBN 978-3-527-40813-9.

[3]   A. F. Kohan, G. Ceder, D. Morgan, C. G. Van de Walle, Phys. Rev. B 16 (2000) 15 019.

[4]   V. Tvarozek, S. Flickyngerova, I. Novotny, A. Rehakova, P. Sutta , M. Netrvalova, P. Gaspierik, L. Prusakova, P. Ballo, E. Vavrinsky: „Influence of spatial sputtering distribution on TCO thin film properties", In ICTF 14: 14th International Conference on Thin Films and Reactive Sputter Deposition, 17.-20.11.2008 Ghett, Belgium. ISBN 978-90-334-7347-0.

[5]   P. Ballo, L. Harmatha, V. Tvarožek, P. Šutta: "First-principles study of p-type doping in ZnO", ČFVK – 3. česká fotovoltaická konference, 10.-13.11.2008 Brno.

978-1-4244-8574-1/10 $26.00 © 2010 IEEE

Synthesis and Doping of Zinc-Oxide Thin Films by RF Sputtering and Ion Implantation

M. Milosavljević[1], D. Peruško[1], V.Milinović[1], P. Gašpierik[2], I. Novotný[2] and V. Tvarožek[2]

[1] VINČA Institute of Nuclear Sciences, Laboratory for Atomic Physics, P.O.B. 522, 11001 Belgrade, Serbia

[2] Department of Microelectronics, Slovak University of Technology, Ilkovičova 3, 812 19 Bratislava, Slovakia

e-mail: momirm@vin.bg.ac.rs

Structural and electrical properties of ZnO:Ga thin films synthesized by RF sputtering and N-doped by ion implantation were studied. The ZnO films with an addition of 2 at% of Ga were deposited on Corning glass substrates to a thickness of 0.5 μm. Nitrogen ions were implanted at 180 keV, to $10^{15} - 10^{16}$ at/cm², at room temperature. It was found that the as-deposited ZnO films have a polycrystalline columnar structure. Incorporation of Ga resulted in their electrical resistivity of the order of 1 Ωcm, electron concentration of 10^{19} cm⁻³ and Hall mobility in the range of $1 cm^2 V^{-1} s^{-1}$. After implantation with nitrogen the polycristalline structure was preserved, though individual columns appeared as discontinued. On the other hand, incorporation of N species into the ZnO lattice can result in both acceptor and donor states, which at this stage of experiments could not be fully clarified.

1. Introduction

Zinc-oxide thin films are a promising transparent conductive oxide (TCO) material, interesting for diverse applications, such as transparent contacts for liquid crystal displays (LCDs), organic light-emitting diodes (OLEDs), solar cells, antistatic coatings, touch display panels, flat panel displays, heaters, defrosters, optical coatings, and various other optoelectronic devices. ZnO is a direct wide bandgap semiconductor (3.37 eV at room temperature), with a large exciton binding energy of 60 meV, and high temperature (melting point of 1975°C) and chemical stability. Undoped zinc oxide exhibits intrinsic n-type conductivity. Turning n-ZnO into p-type is more difficult due to high activation energy of acceptors, low solubility of acceptor dopants and self-compensating process at acceptor doping. Yamamoto et al. [1], proposed a co-doping technique, which involves a simultaneous doping with donors and acceptors, as a way to enhance the solubility of nitrogen and to introduce more NO acceptors into the film. Low solubility of N_2 in ZnO can be increased by co-dopants, group III (Al, Ga), which create complexes with N and enhance the formation of N-accceptors in p-type ZnO. Sputtered ZnO films indicated p-type behavior by single nitrogen doping (ZnO:N) and aluminum-nitrogen (ZnO:Al:N) co-doping [2, 3]

Because of competition between donor´s and acceptor´s in N-built complexes, the manufacturing of p-type ZnO-based films requires to keep the technological parameters within a narrow region of tolerance. Ion implantation is a proven technique for doping of semiconductor materials, but to our knowledge it was not studied yet for nitrogen doping of zinc-oxide thin films. The aim of this work was to synthesize ZnO:Ga thin films by RF sputtering and to perform their N-doping by ion implantation. All ZnO-based thin films were subjected to structural and electrical characterization.

2. Technology and Experimental

RF diode sputtering, ZnO+2% Ga_2O_3 target of 102.4 mm in diameter, and Ar working gas was used for deposition of ZnO:Ga thin films, to a thickness of 0.5 μm on Corning glas substrates, at room temperature. The thin film morphology was studied by a SEM Philips XL30 (cross-section) and a X-ray diffractometer (crystal texture). Electrical properties were measured by Van der Pauw method. The ZnO:Ga thin films were impanted with 180 keV nitrogen ions, to the doses from $10^{15} – 10^{16}$ at/cm^2. The incident ion beam was normal to the sample surface, and the samples were held at room temperature during irradiation. Post-implantation anneals were done at 500°C, for 15-30 min.

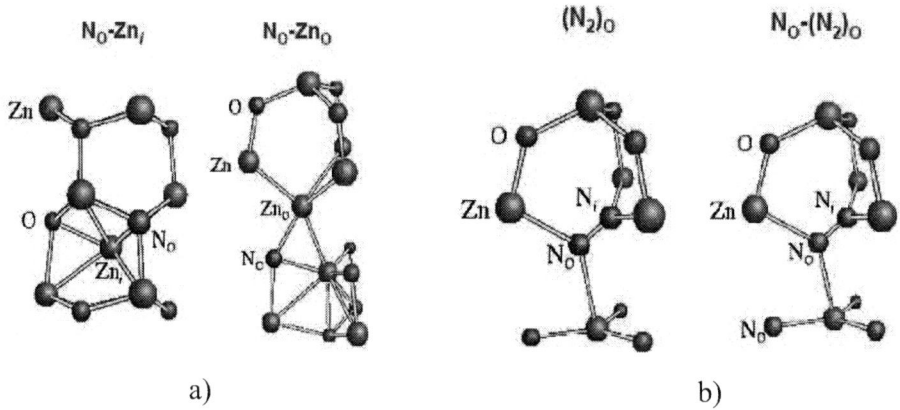

a)   b)

Fig. 1. a) N_O–native defect complexes, b) Defects composed from N

The aim of implanting nitrogen was to study the viability of this technique for doping of ZnO. Incorporation of N in the lattice can be described by the model presented in Fig. 1. [4] The difficulties in achieving a stable p-type ZnO:N are assigned to the compensation of NO acceptors by native and unintentionally introduced defects, the complexes that NO acceptors form with those defects, and purely N composed defects. The most favorable NO-native defect complexes are NO- Zni and NO- ZnO. The deffects purely composed from N are the nitrogen molecules (N2)O and NO- (N2)O complexes.

3. Results and discussion

Fracture cross-sectionial SEM image, taken from an as-deposited ZnO:Ga thin film is shown in Fig. 2. It can be seen that the film grows in form of a polycrystalline columnar structure. The width of individual columns is up to ~ 100 nm, and they stretch from the substrate to the surface. The polycrystalline structure was preserved upon ion irradiation, although with somewhat perturbed structure, similar as we observed earlier in case of metal-nitride thin films [5]. Doping by Ga produced n-type films with resistivities in order of 1 Ωcm, electron concentration of 10^{19} cm^{-3} and Hall mobility in the range of 1 cm^{2}V^{-1}s^{-1}.

978-1-4244-8574-1/10 $26.00 © 2010 IEEE

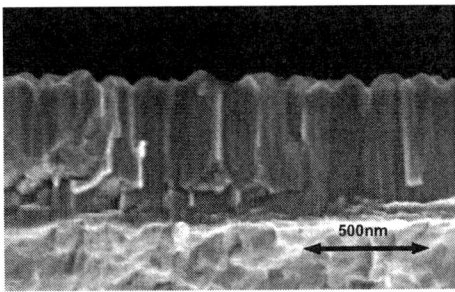

Fig. 2. Fracture cross-sectionial SEM image of ZnO:Ga thin film

The results of Monte Carlo simulations using the SRIM code [6], for ion irradiation of 0.5 μm ZnO:Ga film with 180 keV N^+ ions are plotted in Fig. 3. It is seen that the ion projected range R_p is around 340 nm, but the maximun number of created vacancies around the mid depth of the deposited layer. The maximum of the impact ion energy loss in nuclear collisions also coincides with mid depth of the layer.

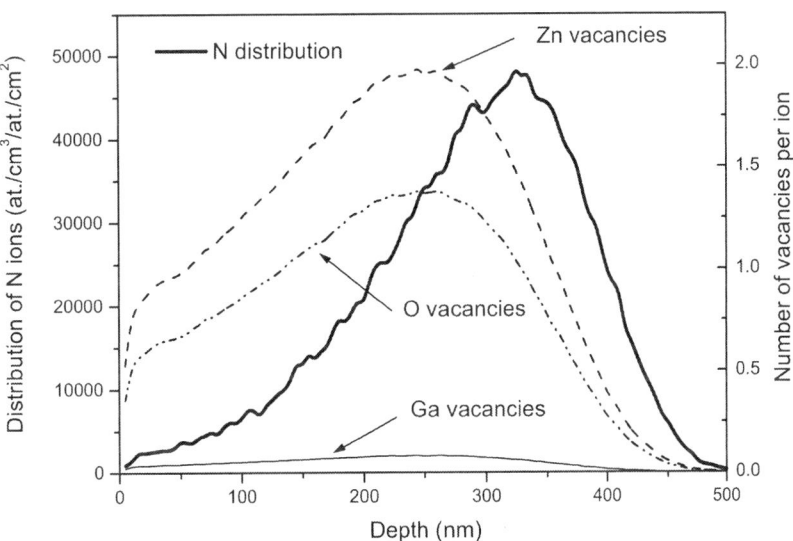

Fig. 3. Simulated plots of the implanted N distribution and of the created Zn, O and Ga vacancies

The total number of created vacancies in the film is 982/ion, total target displacements 1010/ion, and total replacement collisions 28/ion. Based on this it can be concluded that for the applied relatively high N ion doses induce a total rearrangement of the atomic species within the irradiated ZnO:Ga films. Heavier Zn atoms suffer more collisions and displacements compared to the lighter O atoms. Collisions and displacements of Ga are substantialy lower because of their low concentration of 2 at.%. It may be speculated that the impact N ions would tend to replace Zn in the lattice, due to a larger number of created vacancies. On the other hand, they could more easily replace O, because of a closer atomic and mass numbers. However, the competing processes in atomic rearrangements arrise from the chemical driving forces between the thin film constituents. Indeed, chemical recombination is more rapid than ion beam destruction of the crystal lattice, which is evident from the preserved policristalline structure of the films. Hence, it is not so straight forward to predict the dominant displacement/replacement processes, or in such way the incorporation of donor or acceptor states in the ZnO:Ga films.

The results of electrical measurements indicated that N-ion implantation induced some changes in the electrical parameters of the films, but further experiments are needed to clarify between the donor or acceptor states that are created in the films.

4. Conclusions

The presented results have shown the complexity of N-doping of ZnO:Ga thin films. Ion implantation doping is promising, but it requires further studies in order to distinguish between the induced donor and/or acceptor states. New routes of investigations will include ion implantation at an elevated substrate temperature, in order to favor one type of nitrogen inclusions in the chemical bonding with respect to the other. Further attention will also be addressed to post-implantation annealing.

Acknowledgments

This work was supported by the APVV project SK-SRB-0012-09, SK VEGA project 01/0220/09 and MNTR RS project 141013.

References

[1] T. Yamamoto, *Thin Solid Films* **420-421**, 100, 2002.
[2] V. Tvarozek, K. Shtereva, I. Novotny, J. Kovac, P. Sutta, R. Srnanek, and A. Vincze, *Vacuum* **82**, 166, 2007.
[3] K. Shtereva, V. Tvarozek, P. Sutta, and I. Novotny, Chapter 14, 211-234, In: *Micro Electronic and Mechanical Systems*, Publisher: INTECH, 2009.
[4] E.Ch. Lee, Y.S. Kim, Y.G. Jin, and K.J. Chang, *Phisica* B **912**, 308, 2001.
[5] M. Milosavljević, D. Peruško, V. Milinović, Z. Stojanović, A. Zalar, J. Kovač, and C. Jeynes, *Journal of Physics D: Applied Physics* **43**, 065302, 2010.
[6] J.F. Ziegler, J.P. Biersack, and U. Littmark, *The Stopping and Range of Ions in Solids*, Pergamon, New York, 1985.

MO CVD growth of ZnO with different growth rate

D. Nohavica[1,2], P. Gladkov[1,2], J. Grym[1] and Z. Jarchovsky[1]

[1]Institute of Photonics and Electronics, Academy of Sci. of the Czech Republic and
[2]Institute of Physics, Academy of Sci. of the Czech Republic

Low pressure apparatus combining N_2O plasma, DEZn transported in argon and UV irradiation of the deposition zone has been used. ZnO layers were deposited on Si (100) and GaP (111) substrates. Best quality layers were deposited on Si substrates. Growth rate was changed in the range 2 to 90 µm/hour. Surface morphology at smaller growth rate was regular nanowalls type on both Si and GaP substrates. More complex morphology containing longer microrods and pyramids was obtained at high growth rate. Photoluminescence of the samples demonstrated improvement of the ZnO quality when the growth rate increased.

1. Introduction

Zinc oxide is frequently used in several areas of technology including optoelectronics, sensors and catalysis [1–3]. Researchers have demonstrated that ZnO films can be prepared on both single crystal and amorphous substrates [4]. It is worthy to investigate high-quality self-textured ZnO films synthesized on different kinds of substrates. The ZnO films can be also prepared by many methods, such as magnetron sputtering [5], PLD[6], MBE [7], and metal-organic chemical vapor deposition (MOCVD) [8]. MOCVD is a commonly used technique for the epitaxial growth of II–VI materials, such as ZnO. In our laboratory the special research variant of the MOCVD apparatus combining possibility of plasma exciting oxidizer like N_2O, separate branch with DEZn in Ar, UV irradiation of the deposition area, low pressure operation and inertless infrared radiation furnace was developed. In this study we investigate influence of the growth rate on morphology of ZnO deposited on Si (100) and GaP (111) substrates. Photoluminescence study was used for comparison of the growth condition influence on quality of the deposit.

2. Experimental

Schematic view of the reactor and furnace configuration is depicted in Fig 1. Oxidizer input with two electrodes for plasma generation is arranged from the side opposite to

 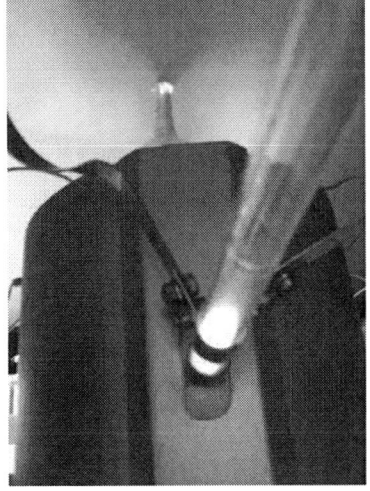

Fig. 1. One side view of the growth apparatus during ZnO deposition and N_2O plasma related illumination in furnace wall.

ultraviolet mercury lamp. Substrates are positioned on quartz holder adjustable parallel or perpendicular to UV irradiation. In Fig.1 the plasma related light in furnace wall is visible. Growth rate was adjusted to changing the vapor phase composition and to obtain the highest growth rates, also the substrate position in gas stream was optimised simultaneously. Surface morphology of the ZnO/Si (100) and ZnO/GaP(111) structures at relativelly slow growth rate app. 1-3 µm/ hour are nanowalls type. The quantity of rods with diameters up to 500 nm in layers morphology increases when the growth rate increases up to 80 µm/hour. In the surface morphology of more than 10 µm thick ZnO layers on Si (100) and GaP (111) dominates the hexagonal pyramids with diameter up to 3000 nm, [9]. Generally the surface morphologies of the thick ZnO layers on both substrates are very similar. First part of this ZnO/GaP structures is polycrystalline with small grains dimension. Low temperature photoluminescence studies of the prepared layers confirm significantly deteriorate optical properties ZnO layers prepared on GaP substrates. Similarity in final morphology of the ZnO on both substrates corresponds to the spontaneous cooperative action of numerous ZnO molecules and long-range spatial correlation under non-equilibrium conditions similarly with CdO [10]. It affects morphology of the second stage of the ZnO deposits. Another evidence of the same mechanisms illustrates Fig 3. were identical morphology of ZnO was observed on two perpendicular planes of the Si substrate. Transition between microrods and hexagonal pyramids (Fig 4. and Fig 5.) is a consequence of the supersaturation increases when micro or nanorods dominated. Increases of supersaturation widen the rods diameter due to anisotropic growth of ZnO crystals or by unidirectional growth of ZnO single crystals due to screw dislocation,[11]. Dimension of the idividual nanoparticles of ZnO prepared in described apparatus at reaction pressure 15 torr approaches 20 nm, (see Fig 7.).

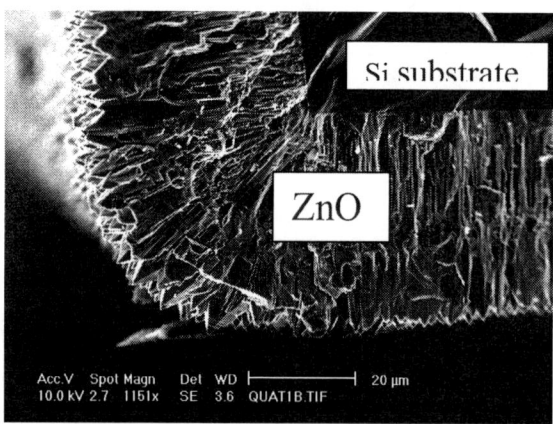

Fig 3. Identical ZnO morfology on two perpendicular plane sof the Si substráte.

Fig 6. a/ and Fig 6. b/ show near band edge (NBE) photoluminescence spectra at 4 K from ZnO/Si films grown at different conditions. Sample in the Fig. 6. a/ was grown with velocity 5µm/hour and sample in Fig 6. b/ with growth velocity 80 µm/hour. The spectra are recorded with 325 nm HeCd laser and excitation densities of 5 W.cm^{-2}. A strong luminescence in the UV region corresponds to the known PL lines for epitaxial and bulk ZnO. The observed NBE PL has a structure corresponding to the lines of bound excitons (BE), two electron satellites (TES) and their LO phonon replica (~72 meV). In our best samples (grown at high growth rate 80 µm/hour) the BE peak at 3.36 eV exhibits FWHM of ~4 meV, while the intensity of the TES peak is substantially lower compared to the BE. Very low PL intensity is observed at energies corresponding to the DAP transitions merging with the BE-2LO peak. All these characteristics are signs of high quality micro-crystalline ZnO. The abnormal high intensity of the TES visible in Fig 6.a/ we ascribe tentatively to a lower quality microcrystalline structure i. e. high concentration of point defects.

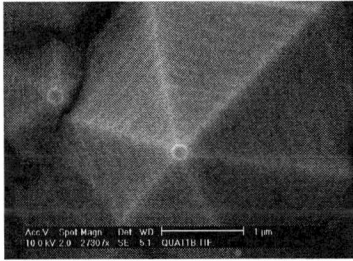

Fig 4. Pyramidal structure with diameter approaching 3000 nm.

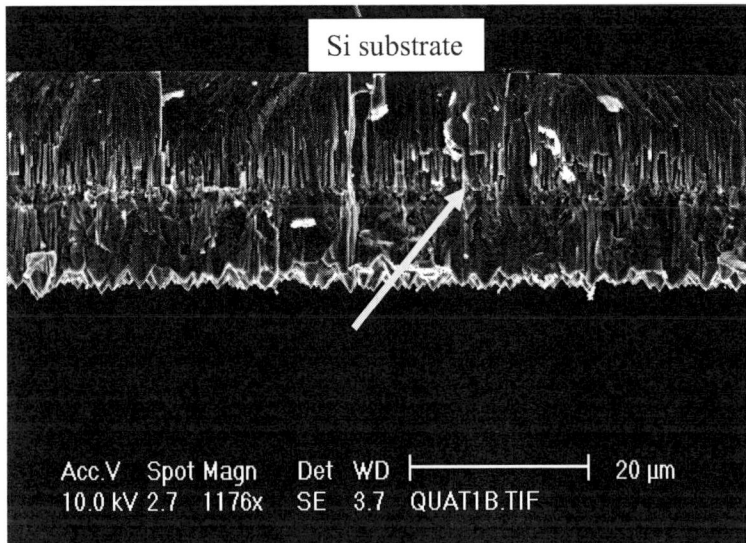

Fig 5. Transition between micro or nanorods and hexagonal pyramids

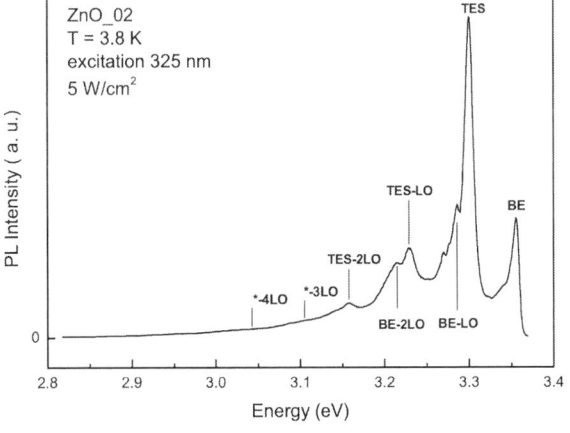

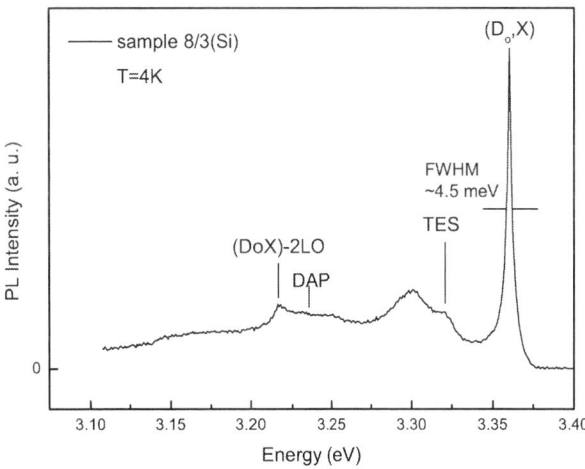

a/

b/

Fig 6. Photoluminescence spectra at 4 K from ZnO/Si(100) films grown at different conditions. Sample a/ was grown at grow rate 5μm/hour and b/ at 80μm/ hour.

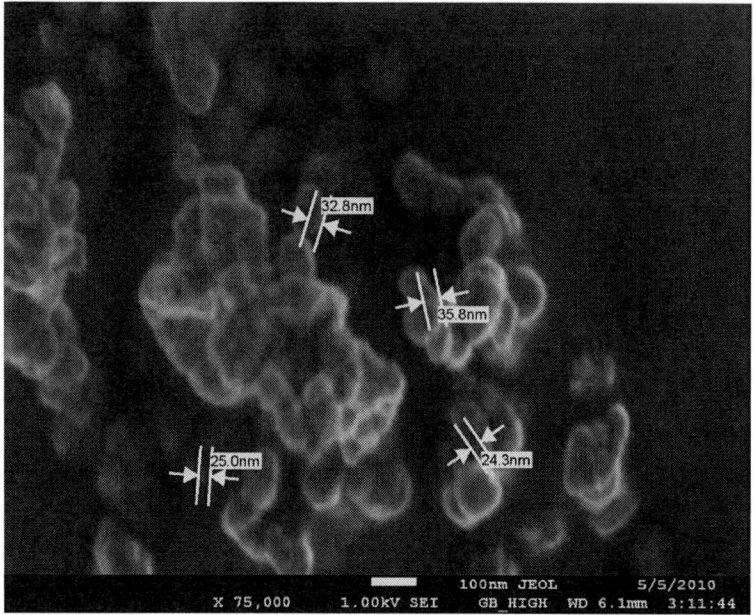

Fig 7. Nanoparticles of ZnO prepared in described apparatus at reaction pressure 15 torr. Dimension of the individual particles approaches 20 nm.

3. Conclusion

Constructed MOCVD apparatus is suitable for growth of ZnO layers with very different growth rate in range 1-90 μm/hour with excelent microstoichiometry. The best samples grown at growth rate 80μm/hour demonstrate FWHM of the bound exciton peak ~4meV. Final morfology of the ZnO on Si and GaP substrates as well as differently oriented Si is interpreted as consequence of spontaneous cooperation action of ZnO molecules. All rods and pyramids have hexagonal symetry. Nanoparticles deposited outside the substráte approach to 20 nm and without polymeric protection produce larger aglomerates.

This work was supported by the Grant Agency of the Czech Republic under contract number P108/10/0253.

References

[1] V.E. Heinrich, P.A. Cox, The Surface Science of Metal Oxides, Cambridge University Press, Cambridge, 1996.
[2] C. Wöll, Prog. Surf. Sci. **82** 55(2007).
[3] C. Klingshirn, Phys. Status Solidi (b) **244** 3027(2007).
[4] Hongtao Yuan, Yao Zhang, J. Cryst. Growth 263 119(2004).
[5] B.J. Kwon, H.S. Kwack, S.K. Lee, Y.H. Cho, D.K. Hwang, S.J. Park, Appl. Phys. Lett. **91** 061903 (2007).
[6] H. Kim, A. Cepler, M.S. Osofsky, R.C.Y. Auyeung, A. Pique , Appl. Phys. Lett. **90** 203508 (2007).
[7] Y.S. Jung, W.K. Choi, O.V. Kononenko, G.N. Panin, J. Appl. Phys. 99 013502 (2006).
[8] Guoqiang Zhang, Atsushi Nakamura, Toru Aoki, Jiro Temmyo, Yoshio Matsui, Appl. Phys. Lett. 89 113112(2006).
[9] D. Nohavica, P. Gladkov and Z. Jarchovsky, in Proceedings of Europeans MOVPE Workshop, Ulm, Germany, (2009), p. 291.
[10] Jizhong Zhang, HuanZhao, Physica B **405** 4116 (2010).
[11] N. Wang et al. Materials Science and Engineering **R 60** 1 (2008).

Optimization of Position of Piezoresistive Elements on Substrate Using FEM Simulations

Pavel Kulha

Department of Microelectronics, Faculty of Electrical Engineering,
Czech Technical University in Prague
Technická 2, CZ – 166 27 Prague 6, Czech Republic
Phone: (+420) 224 352 793, Email: kulhap@fel.cvut.cz

This paper presents an example of optimization of piezoresistive element placement on substrates for different types of deformation transducers (single side fixed cantilever beam and membrane). Modelling of structures was performed by simulator utilizing finite element method (FEM). Modelling and simulation of stress and strain distribution and deformations is practically essential for any design of MEMS structures. The modern simulation tools make the design easier and enable optimization of many different parameters before fabrication of new structure. Designed piezoresistive structures were consequently fabricated and tested.

1. Introduction

The basic function of the strain gauge is based on transforming the strain in certain direction as to change its electric resistance. The key problem in the design of piezoresistive sensors using beam or membrane is finding the right position of the sensing element on the transducer to achieve maximum sensitivity. Thus, it is important to know the distribution of mechanical stresses in particular parts of the transducer under different types of load. Then, the parametric piezoresistive simulation (in the case of piezoresistive sensors) follows with respect to the designed shape of the piezoresistive element. Analytical solution of this problem is very difficult with respect to the complexity of present MEMS devices. Thus, numerical solution and application of finite element method (FEM) is utilized. Models and simulation results presented in the following sections were created in MEMS simulator Coventor. The simulator enables drawing of layout of designed structures, creation of 3d models and mesh generation for FEM and analyzing in different domains, such as mechanical, piezoresistive, electrostatic, piezoelectric, thermal etc.

2. Designed structures

Two different strain gauge topologies were designed and simulated. The first with single resistors or half bridges (of length from 500 μm to 2000 μm and of width from 25 μm to 200 μm,) on single side fixed cantilever beam mechanical transducer (Fig.1), the second with four resistors forming the Wheatstone bridge on membrane mechanical transducer (Fig.2).

The software package Coventor has been used for design of mechanical and thermal characteristics of the structure. The tools enable design, modelling and successive modification of designed MEMS structures. The program enables: drawing of 2D layout and its editing, simulation of production process, generation of 3D model from 2D masks, generation of network by the method of finite elements, solution of mechanical, electrostatic, thermal, piezoresistive, induction, optical, and further simulations.

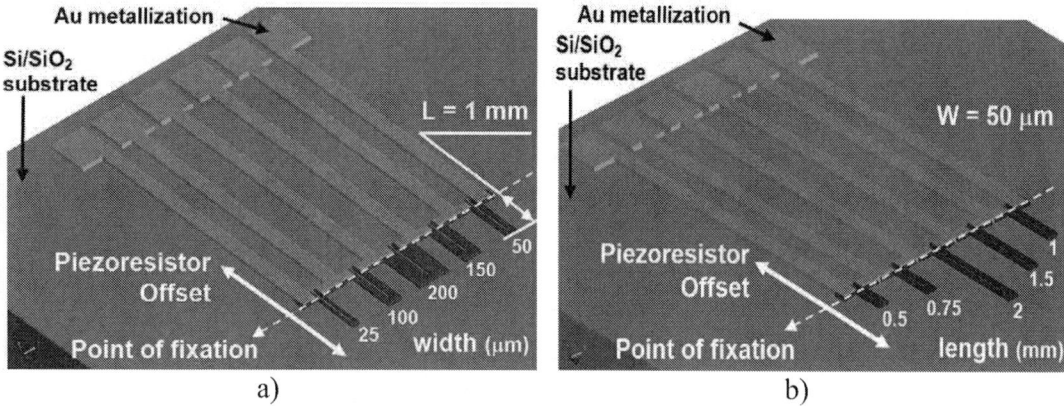

Fig. 1. 3D model of single resistor structures on cantilever beam substrate

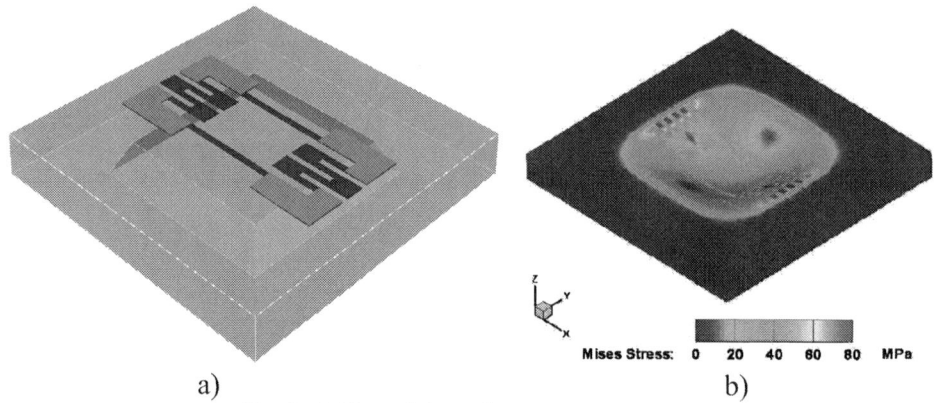

Fig. 2. a) 3D model membrane pressure sensor;
b) results of membrane mechanical simulation under pressure of 0.3MPa

3. FEM simulations and sensitivity analysis

Cantilever beams and membranes are widely used in mechanical transducers that transform measured quantity to stress or strain. The key problem in the design of piezoresistive sensors using beam or membrane is finding the right position of the sensing element on the transducer to achieve maximum sensitivity.

Such sensitivity analysis is available in Coventor. First, mechanical simulation of the stress distribution was performed. Then, piezoresistive simulation used the structural results and calculates changed in applied current while the piezoresistive element was moving along the deformed transducer.

3.1 Optimization of resistor structure with constant length

Fig. 3a depicts the differences in sensitivity and proper position of particular resistors. With increasing width of the resistor slightly decreases relative change in resistance and centre of the wider resistor should be further to the point of fixation.

3.2 Optimization of resistor structure with constant width

Piezoresistors with constant width are even more sensitive to the offset with respect to the point of fixation as is shown in Fig. 3b. The centre of the longest resistor R3 (L = 2mm) should be placed 0.2 mm from the point of fixation, while R4 (L = 0.75 mm) should be 1 mm far from the point of fixation. The longer resistor has also slightly lower sensitivity.

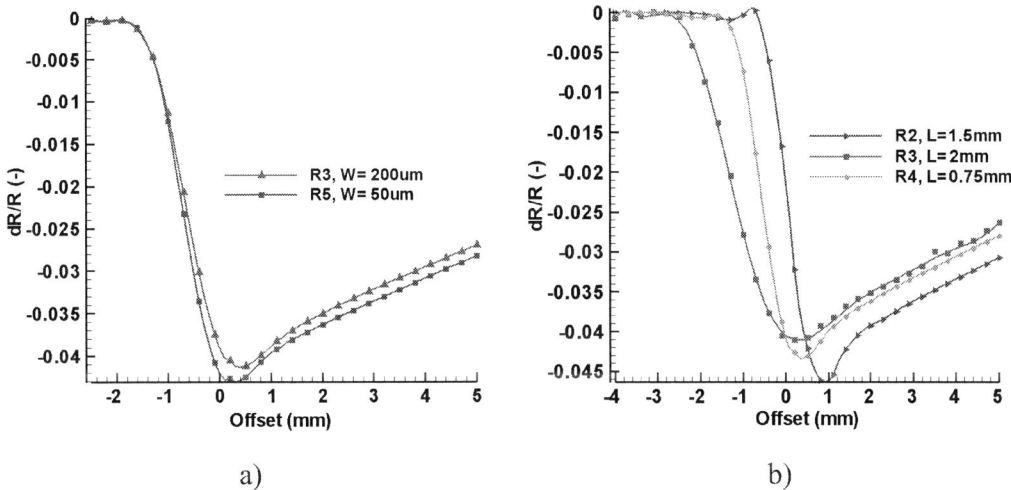

a)   b)

Fig. 3. a) analysis results for single resistors with constant length (L-const)
b) analysis results for single resistors with constant width (W-const)

3.3 Optimization of membrane structure

By the membrane bending, high-stress regions were growing up close to the membrane border and in the middle of the membrane. The resistor placed on the border was subjected to tensile stress while resistor placed in the centre was subjected to compressive strain (Fig. 4a). This feature is very useful and finally, it increases the sensitivity of the sensor.

The sensitivity analysis was performed on two resistive elements, the meander and the straight resistor. The meander was moved along the centre line (y = 0 mm) while the straight resistor along line with position y = 550 μm. The initial position is in the centre of the membrane and the structure is then moved with positive and negative offset (Fig. 4b).

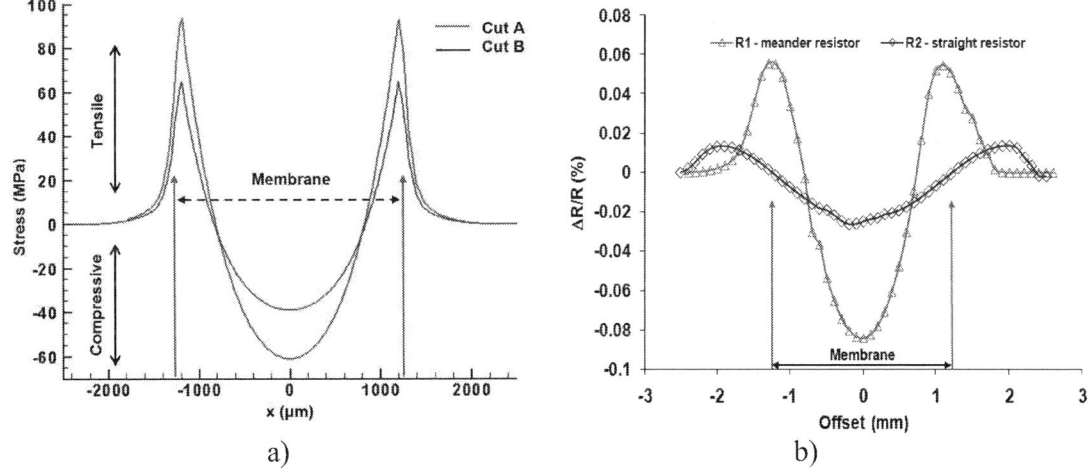

a)   b)

Fig. 4. a) analysis results for single resistors with constant length (L-const)
b) analysis results for single resistors with constant width (W-const)

4. Fabrication and testing

The cantilever beam structures uses piezoresistive boron doped diamond thin films which were deposited on $SiO_2/Si_3N_4/Si$ substrates of size 8x25 mm^2 (Fig.5a.). After the deposition, all samples were lithographically processed and the piezoresistive structures were formed by reactive ion etching in an O_2/CF_4 gas mixture. The thickness of the diamond film was 250 ÷ 300 nm. Finally, Ti and Au thin films were evaporated on piezoresistive structures to form metal contacts of 90 nm in thickness.

The membrane structures were realised on Si/SiO$_2$ substrates (Fig.5b.). Silicon dioxide (2 μm thick) was prepared by thermal oxidation as the basic insulating layer covered by 75 nm layer of PECVD silicon nitride. Piezoresistive layers were prepared by vacuum sputtering of NiCr alloy.

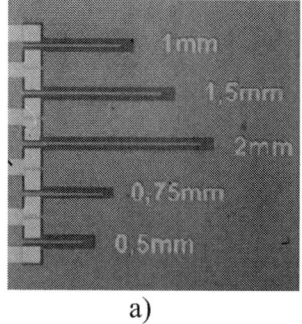
a)

b)

Fig. 5. a) cantilever beam structures with nanocrystalline diamond piezoresistors
b) membrane structure with thin film piezoresistive bridge

The extensive study of fabricated samples was performed. Samples were loaded mechanically and thermally to obtain parameters such gauge factor and its temperature dependence [3] . The simulated value of the gauge factor was 12.4 (-), the measured value 12.6 ± 0.2 (-) [4].

5. Conclusions

Our simulations and experimental results characterizes available technological processes and can be interpreted as basic input data required for design and simulation of more complex MEMS structures working with diamond layers (DMEMS pressure sensors, accelerometers etc.) using FEM software. The simulation results are in very good agreement with the experimental data

Acknowledgement

This research has been supported by the Czech Science Foundation project No. 102/09/1601 "Micro- and nano-sensor structures and systems with embedded intelligence" and partially by the research program No. MSM6840770015 "Research of Methods and Systems for Measurement of Physical Quantities and Measured Data Processing" of the CTU in Prague.

References

[1] O. C. Zienkiewicz, R. L. Taylor, J. Z. Zhu, P. Nithiarasu, "The Finite Element Method", Butterworth-Heinemann, 2005
[2] M. Elvenspoek, "Mechanical Microsensors", Springer, 2001
[3] A. Bouřa,M. Husák, "Universal Test Bench For Sensor Characterization", Electronic Devices and Systems, IMAPS CS International Conference 2010 Proceedings. Brno,2010, vol. 1
[4] P. Kulha, A. Kromka, O. Babchenko, M. Vaněček, M. Husák, "Nanocrystalline Diamond Piezoresistive Sensor", Vacuum. 2009, vol. 84, no. 1, p. 53-56. ISSN 0042-207X

978-1-4244-8574-1/10 $26.00 © 2010 IEEE

A new model of trap assisted band-to-band tunnelling

Miroslav Mikolášek[1], Juraj Racko[1], Ladislav Harmatha[1],
Ondrej Gallo[1], Ján Režnák[1], Frank Schwierz[2], Ralf Granzner[2]

[1] Slovak University of Technology, Ilkovičova 3, 812 19 Bratislava, Slovakia
[2] Technical University of Ilmenau, PF 98684 Ilmenau, Germany
e-mail: miroslav.mikolasek@stuba.sk

The paper describes a new approach to calculate currents in a PN diode based on the extension of the Shockley-Read-Hall recombination-generation model. Presented model is an alternative to Hurkx model of trap assisted tunnelling.

1. Introduction

The presence of defect states in the semiconductor has a significant influence on the carrier transfer and therefore on I-V characteristic. Due to the large width of the semiconductor band gap a high energy is needed for direct band-to-band generation. The presence of traps within the band gap assist in this process, resulting in a split of the electron "jump" from the valence band to the conduction band into two steps, each of them with lower energy demands. Four thermal generation and recombination processes of electrons and holes at the traps involved in this "jump" are described by the Shockley-Read-Hall (SRH) model.

Upon the increase of voltage, the tunnelling of carriers via the band gap becomes also significant. The direct band-to-band tunnelling is, however, energetically "expensive". Therefore, traps present in the semiconductor also in this case assist in the process of carriers transfer. Taking into account the similarities of carrier transfer via the band gap in SRH model and tunnelling process, we present a new model of trap assisted band-to-band tunnelling (TAT), which includes both models and has the ability to simply describe the carrier transfer in semiconductors. In our model, four processes of electron and hole generation and recombination described by SRH model are extended by four electron and hole capture and release processes of tunnelling to and from the traps, giving together eight exchange processes characterized by their escape times (Fig. 1). A detailed description of this model, which can be also an alternative to Hurkx trap-assisted-tunnelling model [1], is given in the sequel.

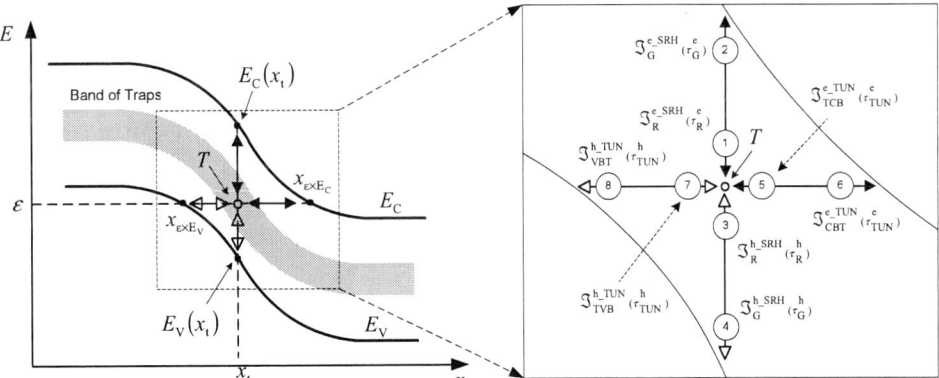

Fig. 1. Eight exchange processes involved in the new trap assisted band-to-band tunnelling.

2. Numerical model

We consider a semiconductor crystal with a band gap, E_G, defined as the width between the valence band edge, E_V, and the conduction band edge, E_C. The traps in the crystal are represented by the band of traps, lie at respective energy levels, ε, and can be described by Gaussian functions for acceptor as well as for donor defects states (Fig. 1). The densities of donor and acceptor trap states, $D_t^{D,A}$, in the text distinguished by upper indices D and A, respectively, are then evaluated as a sum as follows

$$D_t^{D,A}(\varepsilon, x) = \sum_i^N \frac{N_t^{iD,A}(\varepsilon, x)}{\sqrt{2\pi}\, E_G} \exp\left(-\left(\frac{\varepsilon - E_C(x) + E_t^{iD,A}}{2\Delta E_t^{iD,A}}\right)^2\right), \tag{1}$$

where $\Delta E_t^{iD,A}$ is the standard deviation of the Gaussian distribution of i-th band of traps, $E_t^{iD,A}$ is the distance of the middle of the i-th band of traps from the conduction band edge and $N_t^{iD,A}$ is the concentration of traps in the peak of i-th band of traps.

The traps are involved in all eight processes of electron and hole transfer. The rate of this transfer is defined by the probability function of the electron presence at the trap, which can be calculated from following equation

$$f_t(x, \varepsilon) = \frac{\dfrac{1}{\tau_R^e} + \dfrac{1}{\tau_G^h} + \dfrac{f_{F_n}(x_{\varepsilon \times E_c})}{\tau_{TUN}^e} + \dfrac{1 - f_{F_p}(x_{\varepsilon \times E_v})}{\tau_{TUN}^h}}{\dfrac{1}{\tau_R^e} + \dfrac{1}{\tau_G^e} + \dfrac{1}{\tau_G^h} + \dfrac{1}{\tau_R^h} + \dfrac{1}{\tau_{TUN}^e} + \dfrac{1}{\tau_{TUN}^h}}. \tag{1}$$

As one can observe, the probability function is defined not only by the emission-recombination processes, as assumed in the classical SRH model, but also by the capture-release processes involved in TAT transfer of carriers. Four carriers exchange processes between the traps and conduction or valence band are characterized by the change of carriers' energy. These thermal processes are described by the classical SRH model and in Eqn. 1 are represented by electron recombination, electron generation, hole recombination and hole generation escape times, $\tau_R^e, \tau_G^e, \tau_R^h$ and τ_G^h, respectively. A detailed description of these parameters can be found elsewhere [2].

The TAT mechanism of electron and hole transfer from the conduction band to the valence band and vice versa is described by four exchange processes represented by their tunnelling escapes times. In WKB approximation these times are equal for both directions of tunnelling, therefore we can write for electrons and holes (notice that the order of the upper indices distinguishes the equation for electrons and holes)

$$\frac{1}{\tau_{TUN}^{e,h}(\varepsilon, x)} = \frac{m_{TE}^{e,h}\, \sigma^{e,h}}{2\pi^2\hbar^3} \int_{E_C(x_n),\varepsilon}^{\varepsilon, E_V(x_p)} (\pm\varepsilon \mp \varepsilon')\, \Gamma_{WKB}^{e,h}(\varepsilon', x)\, d\varepsilon', \tag{2}$$

where the $\Gamma_{WKB}^{e,h}$ denotes the effective transmission coefficient for trap-assisted tunnelling of electrons and holes

$$\Gamma_{WKB}^{e,h}(\varepsilon) = \exp\left(-\frac{2}{\hbar} \int_{x_{\varepsilon \times E_c}, x}^{x, x_{\varepsilon \times E_v}} \sqrt{2 m_{TUN}^{e,h}\left(\pm E_{C,V}(x) \mp \varepsilon\right)}\, dx\right). \tag{3}$$

The Poisson equation can be then written as

$$-\frac{d}{dx}\left(\kappa \frac{d\psi(x)}{dx}\right) \cong \frac{q}{\varepsilon_0}\left(p(x) - n(x) + N_D^+(x) + N_t^D(x) - N_A^-(x) - N_t^A(x)\right), \tag{4}$$

where the contribution of traps to the carrier transfer is expressed through the concentration of donor traps, $N_t^D(x)$, and concentration of acceptor traps, $N_t^A(x)$. The continuity equations for electrons and holes in this case can be written as

$$\frac{dJ_D^e(x)}{dx} = q\left(R_{\text{SRH-TAT}}^e(x) - G_{\text{SRH-TAT}}^e(x)\right) \quad (5), \quad \frac{dJ_D^h(x)}{dx} = -q\left(R_{\text{SRH-TAT}}^h(x) - G_{\text{SRH-TAT}}^h(x)\right). \quad (6)$$

The first two terms on the right side of Eqns. 5 and 6 characterize the generation and recombination processes involved in the model of carrier transport and can be calculated as

$$R_{\text{SRH-TAT}}^e, G_{\text{SRH-TAT}}^h(x) = \int_{E_V(x)}^{E_C(x)} \frac{\dfrac{1}{\tau_{R,G}^{e,h}}\left(\dfrac{1}{\tau_{R,G}^{h,e}} + \dfrac{1 - f_{F_n}(x_{\varepsilon \times E_c})}{\tau_{\text{TUN}}^e} + \dfrac{f_{F_p}(x_{\varepsilon \times E_v})}{\tau_{\text{TUN}}^h}\right)}{\dfrac{1}{\tau_R^e} + \dfrac{1}{\tau_G^e} + \dfrac{1}{\tau_G^h} + \dfrac{1}{\tau_R^h} + \dfrac{1}{\tau_{\text{TUN}}^e} + \dfrac{1}{\tau_{\text{TUN}}^h}} D_t \, d\varepsilon \;, \quad (7)$$

$$G_{\text{SRH-TAT}}^e, R_{\text{SRH-TAT}}^h(x) = \int_{E_V(x)}^{E_C(x)} \frac{\dfrac{1}{\tau_{G,R}^{e,h}}\left(\dfrac{1}{\tau_{G,R}^{h,e}} + \dfrac{f_{F_n}(x_{\varepsilon \times E_c})}{\tau_{\text{TUN}}^e} + \dfrac{1 - f_{F_p}(x_{\varepsilon \times E_v})}{\tau_{\text{TUN}}^h}\right)}{\dfrac{1}{\tau_R^e} + \dfrac{1}{\tau_G^e} + \dfrac{1}{\tau_G^h} + \dfrac{1}{\tau_R^h} + \dfrac{1}{\tau_{\text{TUN}}^e} + \dfrac{1}{\tau_{\text{TUN}}^h}} D_t \, d\varepsilon \;. \quad (8)$$

If the tunnelling escapes times in this model are set as $\tau_{\text{TUN}}^e \equiv \infty$ and $\tau_{\text{TUN}}^h \equiv \infty$, the TAT process becomes negligible, and the equations are altered to the classical SRH model

$$G_{\text{SRH}}, R_{\text{SRH}}(x) = \int_{E_V(x)}^{E_C(x)} \frac{\dfrac{1}{\tau_{G,R}^e}\dfrac{1}{\tau_{G,R}^h}}{\dfrac{1}{\tau_R^e} + \dfrac{1}{\tau_G^e} + \dfrac{1}{\tau_G^h} + \dfrac{1}{\tau_R^h}} D_t \, d\varepsilon \;. \quad (9)$$

3. Simulation results

New TAT model was employed in simulations of a PN diode with a linear concentration profile (Fig. 2.), which was prepared on the phosphor doped silicon substrate ($N_D = 2.5 \times 10^{18}$ cm^{-3}) with orientation <111> by boron diffusion from an infinite source with surface concentration $N_A = 10^{19}$ cm^{-3} at a temperature of 1020°C for 30 minutes. The structure was contaminated by gold, which forms two bands of traps, one of acceptor type with the distance of the band of traps peak from the conduction band edge, $E_t^A = 0.54$ eV, and one of donor type with the distance of the band of traps peak from the conduction band edge, $E_t^D = 0.83$ eV. We assumed the same width of the band of traps, $\Delta E_t^D = \Delta E_t^A = 0.02$ eV, as well as the same concentration in the peak of the Gaussian distribution with $N_t^D = N_t^A = 10^{14}$ cm^{-3} for both types of the band of traps. The effective cross section for electrons and holes was set constant $\sigma^{e,h} = 10^{-15}$ cm^2. For evaluation of the tunnelling escape times, $\tau_{\text{TUN}}^{e,h}$, effective masses $m_{\text{TE}}^e = 2.19\,m_0$ and $m_{\text{TE}}^h = 0.66\,m_0$ were used. For calculation of the tunnelling probability WKB approximation was used and the effective masses were set as $m_{\text{TUN}}^e = 0.26\,m_0$ and $m_{\text{TUN}}^h = 0.370\,m_0$.

In Fig. 3 is shown the reverse I-V characteristic simulated using our new TAT model with included band-to-band model (BTB) [3] and impact ionization model (II) [4]. To compare the influence of TAT mechanisms the I-V curve calculated only with the SRH model is depicted as well. The trap-assisted-tunnelling results in an increase of the current in the middle voltage region, resulting to a "soft" shape of the I-V curve. At a higher voltage, however, the influence of TAT mechanisms becomes negligible compared with BTB and II processes.

Figure 4 shows I-V characteristics simulated with and without the TAT model for four different widths of band of traps (Fig. 5) and therefore also for four different concentrations of traps in the band gap (the concentration of traps in the Gaussian distribution peak is constant). From the figure we can see the increased role of TAT tunnelling upon the increased voltage.

978-1-4244-8574-1/10 $26.00 © 2010 IEEE

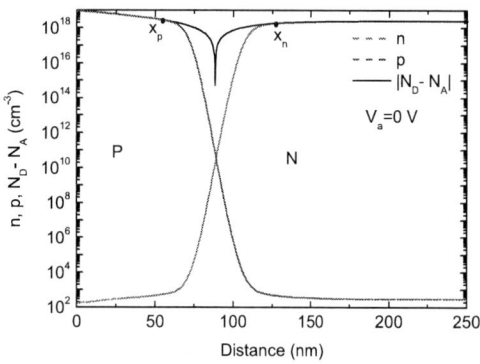

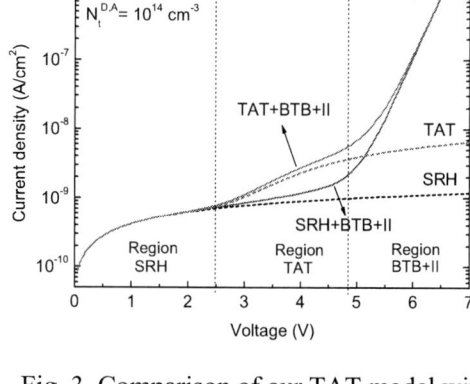

Fig. 2. Concentration profile of simulated PN diode.

Fig. 3. Comparison of our TAT model with classical SRH model.

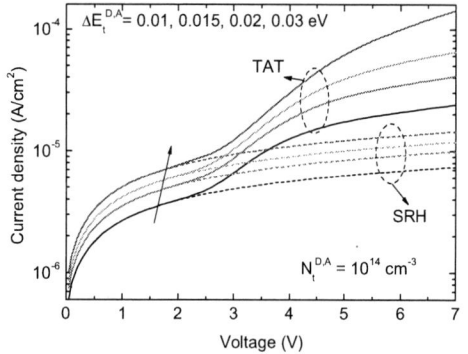

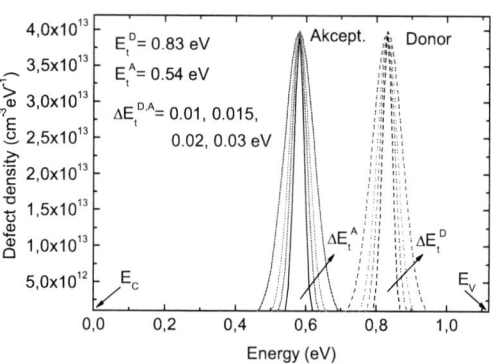

Fig. 4. I-V characteristics in reverse direction for different standard deviations of band of traps.

Fig. 5. The distribution of traps in band of traps with different standard deviation.

At a voltage of 2.5 V the TAT tunnelling process becomes dominant. The increased concentration of traps results in a higher current under increased voltage for I-V curves simulated by TAT model.

4. Conclusion

The new TAT model has the ability to model generation and recombination as well as tunnelling processes of charge transfer in the PN junction. Using this model a real "soft" I-V characteristic usually present in the case of switching diodes and transistors was modelled as a result of high concentration of traps, which assist in the process of tunnelling.

Acknowledgement

The work has been conducted at the KME FEI STU in Bratislava, in the Centre of Excellence CENAMOST (VVCE-0049-07), and supported by the Slovak Research and Development Agency (projects APVV-0133-07) and by the Scientific Grant Agency of the Ministry of Education of the SR (projects VEGA 1/0601/10 and VEGA 1/0507/09).

References

[1] G. A. M. Hurkx et al, *IEEE Trans. on ED*, 39, 1992, p. 331-338.
[2] J. Racko et al, *Central European Journal of Physics*, DOI: 10.2478/s11534-010-0027-7.
[3] S. Selberherr, *Analysis and Simulation of Semiconductor Devices*, Springer-Verlag, Wien-New York, 1984.
[4] G. A. Baraff, *Physical Review* 128, 1962, p. 2507-2517.

Simulation Study of Conduction-state Charge Imbalance in High Voltage Super-junction Power MOSFET

Kondekar Pravin N

ECE Research Group,

Indian Institute of Information Technology, Design and Manufacturing

Dumna Airport Road, Khamaria, Jabalpur (MP) India

e-mail: pnkondekar@iiiitdmj.ac.in

In high voltage super-junction (SJ) power transistor, to achieve high breakdown voltage, exact charge balance between the p–pillar and n-pillar is required. However, if there is a charge imbalance due to doping difference between the n and p pillars, the field profile gets disturbed, and BV is reduced. Effect of the gate voltage variation on the forward blocking capability is investigated using simulation. The electric field profile and the potential contour distribution are investigated during conduction state using simulation tool. Results are summarized in the form of FBV-V_{GS} plots for three different cases of possible drift layer doping variation.

1. Introduction

A relatively high breakdown voltage (BV) and low on resistance (R_{on}) is achieved using a "SuperJunction" (SJ) structure in which alternate "pillars" of p and n materials are employed, as shown in Fig.1. Exact charge balance (i.e., $N_aW_p = N_dW_n$) is required in order to achieve the highest BV in the SJ structure [1-2]. When this condition is satisfied, a relatively flat or constant electric field profile results. However, if there is a charge imbalance due to doping difference between the n and p pillars, the field profile gets disturbed, and the BV is reduced. The off state charge imbalance is caused due to the inherent doping difference (limitation of fabrication process) in two pillars has been explained in detail [3].

Here, we study the effect of charge imbalance on the forward blocking voltage (FBV) when the device is turned on by applying a gate voltage greater than the threshold voltage. Interestingly, the

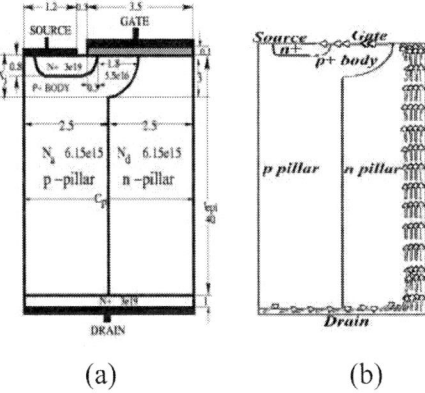

(a)   (b)

Fig.1 (a) Dimensions (in μm) and doping densities (/cm3) of the simulated SJ-MOSFET structure. (b) A vector plot of current density for $V_{GS}=10$ V and $V_{DS}=315V$

presence of carriers in the on state changes the electric field profile in the device very significantly, and this has a major effect on the FBV of the device. We present device simulation results for three cases: (i) the doping densities in the n and p pillars (N_d and N_a, respectively) are equal, i.e., $N_d=N_a$, (ii)

$N_a > N_d$ by 10.5 %, and (iii) $N_a < N_d$ by 10.5 %. The results are explained in terms of the potential contours and electric field profile. Finally, the breakdown voltage is plotted as a function of the gate voltage which shows comprehensively the effect of charge imbalance on the FBV in the SJ MOS transistor.

2. Simulation Results

The analytical design procedure for super-junction MOSFET is explained in detail in [4]. The SJ MOS transistor structure for 600V is simulated as shown in Fig.1 (a) the "nominal" value of N_a or N_d is chosen to be 6.15×10^{15} /cm^3 with an epi-layer thickness of $t_{epi}= 40$ μm. The widths of the n and p pillars are assumed to be equal (i.e., half of C_p, the "cell pitch"). This device is simulated with ISE-TCAD tools [5]. Breakdown is said to take place when the ionization integral becomes 1 (unity) along a suitable path in the device. As our goal here is to study the effect of an unintentional "charge imbalance" created by the condition, $N_a \neq N_d$, we have simulated the structure with three different sets of N_a and N_d, as mentioned in the Introduction

In general, there is a major difference between the charge density distribution within the device in the on-state and in the off-state. An example of the vector plot of the current density in the on state is shown in Fig. 1 (b). Note that, in the n pillar, the region where the current density is significant is almost charge neutral (this is not explicitly shown in the figure), which must be contrasted with the off-state situation in which a relatively small V_{DS} causes the entire n and p columns to be depleted, thereby creating space charge throughout. This gives rise to very different potential distributions in the on and off state, as we shall see in the following. Note that, in the on state, the extent of the neutral region in the n pillar which is equivalent to a JFET channel [6] would depend on various factors. The current at a position y in the JFET channel (i.e., the n pillar) is given by,

$$I = qn(y)Wb(y)v(y). \tag{1}$$

Here, W is the device width (perpendicular to the cross sectional view), $v(y)$ is the drift velocity, and $b(y)$ is width of the neutral part of the n pillar, and $n(y)$ is the electron concentration. Here y is the vertical distance along the JFET channel. We will now describe qualitatively the different charge imbalance situations.

Equation. 1 gives the drain current which should be constant throughout the device for a given bias condition. It is observed that, for the full FBV applied at the drain, the drift velocity of the electrons is saturated for all the gate bias conditions above the threshold voltage. At low gate voltage, the width of the conducting channel $b(y)$ is small, and it is almost uniform throughout the device. For higher gate voltages, $b(y)$ increases and becomes non-uniform due to the JFET action and it affects the charge distribution.

The I_D-V_D characteristics for $N_a=N_d =N_0=6.15 \times 10^{15}$ /cm^3. The drain current shows very good saturation behavior. At high gate voltages, the transconductance is rather poor, i.e., the gate voltage does not have significant effect on I_{Dsat}. This is, of course, the well-known phenomenon of quasi-saturation of the I_D-V_G characteristics as described in, for example, [7]

We are particularly interested in the breakdown behavior of the device. The drain voltage at which a given I_D-V_D curve terminates (for a particular V_{GS}) is the FBV of the device. It can be easily seen, that the gate voltage has a significant influence on the FBV. The simulated I_D-V_D characteristics for the $N_a>N_d$ (10.5% Variation from Nominal optimum value N_0 for inherent fabrication limitation) is somewhat similar to the $N_a=N_d$ case (in terms of the saturation current levels) [8]

For the off state ($V_{GS}=0$), $N_d > N_a$, we can note that, the BV of the device is 356 V due inherent doping imbalance. The important observation here is that, the BV first increases as V_{GS} increases, reaches a maximum value and then decreases, as V_{GS} is increased further.

978-1-4244-8574-1/10 $26.00 © 2010 IEEE   200

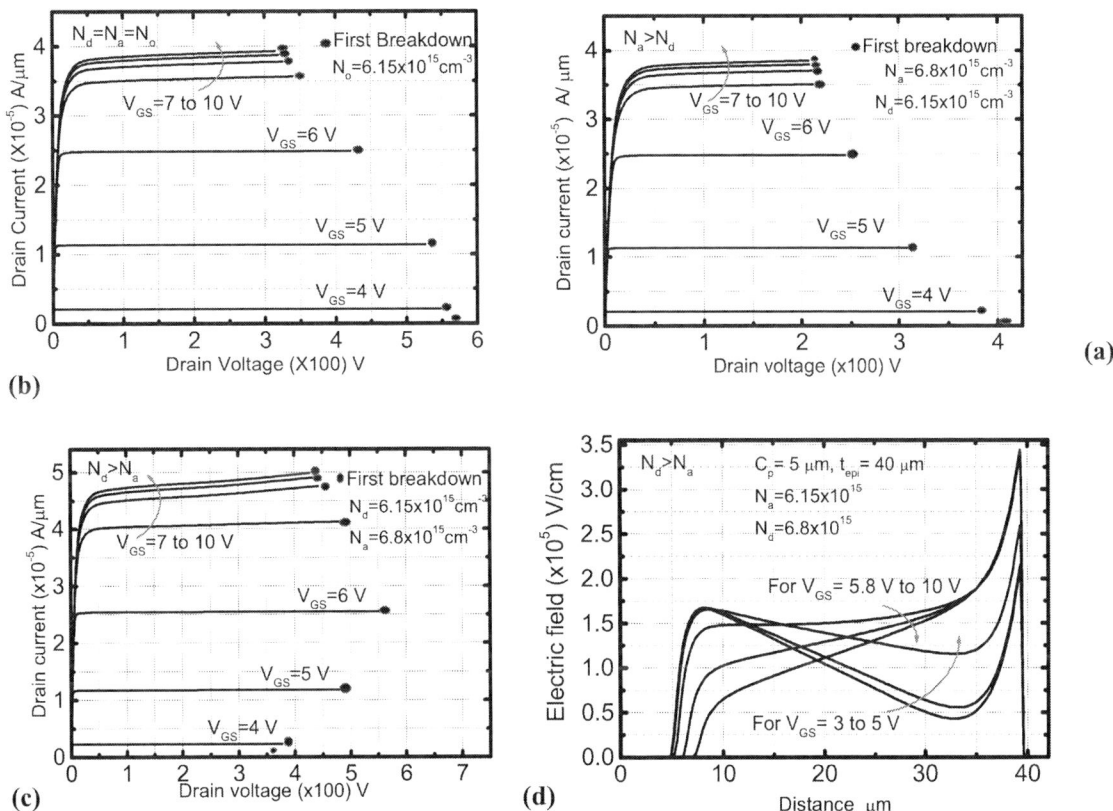

Fig. 2 (a) and (b) I_D-V_D Curves for drift layer with $N_a=N_d$ and $N_a>N_d$ respectively and (c) $N_d>N_a$ (d) the y-component of the electric field profiles along the left edge of the p pillar of SJ-MOSFET

It is instructive to look at the y-component of the electric field profiles as shown in Fig. 2 (d) along the left edge of the p pillar of SJ-MOSFET structure given in Fig.1. It can be observed here that, the y-component in the middle of the device is increasing for $V_{GS}=3$ to 5.8 V for which, the case of $N_d>N_a$ is still valid and then it becomes almost flat at $V_{GS}=5.8$ V giving full BV as FBV (a case effectively similar to off state $N_d=N_a$ case). Further increase in V_{GS} above 5.8 V will establish a case effectively similar to the $N_a>N_d$ case. The electric field profile will tilt toward triangular, increasing its peak near the bottom of SJ-MOSFET affecting BV

3. FBV versus Gate Voltage

The results of the off-state [4,10] and the on-state charge imbalance can easily be summarized looking at I_D-V_D characteristics up to breakdown for all three cases (i) $N_d=N_a$, (ii) $N_a>N_d$ by 10.5 %, and (iii) $N_d>N_a$ by 10.5 %. Forward blocking voltage as a function of gate voltage is plotted for all these cases as shown in Fig. 3. It is now simple to understand the behavior of SJ MOSFET at one glance as follows

1. Consider the $N_d=N_a$ curve first, with $V_{GS}=0$V, the BV is 555 V [4]. As V_{GS} increases above the threshold voltage (3.5V), current starts flowing through the device. The injection of electrons in the n drift region creates charge imbalance similar to the $N_a>N_d$ case, causing impact ionization at a lower drain voltage affecting BV. At gate voltage more than 8 V, the drain current does not increase appreciably and control from gate is lost due to quasi-saturation and FBV settles at 315 V.

2. There is a difference in the off-state ($V_{GS}=0$V) BV between $N_a>N_d$ and $N_d>N_a$ due lack of symmetry in the structure

3. In the case of $N_d > N_a$ by 10.5 %, initially as V_{GS} increases to 5.8 V, FBV increases and reaches maximum similar to off-state $N_d = N_a$ and then it starts decreasing similar to $N_a > N_d$ type, as V_{GS} increases further. This is because, at a larger V_{GS}, the number of electrons in the n pillar is more, which neutralize the donor ion charge. On the other hand, the acceptor ion charge in the p pillar does not change. This amounts to creating an effective charge imbalance similar to the $N_a > N_d$ case.

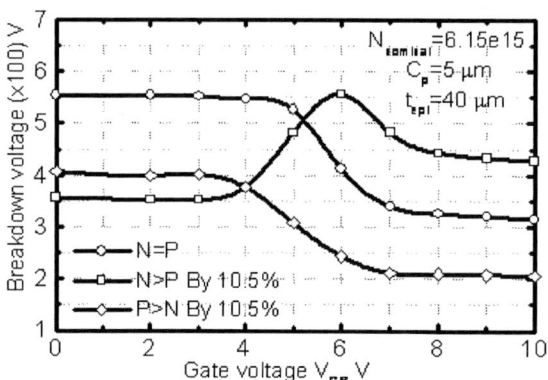

Fig.3 BV-V_{GS} curves for all the three cases of SJ MOSFET

4.   Conclusions

We have studied the effect of the gate voltage on the forward blocking capability of the super-junction high voltage power transistor. As V_{GS} and therefore the drain current changes, the charge distribution within the device is affected, this in turn, gives rise to a significant change in the nature of potential contours. We have clearly shown, with the help of simulation results, that the maximum FBV for a given V_{GS} occurs when the potential contours are evenly distributed, i.e., the electric field profile is relatively flat. Also, the results are summarized in the form of BV-V_{GS} plots for three different cases, $N_a = N_d$, $N_a > N_d$, and $N_d > N_a$ (lowest on resistance)

Acknowledgement

I am thankful to Director, IIITDM for providing facilities and moral support for carrying out this research and MHRD-Govt of India for providing financial assistance to present my work at this conference

References

[1]   M. N. Darwish and K. Board, *IEEE Trans. Electron Devices,* vol. ED-31, no. 12, pp. 1769-1972, Dec. 1984.

[2]   G. Deboy, M. Purschel, M. Schmitt, and A. Willmeroth, *Proceedings of ESSDERC 2001*, pp. 61-68.

[3]   Pravin N. Kondekar,Oh Hwan Sool, *Journal of Korean Physical Society*, vol.48, no.4, pp.624~630 April 2006.

[4]   Pravin N. Kondeka, *Journal of Korean Physical Society*, vol.44, no.6, pp. 1565~1570, June 2004

[5]   *Integrated System Engineering, ISE-TCAD Manuals*, AG, Zurich, Switzerland

[6]   K. Lehovec and R. S. Miller, *IEEE Trans. Electron Devices*, vol. ED-22, no. 5, May 1975.

[7]   M. N. Darwish, *IEEE Trans. Electron Devices*, vol. ED-33, no. 11, pp. 1710-1716, 1986.

[8]   K. Sheng, F. Udrea, and G. A. J. Amaratunga, *Proceedings of ESSDERC 2001*, pp. 251-254, Nuremberg, Germany, Sept. 2001.

[9]   P. Shenoy, A. Bhalla, and G. Dolny, *Proceedings of ISPSD '99*, pp. 99-102, Toronto

Electrode configuration for EMG measurements

E. Vavrinsky, K. Rendek, M. Daricek, M. Donoval, F. Horinek, M. Horniak, D. Donoval

Department of Microelectronics, Slovak University of Technology in Bratislava
Ilkovicova 3, SK-812 19 Bratislava, Slovakia
e-mail: erik.vavrinsky@stuba.sk

This paper explains problematic of EMG (electromyography) signal measurements. Described experiments are partition of complex project "MEDISYS" dedicated to design of modern medical equipment, consisting the latest technologies and IC. The design is based on low noise wireless measurement block (for ECG, EMG and EEG monitoring) followed by actuator block with mechanical constructions and control electronic.

1. Introduction

EMG is a technique for evaluating and recording the electrical activity produced by skeletal muscles. An EMG detects the electrical potential generated by muscle cells when these are electrically or neurologically activated. The signals can be analyzed to detect medical abnormalities, activation level, recruitment order or to analyze the biomechanics of human or animal movement. Single muscle membrane generates electric potential. The EMG potentials range between as low as 50 µV up to 30 mV, depending on the muscle type and conditions during the observation process. EMG signals are used in many clinical and biomedical applications. In modern medicine processes is EMG used as a diagnostics tool for identification of neuromuscular diseases, assessing low-back pain, kinesiology, and disorders of motor control. Generated signals are often used as control signals for prosthetic devices such as prosthetic hands, arms, and lower limbs. With new low noise technologies can be the method used to sense even isometric muscular activity where no movement is produced with significant impact on neural and neuromuscular diseases. The detected signals can be in research used to determine and treat many other kinds of disabilities, or to control electronic devices such as assisting robotic hands, mobile phones or PDA. [1]

2. Experiments

In our first experiments we measured EMG on left forearm by disposable pre gelled silver/silver chloride surface bar electrodes (size 22x23 mm). For our experiments we have used amplifier, designed and constructed for the purposes of wireless low noise signal sensing and processing. The core of the EMG amplifier is based on precise instrumentation amplifier designed for bipolar detection arrangement. Block diagram of experimental set-up is in Fig 1. All power tests (total time: 120 s) was divided in four phases: relaxation (0-30 s), periodic power test (30-60 s), relaxation (60-90 s) and balanced power test (90-120 s).

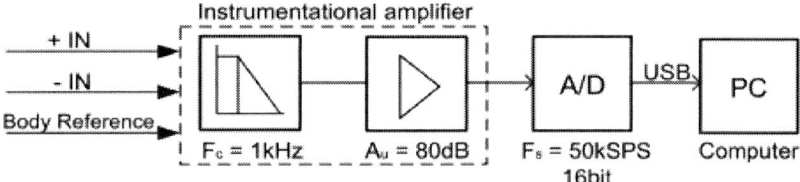

Fig. 1: Block diagram of experimental set-up

In 1st application experiments we have tested functionality and main parameters of dedicated amplifier (Fig. 2a) by comparison with commercially used "ECG/EMG sensor" from Vernier (Fig. 2b) [2] and precise 24-bit ±5 V measuring instrument "NI 9234" with "Hi-Speed USB Carrier NI USB-9162" (Fig. 2c) connected to personal computer and controlled by LabView 8.6, from National Instruments [3]. In all experiments we used 3-electrode configuration, except of NI 9234, and the output signal was measured in full EMG spectrum (Band pass filter (BPF) set-up: 2 – 300 Hz).

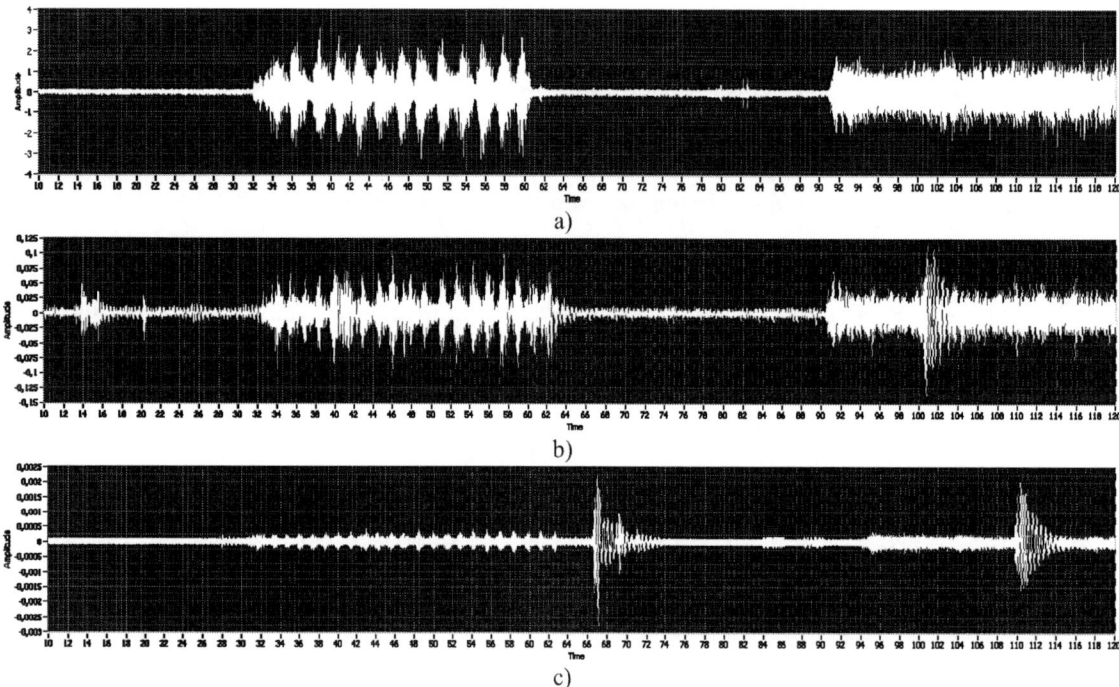

Fig. 2: Amplifiers comparison: a) Medisys (new designed amplifier), b) Vernier, c) NI 9234

As 2nd we have investigated influence of using ground electrodes (2-electrode vs. 3-electrode configurations) in our designed "Medisys" amplifier. We found that 3-electrode configuration offers more stabile signal (noise level is lower) in comparison to 2-electrode configuration (Fig. 3).

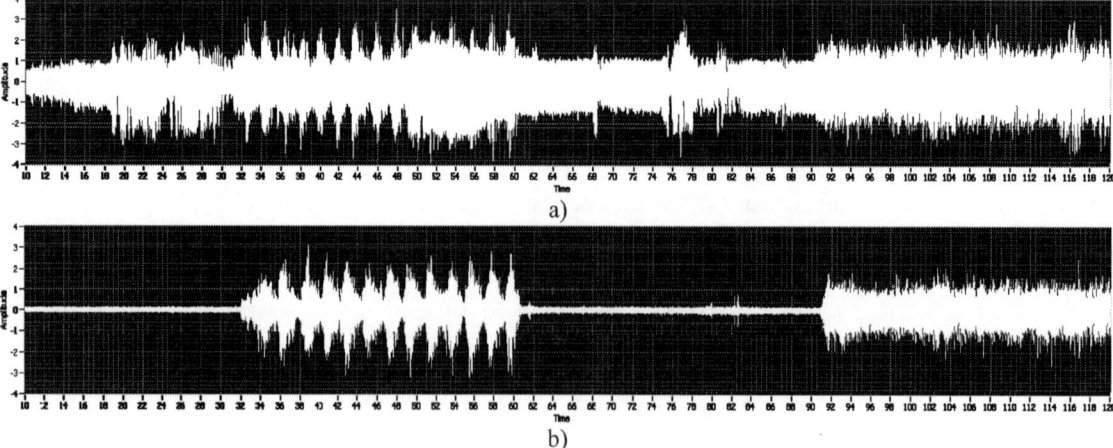

Fig. 3: Electrode configurations comparison: a) 2-electrode, b) 3-electrode

Next we found that for forearm EMG measurement is ideal to use BPF: 55-95 Hz and this filter set-up we used in all next experiments (Fig. 4).

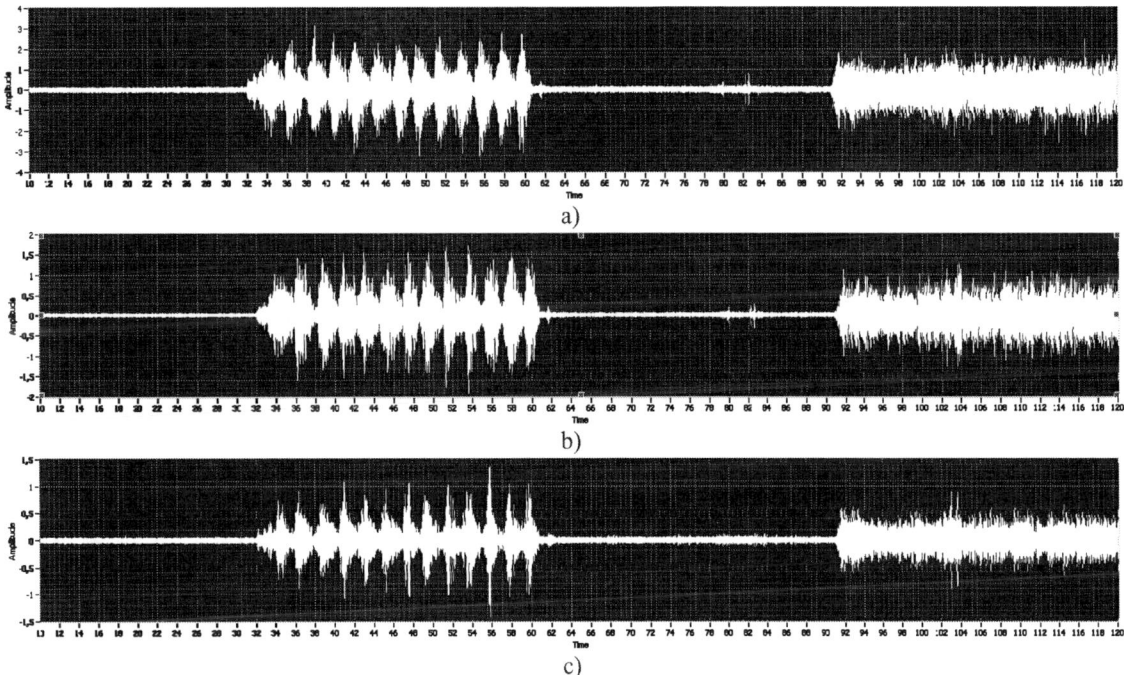

Fig. 4: Band pass filters set-up: a) 2-300 Hz, b) 55-95 Hz, c) 150-200 Hz

In next step we examined influence of ground electrode separation distance (Fig. 5). We can see that electrode distance is not a critical parameter and it is possible to put our electrodes relative close together.

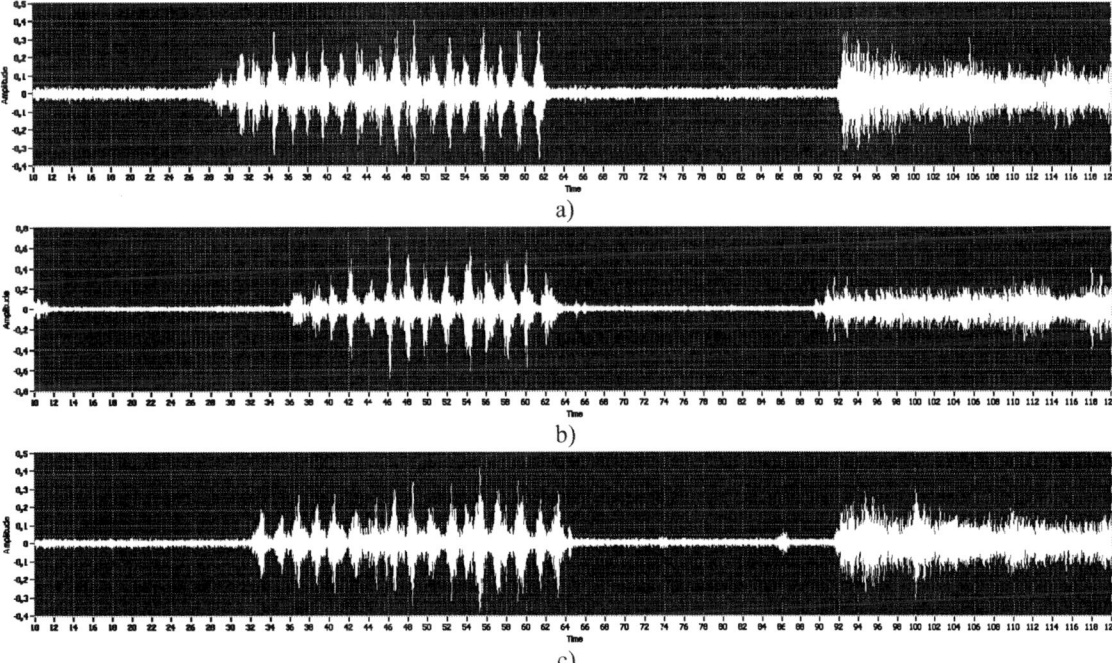

Fig. 5: Ground electrode separation distance: a) 2500 mm, b) 30 mm, c) 5 mm

As last and we investigated influence of size of active electrode area on output signal. We found, that the size of active area is very important input factor for output signal quality (Fig. 6) and by using microelectrodes (active sensing area under 300 mm^2) is better to stop using of conductive gels.

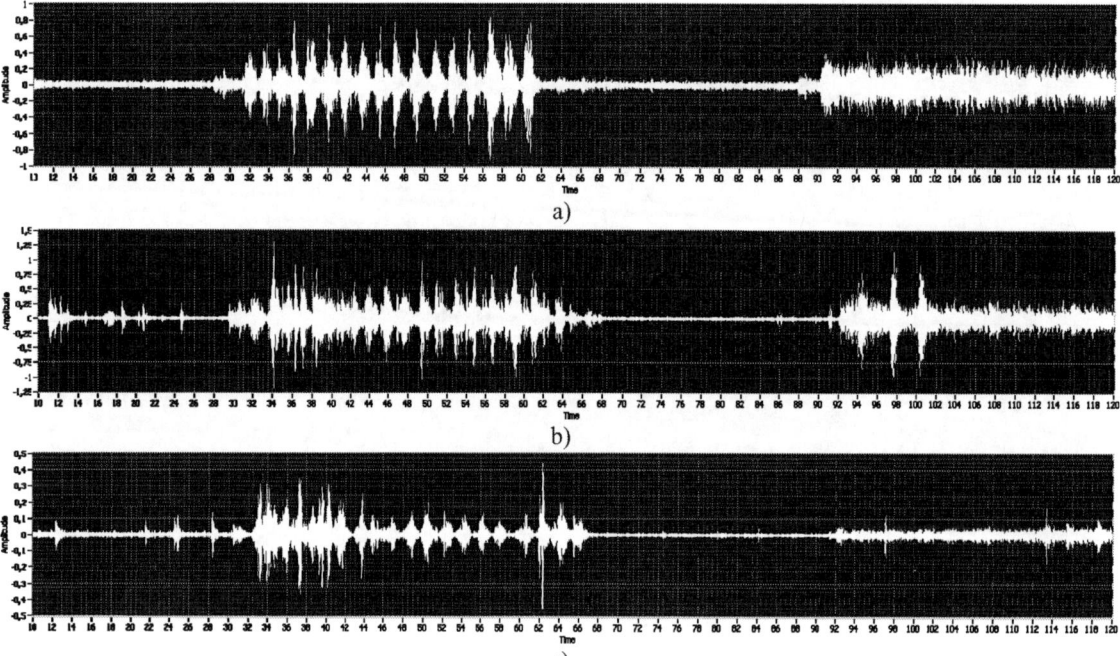

Fig. 6: Size of active electrode area: a) 1700 mm^2, b) 450 mm^2 c) 100 mm^2

Conclusions

The obtain results very clearly proofed high quality and well noise cancellation of dedicated amplifier with better measurement results in comparison to other conventionally used devices.

Overall short-term goal is maximal miniaturization of electrodes with analysis of the potential of integration together with the dedicated amplifier and supporting electronics, including wireless data transfer. The obtained results indicated that mostly the area of active electrode is very important factor. The overall size of electrodes is very critical in signal-to-noise ratio. To improve quality of output signal we propose digital narrowband filters (mostly in range of 55-95 Hz), but for lab testing purposes we considered wider frequency range: 2 Hz – 300 Hz. As next research step another electrode materials (like Au, Pt, Al, ZnO(Al), ITO) and different geometrical configurations (IDA etc.) will be investigated.

Acknowledgement

Presented work was supported by the Excellence Centre „CENAMOST", by the APVV project VMSP-P-0127-09 and project VEGA grant 1/0220/09.

References

[1] Electromyography, http://en.wikipedia.org/wiki/Electromyography, 2010.
[2] Vernier - EKG-BTA sensor, http://www2.vernier.com/booklets/ekg-bta.pdf, 2010.
[3] National Instruments, http://www.ni.com/, 2010.

Semi-insulating GaAs radiation detectors: PICTS study of neutron-induced defects

F. Dubecký[1], M. Ladziansky[2], D. Kindl[3] and V. Nečas[2]

[1] Institute of Electrical Engineering, Slovak Academy of Sciences,
Dúbravská cesta 9, SK-841 04 Bratislava, Slovakia
[2] Department of Nuclear Physics and Technology, Slovak University of Technology,
Ilkovičova 3, SK-812 19 Bratislava, Slovakia
[3] Institute of Physics, Academy of Sciences of the Czech Republic, Cukrovarnícka 10,
CZ-162 53 Praha 6, Czech Republic
e-mail: elekfdub@savba.sk

Influence of damage by neutrons introduced in semi-insulating GaAs detectors is studied by current-voltage measurement and Photo-Induced Current Transient Spectroscopy (PICTS). Significant rise of the reverse current is observed at neutron fluencies exceeding 10^{13} ncm^{-2}. The PICTS is used for evaluation of deep-level states in detector structures prior and after neutron bombardment. Formation of a new significant neutron-induced acceptor-like deep level with apparent energy position (E_C-E_t) 1.02 eV was observed for fluencies >10^{13} ncm^{-2}.

1. Introduction

Higher cross-section of neutrons in comparison with photons in light element materials makes neutrons suitable for imagining of such materials. Hence digital neutron imaging seems to be a powerful tool in industrial and security applications (material defectoscopy, exposure of plastic weapons, drugs, explosives, etc.). Semi-insulating (SI) GaAs proved relatively high resistance against damage by neutrons [1], which makes this material suitable candidate for fabrication of detectors applicable in neutron digital imaging.

This paper, concerns with the neutron damage study of fabricated SI GaAs detectors bombarded by various fluencies of neutrons. Current-voltage (I-V) measurement and Photo-Induced Current Transient Spectroscopy (PICTS) are used for characterization of investigated SI GaAs detectors.

2. Experimental results

Set of large area (2×2 mm^2) detectors (Fig. 1) based on bulk SI GaAs (CMK Ltd., Žarnovica, Slovakia) grown by the liquid encapsulated Czochralski (LEC) with thickness of

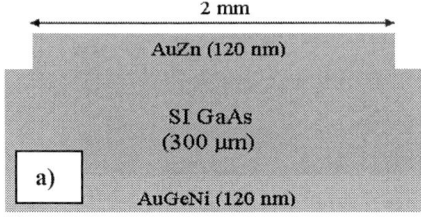 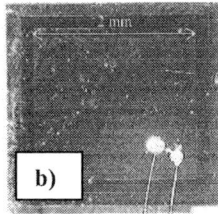

Fig. 1. *SI GaAs detector: a) cross section, and b) photo. Area of AuZn Schottky contact is 2x2 mm^2 and thickness of the LEC SI GaAs substrate is 300 μm.*

300 μm was fabricated. The Schottky barriers were for-med on the wafer topside by evaporation of AuZn eutectic metallization with thickness of 120 nm using double-sided optical photo-lithography and lift-off process. Full area backside eutectic alloy metallization

(thickness of 120 nm) was deposited by evaporation creating a quasi–ohmic non-injecting contact. Sample surface was passivated by the plasma enhanced CVD silicon nitride (thickness of 100 nm) deposited at a temperature of 290 °C with following tempering at 100 °C for 10 hours.

Neutron bombardment of SI GaAs detectors was performed in the Cyclotron Centre in Řež (Czech Republic). Beryllium target was bombarded by 37 MeV primary protons. The result of $^{9}Be(p,n)^{9}B$ reaction provided constant continuous energetic spectrum of fast neutrons in the range from 2 MeV to 30 MeV with the exponential flux decrease down to 37 MeV at the end of the spectrum. Total integral neutron fluencies impacted to detectors varied from 10^{11} to 10^{14} ncm^{-2}.

I-V characteristics of detectors were measured by the automatic pA meter with internal voltage source from DS Lab, Ltd. The measurement was performed at room temperature in the dark. I-V characteristics measured prior and after neutron bombardment are depicted in Fig. 2. Significant rise of the reverse current is observed at neutron fluencies exceeding 10^{13} ncm^{-2} (Fig. 2b). Deep-level defect states formed in bulk SI GaAs detectors were evaluated by the PICTS technique with the aim to evaluate changes in the material electron structure caused by neutrons. Measurement was performed by Polaron S4600 equipment with the choice of optical excitation. Detector was placed inside of cryostat and biased by controllable constant voltage source. Decrement of the voltage at series resistor R_M allows measurement of current transients through the detector. Signal is amplified by DPLVA-100-B-S amplifier, correlated by the PICTS apparatus and processed by a personal computer. For optical excitation was used GaAs/GaAlAs semiconductor laser with the output wavelength of 854 nm. Bulk LEC SI GaAs contains several native defects introduced during the crystal growth. Its relatively high resistivity at room temperature ($\sim 10^7$ Ωcm) is achieved under As-rich conditions during the growth process. Semi-insulating character of the material [2] is caused by native deep-level donor-like defect labelled EL2 following previous studies and observations of many authors. Some additional deep-level defect states additionally participate in the overall compensation mechanisms. Deep-level states in SI GaAs were investigated by various deep-level transient spectroscopy techniques such as TSC [3], MCTS, IDLTS [4], etc. The original status of deep-levels present in used SI GaAs material was deduced from investigated reference detector (labelled D0). Changes in the defect electronic structure due to the neutron bombardment are evaluated from the fabricated SI GaAs detectors bombarded by different fluencies of neutrons.

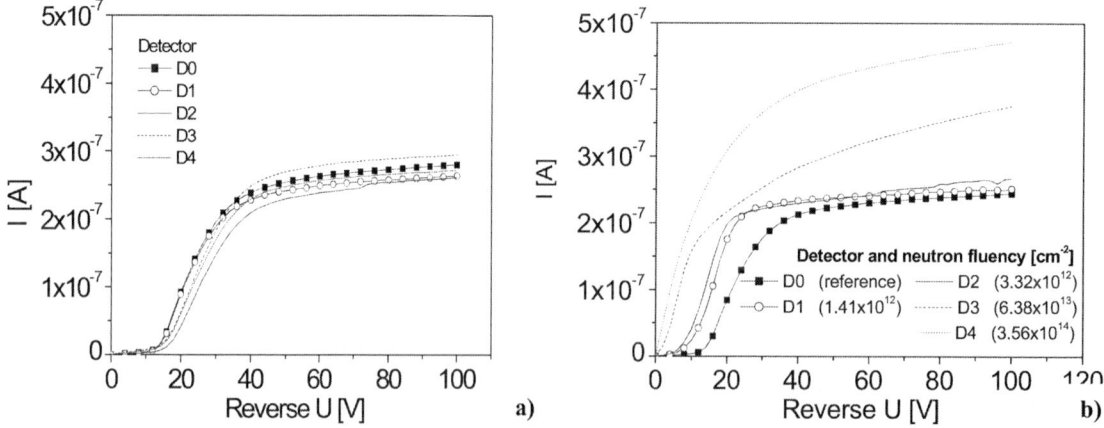

Fig. 2. *Measured reverse I-V characteristics of SI GaAs detectors: prior (a) and after (b) bombardment by different neutron fluencies. Measurement was performed at room temperature in the dark.*

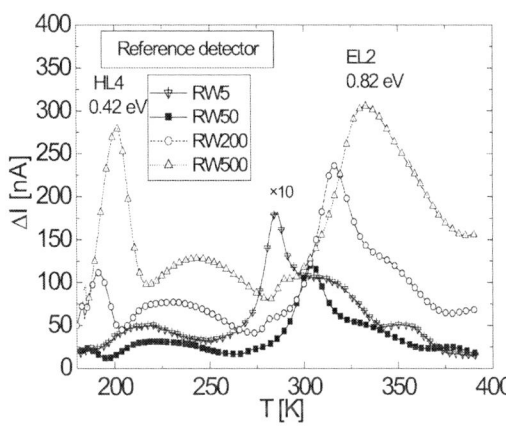 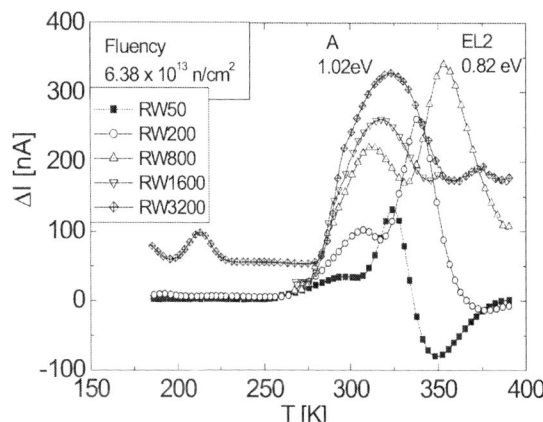

Fig. 3. *PICTS spectra of the reference SI GaAs detector. Emission rate from the native EL2 defect in GaAs dominates.*

Fig. 4. *PICTS spectra of SI GaAs detector irradiated by neutron fluency of $6.4 \times 10^{13} cm^{-2}$. Emission rate from neutron-induced trap A dominates.*

Obtained PICTS spectrum represents two dominant emission centres (Fig. 3) measured at various "rate windows" (RW), presenting frequency of optical excitation pulses. Calculated apparent energy positions of detected deep levels in GaAs band gap give 0.82 eV (EL2) and 0.42 eV (HL4) with respect to the conduction and valence bands, respectively. These values are in good agreement with the published data [5] and results obtained by the MCTS (Minority Carrier Trap Spectroscopy) and the I-DLTS (Isothermal Deep-Level Transient Spectroscopy) using optical excitation [4]. Measurement of detectors irradiated by neutron fluencies lower than 10^{13} ncm^{-2} shows non-significant change in the measured PICTS spectra. However, at neutron fluencies exceeding 10^{13} ncm^{-2}, a huge deep-level defect (labelled A) occurred (Fig. 4). Formation of this neutron-induced defect is explained by shifting of Ga and As atoms into the interstitial positions, thus forming a new complex defect (along with the original defects) with tendency to form areas of disturbances. Comparable concentration of such defects with respect to effective concentration of free charge carriers causes degradation of the SI GaAs material and consequently detector characteristics [6]. Apparent energy position of this newly observed acceptor-like defect induced by neutrons was calculated to 1.02 eV under the conduction band (Fig. 5). This defect causes degradation of electrical and detection performance of detectors and significant decrease of their effective dc resistance.

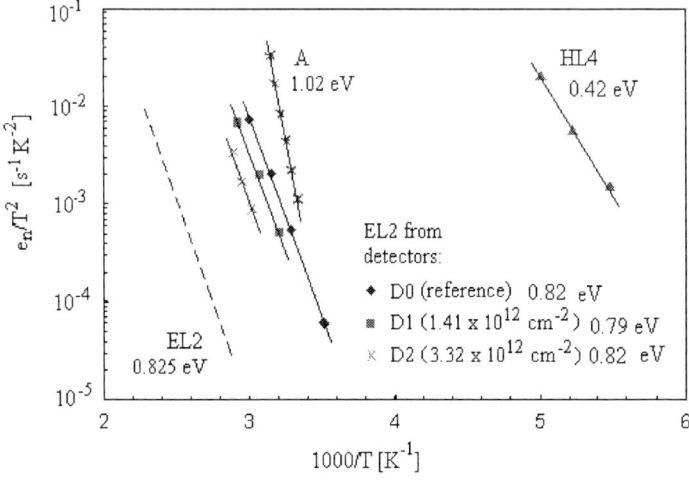

Fig. 5. *The Arrhenius plot of deep levels obtained in SI GaAs structures by the PICTS: EL2 level according to the literature [4] - dashed line and neutron induced deep level (A) calculated using data of detector irradiated by fluency of 6.4×10^{13} ncm^{-2}. Lower fluencies (D1, D2) are also shown.*

3. Conclusions

Degradation of fabricated SI GaAs detectors after fast neutron bombardment was investigated by means of I-V and PICTS measurements. Significant rise of detectors reverse current is observed at neutron fluencies exceeding 10^{13} ncm^{-2}. Energy positions of deep-level defects in the band gap of GaAs prior and after the neutron bombardment was studied using the PICTS. Calculated apparent activation energies of the dominant deep levels in SI GaAs prior exposing to neutron flux give 0.82 eV for EL2 and 0.42 eV for HL4. Formation of dominating neutron-induced deep-level complex defect is indicated at fluencies exceeding 10^{13} ncm^{-2}. Calculated apparent activation energy of this acceptor-like trap gives 1.02 eV with respect to the conduction band.

Acknowledgement

Authors acknowledge V. Linhart (IAEP CTU, Prague, Czech Republic, P. Bém (NPI ASCR, Řež, Czech Republic) for neutron bombardment of the samples, P. Boháček, M. Sekáčová and J. Huran (IEE SAS Bratislava) for assistance in detector technology. This work was done in the Center of Excellence CENAMOST (Contract No. VVCE-0049-07) with partial support from grants No. APVV 0655-07 (Slovak Research and Development Agency), 2/0163/09, 1/0689/09, 2/0192/10 (Grant Agency for Science), EURATOM/CU (Project P3), and by the IP ASCR, v.v.i. in the frame of the Institutional Research plan AV0Z10100521.

References

[1] M. Morovič, et al.: *Nucl. Instr. and Meth. in Phys. Res.* **B 197**, 240, 2002.
[2] O. Oda: *Compound semiconductor bulk materials and characterisation,* World Sci. Publ. Co., 2007.
[3] M. Pavlovič, U.V. Desnica, *J. Appl. Phys.* **88**, 4563, 2000.
[4] Ľ. Stuchlíková, et al., *Phys. Stat. Sol.* **a 138**, 241, 1993.
[5] M.G. Martin, et al., *Electr. Lett.* **13**, 191, 1977.
[6] A. Janotti, et al., *Brasil. J. Phys.* **27/A**, 110, 1997.

Wireless Sensor System for Overhead Line Ampacity Monitoring

Jakub Frolec and Miroslav Husak

Department of Microelectronics, Czech Technical University in Prague,
Technicka 2, 166 27 Prague, Czech Republic
e-mail: frolejak@fel.cvut.cz

Due to the economic pressures of internationalized energy market together with rising demand for electrical power, a growing need has been observed in recent years to increase ampere capacity (ampacity) of high voltage overhead transmission line infrastructure. Long construction time, difficult legal procedures and cost considerations encourage the line operators to exploit the wide margins of the traditional worst-case condition oriented overhead line design. To operate the transmission lines closer to the limits of their instantaneous weather dependent ampacity, live monitoring of line state must be employed. In this paper we propose an extensible sensor system for overhead conductor temperature and current measurement. The system comprises of sensor units mounted directly on the high voltage conductors, which acquire the current and temperature data and transmit it to a ground interfacing unit connected to line operator's control system computer network.

1. Introduction

Recent liberalisation and internationalisation of European energy markets in conjunction with constantly growing demand for electrical power challenge the overhead transmission line operators to increase the ampacity – or ampere capacity – of the transmission network. The vast investment costs and lengthy legal procedures involved in building new high voltage overhead lines lead the transmission network operators to seek ways to utilise the existing lines more efficiently. Traditional high voltage transmission line design has aimed to ensure that certain temperature (specific to conductor material and construction) never be exceeded even under the worst-case combination of line load and weather conditions. Exceeding the maximum temperature results in reduced tensile strength of the conductor, increased corrosion and safety risks brought about by excessive conductor sag. While the traditional approach has ensured reliable operation of the transmission infrastructure by what may seem as overdesign in the majority of times, it also presents the opportunity to utilize the transmission lines closer to the limits of their instantaneous ampacity – given that the state of the lines is monitored in real time and transmitted power regulated accordingly. The line monitoring and dynamic ampacity rating problem has been approached by several researchers in different ways. The Liège University team around J. Lilien developed an accelerometer equipped sensor system that calculates conductor sag based on conductor vibration analysis [1]. Hinrichsen et al proposed a system where temperature of conductor-mounted passive SAW (Surface Acoustic Wave) component is evaluated based on reflected microwave signal properties [2]. Another work proposes conductor sag evaluation based on PLC signal attenuation monitoring [3].

In this work the authors propose an extensible sensor system for current and temperature measurement usable both for transmission line state monitoring and ampacity model verification. The system comprises of several sensor units mounted directly on phase conductors that acquire and transmit the temperature and current data to the ground unit located in an adjacent power substation.

2. Ampacity

Ampacity of a conductor can be defined as the maximum current in the conductor such that the conductor temperature does not exceed its material and construction dependent maximum temperature. While the maximum operating temperature for the most commonly used ACSR (Aluminium Conductor Steel Reinforced) conductors is 80 °C, the high voltage transmission lines are often designed to achieve the minimum safe clearance above ground at temperatures as low as 60 °C. There are several thermal behaviour models used for ampacity calculation. They usually treat the major thermal effects in similar way differing often only in treatment of the less significant effects. In the most commonly used general form ampacity can be written as:

$$I = \sqrt{\frac{P_c + P_r - P_i}{R_{AC}}}, \tag{1}$$

where P_c is the power dissipated from the conductor due to air convection, P_r represents its thermal radiation, P_i is the solar irradiation $\left(Wm^{-1}\right)$ and R_{AC} $\left(\Omega m^{-1}\right)$ represents the AC resistance. All the quantities pertain to unit length of the conductor. By evaluating Eqn. (1) according to the selected ampacity model, one can find the instantaneous ampacity for given weather conditions.

3 Description of the sensor system

The wireless sensor system the authors propose consists of several units: Up to six sensor units mounted on the phase conductors of the overhead transmission line and one ground unit located in a power substation where it interfaces with line operator's control system computer network.

Wireless communication and networking between the sensor units and the ground unit is implemented using ZigBee in the 2.4 GHz band. This low data rate secure wireless networking standard was chosen for ease of use and availability of both hardware and software components. Both the sensor units and the ground unit are built around a ZigBee-enabled microcontroller responsible for measurement, data preprocessing, networking and communication. The temperature is measured using an intelligent sensor mounted directly on the conductor. Current measurements are accomplished using a flexible Rogowski coil sensor with built-in precision integrator. Provisions have been made for future inclusion of other sensors such as an accelerometer for conductor vibration analysis.

The sensor units are equipped with a low-gain patch antenna. Its wide beam suitably compensates for its fixed position and orientation and its low flat profile does not significantly increase corona discharge incidence. In order to increase the operational range of the system, the ground unit utilizes a high-gain reflector antenna. The sensor unit is supplied directly from the measured conductor using a current transformer.

3.1 The sensor unit – temperature measurement

To measure the conductor temperature, various temperature sensors can be used. These range from the simplest two-pin devices with temperature-dependent resistance through more intelligent sensors where the temperature information is presented in the form of output pulse width modulation to fully digital parts that provide the temperature information in digital form over one of the industry standard microcontroller peripheral buses. The two-pin analog sensors are often plagued with significant nonlinearity and typically will have to be used in bridge configuration. The more precise PWM devices are often unable to measure temperatures below 0 °C or require negative bias voltage to do so. The fully digital devices with SPI bus data output chosen by the authors require no difficult hardware interfacing or software processing and do not occupy AD converter channels. In order to achieve a good thermal contact, the temperature sensor case is clamped to the conductor.

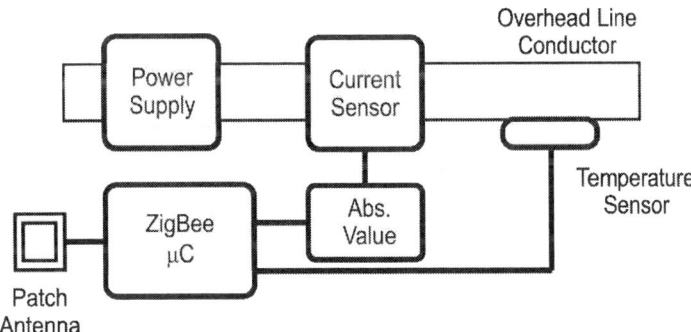

Fig. 1. The sensor unit.

3.2 The sensor unit – current measurement

Due to the very high currents involved – up to 1000 A – and the requirement that it must be possible to install the sensor units without disconnecting or damaging the high voltage conductor, it is obvious that a contactless measurement method must be employed. Classical closed core current transformer cannot be used for the same reasons. The practical choices of current sensor include a split core current transformer, flexible Rogowski coil or Hall sensor used with split magnetic core. Flexible Rogowski coils generally show very good linearity and precision over the widest current range, however, as their output voltage is proportional to the derivative of the measured current, an integrator has to be employed. Based on excellent precision parameters, a flexible Rogowski coil sensor supplied with a calibrated integrator was selected. The current sensor is followed by an absolute value circuit which passes the signal to the AD converter of the microcontroller. RMS value of the current is calculated and averaged over several signal periods.

3.3 The sensor unit – power supply

Due to the very wide range of operating temperatures at the high voltage conductor, backup battery cannot be reliably included as a core component in power supply design. Prolonged exposures to high temperatures would drastically degrade battery capacity, necessitating too frequent maintenance. Split core current transformer eventually chosen as the power supply method for its straightforwardness would benefit from the backup battery only under low current conditions where ampacity monitoring is no longer useful. The wide range of currents carried through the conductor (from several tens of amperes to several hundred) makes it is necessary to utilize a regulator circuit to keep the supply voltage constant.

3.4 The ground unit

The ground unit employs the same ZigBee-enabled microcontroller as the sensor units. Interfacing to the transmission line operator's computer system is for the sake of simplicity implemented using a USB virtual serial port device connected to microcontroller UART peripheral. One of the advantages of this solution is that it is not required to write any drivers for the host computer, as serial port application programming interface tends to be accessible from user level applications. The ground unit is supplied from the +5 V line of the host computer USB port.

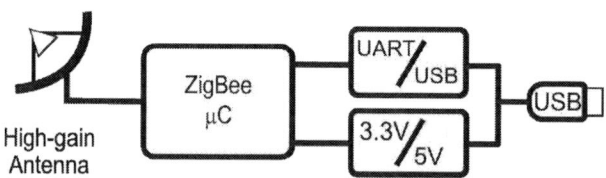

Fig. 2. The ground unit.

4. Conclusions

This paper presents a wireless sensor system for overhead transmission line ampacity rating capable of monitoring current and temperature of several high voltage phase conductors. In the first part the authors give reasoning for dynamic ampacity rating and outline the presented sensor system. The second part briefly defines ampacity. The third part of the paper describes individual building blocks of the system with short discussion of advantages of the chosen implementations.

The presented sensor system is intended as a basis for future dynamic ampacity rating system usable as a part of transmission network management and control system and for verification of ampacity model properties. At first, only operation in vicinity of a power substation is considered due to the short range nature of the presented sensor system. Subsequently a modified system with GPRS data capability and broader variety of sensors will be developed for use anywhere in the field.

Acknowledgement

This work was supported by the Grant Agency of the Czech Technical University in Prague, grant No. SGS 10/802800.

References

[1] L. Renson, C. Jamar, S. Guérard, J. Lilien and J. Destiné in *Proceedings of the 1st International Operational Modal Analysis Conference (IOMAC)*, Copenhagen, Denmark, 2005

[2] V. Hinrichsen et al. in *Proceedings of the 19th International Conference on Electricity Distribution (CIRED 2007)*, Vienna, Austria, 2007, paper 0788.

[3] W. de Villiers, J. H. Cloete, L. M. Wedepohl and A. Burger, *IEEE Trans. on Power Delivery.* **23**, p. 389, 2008.

Wireless Sensor Network Control System

M. Husak, A. Boura and J. Jakovenko

Department of Microelectronics, Czech Technical University in Prague,
Technicka 2, 166 27 Prague 6, Czech Republic
e-mail: husak@fel.cvut.cz, bouraa@fel.cvut.cz and jakovenk@fel.cvut.cz

The core of the paper is the control system with a few electronic blocks (control, actuator, software, and RF wireless communication). The control block drives operations in the system, the wireless block ensures communication, and data transfer, the actuator part is drived by control block. The software drives all operations in the system. Sensors ensure basic information about environment. The number of sensor can be variable. The control block cooperates with wireless sensors, and wireless actuators. The heart of the system is the control microprocessor. The system communicates with PC, mobile phone etc. There was used a new architecture of a multisensor system for physical measurement using wireless data transfer in the paper. Different control software was designed for wireless parts.

1. Wireless sensor network system

The system is used as a support for wireless transfer of sensor data. The system was designed as modular with possibilities of number components extending. The system was designed for measurement and control use. There is described a new architecture of a multisensor system for remote temperature measurement using wireless communications in the paper. There are used sensors with digital outputs in the system [1]. The number of sensor can be variable. The control software of the whole system has been designed. Partial control programmes were designed for wireless unit control. There are many program functions implemented. The system contains units: wireless, control, actuator, software equipment. RF wireless unit contain integrated RF chip nRF9E5. One ensures wireless communication between control unit and sensors as well as wireless switch unit. The control unit controls system operation, i.e. communication transfer, sensor data processing as well as switching of actuator unit. Actuator switch unit is wireless controlled by control unit. There were hardware and software realized and tested in the designed system. Wireless communication is ensured in the range of 300 m in the free space. The system was designed to operate with different type of physical sensors. The system can used PC, PDA or mobile phone to communication with control unit as well as signal processing [2].

The wireless system consists of several main blocks. The control block is the basic part, securing communication in the system. The unit communicates with wireless physical sensors and wireless actuators. It is provided with custom-set programs, controlling the whole system. Second part of the system is represented by the wireless physical sensors. The sensors are placed in required localities. The sensors measure physical quantities periodically and convey the measured data to the control unit. The third part is the wireless actuators. The blocks communicate with the control part periodically. A block schematic diagram of the whole system is shown in Fig. 1.

978-1-4244-8574-1/10 $26.00 © 2010 IEEE

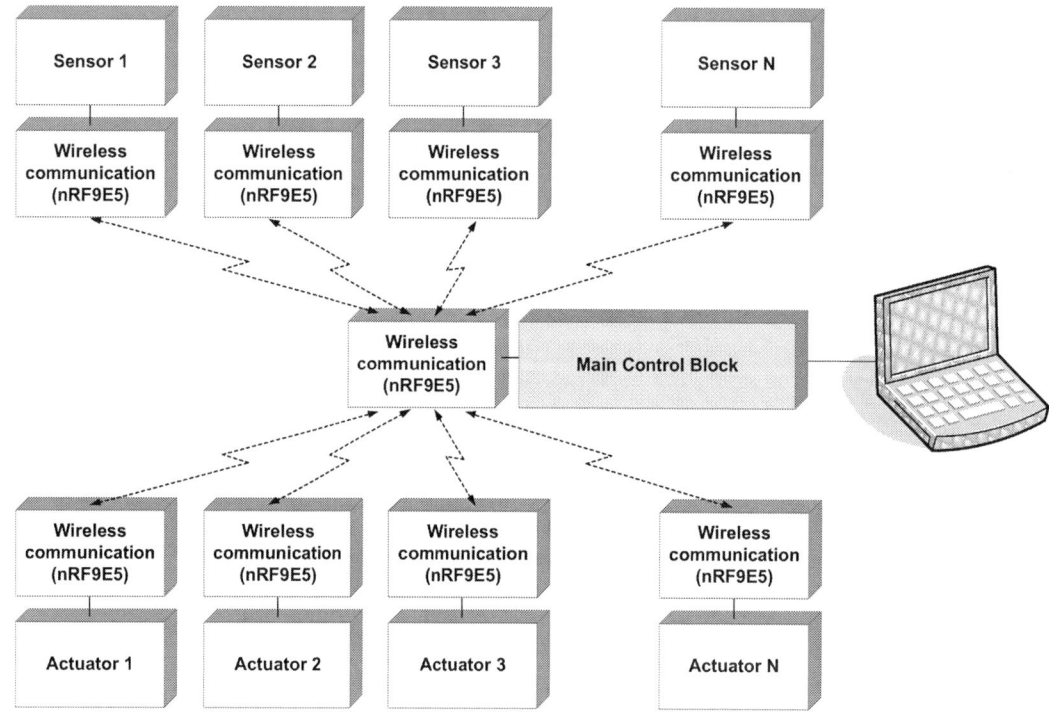

Fig. 1: Wireless sensor network system

The physical sensors and actuators can be placed at random individually and independently according to the user's needs. The only condition they have to fulfill is that they must lie within the wireless communication range. During normal operation, information from individual zones, are displayed periodically. They indicate the set and measured physical quantities in a particular zone, sensor and control unit battery voltage.

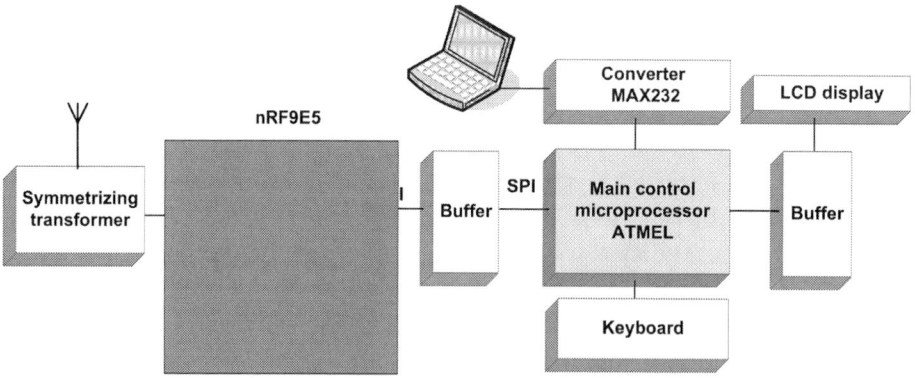

Fig. 2: Wireless control block

2. Control block and sensor block

The block diagram of the control part is in Fig. 2. The core of that part is the control microprocessor, through which all remaining parts of the system are connected. A type HD4478U controller, common with character type LCDs is used to control the display [3]. The real time circuit communicates by means of a serial bus. The bus consists of the SCLK

978-1-4244-8574-1/10 $26.00 © 2010 IEEE   216

signal for clock frequency transmission, the I/O signal for data and chip select signal transmission. The transmitting and receiving parts are very similar to those of the wireless temperature sensor and the switch unit. The microprocessor, memory and antenna connections are the same, the SPI bus is used for connection with the controlling microprocessor (MISO, MOSI, SCK signals). The slave select signal is software on one of the microprocessor ports. Two voltages are needed for control unit supply, namely 3.3 V and 5 V. The 5 V supply is used for of the LCD display and its buffer circuit. All other circuits are supplied by the 3.3 V. A type HD4478U controller, common with character type LCDs is used to control the display [3]. A type AT89S8253 circuit is the core of the control unit. Since the nRF9E5 circuit has the SPI hardware as master only, the control processor must be operated in slave mode. A simple converter serves for communication with a PC over RS232 line. It is based on a type MAC3232 circuit. The circuit contains a doubler and voltage inverter in addition to the RS232 line drivers/converters [4].

The main microprocessor is 8051 architecture compatible. It includes 4 kB program memory, 256 bytes of data memory and special function registers. The upper 128 bits are accessible by indirect addressing since they are shared with special function register addresses. The program memory is a RAM type and the program is recorded in it by the Bootloader after SPI from the EEPROM memory after resetting. A header must be present in front of the program in the memory, containing the memory speed, crystal frequency and user data.

An ATMEL 89S8253 type was selected as the control processor. It is a 8051 architecture compatible microprocessor, containing additional 2 kB data EEPROM, 12 kB FLASH program memory, SPI interface and further hardware. Both the program and data memories are In System Programmable (ISP), by series programming through the included SPI interface. Further the circuit contains a 256 byte ARM memory, whose upper 128 bits are accessible by indirect addressing since they are shared with SFR special function register addresses. The lower 32 bytes of RAM are four register banks. Instructions are fully compatible with the 8051 architecture and operate identically [5]. The instruction timing is the same as in the preceding case.

The nRF9E wireless chip is the core of a every sensor block. The sensor used contains an AD converter and a series interface. The DS620 sensor as example can be used for the measurement of temperature. The circuit contains a temperature sensor, A/D converter, comparator and a series interface. It does not need any external components for its operation [6]. The temperature measurement range is -55 ^{0}C to +125 ^{0}C. The A/D converter resolution can be adjusted from 10 to 13 bits, corresponding to 0.5 ^{0}C to 0.0625^{0}C resolution. The conversion duration depends on resolution and takes between 25 ms and 200 ms. A twin lead I$_2$C bus is used for communication. The SCL terminal serves for reception of clock pulses and the SDA terminal for data reception or transmission. The circuit also contains an EEPROM memory to which a part of registers can be copied and so preserve the setting even when the power supply is disconnected. The communication with the microprocessor takes place over an I^{2}C bus.

3. Software of measurement instrumentation hardware

The run of the system is supported by the sensor block, control program, control program of the wireless transmission and reception section of the control block. The control program of the Atmel processor in the control unit is the largest program in the system. The control program is divided to different functions. Individual functions control behaviour of the corresponding instrumentation parts. The program for communication with AD converter is simple implementation of the SPI protocol. Another function performs data read-up and saving from/to the EEPROM inner memory of the processor. Another function controls character LCD display. 8-wire bus is used for data transfer. Further function performs data transfer using serial bus and communication with the sensor. The PC instructions are loaded form the computer in the loop. Individual functions might be called from these instructions. All functions are periodically running in the PC interruption. Basic functions are displayed in the flow diagram – see Fig. 3.

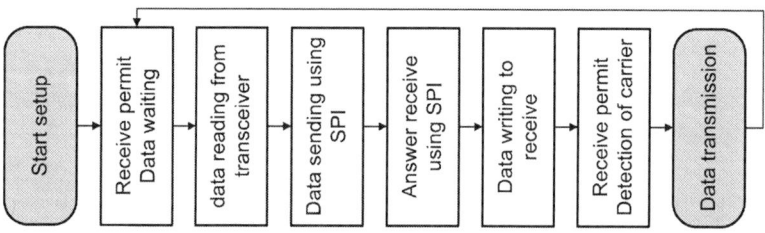

Fig. 3: Control block program

4. Results and conclusions

The highest RF, output power +10 dBm, or 10 mW, means a free space more than 300 m. During tests, the system worked flawlessly even in close presence of common interference sources like TV receivers, PCs, microwave ovens etc. The system can be used with minor modification in the all energy domains for processing of physical as well as biochemical quantities. In the work was used example of the temperature domain. Where is made design with other sensor type, it is necessary adjust output parameters of the sensor to input parameters of the wireless unit. Sensor with analog output must used amplifier and AD converter. The system is designed like multisensor system to be able to work with independent thermal sensors and switch units. The communication is two bidirectional among parts of system, i.e. the data transmission to the control unit as well as also to the wireless sensors and switch units. The control unit is patchable with computer and one makes it possible extract user's programmes. The system can be extended also alongside hardware.

Acknowledgement

This research has been supported by the Czech Science Foundation project No. 102/09/1601 "Micro- and nano-sensor structures and systems with embedded intelligence" and partially by the research program No. MSM6840770015 "Research of Methods and Systems for Measurement of Physical Quantities and Measured Data Processing" of the CTU in Prague.

References

[1] Varadan,V.K. et al., RF MEMS and Their Applications, Wiley, 2003
[2] Kirianaki,N.V.,et.al., Data Acquisition and Signal Processing for Smart Sensors, Wiley, 2002
[3] http://pdf1.alldatasheet.com, 2007
[4] http://datasheets.maxim-ic.com, 2007
[5] http://www.atmel.com, 2007
[6] http://datasheets.maxim-ic.com/en/ds/DS620.pdf, 2007

Detection of soft X-rays using semi-insulating GaAs detector

B. Zaťko[1], F. Dubecký[1], P. Boháček[1], V. Nečas[2], L. Ryć[3]

[1]Institute of Electrical Engineering, Slovak Academy of Sciences,
Dúbravská cesta 9, SK-841 04 Bratislava, Slovakia
[2]Faculty of Electrical Engineering, Slovak Technical University of Technology,
Ilkovičova 4, SK-812 19 Bratislava, Slovakia
[3]Institute of Plasmas Physics and Laser Microfusion, Hery Str. 23,
P. O. Box 49, PL-00 908 Warsaw, Poland

e-mail: elekbzat@savba.sk

This work deals with the performance of thin semi–insulating GaAs–based detectors of ionizing radiation DC–coupled to the low noise readout electronics. Fabricated detector have square shape Ti/Pt/Au Schottky blocking contact with dimension of 0.5 mm and full area AuGeNi eutectic alloy quasi–ohmic contact on the opposite site. The current–voltage characteristic and ^{241}Am pulse–height spectra of the detectors were measured and evaluated. The best energy resolution of 6.7 % for 59.5 keV gamma photons was achieved at lowered temperature of 255 K. The detector and the first stage of preamplifier (JFET) was cooled using one stage Peltier cooler.

1. Introduction

This work is dealing with the study of SI (semi-insulating) GaAs radiation detectors that can be applicable in diagnostics of hot tokamak plasmas, which mainly generates X- and γ-rays and fast neutrons. These products and especially the spectrum of soft X-rays represent an important source of information about the condition of the plasmas as are the concentration of contaminant atoms, transport coefficients of injected atoms and temperature of electrons.

The fundamental request for detectors used in the hot plasmas diagnostics relates to high resistance to the damage by neutrons and gamma radiation. Detector based on bulk SI GaAs offer such properties [1-3]. Their application as a neutron detector was demonstrated in previous papers (e.g. [4]). One of the most important task in hot plasmas diagnostics presents soft X-rays spectrometry in a region of 1÷10 keV. There are currently used Silicon drift detectors (SDD) performing very short lifetime in the hot plasmas environment. Optimized GaAs detectors based on detector-grade materials with the active region width of about 100 μm could be potentially more damage resistant in comparison with the SDD.

In the present paper, we are continuing our study [5] and optimizing read-out electronics and bias voltage of the detector. The evaluation of the fabricated bulk SI GaAs-based radiation detector in terms of its spectrometric ability for the detection of soft X-rays was accomplished.

2. Detector fabrication

The detector was prepared from 2" bulk undoped SI GaAs wafer with low dislocation density (< 3000 cm^{-2}) grown by the vertical gradient freeze method (producer CMK Ltd., Žarnovica, Slovakia). The wafer with (100) crystallographic orientation was polished by the producer from both sides to (250±10) μm. Resistivity and the Hall mobility measured by the van der Pauw method at 295 K in our laboratory give values of 1.82×10^7 Ωcm and 7062 cm^2/Vs, respectively. The measured values fulfill key requirements for a "detector-grade" bulk SI GaAs material [6].

Square surface Schottky electrodes of Ti/Pt/Au (10/40/70 nm) with 0.50 mm size and 2 mm pitch were formed using photolithography masking onto one side (top) of the wafer. Just before evaporation the surface oxides were removed in a solution of HCl:H$_2$O = 1:1 at room temperature (RT: ~300 K) for 30 sec. Wafer fragment of about 1×1 cm^2 was lapped and chemo-mechanically polished down to (130±10) μm from the reverse (back) side. The quasi-ohmic electrode was formed by evaporation of AuGeNi/Au (50/70 nm) on the backside of the sample just after the surface oxides removing (as on the topside). The metal contacts were evaporated in a dry high-vacuum system. Finally, the wafer fragment was cleaved onto individual detector chips.

3.   Experiment and discussion

Current-voltage characteristic of the fabricated SI GaAs detector measured at RT (room temperature, ~295 K) is depicted in Fig. 1. The operation bias ranges between 50 V and 130 V. The detector was directly coupled to the laboratory-made spectrometric preamplifier illustrated in Fig. 2. The spectrometric system consists of the charge sensitive preamplifier Eurorad PR307SF (hybrid circuit). The fabricated SI GaAs chip was contacted onto a small PCB (printed circuit board) holder close to the input JFET and feedback elements (500 MΩ, 0.2 pF) using conductive silver epoxy. The PCB holder was glued using thermally conductive silicone paste onto the cooled side of the single stage Peltier cooler packaged in the TO8 transistor holder. During testing and operation, the holder was hermetically closed by cup with thin Al window. Signals from preamplifier were led to the shaping amplifier based on the set of Cremat CR-200 (hybrid circuits). Shaped signals continue to the ORTEC 800 analog-to-digital converter and M2D multichannel analyzer connected with a PC.

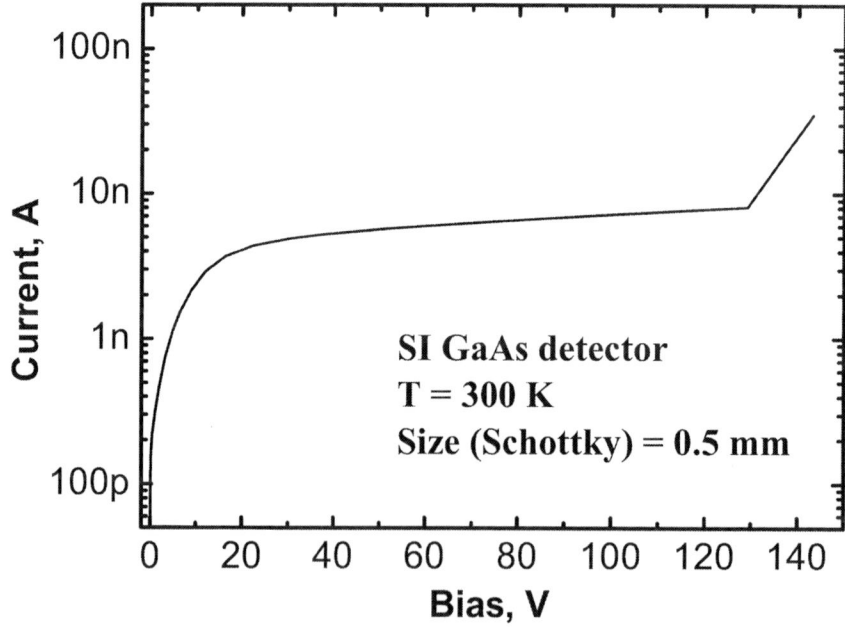

Fig.1: Current-voltage characteristic of fabricated SI GaAs detector.

Fig.2: The charge sensitive preamplifier with cooled SI GaAs detector and external JFET (magnified picture).

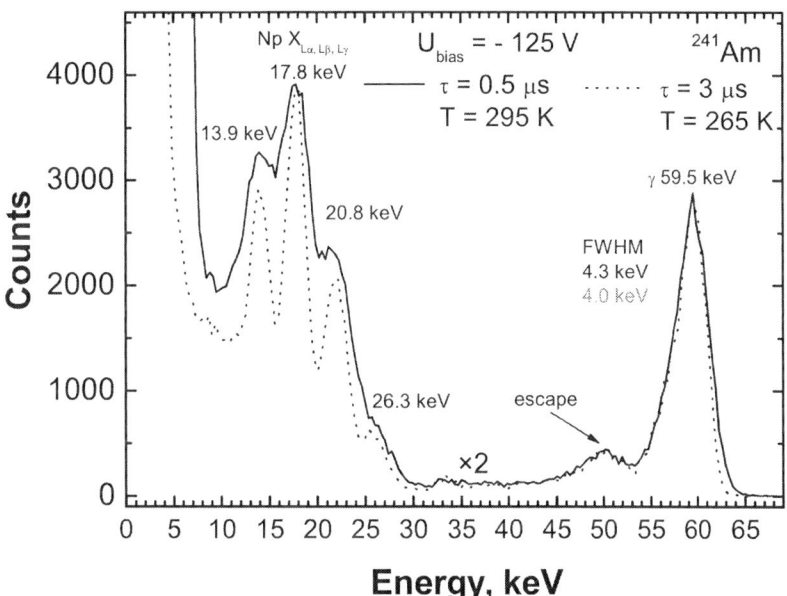

Fig.3: The calibrated pulse height spectra of ^{241}Am radioisotope measured temperature of 295 K and 255 K.

The developed system was applied to detection of X- and γ-rays from the ^{241}Am radioisotope. Fig. 3 shows measured pulse-height spectra of the evaluated detector at a bias voltage of -125 V and temperature of 295 K and 265 K. The lowest detectable energy of about 7 keV and 4 keV was observed at 295 K and 265 K, respectively. Calculated energy resolution for 59.5 keV and 17.8 keV peaks is 4.3 keV and 3.6 keV at RT and 4.0 keV and 3.0 keV at 265 K, respectively. The values of the preamplifier noise obtained by testing with the KFKI mercury pulse generator were 2.0 keV and 1.65 keV at 295 K and 265 K, respectively.

It is evident that decreasing the temperature improves the spectrometric performance of the preamplifier and the energy resolution of the detector mainly in low energy region (< 30 keV). However, the energy threshold (4 keV at 265 K) is not low enough for detection of soft X-rays emissions from hot plasmas. Hence, further improvements are necessary. The key task concerns the front-end electronics, mainly by the selection of the proper separated input JFET and optimization of the circuitry. The noise lowering will be possible by the cooling of the input JFET down to lower temperature using two- or three-stage Peltier cooler. The SI GaAs detector needs additional improvement modification: (i) thinner active base (≤ 100 μm) for sandwich electrodes as the detection efficiency is sufficient for low energy of X-ray photons, (ii) further optimization of electrodes technology and topology, and (iii) a collimator and ultra thin Be window will be necessary for the final detector design. Decreasing of the detector operational temperature will improve the detector performance, particularly signal-to-noise ratio, and the detectable energy threshold.

4. Conclusions

We evaluated spectrometric ability of SI GaAs detector in soft X-rays region. The ultra low-noise front-end electronics presents the task of the key importance in such application. We developed and tested the charge sensitive preamplifier with the separated JFET operated at RT and lowered temperature. SI GaAs detector coupled to the developed front-end electronics gives detectable energy threshold of about 7 and 4 keV at 295 K and 265 K, respectively. The fabricated and evaluated surface Schottky barrier detector has an active base of 130 μm width and gives energy resolution of about 3.0 keV FWHM for the 17.8 keV photopeak from [241]Am source at 265 K. The results show that first of all the noise of the front-end electronics chain (currently about 1.6 keV) has to be reduced. Following this, the optimization of GaAs detector should continue toward thinner base and/or improved overall electrode topology and technology.

Acknowledgement

This work was partially supported by the Slovak Grant Agency for Science through grant Nos. 2/0192/10 and 2/0153/10 and by the Slovak Research and Development Agency under contract No. APVV-0713-07. The work was also supported by the EURATOM/CU Slovak association.

References

[1] Nečas V., et al.: NIM in Phys. Res. A **458** (2001) 348.
[2] Morvic M., et al.: NIM in Phys. Res. B **197** (2002) 240.
[3] Ly Anh Tu, et al.: Nuclear Physics B (Proc. Suppl.) **150** (2006) 402.
[4] Dubecký F., et al.: Proc. ASDAM 2008, Smolenice Castle, Slovakia, (2008) 299.
[5] Zaťko, B. et al.: NIM in Phys. Res. A (2010) in print.
[6] Dubecký, F. et al.: NIM in Phys. Res. A **576** (2007) 27.

Use of Barometric Sensor for Vertical Velocity Measurement

M. Husak, and J. Jakovenko

Department of Microelectronics, Czech Technical University in Prague,
Technicka 2, 166 27 Prague 6, Czech Republic
e-mail: husak@fel.cvut.cz and vitekt1@fel.cvut.cz, bouraa@fel.cvut.cz

The measure system is based on the principle of evaluating the change of an atmospheric pressure. The vertical velocity is derived from the exponential form of the barometric equation which relates the air pressure versus the altitude. The relatively simple method has some drawbacks (a nonlinearity of the exponential dependency of the pressure versus altitude, the air temperature plays a significant role). There are used the compensated methods for measuring the vertical velocity. The correction is calculated from the horizontal velocity change which induces another vertical velocity change. The solution is based on the altitude equation, including the effects of temperature. The acceleration is evaluated in the horizontal direction using the information about the dynamic pressure. The possibility of nonlinearity compensation and temperature compensation are used.

1. Introduction

Information about the vertical acceleration is necessary for different use. We can use a several ways how to calculate the vertical velocity. The simple solutions are based on principle of calculation the change of a barometric pressure. The measurement is relatively simple method with some drawbacks, for example a nonlinearity of the exponential dependency of the pressure versus altitude. A significant role plays also the air temperature. The vertical acceleration measurement is typically method used in devices that measure the air planes rising or falling.

Compensated methods for measuring the vertical acceleration are used. The compensation and correction are calculated from the horizontal velocity change. The acceleration is evaluated in the horizontal direction using the information about the dynamic pressure [3].

2. Principles

The principle is described in [1]. The vertical velocity is derived from the exponential shape of the barometric equation (one relates the air pressure versus the altitude)

$$p(z) = p_o e^{-\frac{z}{z_o}}, \qquad z_o = \frac{kT}{gm_o} \tag{1}$$

where p_o is the sea level altitude air pressure, $p(z)$ is the barometric pressure in altitude z, T (K) is the temperature, other elements are constants, m_o (-), g (m·s^{-2}), k (J·K^{-1}). The element e^{-z} can be written in the Taylors polynomial shape. Omitting the powers from this formulation bigger then one, it can be written

$$\frac{dz}{dt} = -\frac{z_o}{p_o} \frac{dp(z)}{dt} \tag{2}$$

If the air pressure sensor with the voltage output v(p) is used and if the output voltage is direct proportional to the barometric pressure p(z) it can be obtained simple formulation for the vertical velocity in the shape

$$\frac{dz}{dt} = -\frac{z_o}{p_o}\frac{dv(p)}{dt} \qquad (3)$$

where v(p) is the sensor voltage output. Dependency according the equation can be realized using the differentiator. Non-linearity compensation can be done using the circuit with the inverse characteristic (logarithmical circuit). The temperature compensation is essential to solve as well. The circuit solution can be done by temperature compensated (pn junction in the shape of diode) logarithmic circuit which realizes linearization of the input exponential voltage. Following circuits are the amplifier and the differentiator. There can be expressed the output voltage from the simple differentiator with the feedback diode in the shape

$$v_2 = -\frac{kT}{q}\,ln\frac{v_1}{Ri_s} \qquad (4)$$

where i_s represents the current given by the diode technological parameters and R is the input resistor of the differentiator. Transfer characteristic according (4) depends on the temperature. The characteristic have the voltage shift c for different temperatures. The temperature dependency for the current i_s causes small distortion. Output voltage can be expressed in the shape

$$v_2 = f(v_1) + c \qquad (5)$$

Differentiating this equation the constant c disappears. This allows to use the feedback diode (distortion is thus given only by the temperature dependency of the technological factor i_s). The transfer characteristic according the equation (4) is very flat for higher input voltages v_1, so the sensitivity is small. The sensitivity can be increased using the amplifier which multiplies the characteristic by the constant. To obtain more precise calculation of the altitude the equation (1) can be modified using Babinet formula to the form (6)

$$z(p) = \frac{T_o}{T_r}\left(1 - \left(\frac{p}{p_o}\right)^{\frac{T_r \cdot R}{Mg}}\right) \qquad (6)$$

where z(p) is the altitude, p is the air pressure, p_o=101.325 kPa is the sea level barometric pressure according the ISA, Tr=0.0065 K·m^{-1} is the temperature gradient according the ISA, R=8.3 J·K^{-1}mol^{-1} is the universal gas constant, M=0.02894 kg·mol^{-1} is the air molar mass, T_0=288.15 K is the temperature at the sea level according the ISA, g=9,81 m·s^{-2} is the gravitational constant. Software solution and microcontroller were used for the compensation of this no linearity.

3. Hardware design

The system consists of several blocks. The block diagram of the design system is depicted on the figure 1. [2]. The system consisted from a few parts (main parts, audio parts, altitude part, power supply, output indicators etc.). The analog differentiating network delivers the differentiation peaks at its output. The peaks are read off by the microprocessor. At the same time, the altitude above sea level is measured and its value applied for correction of the measured peak. The design must care for a quick and accurate measurement of the differentiation peaks. The parameters given in [4] were taken as the design basis.

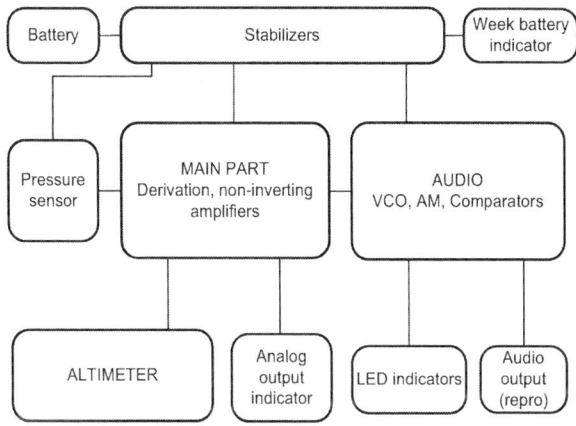

Fig. 1: Block diagram of the realized system

The barometric system works with the MPX4115 pressure sensor, the principle is described in [1]. The data sheet gives the output voltage dependence for that type of pressure sensor as [5]

$$v_{out} = V_{cc}(c_1 p - c_2)$$ (7)

where V_{cc} is the supply voltage and p is the pressure, c_1=0.009 a c_2=0.095. By arrangement of the equation we can obtain the formula for pressure

$$p = \frac{\dfrac{v_{out}}{V_{cc}} + c_2}{c_1}$$ (8)

By setting the constants into (6) and arrangement we can obtain the formula

$$z(p) = \frac{T_o}{T_r}\left(1 - \left(\frac{p}{p_o}\right)^{\frac{T_r \cdot R}{Mg}}\right) = c_3\left(1 - \left(\frac{p}{p_o}\right)^{c_4}\right)$$ (9)

where the constant c_3=44330.8 and c_4=0.190261.

Derivation of the correction data. The values read for altitude are used for the differentiation peaks correction. Under the assumption of direct proportionality, the differentiating network output (the differentiation peaks value) can be written as

$$v_{peak} = c_5 \frac{dv_{out}}{dt}$$ (10)

where v_{peak} is the output voltage of the differentiating network and c_5 is a constant. The equation (7) applies for the pressure sensor output voltage, and (12) can then be arranged to the form

$$v_{peak} = c_6 \frac{dp}{dt}$$ (11)

where c_6 is a constant. The following formula applies to the vertical velocity

$$\frac{dz}{dt} = \frac{dz}{dp}\frac{dp}{dt}$$ (12)

where the dp/dt term corrensponds to the v_{peak} measured voltage value. To express the dz/dp term, we start with the equation (6), and differentiate it over pressure

$$\frac{dz}{dp} = c_7 p_o^{c_8 p^{-c_8}}$$ (13)

where the values of the calculated constants after setting in are $c_7=83.241276$ and $c_8=0.809739$. By setting the converted pressure formula into (15) we get

$$\frac{dz}{dp} = c_7 \left(1 - \frac{z}{c_3} \right)^{-c_9} \tag{14}$$

After setting of equation (6) into (14) we get the formula for the actual dependence of vertical velocity on the differentiation peaks and on the altitude as

$$\frac{dz}{dt} = \frac{v_{peak}}{c_6} c_7 \left(1 - \frac{z}{c_3} \right)^{-c_9} \tag{15}$$

Marking one part of the formula (16) as a function $F(z)$, we can use this function to generate a table with altitude values and values of the $F(z)$ function as the corresponding corrections

$$F(z) = \left(1 - \frac{z}{c_3} \right)^{-c_9} \tag{16}$$

The resulting formula for the real vertical velocity has the form

$$\frac{dz}{dt} = c_7 \frac{v_{peak}}{c_6} F(z) \tag{17}$$

where the $F(z)$ values relating to the particular altitude are stored in the table.

6. Conclusions

The designed simple system is capable to measure a vertical velocity from 2 m·s^{-1}, the altitude measurement is auxiliary information for the vertical velocity calculation, the measurement accuracy is 2 m. The system is necessary to calibrate, one also contains a calibration section. The system designed includes distortion compensation. The calibration of the designed system is performed by means of the SPICE program.

An electronic circuit was designed, taking care of the mathematical functions including compensation functions. The transient analysis of the electronic circuit connection was used for vertical velocity simulation. The type MPX4115 pressure sensor was used in the system. Problem of nonlinearity due to the exponential function of barometric pressure versus altitude is discussed in detail. Using a linear interpolation, the altitude is calculated from the measured value.

Acknowledgement

This research has been supported by the Czech Science Foundation project No. 102/09/1601 "Micro- and nano-sensor structures and systems with embedded intelligence" and partially by the research program No. MSM6840770015 "Research of Methods and Systems for Measurement of Physical Quantities and Measured Data Processing" of the CTU in Prague.

References

[1] Husak,M. - Jakovenko,J. – Stanislav,L., Pressure Sensor Data Processing for Vertical Velocity Measurement. Nanotech 2008.
[2] Vrkoc,J: The simple analogue variometer. The final thesis, CTU 2009.
[3] http://home.att.net/~jdburch/systems.htm
[4] www.sensair.com
[5] Freescale Semiconductor. MPX4115A, MPXA4115A, 2006. www.freescale.com.

Monitoring of Car Driver Physiological Parameters

E. Vavrinský [1], V. Tvarožek [1], V. Stopjaková [1], P. Soláriková [2], I. Brezina [2]

[1] Department of Microelectronics, Slovak University of Technology
Ilkovicova 3, SK-812 19 Bratislava, Slovakia
[2] Department of Psychology, Comenius University
Gondova 1, SK-818 01 Bratislava, Slovakia
e-mail: erik.vavrinsky@stuba.sk

Miniaturization of biomedical sensors has increased the importance of microsystem technology in medical applications, particularly microelectronics and micromachining. This paper presents a new approach to biomedical monitoring of car driver physiological parameters. The system is based on our preliminary described theory and experiments [1].

1. Test set-up

New thin film multipurpose microsensors (IDAT) (Fig. 1), have been developed, where an interdigital array (IDA) of microelectrodes is integrated together with temperature sensor (T) on a single chip. The developed microsensor allows measurement of psychogalvanic reflex (PGR) by IDA structure and body temperature by T meander, locally from "one place". Moreover, it can be found in previous experiments on the heart rate monitor. Therefore, the microsensor allows continual monitoring and analysis of complex physiological, pathophysiological, and therapeutic processes.

The microelectrodes were fabricated by a standard thin film technology: Pt (Au) films (150 nm in thickness) underlaid by Ti film (50 nm) were deposited by rf sputtering on Al_2O_3 substrates, and microelectrodes were lithographically patterned by lift-off technique. The total size of the microelectrode chip is 10 x 13 mm. IDA structure was made in symmetric configuration, 200 μm / 200 μm (finger/gap) dimensions. Total resistance of thermal resistive meander by using Pt is between 530 and 540 Ω.

The car driver monitoring system consists of:
- 2 local IDAT microsensors on a driving wheel for psychogalvanic reflex (PGR), body temperature and heart rate monitoring
- 1 global PGR and 1 global ECG sensors for monitoring of conductance and ECG between left and right hand
- 1 smart pressure sensor placed in the car seat for heart pulse and respiration frequency monitoring [2]. It can be partially used for the driver weight measurement.
- The system also includes an infrared camera for monitoring of face mimic representing different psychological emotions, which are visually recognized and diagnosed using software "eMotion" [3].

The system has been tested using "Compact RIO system" and controlled by NI Labview 8.6, where the whole mathematical apparatus for signal processing, filtering and analysis is implemented [4].

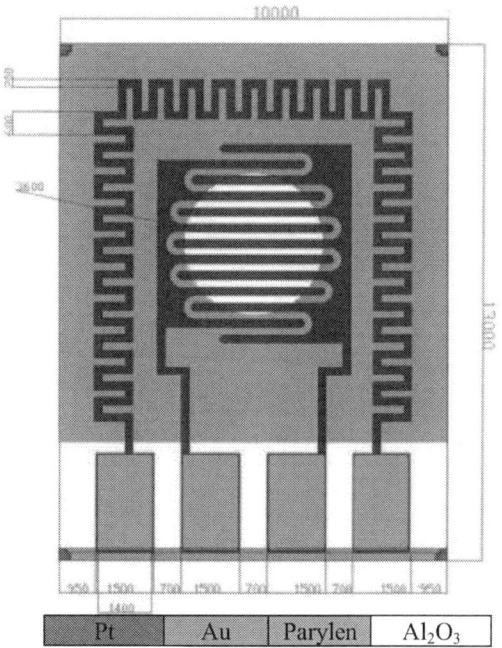

| Pt | Au | Parylen | Al₂O₃ |

Fig. 1. IDAT microelectrodes

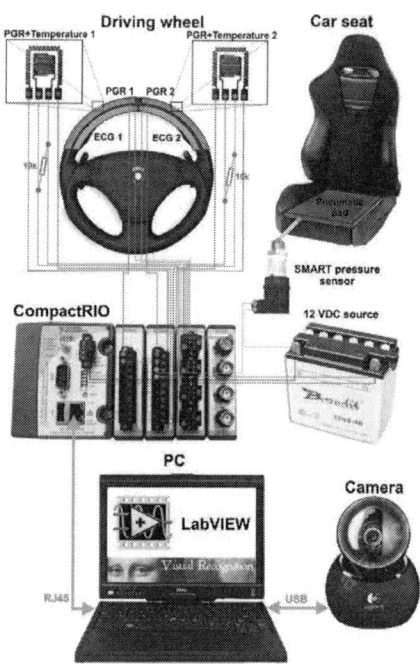

Fig. 2. Car driver monitoring system

2.   Results and discussion

A) Driving wheel sensors
-   ECG and PGR by macroelectrodes

For global monitoring of ECG and PGR by driving wheel we used aluminum macro electrodes. ECG electrodes were connected to NI 9234 (25,6 S/s, IIR Bandpass filter (1 - 130 Hz) and Bandstop filter: (48 - 52 Hz). Conductivity (PGR) electrodes were serial connected to NI 9263 (V_{OUT} = 3V, f = 1 kHz) and NI 9203 (100 S/s. Typical result for ECG monitoring is shown in Fig. 3. The PGR response of macroelectrodes (skin conductivity) corresponds to typical signal of commercial PGR sensors like in old results [1]. Macroelectrodes offer very fixed contact between human skin and electrodes and the total reliability is very good.

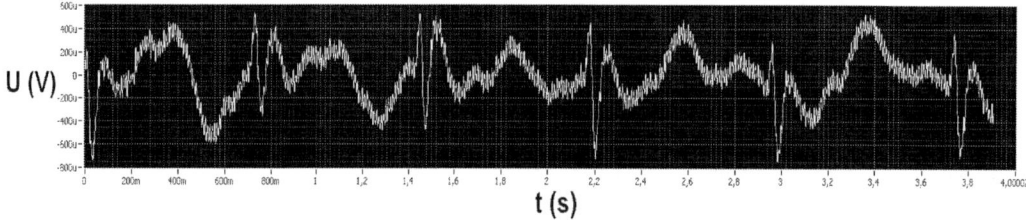

Fig. 3. ECG signal from a driving wheel

-   PGR, body temperature and heart rate by multipurpose IDAT microsensor

In this set-up we used multipurpose IDAT microelectrodes (Fig. 1) placed up on driving wheel and connected to NI 9219 card (100 S/s). The PGR and temperature output signal corresponds to preliminary experiment and the heart rate was easily read out by derivation of the measured waveform. Standard psychotests also showed that the response signals of IDAT microelectrodes and macroelectrodes were similar. IDAT microelectrodes signals were more stabile with shorter response time, but for better reliability in real praxis, we need to place microsensors on several positions of driving wheel – to obtain more fixed contact. In real praxis is ideal to combine macro and micro-sensors results [1].

978-1-4244-8574-1/10 $26.00 © 2010 IEEE   228

B) Seat sensor

In this set-up we used smart pressure sensor Treston DMP 331 [5] to monitor pressure in air filled seat cushion via serially connected NI 9219 card. As power supply we used 12 DC batteries. Typical measured signal is shown in Fig. 4a. Period designated as T_{heart} corresponds to heart pulse signal, and period $T_{respiration}$ corresponds to the respiration frequency. For better readability and reliability of heat rate, we can use IIR Highpass filter (700 mHz) (Fig. 4b).

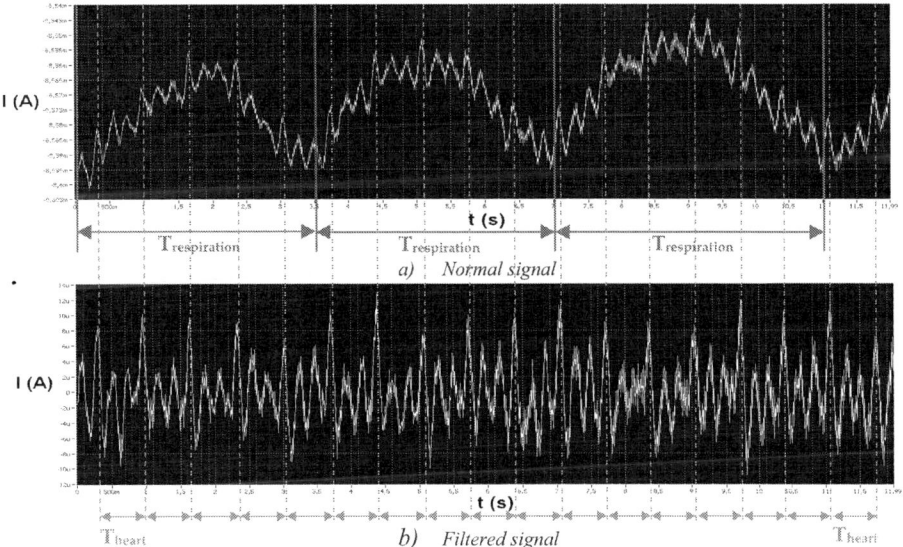

Fig. 4 Heart pulse and respiration frequency measured by the seat sensor

C) Visual emotion recognition

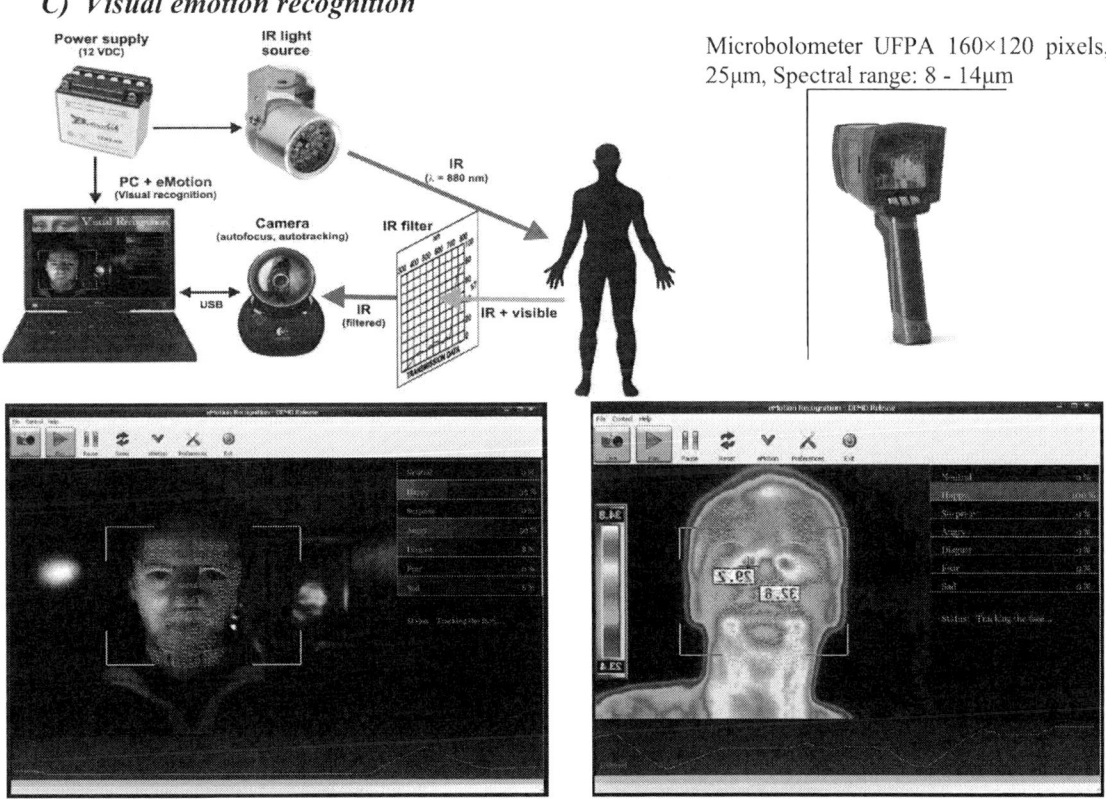

Microbolometer UFPA 160×120 pixels, 25μm, Spectral range: 8 - 14μm

Fig. 5. Visual emotion recognition

For visual emotional recognition we used cameras and eMotion software. To improve reliability in real conditions (daylight, night), and to minimize the influence of unwanted optical effects like shadows and reflex from the outer sources, we used an infrared (IR) modified (active method) and thermal camera (passive method).

- *Active method* (Fig. 5 left): To keep cost of the system down and make the system widely shareable, we have modified web camera Logitech QuickCam® Orbit AF [6] for near IR spectra (0,8 – 1,3 µm) using optical filters [7].
- *Passive method* (Fig. 5 right): In this set-up thermal camera EasIR 4 [8] for middle IR (8 - 14µm) was connected to a personal computer via S-Video. This experimental equipment is not one of the low-cost versions, however, it can be additionally used for "contact-free stress monitoring for drivers divided attention" like in [9].

3. Conclusion

Used set-up offers continuous biomonitoring and analysis of different electrophysiological aspects of human physiology in a completely safe and non-invasive manner. This technique also has no undesired influence on natural physiological processes. Motivated by the promising results achieved so far, the research will go on by the next step that is integration of the whole biomonitoring system into a real car conditions. We prove that ideal and reliable car monitoring system needs often using multiple measurements methods and the final product will need very robust and smart analyzing software. More results will be found in [10].

Acknowledgement

Presented work was supported by the Excellence Centre „CENAMOST", project VEGA grant 1/0220/09, V-10-025-00 and Project ENIAC No. 120 228 (Project MAS - Nanoelectronics for Mobile Ambient Assisted Living). I would like thank to Dr. I. Novotny and Dr. S. Flickyngerova for production and analysis of microelectrodes.

References

[1]   Vavrinsky, E.; Stopjakova, V.; Brezina, I.; Majer, L.; Solarikova, P.; Tvarozek, V. (2010) Electro-Optical Monitoring and Analysis of Human Cognitive Processes, *Semiconductor Technologies*, IN-TECH, Vienna, Austria, ISBN 978-953-307-080-3

[2]   Partin, D. L.; Sultan, M. F.; Trush, Ch. M.; Prieto, R.; Wagner, S. J. (2006). Monitoring Driver Physiological Parameters for Improved Safety, *SAE World Congress*, Detroit, USA, ISBN 0-7680-1633-9

[3]   eMotion, Faculty of Science, University of Amsterdam, Netherlands (2009). http://www.visual-recognition.nl/eMotion.html.

[4]   National Instruments, (2009). www.ni.com

[5]   Treston. (2010). Pressure sensors with internal transmitter, http://www.treston.cz

[6]   Logitech. (2009). Logitech Quickcam Sphere AF, http://www.logitech.com/en-in/webcam-communications/webcams/devices/3480

[7]   Apollo Design Technology. (2008). http://www.internetapollo.com

[8]   Guide Infrared. (2010). Thermal camera EasIR 4. http://www.wuhan-guide.com/Content.aspx?lang=en&id=195

[9]   Shastri, D. (2008). Contact-free Stress Monitoring for User's Divided Attention, Human Computer Interaction, IN-TECH, Vienna, Austria, ISBN 978-953-7619-19-0, pp. 127-134

[10] Vavrinsky, E.; Solarikova, P.; Stopjakova, V.; Tvarozek, V.; Brezina, I. (2010) Implementation of Microsensor Interface for Biomonitoring of Human Cognitive Processes, *Biomedical Engineering, Trends, Researches and Technologies*, IN-TECH, Vienna, Austria, ISBN 978-953-7619-X-X

Broadband amplitude-stabilized oscillator

Julius Foit, Jan Novák. Department of Microelectronics, CTU Prague

Abstract: - *In many applications we need oscillators tunable over a considerable frequency range by simple LC circuits with broadly varying dynamic resistances (quality factor Q of the reactances), while insisting on keeping a rather constant amplitude of the A.C. voltage across the LC circuit, or at least of some output voltage (not necessarily of a harmonic waveform) somewhere in the circuit. If we manage to arrange the circuit in such a way that a variable external load will not cause appreciable frequency shifts of the oscillations, it will be felt as a particular benefit. Another particular benefit will be strongly appreciated: if the circuit does not require any taps (inductive or capacitive) or any transformer coupling in the frequency-determining LC circuit. Perhaps even more appreciated will be a possibility to have one side of the LC circuit grounded. The following paper shows a surprisingly simple circuit capable to fulfill these seemingly conflicting requirements, all at the same time.*

Keywords: - *oscillators, distortion, LC tuning, amplitude stability*

I. REQUIREMENTS

The requirements specified above indicate that we need a circuit possessing adequate internal automatic adjustment of the basic oscillation condition – the loop gain. The circuit must have sufficient reserve of gain in order to oscillate with poor-Q LC circuits, and at the same time have enough regulation capability to keep the amplitude of LC circuit oscillation voltage within a rather narrow tolerance band. Keeping the oscillation amplitude within reasonable limits also improves the frequency stability of the oscillator since it keeps the THD (total harmonic distortion) of the generated waveform relatively low. Theoretically, there exist several methods for obtaining such a behavior in oscillator circuits; unfortunately, however, most of them are rather complex and not at all suited for LC tuning in a broad frequency range. Yet it can be shown that exploiting the basic properties of semiconductor active devices, it is possible to obtain a quite impressive behavior in a rather simple circuit.

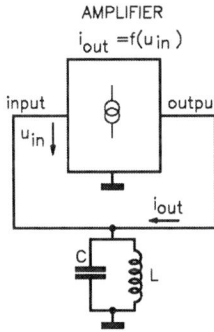

Fig. 1 Oscillator principle

II. BASIC PRINCIPLE

An example of one of the well-known basic arrangements of an LC oscillator is shown in *Fig. 1*. The amplifier in this case is operating as a non-inverting voltage-controlled current source. The LC circuit serves as a converter, converting the output current i_{out} of the amplifier back again to voltage u_{in}, serving as input to the amplifier. The oscillation condition can be formally written as

$$A_O = \frac{i_{out} R_d}{u_{in}} \geq 1 \qquad (1)$$

where:
A_O is the overall voltage amplification,
R_d is the dynamic resistance of the LC circuit at its resonant frequency.

Of course, in real circuits the value of R_d depends on the particular properties of the LC circuit and it can fall anywhere within a range of several orders of magnitude. Also, equation (1) assumes that the amplifier itself is ideal, i.e. frequency-independent which is not quite true frequently. Nevertheless, *Fig. 1* and equation (1) give a simple insight into the basic problem: in order to

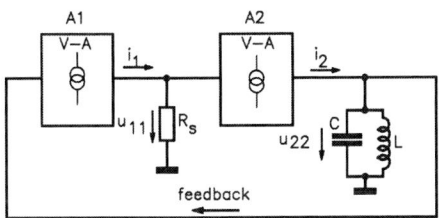

Fig. 2 Two-stage LC oscillator

secure proper operation, the amplifier properties need to be adjustable in a broad range if we assume that very different LC circuits are going to be used. It is certainly possible to adjust the amplification so that the condition (1) is securely fulfilled for the worst-case LC circuit and then just rely on the fact that excessive amplification will finally be cut down by the nonlinearity of the amplifier under conditions of severe overdrive. This, however, is far from the optimum oscillator operating conditions. A heavily overdriven amplifier frequently loses most of its good properties. First of all, its input and output differential impedance can drop to a fraction of the optimum (infinity) value. Second, a large degree of nonlinear distortion can appear in the waveform of the current driving the

978-1-4244-8574-1/10 $26.00 © 2010 IEEE   231

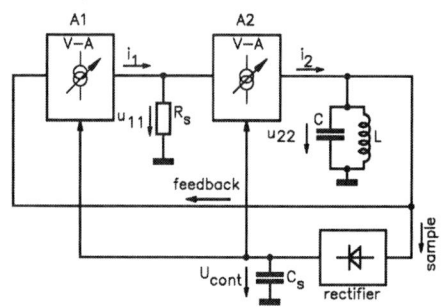

Fig. 3 Oscillator with automatic gain control

LC circuit, further impairing frequency stability. Moreover, all these effects are strongly dependent on the momentary value of the amplifier power supply voltage, causing a further deterioration of frequency stability.

III. SOLUTION

Different kinds of real LC oscillator circuits use very different internal design structures of the amplifier block (Fig. 1). The popular common emitter or common source transistor stage has two important drawbacks: it is an inverting amplifier, and its output does not behave as a really good current source, especially under the conditions of heavy overdrive. Various well-known oscillator circuits try to avoid these problems either by transformer coupling or by taps on the LC circuit. Both methods complicate the LC circuit design and remove the problems only partially.

Moreover, this solution does not at all address the wide variability of LC circuit parameters appearing in real life; as a result, oscillators with tapped or transformer-coupled LC circuits must be individually designed for each particular LC circuit in order to obtain optimum operating conditions.

In the following we will show that most of the problems mentioned above can be solved in a very efficient way by choosing a cascaded pair of two amplifier stages, both non-inverting and operating as voltage-to-current converters, instead of a single amplifier stage. The general idea of the circuit is shown in *Fig. 2*. It consists of two amplifying stages *A1, A2,* each operating as a non-inverting voltage to current converter (voltage controlled current source). The output current i_1 of stage *A1* is converted to voltage u_{11} by passing it through the coupling resistor R_s. The u_{11} voltage serves to drive the second stage *A2*. The output current of stage *A2* is converted to voltage once again, but this time the output voltage u_{22} is generated over the dynamic resistance of an LC tuned circuit. The voltage u_{22} is fed back to the input of stage *A1*, completing the positive-feedback loop. The over all loop amplification can be written as

$$A_{tot} = A1 . A2 = R_s . R_d . |y_{21s1}| . |y_{21s2}| \qquad (2)$$

where

A1, A2 are the equivalent voltage amplifications of both amplifier stages,

$R_d = Q\omega L$ is the dynamic resistance of the LC circuit at resonance (at the resonant frequency ω), is the quality factor of the LC circuit,

$|y_{21s1}|$, $|y_{21s2}|$ are the real parts of common-source differential forward transfer admittances of both amplifying stages.

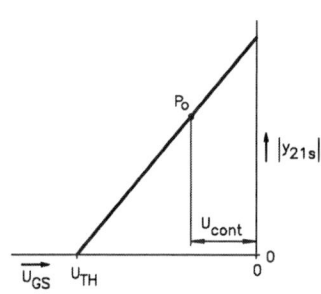

Fig. 4 The control characteristic of a JFET [2]

The basic condition (1) for self-sustained oscillations, $A_{tot} > 1$ must be fulfilled for all expected R_d values of the LC circuit involved. Theoretically, the fulfillment of this condition presents no problem; a totally different situation, however, appears in real life when we insist on having the circuit operating as an LC oscillator with:

a) broad range of tuning inductances and capacitances (several orders of magnitude for each of them),

b) a broad range of quality factors Q of the LC circuit (mostly determined by the inductor),

c) obtaining the same signal amplitude from the oscillator at any of the a) and b) conditions, and at the same time, keeping the best possible frequency stability versus supply voltage and load.

Most of the common simple LC oscillator circuits are just unable to fulfill all these requirements at the same time. Indeed there do exist oscillator circuits capable of fulfilling some of the mentioned requirements one by one, but none of them is able to fulfill all of them without raising the circuit complexity beyond reasonable limits. If we can arrange the voltage-to-current converting stages *A1, A2* in such a way that their conversion coefficients (i.e. amplification) can be controlled by an external D.C. signal, and add a circuit deriving such a control signal from the u_{22} voltage amplitude, the block diagram of *Fig. 2* will result. In fact, we have added one more feedback to the oscillator circuit in

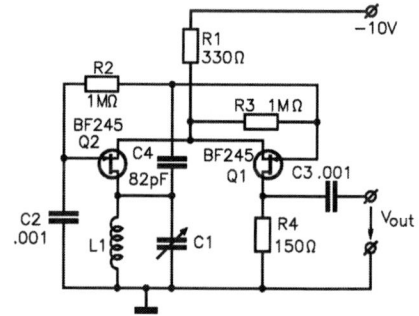

Fig. 5 Basic oscillator circuit [1]

978-1-4244-8574-1/10 $26.00 © 2010 IEEE   232

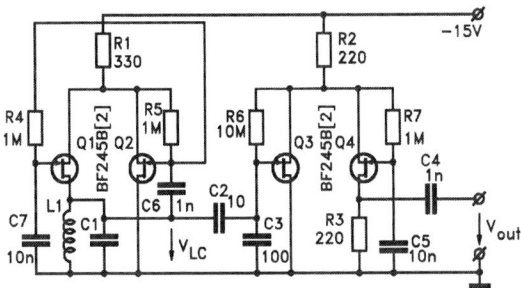

Fig. 6 Oscillator with linear output buffer

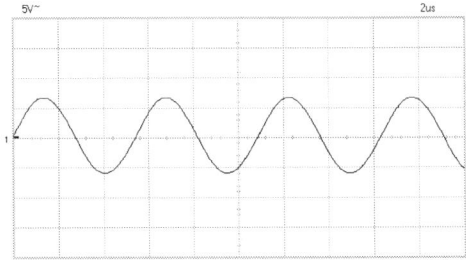

Fig. 7 V_{L1C1} waveform at 180 kHz(circuit Fig. 6).

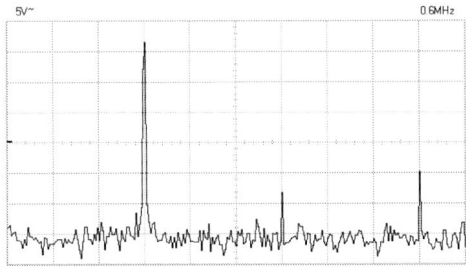

Fig.8 V_{L1C1} (Fig.6 circuit) spectrum at 180 kHz
(vertical scale 10 dB/division)

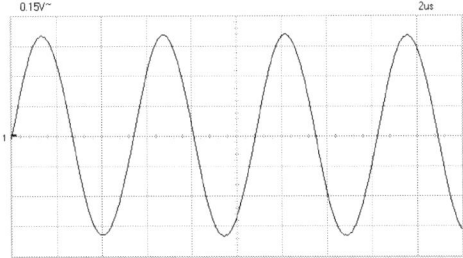

Fig. 9 V_{out} waveform (circuit Fig. 6)

addition to the original positive feedback needed for oscillations, this time a negative one, limiting the amplitude of output voltage u_{22} across the LC circuit.

Now, let us try to find the best possible active devices to be used in the amplifier blocks $A1$, $A2$. For best operation they should fulfill five basic requirements:
- their inputs should be voltage-controlled,
- the amplification should best be a proportional (or inverse) function of a D.C. control voltage.
- the output of $A2$ should have the highest possible differential internal impedance (current source),
- both stages should be non-inverting,
- the input currents of both stages should be as small as possible (preferably zero).

In order to obtain best characteristics of the gain control, the parameter with dominant effect on gain should be proportional to the control voltage. A device fulfilling this requirement very accurately is a JFET [2], as shown in *Fig. 4*.

The result of these deliberations is clear: the best active devices for both amplifier stages are JFETs, $A1$ operating as a common drain amplifier, $A2$ operating in common gate connection. The resulting final circuit schematic diagram is shown in *Fig. 5*. Type BF245B JFETs were chosen for both stages, selected units with drain current 5 mA at $V_{GS} = 0$ and at $V_{DS} = 15V$.

As can be seen from *Fig. 5*, the transistor $Q1$ operates in common drain connection as the amplifier stage $A1$, the transistor $Q2$ as the amplifier stage $A2$. The rectification of an LC circuit A.C. voltage "sample" (see *Fig. 3*) is performed by the gate junction of Q1. The coupling capacitor C4 serves also as the smoothing capacitor C_S (see *Fig. 5*) and the D.C. control voltage is brought to the gate of $Q2$ through the resistor R2. C2 connects the gate of $Q2$ to the common conductor for A.C. signal components, i.e. to operate $Q2$ in common gate connection, driven into its source from the source of $Q1$. A rather small-value resistor R4 is added in series with the drain of $Q2$ to provide an output waveform well insulated from the frequency-determining components of the circuit. In this way, the effects of variations of load at the output terminals on the operating frequency are minimized. Not negligible advantages of the circuit in *Fig. 5* are the facts that one side of the LC circuit is directly connected to the common conductor (grounded), and that no capacitive or inductive tapping or transformer coupling is needed in the tuning circuit, just a plain parallel LC combination.

IV. RESULTS

In certain applications, we may need to have the output waveform of the circuit spectrally clean, i.e. a harmonic (sine) wave, and at the same time to keep the influence of variable load impedance as low as possible. In such a case we must add a buffer circuit. A very efficient buffer fulfilling these requirements can be built using a similar circuit arrangement as the oscillator itself – two JFETs operating as a two-stage source-coupled amplifier. The resulting circuit connection is shown in *Fig. 6*. The graphs in *Fig. 7* and *Fig. 8* show that the principal output voltage is a rather good harmonic wave; the 3rd harmonic relative amplitude is about 45 dB below the fundamental harmonic, the 2nd harmonic amplitude is even lower. A salient property of the circuit is that the signal cleanness is preserved even when widely different values of L1 and C1 are used; as a matter of fact, the same output voltage and spectral purity is obtained even at frequencies as low as 5 kHz and as high as 50 MHz with no adjustment of any passive component values of the circuit (except the L1, C1 values, of course). The amplitude shown in *Fig. 7* remains constant within 3% for supply voltages U_{DD} from 8 to 30V [1]. The same amplitude stability holds for the output voltage V_{out} (1.8 V_{P-P}, half-sine pulse, in the *Fig. 5* circuit).

978-1-4244-8574-1/10 $26.00 © 2010 IEEE   233

The extremely small value of the common drain JFET input admittance permits to use a very small coupling capacitance (see *C2* in *Fig. 6*), thereby keeping the overall tuning capacitance of the LC circuit essentially concentrated in *C1* alone. *C3* and *C2* form a voltage divider which keeps the *Q3* input voltage low enough to avoid nonlinear distortion in the buffer amplifier, and further decreases the already small effect of load variations reverse transfer through the buffer amplifier. The buffer output voltage amplitude can be controlled (within certain limits) by the values of *C3* and/or *R3*. The *R3* value shown in *Fig. 6* yields an overall buffer voltage amplification approximately unity, providing a harmonic output voltage $V_{out} \approx 1$ V_{pp} with a relatively low distortion (see *Fig 9* and *Fig. 10*), with second harmonic relative amplitude almost 40 dB below the fundamental.

If we do not insist on low distortion of the output voltage and prefer a larger V_{out} amplitude, *C3* value can be decreased to diminish the *C2-C3* voltage division ratio. The result of such an alteration is shown in *Fig. 11* and *Fig. 12*.

The output amplitude in this case is about $V_{out} \approx 2$ V_{p-p}, with a slight flattening beginning at the negative peak, resulting in the second harmonic relative amplitude of about -28 dB and third harmonic at about −36 dB.

One important property of the circuit is its low sensitivity to power supply voltage variations. With the *L1C1* circuit set to *C1* = 800 pF and *L1* = 3,16 mH, i.e. oscillation frequency 100 kHz, upon variation of U_{DD} from 15 to 35 V the oscillation frequency varied by less than 5 Hz, yielding a relative frequency-to-supply-voltage sensitivity of $2{,}5 \times 10^{-6}$ V^{-1}.

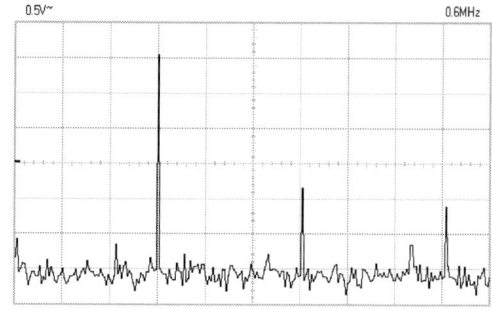

Fig. 10 V_{out} spectrum (circuit Fig. 6)
(vertical scale 10 dB/division)

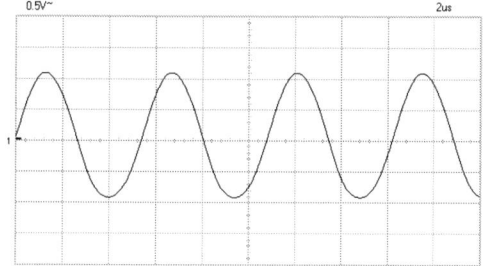

Fig. 11 V_{out} (circuit Fig. 6, C3=47pF)

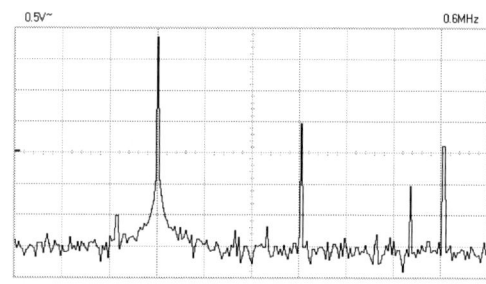

Fig. 12 V_{out} spectrum (circuit Fig. 6, C3=47pF)

V. CONCLUSION

The target goal – a widely tunable oscillator, with stable amplitude and frequency, yet as simple as possible, was reached. All circuits shown use N-channel FETs; of course, P-channel FETs can be used just as well, with opposite U_{DD} polarity.

ACKNOWLEDGMENT

This research was partly supported by the research program No. MSM6840770015 "Research of methods and systems for measurement of physical quantities and measured data processing of the CTU Prague and partly by the Czech Science Foundation Project No. 102/09/1601 "Micro- and nano-sensor structures and systems with embedded intelligence".

REFERENCES

[1] Foit, J.: *LC oscillator has stable amplitude.* Electronic Design News, Vol. 50 (2005), No. 10, pp. 93 – 96, Reed Elsevier Electronics Group, USA, ISSN 0012-7515

[2] Vaníček, F.: *Elektronické součástky.* ČVUT, Praha 1999 (monograph, in Czech, ISBN 80-01-01897-0)

[3] Foit, J.: *Přeladitelný LC oscilátor s konstantní amplitudou,* Patent application PV 2010/190, Úřad průmyslového vlastnictví, Praha 2010 (in Czech)

[4] Foit, J.: *Přeladitelný LC oscilátor s konstantní amplitudou,* Registered invention: Užitný vzor 20874, Úřad průmyslového vlastnictví, Praha 2010 (in Czech)

978-1-4244-8574-1/10 $26.00 © 2010 IEEE

Potentiality of the Inductive Powering for Measurement in the Enclosed Systems

Adam Bouřa, Miroslav Husák

Department of Microelectronics, Faculty of Electrical Engineering, Czech Technical University in Prague,
Technická 2, CZ – 166 27 Prague 6, Czech Republic
e-mail: bouraa@feld.cvut.cz, husak@feld.cvut.cz

Paper presents wireless powering and signal transfer solution using the inductive coupling. It can be suitable for enclosed systems that are isolated from the surroundings and where the batteries cannot be used. It can be used for biomedical probes, probes bricked in to the wall for long-term monitoring, extreme temperature environment measurement etc. The powering is provided by the near magnetic field. The serial and parallel resonances are considered in the design in order to increase the voltage transfer. Voltage levels, power transfer efficiency and signal modulation effects are studied up to distance of 30 cm. Presented principle of powering and communication is similar to the RFID systems (Radio Frequency Identification). The main contribution of this paper is powering potentialities and distance limits prediction of this powering strategy. The simple coils are considered. Special converter is presented for this purpose.

1. Introduction

Wireless measurements are important in situations where the standard techniques cannot be used. Wireless solution requires the special way of powering. Most used technique is to power the circuits using some kind of battery. The battery operation of the system has its limits caused namely by the limited lifetime, toxicity of the battery or an extreme temperatures environment that is improper for the batteries. There are several alternative possibilities how to power the isolated system but sometimes the energy must be wirelessly injected to the system. Method, which is described in this paper, injects the energy to the system and it is also proper for the data transfer. Most known application of this method is RFID (Radio Frequency Identification) systems, where the chip containing the identification tag it is powered. This paper presents possibilities and limits of the powering via the inductive coupling where the serial and parallel resonance is used for maximization the transfer efficiency. Special kind of converter is presented which allows measurement and direct modulation of the transfer signal.

2. Model of the Inductive Coupling

When two or more inductances partly share the magnetic field, the coupling is presented. Voltage transferred to the second inductance can be calculated using the Faraday's law. The secondary voltage thus depends on the secondary coils geometry and on the magnetic field distribution of the primary inductance.

2.1 Coupling Coefficient

In terms of the circuit theory the inductance coupling can be described using the mutual inductance or using the coupling coefficient. This approach is suitable because the standard simulation tools can be used for a description of the powering. As well as the voltage transfer also the coupling coefficient is given by the coils geometry and the magnetic field distribution. Most important topology of the inductances is the axial orientation. Coefficients dependency versus the distance is given by the magnetic field intensity distribution.

Circular coil of radius r, winding number N and current carrying I, induces the magnetic field inductance B (in environment of permeability μ_0). The inductance versus the axial distance x can be expressed using the equation (1) [1].

$$B = \frac{\mu_0 \cdot N \cdot I \cdot r^2}{2 \cdot \left(r^2 + x^2\right)^{\frac{2}{3}}} \approx \frac{\mu_0 \cdot N \cdot I \cdot r^2}{2} \frac{1}{x^3}\Big|_{x \gg r} \qquad (1)$$

The field drop is proportional to $1/x^3$ for distances much bigger than the coils diameter and thus also k is dependent this way. For smaller distances the secondary coil is embracing also the non-axial magnetic field. This is why the coupling coefficient is changing faster for the smaller distances. This change is approximately exponential. Fig. 1 shows the measurement result of the coupling coefficient between the circular coil (diameter 135 mm, winding number 10, inductance 39 μH and serial resistance 1.4 Ω) and the rectangular surface coil (maximal side length 90 mm, minimal side length 4 mm, winding number 45, inductance 82 μH and serial resistance 22 Ω). This dependency is approximated using the exponential and the $1/x^3$ functions.

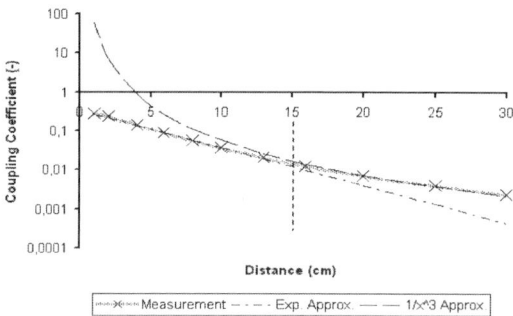

Fig. 1. Example of approximation by the exponential and $1/x^3$ functions for the coils

Character of this dependency is changing approximately at the distance equal to the diameter of the bigger coil (vertical dash line in the picture). This knowledge of the coefficients character can be used for a prediction of the coefficients value for the other coils and distances.

2.2 Equivalent Circuit

For an understanding the effects in the circuit it is helpful to derive an equivalent circuit (Fig. 2) of the coupled inductances. Derivation of this circuit can be realized by observing the circuit behaviour under the loading impedance change [3]. The mutual inductances split into the independent leakage and magnetizing inductances [2].

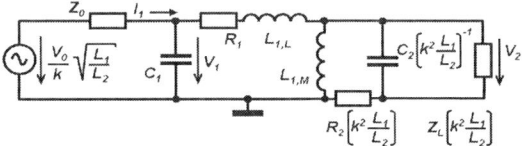

Fig. 2. Equivalent circuit respecting the parasitic properties of the inductances

It is evident from the Fig. 2, that the circuit exhibits some resonant behaviour. Omitting the parasitic effects, the parallel resonance frequency f_{rp0} can be expressed by the (2). The inductance L_e is a parallel combination of the $L_{1,L}$ and $L_{1,M}$. Parallel capacitance C_p is the secondary side capacitance transformed according the equivalent circuit from the fig. 2.

$$f_{rp0} = \frac{1}{2 \cdot \pi \sqrt{L_e C_p}} = \frac{1}{2 \cdot \pi \sqrt{L_2 \cdot C_2 \cdot (1 - k^2)}} \tag{2}$$

Also serial resonance f_{rs0} on the input can be present (impedance Z_0 from the fig. 2 must be capacitance) in order to increase the voltage transfer.

Both resonances, the serial on the primary side and the parallel on the secondary side can be present simultaneously. There are two tank circuits. The serial tank - C_s-$L_{1,L}$ and the parallel tank C_p-$L_{1,M}$. General properties of this configuration is quite complex - depending on the tank circuits resonation frequencies. Behaviour of this circuit depends namely on the coupling coefficient. The resonance frequencies can be changed independently for small coupling coefficients. For high coupling between the inductors the resonance frequencies are also coupled. Fig. 3 presents the change of the voltage transfer for different coupling coefficients.

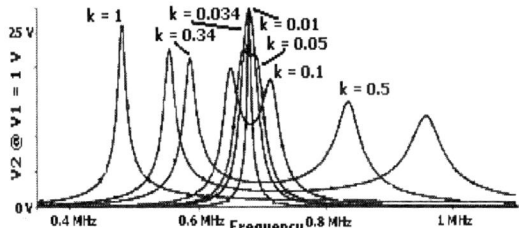

Fig. 3. Simulation according the fig. 3; $L_1 = 55$ µH, $L_2 = 82$ µH, $C_s = 1$ nF, $R_1 = 1.2$ Ω, $R_2 = 21.4$ Ω, $C_2 = 671.5$ pF

2.3 Powering and Communication

The voltage transfer in resonance can be used for maximizing the voltage level of the power transfer (it can increase the transfer several hundred times) and it is also very sensitive on the loading impedance because it is changing the quality factor of the resonance tank circuit. Also the input current of the inductance L_1 is changing according the load. This effect can be used for communication between the powered device and the transmitting device. Simple powering and communication scheme is presented on the Fig. 4.

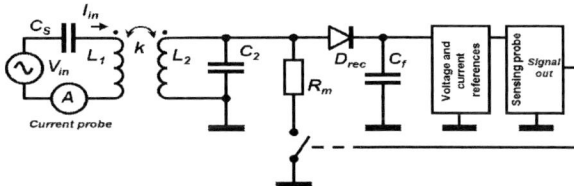

Fig. 4. Simplified powering and communication scheme

3. Converter

According the presented communication scheme it is necessary to have a switched output from the sensing probe in order to active the modulating resistor on the secondary inductance. Standard analog to digital converters (ADC) cannot be used for this purpose because of the power consumption and complexity of data output. It is necessary to design a simple converter with specific signal output. Possible solution is presented on the fig. 5 which represents current to pulse-ratio converter.

Principle of this converter is based on a dual-slope ADC while the final digitalization is performed on the receivers side. Input current I_{in} is integrated by a capacitor and when its voltage exceeds the specific level it changes the current flow and the capacitor is discharged by the reference current I_o.

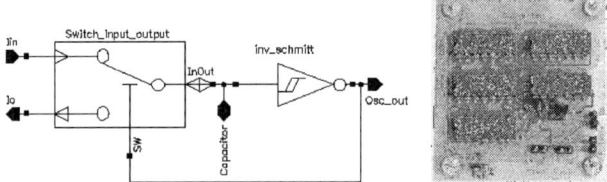

Fig. 5. Block scheme of the converter (left) and its realization using the HEF 4007 circuits (right)

4. Conclusion

Presented paper summarizes basic theory of the no-ideally coupled inductances, which can be used for powering the micro-systems (e.g. [4]). Measurements in real application were performed for surface rectangular coil on the secondary side and for the simple winded coil on the primary side. To maximize the voltage transfer with respect to the possible power reliability the frequency must be tuned properly. Different resonances can be used. The power transfer can be up to several tens of miliwats for distance up to 15 cm (at one volt on the primary side). Special kind of converter can be design for the signal transfer.

Acknowledgement

Research described in the paper has been supported by the Czech Science Foundation project No. 102/09/1601 "Micro- and nano-sensor structures and systems with embedded intelligence" and partially by the research program No. MSM 6840770012 "Transdisciplinary Research in Biomedical Engineering II" of the CTU in Prague.

References

[1] T. Prochazka, Antennas for RFID systems, Elektrorevue, 2002, vol. 22, electronic file available at http://www.elektrorevue.cz/clanky/ 02022/index.html, 2009.

[2] A. Boura, P. Kulha, M. Husak, Simple Wireless A/D Converter for Isolated Systems, IEEE ISIE 2009 Proceedings [CD-ROM], Seoul, Korea, p. 323 – 328, ISBN: 978-1-4244-4349-9

[3] Wu, J., Quinn, V., Bernstein, G. A simple, wireless powering scheme for MEMS devices, MEMS Components and Applications for Industry, Automobiles Aerospace, and Communications, Proceedings of SPIE, 2001, Vol. 4559, URL: http://www.ece.utk.edu/~jaynewu/Papers/SPIE%202001%204559_7.pdf

[4] Kulha, P. - Kromka, A. - Babchenko, O. - Vaněček, M. - Husák, M. - et al.: Nanocrystalline Diamond Piezoresistive Sensor. Vacuum. 2009, vol. 84, no. 1, p. 53-56. ISSN 0042-207X.

Influence of Conductor Systems on the Crosstalks in Integrated Circuits

J. Novak, J. Foit, V. Janicek

Department of Microelectronics, Czech Technical University in Prague
Technicka 2, CZ-166 27 Prague 6, Czech Republic,
novakj2@fel.cvut.cz, foit@fel.cvut.cz

The advent of novel sub-micron technologies of IC fabrication led to such a decrease in lead-to-lead separation that it is not possible any more to neglect the influence of these leads on the reliability of the system operation. Both the small lead separation and the application of multilayer interconnecting systems cause parasitic electromagnetic couplings [1]; in the case of a unipolar CMOS technology, the capacitive coupling is the dominant effect. It is impossible to measure direct the rapid variations voltage between leads inside the IC.

1. Introduction

The capacitive coupling is the predominant type of coupling in high-impedance circuits, like NMOS and CMOS IC's. Fig. 1 shows the way of transforming the capacitive coupling between conductors inside an IC to ideal capacitors [2]. The C_S capacitors represent the capacitive coupling between particular leads and the substrate; the substrate is usually connected to common ground or a power supply lead, depending on the particular IC fabrication technology. The capacitor C_V represents the capacitive coupling between individual leads inside the IC. In Fig. 1 the IC1 is the interference transmitter and the IC4 is the interference receiver.

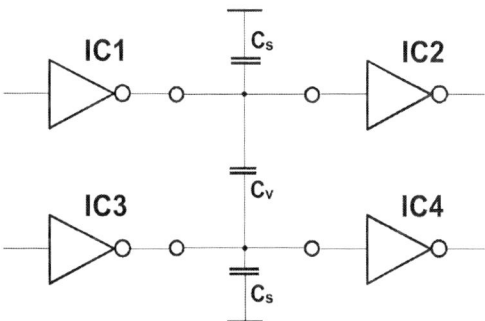

Fig. 1: Equivalent circuit of a capacitive coupling between leads inside an IC

2. A. C. analysis of capacitive couplings

In order to make possible a quantification of the frequency characteristic of the capacitive coupling, it was necessary to rearrange the capacitive coupling equivalent circuit (Fig. 1) to a form from which it was possible to derive the coupling transfer function (Fig. 2) [3]. The resistor R_I represents the internal resistance of the interference source, in this case the internal resistance of the gate IC1 (Fig. 1).

978-1-4244-8574-1/10 $26.00 © 2010 IEEE   239

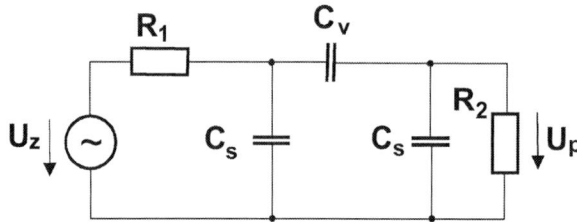

Fig. 2: Equivalent circuit of capacitive coupling rearranged for A.C. analysis

The resistor R_2 represents the internal resistance of the driver of the inactive part of the line, i.e. IC3. The input impedances of the gates IC2 and IC4 were neglected. The U_Z voltage represents the source of interference, and U_P is the voltage across the interference receiver. By an analytical solution assuming $R = R_1 = R_2$ we found the coupling transfer function P [3]:

$$P = \frac{U_P}{U_Z} = \frac{j\omega R C_V}{-\omega^2 R^2 C_S (C_S + 2C_V) + 2j\omega R (C_S + C_V) + 1} \tag{1}$$

The transfer function P has the character of a bandpass filter with characteristic frequency ω_0 and maximum transfer P_0.

$$P_0 = \frac{1}{2} \frac{C_V}{C_V + C_S}, \qquad \omega_0 = \frac{1}{R\sqrt{C_S^2 + 2C_S C_V}} \tag{2, 3}$$

2.1 Transfer function of the coupling

Figure 3 displays the effect of coefficient $k=R_2/R_1$ on the position of the transfer function maximum. The graph in Fig. 4 shows the transfer function P_m versus the coefficient k. The function P_m was normalized for $k = 1$. From the graph in Fig. 4 it follows that with increasing k the transfer P_m grows up to twice its initial value.

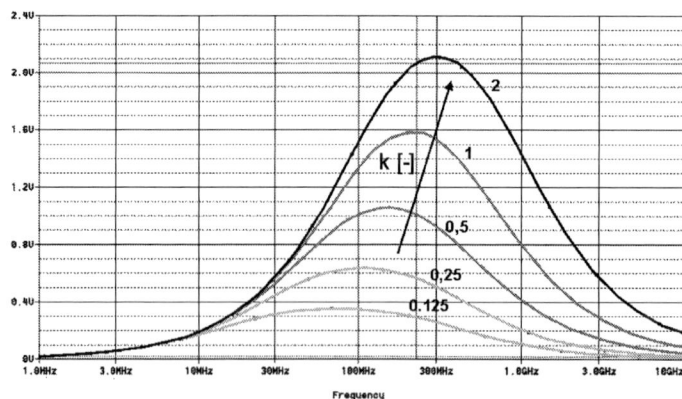

Fig. 3: Influence of coefficient k on the frequency characteristics of the coupling transfer function

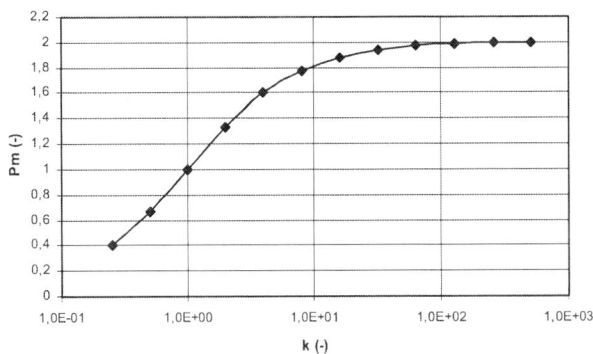

Fig. 4: Normalized maximum coupling transfer P_m versus coefficient k

3. Conductor system in the integration circuit

Applying the formula (2) and the results of table 1 it is possible to find the maximum transfer P_0 caused by parasitic capacitive coupling.

Tab. 1: Influence of conductor systems on the maximum transfer P_0, the lead dimensions correspond to the metallization layer in the AMI [4] process technology

no.	Cross-section of a conductor system	Conditions	Maximum transfer P_0.[-]
1		$w = $ min $\quad$ min $\leq s \leq 5w$	0,28 - 0,07
2		$w_1 = $ min, $s = $ min $\quad$ min $\leq w_2 \leq 5w_1$	0,022 - 0,008 (between 1-3)
3		$w_1 = $ min, $w_2 = $ min $\quad$ min $\leq s \leq 5w_1$	0,42 - 0,02 (between 2 - 1+3)
4		$w_1 = $ min, $w_2 = $ min $\quad$ min $\leq s \leq 5w_1$	0,13 - 0,01 (between 1-2)
5		$w_1 = $ min, $s = $ min $\quad$ min $\leq w_2 \leq 5s$	0,10 - 0,08 (between 1-2)
6		$w = $ min $\quad$ min $\leq s \leq 5w$	$5,1.10^{-4}$ - $-3,8.10^{-4}$ (between 1-2)
7		$w = $ min $\quad$ min $\leq s \leq 5w$	$2,2.10^{-4}$ - $-1,7.10^{-4}$ (between 1-2)

Table 1 shows examples of conductor systems inside an integrated circuit in AMI C05M-A 3M/2P/HR CMOS technology. The shielded conductor systems can be divided to systems using direct electric field shielding and system using indirect shielding. The systems utilizing direct shielding put an additional conductor, connected to common ground, in the same plane, between the interfering and the interfered conductor. An example of a directly shielded system is the three-lead system (Tab. 1 – No. 2), where the central conductor is located directly between the two interacting conductors, and is connected to the common ground.

In indirectly shielded systems, the shielding conductors are located outside the plane of the interacting conductors, in other metallization layers (Tab. 1 – No. 4, 5). The indirect shielding is based on the principle of deformation of the electric field by the shielding conductor. So, the presence of the shielding conductor changes the electric induction flux between the interacting conductors, draining away part of it into the shielding conductor.

The shielding systems mentioned above make use of either direct or indirect electrical shielding (Tab. 1 – No. 6, 7). If an extra high degree of separation is required in the integrated circuit, in other words especially low coupling capacitances between signal conductors, both direct and indirect shielding must be combined in a single unit.

4. Summary

The application of table 1 in IC design makes possible, using the C_V and C_S capacitances, a simple quantitative calculation of the maximum value of the capacitive coupling. By application of this method it is possible to forecast the maximum value of crosstalk between connecting lines inside an IC without the need to resort to time-consuming analog simulations respecting the parasitic coupling effects.

The accurate calculation of parasitic coupling of the signal interconnections helps us to design detailed circuit models of the interconnecting systems. Critical spots in an electronic system, where parasitic couplings cause failures, can be found by simulation of the circuit models. Then, it is possible to take appropriate measures to decrease the parasitic capacitances of signal conductors, like adjusting the integrated circuit layout or by applying shield conductors.

Acknowledgement

This research has been supported by the following research programme of the Czech Technical University in Prague: No. MSM 6840770012.

References

[1] Thomas Kropewnicki, Process Integration Challenges for Copper Dual Damascene Interconnects, Georgia Tech MiRC Seminar, 2002.

[2] C. S. Walker, *Capacitance, Inductance and Crosstalk Analysis*, Norwood, Artech House 1990.

[3] J. Novák, J. Foit, V. Janíček, Frequency characteristic of the capacitave coupling in CMOS integrated circuits, *Proceedings of the ASDAM'04*, Smolenice, Slovakia, pp. 267-270, 2004.

[4] *Design rule manual C05M-D*, DS13315_V5.0.pdf

Simulation of a planar Micro Ion Mobility Spectrometer for Security Applications

R. Cumeras, I. Gràcia, E. Figueras, L. Fonseca, J. Santander,
M. Salleras, C. Calaza, N. Sabaté, and C. Cané

Department of Micro and Nano Systems, Instituto de Microelectrónica de Barcelona, IMB-CNM (CSIC),
Campus UAB s/n, Bellaterra, E08193 Barcelona, Spain

e-mail: raquel.cumeras@imb-cnm.csic.es, isabel.gracia@imb-cnm.csic.es,
eduard.figueras@imb-cnm.csic.es, luis.fonseca@imb-cnm.csic.es, joaquin.santander@imb-cnm.csic.es, carlos.calaza@imb-cnm.csic.es, neus.sabaté@imb-cnm.csic.es and
carles.cane@imb-cnm.csic.es

Simulations of a planar micro Ion Mobility Spectrometer (µIMS) for sensing DMMP and Acetone vapours for security applications are performed. In µIMS ions are discriminated by the application of the proper separation voltages to the electrodes of the system. By simulation, optimum voltages for achieving the proper sensitivity have been obtained.

1. Introduction

Security has become a very attractive application for gas chemical sensing systems as many of the potentially risky substances may be detected by sensing their vapours at low concentrations. Nowadays, the main concern is the detection of chemical warfare (CW) agents, explosives and health risky volatiles. Systems for detecting these volatiles must be fast, and selective while working at very low concentrations (ppb's). Ion Mobility Spectrometers are a good option due to their fundamental advantages: high resolution (~ppb) and fast measurements (~ms), being already used in airports, public spaces and industries [1]. Thus, the development of such systems using micro-technologies for smaller systems integration is very challenging.

In this work the modelling, with COMSOL Multiphysics software, of a planar micro Ion Mobility Spectrometer (µIMS) for security applications is presented. As representative example of possible cohabitation of various target vapours in Security applications, simulations are done with the aim of being used for the detection and discrimination between $DMMPH^+$ and Ac_2H^+. IMS shows good separation capability for multiple ions with short response time, and especially µIMS's allow low temperature operation with higher sensitivity compared to other IMS implementations [2].

2. µIMS working principle

In figure1 a simplified diagram of the proposed micro planar IMS is presented. Simulated vapours are fully ionized by an external source prior to enter the IMS and flows through the gas chamber defined by two electrode plates on which a voltage is applied. The mobility K (cm^2/V·s) of a given ion at constant temperature and pressure through a drift gas with gas density N (m^{-3}) under the influence of a high electric field E (V/cm) it is not constant and can be expressed by [3]:

$$K\left(E/N\right) = K\left(0\right) \cdot \left[1 + \alpha\left(E/N\right)\right] = K\left(0\right) \cdot \left[1 + \alpha_2 \cdot \left[E/N\right]^2 + \alpha_4 \cdot \left[E/N\right]^4 + \ldots\right] \qquad (1)$$

being $K(E/N)$ the mobility coefficient of the ions within a carrier gas with density N and under a electrical field E; $K(0)$ the mobility coefficient under low or zero field; and $\alpha(E/N)$ is a normalized function describing ion mobility dependence on electric field and it is specific for each type of ion. $K(E/N)$ is characteristic of each ion and each medium, and is the basis for its identification. E/N is the electric field in Townsend units ($1\,Td = 10^{-17}\,Vcm^2$). Equation 1 is a convenient mathematical expression for the alpha function [4]. $K(0)$, α_2, and α_4, are characteristic of each ion and are obtained experimentally. When electric field exceeds 10kV/cm ($E/N \sim 40Td$) the mobility of some ions increase, decrease or remains unchanged.

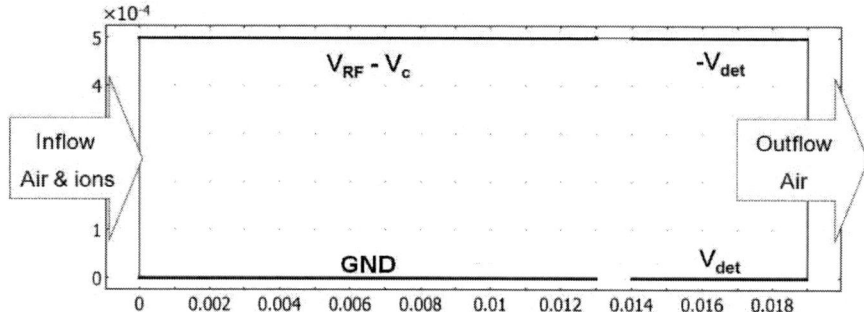

Figure 1: micro Ion Mobility Spectrometer schematics.

While ions are carried forward between the electrodes by gas flow, the use of an AC voltage (V_{RF}) between the plates will induce a vertical deviation that is characteristic of each type of ion, and due to the non linearity of his mobility there is a net displacement for each period. One of the plates is biased at high voltage with an asymmetric waveform, V_{RF} (t) and the other is grounded. This waveform must satisfy that its integration over a period has to be zero. While all ions interact with the applied RF field and are drawn towards the electrodes, applying the convenient DC voltage (V_C) the desired ions will be allowed to drift to the detector, because it compensates the displacement and prevent the ion migration towards either electrode. So selected ions can be kept in the gas stream can pass through the filter electrodes and reach the detector being this V_C voltage a characteristic of each ion species.

3.   Modelling a 2D planar μIMS

Finite element method (FEM)-based models have been implemented in COMSOL Multiphysics using a two-dimensional geometry, to simulate the behaviour of ions of two different vapours in a μIMS. Created model takes into account nonlinear combined effects of different forces and concentrations fields and combines fluid dynamics and electric field which have been found to be the most significant effects. Other effects not included in the simulations presented, such as electric repulsion in ion cloud due to space charge have been found to be considerably less significant (for the low concentration level simulated, 1ppm)[5].

Model assumptions are resumed in figure2. Drift gas (N_2) velocity in the μIMS gap has been calculated using the Navier-Stokes dynamics. Electric potentials applied to the μIMS and detector electrodes are calculated using the conductive media DC and, the movement of ions is calculated with electrokinetic flow, which takes into account of ions behaviour.

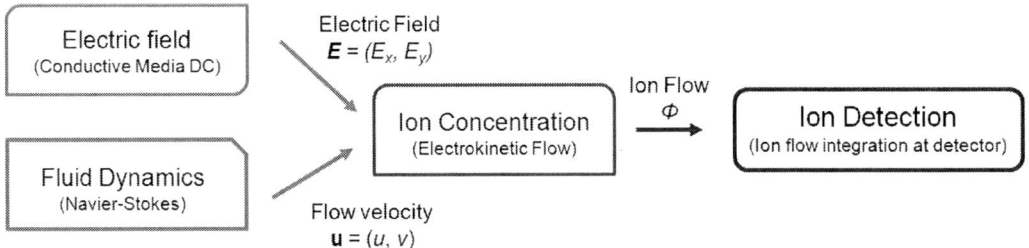

Figure 2: Block diagram of key computational steps involved in modelling µIMS with COMSOL Multiphysics software. Straight squares indicate main modules and dashed squares indicate variables needed for the modules.

A two-harmonic waveform voltage with amplitude ranging between 0 and 1500 V and 2MHz frequency has been applied as separation voltage. Compensation voltages in the range of -6V to 2V have been considered. The target vapour is assumed to be 100% ionized by an external source prior to enter the IMS and flows through the gas chamber defined by two electrode plates on which a voltage is applied. The dimensions of the IMS considered chamber for simulations are: $13 \times 5mm^2$ for electrodes in the drift region, $5 \times 5mm^2$ for electrodes in the detection region with a separation of 0.5mm between plates and 1mm between regions. Concentrations have been fixed to 1ppm in all studied cases.

Numerical simulations of the vapour ions in the µIMS gas chamber have been performed for the vapour phase compounds of DMMP (a chemical warfare agent stimulant) and Ac (a health risky volatile) gases in N_2. Ions modelled are listed in table1 and have been selected because their main properties: K_0, α_2 and α_4 are available in the literature [4]. Following assumptions are made to simplify numerical simulations: 1) All ions are singly charged and do not have dipolar moment 2) Ions are assumed to be free from clusters -from water vapour and nitrogen in the ionization process- 3) Ions do not interact with one another, so that interactions resulting from space charging do not occur 4) Ions are created immediately upon entering the analyzer, and the type of ionization is not considered 5) Reactant and product ions are not present in the system.

Table 1: Parameters used on simulations for the studied compounds [6-7].

Chemical	Ion Acronym	K_0 (10^{-4} m^2/V·s)	α_2 (Td^{-2})	α_4 (Td^{-4})
2-propanone (Ac)	Ac_2H^+	1.88 [6]	$1.34 \cdot 10^{-5}$	$1.77 \cdot 10^{-9}$
Dimethyl methylphosphonate (DMMP)	$DMMPH^+$	1.94 [7]	$5.09 \cdot 10^{-6}$	$-1.58 \cdot 10^{-10}$

4. Results and Discussion

Figure3 shows the distribution of the ions concentration through the drift channel for Ac_2H^+ and $DMMPH^+$ and for two different separation voltages. $DMMPH^+$ ion concentration is shown as line contour plots whereas Ac_2H^+ ion concentration is drawn by a continuous surface plot. Concentrations of the Ac and DMMP gases are presented for two different situations: A) for $V_{RF} = 875$ V ($E_{RF}/N = 73$ Td) shows interferences in results: Ac and DMMP ions reach the detector for the same value of $V_C = -2.3V$ ($E_C/N = 0.19$ Td) so, for just certain values of the applied voltage ion with the same DC voltages will not be separated. B) for $V_{RF} = 1000$ V ($E_{RF}/N = 79$ Td) shows separated measurement for a $V_C = -1.9$ V ($E_C/N = 0.15$ Td) only Ac_2H^+ detection is obtained: the V_C that allows the detection of each component is well differentiated. Obtained current intensities, with initial concentrations of 1ppm, are in all cases in the range of nA.

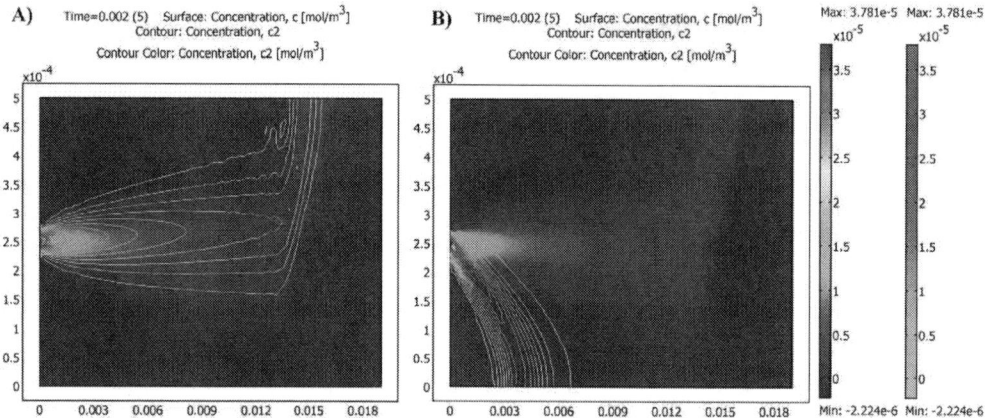

Figure 3: Simulation concentrations of dimer ion Ac_2H^+ -represented by continuous surface plot-mixed with monomer ion $DMMPH^+$ -represented by line contour plots-. For a separation voltage of A) $V_{RF} = 875V$ ($E_{RF}/N = 73$ Td), both ions have the same $V_C = -2.3V$ ($E_C/N = 0.19Td$) and are attracted to the same detector electrode. No differentiation is obtained. B) $V_{RF} = 1000V$ ($E_{RF}/N = 79$ Td), and for a $V_C = -1.9V$ ($E_C/N = 0.15Td$) only Ac_2H^+ detection is obtained. Differentiation is achieved.

From our results for low RF electric fields: $E/N < 40$ Td there is no dependence of mobility with electric field, therefore $V_C = 0$ V for all ions, while from $E/N \geq 40$ Td ions are differentiated. Increasing electric field they can be separated due to their mobility differences. Ions with the same rate E/N also have the same V_C. For a rate of $E/N \sim 73$ Td, compensation field of Ac_2H^+ and $DMMPH^+$ is the same: E_C/N (73 Td) $\sim 0,19$ Td, yielding a compensation voltage of V_C (0,19 Td) $\sim -2,3$ V. At this specific compensation voltage, no differentiation is expected to be achieved for ions with the same charge.

5. Conclusions

Simulations of a planar micro IMS (μIMS) have been done with COMSOL Multiphysics software for two compounds in vapour phase that could be considered representative for security applications. Simulations have shown that Acetone ions can be separated from ions of dimethyl methylphosphonate for an $E/N \geq 79Td$ applying a determinate V_C that makes only selected ions to reach the detector.

Acknowledgements

This work and Thesis grant of Ms. R. Cumeras have been financially supported by the Spanish Ministry of Science and Innovation MICINN-TEC2007-67962-C04-01 project.

References

[1] B. M. Kolakowski, Z. Mester, *Analyst* **139**, 96, 2007.

[2] R. Guevremont, R. W. Purves, *Rev. Sci. Instrum.* **70**, 1370–1383, 1999.

[3] A. A.Shvartsburg, *Differential Ion Mobility Spectrometry*, CRC Press, BocaRaton, 2009

[4] E. V. Krylov, E. G. Nazarov, R. A. Miller, *Int. J. Mass Spectrom.* **266**, 76–85, 2007.

[5] D. A. Dahl, T. R. McJunkin, J. R. Scott, *Int. J. Mass Spectrom.* **266**, 156-165, 2007.

[6] E. Krylov, E. G. Nazarov, R. A. Miller, B. Tadjikov, G. A. Eiceman, *J. Phys. Chem. A* **106**, 5437-5444, 2002.

[7] N. Krylova, E. Krylov, G.A.Eiceman, J.A.Stone, *J.Phys.Chem. A* **107**, 3648-3654, 2003.

Characterization of high permittivity GdScO₃ films prepared by liquid injection MOCVD

Jurkovič M[1], Hušeková K[1], Čičo K[1], Dobročka E[1], Nemec M[2], Fedor J[1], Fröhlich K[1]

[1] – Institute of Electrical Engineering, Slovak Academy of Sciences, Dúbravská Cesta 9, 841 04, Bratislava, Slovakia

[2] – Faculty of Electrical Engineering and Information Technology, Slovak University of Sciences, Ilkovičova 3, 812 19 Bratislava, Slovakia

We present electrical characterization of Si(p)/GdScO₃/Ru metal-oxide-semiconductor structures prepared by liquid injection metal organic chemical vapour deposition. Capacitance-voltage measurement revealed dielectric constant κ=22. Density of interface states was determined using conductance measurement. Annealing in forming gas resulted in decrease of the interface state density in the middle of the Si band-gap to 5 * 10¹¹ eV⁻¹cm⁻².

Introduction

GdScO₃ thin films attracted recently attention as promising candidate for next generation of high-κ dielectrics in CMOS technology. Hf-based dielectrics, that are currently implemented in a production, suffer either from low recrystallization temperature (~500 °C for pure HfO₂) or from lowering of dielectric constant when introducing Si to enhance amorphous phase stability (κ ~ 14÷16 for HfSiO, depending on Si content). Dielectric constant of gadolinium scandate is higher than 20. In addition, GdScO₃ films can be prepared in amorphous form that persists under post-deposition annealing up to high temperature. However, advantages of the GdScO₃ thin films over the Hf-based high-κ dielectrics should be still clearly identified.

Until now, the GdScO₃ films were prepared by pulsed laser deposition, PLD, (1), electron beam evaporation (2), and atomic layer deposition, ALD, (3, 4). In our contribution we present electrical properties of GdScO₃ films prepared by liquid injection metal organic chemical vapor deposition.

Experimental methods

The GdScO₃ films were deposited at 600 °C in a low-pressure hot-wall quartz reactor. The solution of Gd(thd)₃ and Sc(thd)₃ precursors dissolved in toluene in a concentration of 0.02 M was introduced using computer controlled microvalve into the evaporation part heated at 250 °C. Opening time of the micro valve was 3 ms, resulting in the droplet mass of 5.7 mg. The reaction atmosphere was composed of O₂ (170 sccm flow) and Ar (21 sccm flow) with a total pressure of 200 Pa. Injection frequency was 0.33 Hz. Employing these parameters the film growth rate was adjusted to 0.8 nm/min. Si (100) oriented wafer slices were used as substrates. The substrates were cleaned in acetone, isopropyl alcohol and dipped in buffered HF before the deposition.

Dielectric constant of the GdScO₃ films was determined from capacitance-voltage measurements performed at the frequency of 1 MHz. For the capacitance-voltage measurement Ru films were deposited by liquid injection MOCVD at 300 °C.

Subsequently, Ru electrodes were patterned by standard optical lithography followed by Ar ion beam etching.

Post deposition annealing was performed in forming gas (90% N_2 + 10% H_2) at temperature 430 °C during 30 min.

Conductance technique measurements were performed on MOS structure with $GdScO_3$ insulating layer with thickness of 10 nm. First, the energy distribution has to be calculated from low frequency C-V curve. From calculated energy distribution, each gate voltage is corresponding to energy level in semiconductor band-gap. To calculate interface state density, conductance as a function of frequency was measured at each desired energy level. Conductance was measured in frequency range from 20 Hz to 1 MHz. Each conductance curve consisted of 183 points measured in range of 20 Hz – 1 MHz, 40 points were measured in range of 20 Hz – 5 kHz, 34 points in range of 5 – 10 kHz and also in range of 10 – 20 kHz, 63 points in range of 20 – 250 kHz and 6 points in ranges of 250 – 500 kHz and 500 kHz -1 MHz. To determine density of interface states from measured conductance curves, we simulated conductance curves with Nicollian and Goetzberger model (5) and fitted the simulated curve to the measured one.

Results and discussion

Dielectric constant of oxide layer in metal-oxide-semiconductor structure was determined from the set of samples with various oxide thicknesses. Equivalent oxide thickness (*EOT*) of the metal-oxide-semiconductor can be expressed as

$$EOT = \left(\frac{\kappa_{SiO_2}}{\kappa_{ox}} \right) \cdot t_{ox} + t_{SiO_2} \qquad [1]$$

where κ and t are dielectric constants and thicknesses of SiO_2 and oxide films, respectively. Plotting the *EOT* versus the t_{ox} we can determine dielectric constant κ_{ox} as well as physical thickness of the SiO_2 interfacial layer t_{SiO2}. Capacitance-voltage characteristics for set of $GdScO_3$ films with thickness ranged from 7.6 to 22.7 nm are shown in Figure 1. Dielectric constant of 22 was extracted using Equation [1], Figure 2. Determined dielectric constant is comparable to those obtained by PLD (1) or ALD (2, 3)

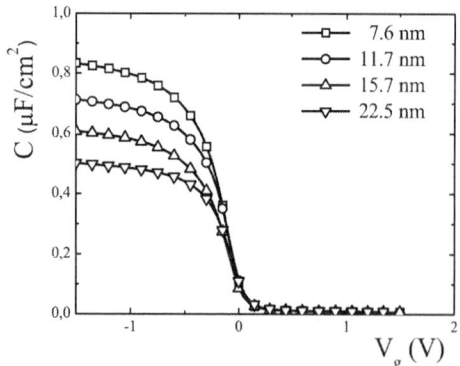 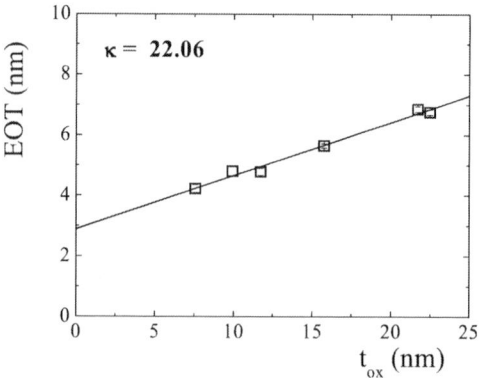

Figure 1. Capacitance-voltage characteristics of the $GdScO_3$ films with various thicknesses.

Figure 2. Determination of dielectric constant κ.

techniques. However, interface layer equal to ~ 3 nm is relatively high. We suppose that the interface layer is formed during high deposition temperature (600 °C) of the GdScO$_3$ film.

We have examined energy distribution of interface states density by conductance method measurements and simulations. Energy distribution is calculated from low-frequency C-V curve using Equation [2] (5)

$$\varphi_S = \int_{V_{FB}}^{V_G} \left(1 - \frac{C(V_G)}{C_I} \right) dV_G \qquad [2]$$

where φ_S is surface potential, V_G is gate voltage, V_{FB} is flatband voltage and C_I is capacity of MOS structure in accumulation. Low-frequency C-V curve was measured at 1 kHz. We choose 10 energy levels in band-gap, on each energy level we measured conductance characteristics. Measured conductance curves were corrected for influence of series bulk resistance and oxide capacity estimated by measurement in deep accumulation.

Simulated conductance characteristics were done using Niccolian-Goetzberger model (6) described with Equation [3]:

$$\frac{G_p}{\omega} = qD_{it} \int_{-\infty}^{\infty} \frac{\exp\left(\frac{-v^2}{2\sigma^2} \right)}{\sigma\sqrt{2\pi}} \frac{\ln\left[1 + \omega^2 \tau_m^2 \exp(-2v) \right]}{2\omega\tau_m \exp(-v)} dv \qquad [3]$$

where G_p is parallel conductance, ω is radian frequency, σ is standard deviation of surface potential, τ_m is time constant of interface state and v is surface potential. Changing the σ, τ_m and D_{it} parameters in the equation, we were able to fit the simulated curve to measured.

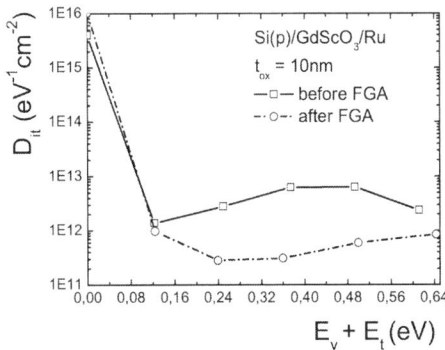

Figure 3. Energy distribution of density of interface states on 10 nm GdScO$_3$ film.

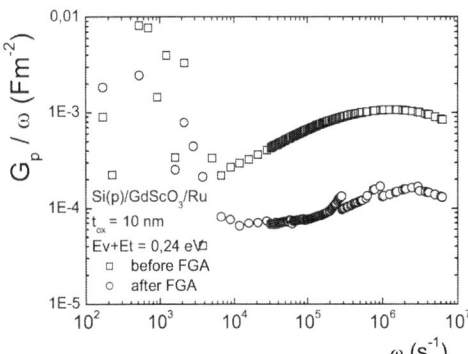

Figure 4. Effect of FGA on measured conductance curves.

From measurement analysis, we have calculated density of interface states of 5×10^{11} eV^{-1}cm^{-2} in the middle of band gap. Results also prove positive effect of forming gas annealing on level of D_{it}, when density of states has lowered after annealing by approximately half order to one order of magnitude, Figure 3. The positive effect of forming gas annealing can be clearly visible on conductance curves, Figure 4.

Conductance lowers after annealing approximately by one order of magnitude, shape of the conductance curve remains.

However, the level of interface states is still high and should be at least around $1x10^{11}$ eV^{-1}cm^{-2} and lower. Further experiments are required to find the best possible annealing settings.

Conclusion

In conclusion, GdScO$_3$ thin dielectric films grown by liquid injection MOCVD exhibit dielectric constant $\kappa = 22$. This value is similar to those obtained by PLD and ALD growth techniques. Analysis of conductance measurements revealed relatively high density of interface states. These values can be decreased by post-deposition annealing in forming gas. Further optimization of the deposition process and also of the post-deposition processing is needed to achieve density of interface states below 10^{11} eV^{-1}cm^{-2}.

Acknowledgments

This work was supported by the project of the structural funds of the European Union entitled: "Centre of excellence for new technologies in electrical engineering", ITMS code 26240120011 (Device for Metal Coating, ATC-ORION-8E).

References

1. C. Zhao, T. Witters, B. Brijs, H. Bender, O. Richard, M. Caymax, T. Heeg, J. Schubert, V. V. Afanasev, A. Stesmans and D. G. Schlom, *Appl. Phys. Lett.,* **86**, 132903 (2005).
2. M. Wagner, T. Heeg, J. Schubert, St. Lenk, S. Mantl, C. Zhao, M. Caymax and S. De Gendt, *Appl. Phys. Lett.,* **88**, 172901 (2006).
3. K. H. Kim, D. B. Farmer, J.-S. M. Lehn, P. V. Rao and R. G. Gordon, *Appl. Phys. Lett.,* **89**, 133512 (2006).
4. P. Myllymäki, M. Roeckerath, M. Putkonen, S. Lenk, J. Schubert, L. Niinistö and S. Mantl, *Appl. Phys. A,* **88**, 633 (2007).
5. Schroder, D.K.: *Semiconductor Material and Device Characterization.* 2nd edition., New York : Wiley, 1998, 784 s., ISBN 0-471-24139-3.
6. E. H. Nicollian, A. Goetzberger in *Bell Syst. Tech. J.* **46**, 6, 1967.
7. K. Fröhlich, A. Vincze, E. Dobročka, K. Hušeková, K. Čičo, F. Uherek, R. Lupták, M. Ťapajna and D. Machajdík, in *CMOS Gate-Stack Scaling — Materials, Interfaces, and Reliability Implication,* Mater. Res. Soc. Symp. Proc., **1155**, 2009.
8. C. Adelmann, S. van Elshocht, A. Franquet, T. Conard, O. Richard, H. Bender, P. Lehnen and S. De Gendt, *Appl. Phys. Lett.,* **92**, (2008).

Biomedical signal amplifier for EMG wireless sensor system

K. Rendek[1], M. Daříček[2,1], E. Vavrinský[1], M. Donoval[2] and D. Donoval[1]

[1]Faculty of Electrical Engineering and Information Technology,
Slovak University of Technology, Bratislava, Slovakia
[2]Nanodesign Bratislava
E-mail: karol.rendek@stuba.sk

This work presents differential amplifier designated as a part of modular biomedical sensor system (MBSS) used for EMG (electromyography) signal measurements. Except the standard features which characterize this type of device (high gain, low noise), some special requirements comprising very low supply voltage and low power consumption could be reached by application of the described amplifier. The amplifier provides switch-able 60/80dB gain at 15 Hz - 1kHz frequency bandwidth. A fourth order anti-aliasing filter for next processing block (AD converter) is included as well.

1. Introduction

The application of the portable instruments intended for measuring and evaluating of the electrical activity produced by skeletal muscles brings a lot of advantages to biomedical electronic and medical care. Our application of such portable system is focused on measurement of the electromyography (EMG) signals used for further development of SMART biomedical instruments. The design consists of the portable battery powered EMG sensors and computer (PC) extended by RF communication module used for transfer the EMG signals measured on separate body places. The wireless EMG sensor contains the electrodes placed on human body, analog amplifier, analog to digital converter, microcontroller containing complete radio block and power source. Measured signals at various body locations are sent via RF communication to computer for advanced analysis. Principal schematic of designed portable modular system is in Figure 1.

Activity of skeletal muscles is controlled by neural pulses/commands from the brain. Single muscle membrane generates electric potential. Detection of the electric potential generated by muscle cells when these are electrically or neurologically activated represents the EMG signal. The amplitude depends on the muscle type and conditions during the observation process and ranges from μV to mV. Subsequently the EMG signal can be used in many clinical and biomedical applications[1].

This paper describes battery powered amplifier for EMG measurements as a heard of wireless modular system. The amplifier is part of the first stage of the wireless measurement system. The aim is to analyse biomedical signals at several locations of human body simultaneously at the same time and create complex EMG model.

2. The RF EMG sensor circuit concept

The EMG sensor is based on precise instrumentation amplifier designed for bipolar detection arrangement. The amplified bio-signals are sampled by 16bit analogue-digital converter (ADC), pre-processed by microcontroller and sent via RF-bridge to computer for further digital processing.

Input amplifier for EMG measurement is represented by INA326 [2]which is a good choice for portable medical instruments in terms of its small size, low supply voltage, high common mode rejection (CMR) and rail-to-rail output. The INA326 solves many of the typical challenges of measured EMG signals. Potentials measured on the skin may vary from µV to mV. The measured voltage depends on the type of the muscle and also on the position of the sensors. For this reason EMG amplifier is designed with variable gain 60 dB or 80 dB. Although EMG signals generally lies in the range 20 – 200 Hz, amplifier was designed for frequency range 15 Hz - 1 kHz, because wider sampling range improves DSP processing methods. Common-mode signal coming from ambient electromagnetic energy and electrical lines helps to reject high CMR 100dB of INA326. The human body behaves like huge antenna receiving ambient electromagnetic signals, particularly 50 Hz from main power supply as well. These signals are in the EMG frequency range and have to be suppressed to minimum. The rail-to-rail output, provided 20 dB gain, offers sufficient dynamic range at battery power supply. Output of the instrumentation amplifier is equipped by an active band pass filter, with pass band 15 Hz – 1 kHz. The high pass part of the filter works as the feedback to body reference electrode and effectively suppress ineligible frequencies. Corner frequency of low pass part is set to 1 kHz[3][4][5]. After this analogue processing is [6]signal prepared for 16 bit ADC working at 100 kHz sample rate.

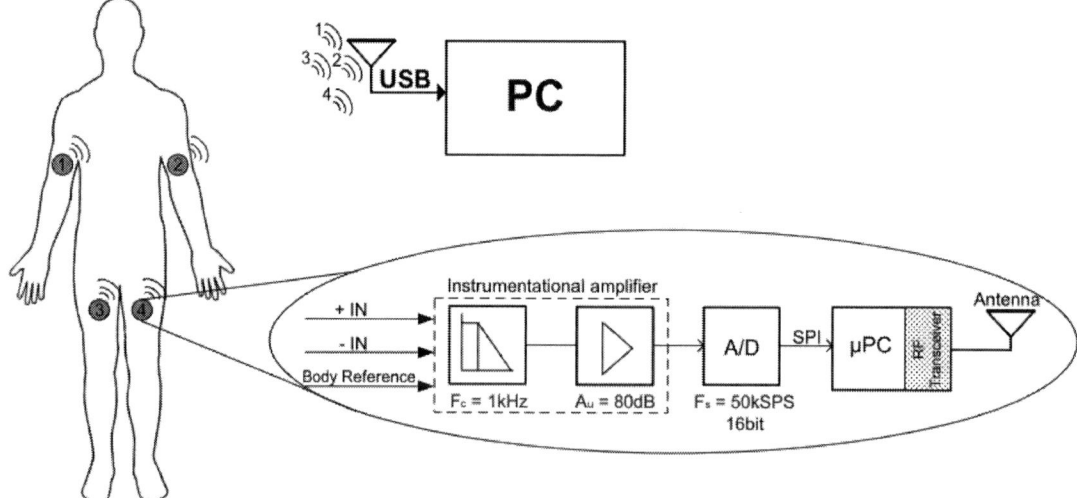

Figure. 1. Block diagram of wireless sensor system

Data from ADC are sent to microcontroller containing complete RF communication block, where they are pre-processed and via RF-bridge sent to main CPU of the system for further DSP processing. Further OPA2335 buffer, inverts and gains the common-mode voltage taken at the midpoint of the INA326 gain setting resistor. Common-mode signals are rejected by active reference drive circuit applied to body reference electrode. The whole portable RF amplifier is battery powered by Li-Pol accumulator. The block diagram is shown in the Fig. 1.

3. Results and discussion

The simulated magnitude-frequency characteristics of designed biomedical amplifier are shown in Fig. 2. They have a characteristic band-pass shape with lower cut-off frequency around $f_L = 15$ Hz and upper cut-off frequency around $f_U = 1$ kHz.

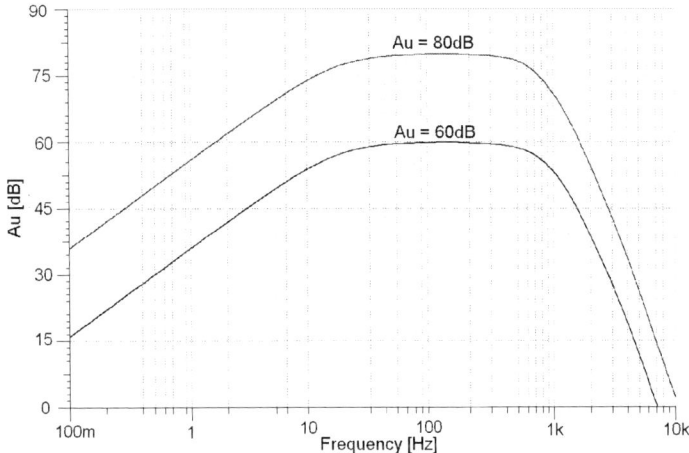

Figure. 2. Simulated magnitude-frequency characteristics of designed biomedical amplifier

Example of EMG signal measured by designed biomedical amplifier with 60 dB gain is shown in Fig. 3. It is then further processed by signal acquisition module NI9234 and digital band pass filter with lower cut-off frequency $f_L = 55$ Hz and upper cut-off frequency $f_U = 95$ Hz. There are 4 characteristic regions of muscle activity: 1- no activity, 2- alternating load, 3- idle and 4- continuous tension.

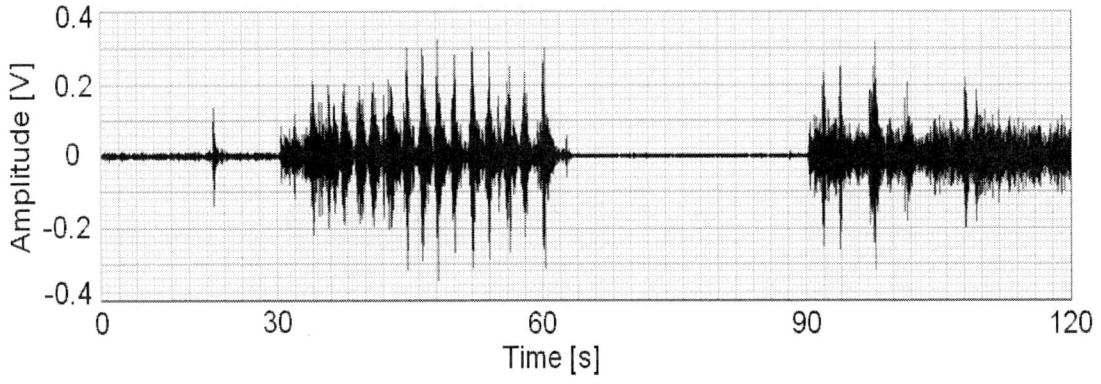

Figure. 3. EMG signal measured by designed biomedical amplifier with Au = 60 dB. It si further processed by signal acquisition module NI9234 and digital band pass filter with lower cut-off frequency $f_L = 55$ Hz and upper cut-off frequency $f_U = 95$ Hz

The power consumption of designed prototype of RF EMG sensor is close to 10 mA. The noise immunity helps to improve small size of electronic parts.

4. Conclusion

The new, original results were obtained in presented research, which verify the desired hardware requirements on the system. Sampled and digitally processed data shows different EMG signal characteristic for each type of muscles activity. These signals are taken from passive gold plated electrodes, which represent large potential for further improvement of global MBSS. In our future work some modified electrodes will be analyzed and verified. Active electrodes and excluding body reference electrode are the most important expected future changes of this part of MBSS.

Acknowledgement

The work has been supported by APVV project number VMSP-P-0127-09.

References

[1] De Luca C. J.: Electromyography. Encyclopaedia of Medical Devices and Instrumentation, (John G. Webster Ed.) John Wiley Publisher. 98 – 109, 2006
[2] Texas Instruments, "INA326 Datasheet" http://www.ti.com/, 2004.
[3] Kugelstadt T.: Getting the most out of your instrumentation amplifier design. http://www.ti.com/, 2005
[4] Kitchin Ch., Counts L.: A Designer's guide to instrumentation amplifiers 3RD Edition. http://www.analog.com/, 2006
[5] Abad S. L. M., Maghooli K.: Low supply voltage electrocardiogram signal amplifier. IEEE, 799-801, 2007
[6] Texas Instruments, "OPA2335 Datasheet" http://www.ti.com/, 2003
[7] Texas Instruments, "ADS8320 Datasheet" http://www.ti.com/, 2007
[8] Nordic semiconductor, "nRF24LE1 Datasheet" http://www.nordicsemi.com/, 2010

Resistive switching in $RuO_2/TiO_2/RuO_2$ MIM structures for non-volatile memory application

B. Hudec[1], M. Hranai[1], K. Hušeková[1], J. Aarik[2], A. Tarre[2] and K. Fröhlich[1]

1 – Department of Thin Oxide Films, Institute of Electrical Engineering, Slovak Academy of Sciences, Dúbravská Cesta 9, 841 04 Bratislava, Slovak Republic
2 –Institute of Physics, University of Tartu, Riia 142, 51014 Tartu, Estonia
e-mail: boris.hudec@savba.sk

In this paper we describe resistive switching in $RuO_2/TiO_2/RuO_2$ structures. Electrodes (RuO_2) were grown by metal organic chemical vapor deposition and dielectric TiO_2 switching layers with rutile structure were prepared by atomic layer deposition. After proper nitrogen annealing of as-grown samples followed by an electro-forming procedure and forming procedure bipolar resistive switching was observed. For various switching parameters 100 switching cycles were performed to test retention characteristics. Ratio of high to low resistance up to 10^3 was obtained for reading voltage of -0.3 V.

1. Introduction

Non-volatile memory technologies are getting more and more attention every year. In the last edition of International Technology Roadmap for Semiconductors from 2009 [1], more attention is paid to the non-volatile memory technology than ever before. ITRS defines several different technological approaches, including nanoionic memories, charge trapping memories, etc. Nanoionic resistive switching memories are very promising because of their high speed, easy manufacturability in the case of using binary oxide materials and long retention time (10 years).

Resistive switching (RS) phenomena comprise reversible change of a sample´s resistance, induced by voltage/current exceeding a threshold level. We distinguish low resistance state (LRS) and high resistance state following SET and RESET sweep, respectively. Resistance state changes occurring in the same polarity are typical for unipolar RS (URS), whereas bipolar RS (BRS) requires sweeping in opposite polarities. The exact mechanism behind resistive switching is still not completely clarified. However, barrier height collapse caused by ionic migration to and from a metal/dielectric interface [2], [3] and filament formation from oxygen vacancies or ionic interstitials are believed to be the cause [2], [4].

Many studies are now being performed on the resistive switching capabilities of TiO_2 [5-7]. Majority of the papers are dealing with resistive switching in anatase TiO_2, because rutile TiO_2 is difficult to grow at temperatures lower than 700 – 800 °C.

In this work, we investigated switching properties of the rutile type TiO_2 dielectric film. Using RuO_2 bottom electrode with rutile structure as a seed layer, we obtained pure rutile TiO_2 crystallographic phase at the deposition temperature of 400°C. Details of the template effect have been reported earlier [5].

Samples have been characterized by *I-V* sweeps in order to analyze their resistive switching capabilities. It is worth to mention that the switching on the samples with good oxygen stoichiometry of TiO_2 layer was not present or was unstable. As expected, better results were obtained on samples with oxygen deficient TiO_2 layer (TiO_{2-x}).

978-1-4244-8574-1/10 $26.00 © 2010 IEEE

2. Experimental

Thin RuO_2 films were deposited by MOCVD in a low-pressure, hotwall quartz reactor operated at a pressure of 2 Torr. Precursor bis(2,2,6,6-tetramethyl-3,5-heptanedionato)(1,5-cyclooctadiene) ruthenium was dissolved in iso-octane (concentration 0.035 M) and injected into the evaporation chamber using solenoid microvalve. The injector was open for 3 ms with a frequency of 0.33 Hz. Oxygen was used as a reactant gas with flow rate of 170 sccm, while argon was used as a carrier gas with a flow rate of 21 sccm. The films were grown at a deposition temperature 290°C on Si(100) substrates covered by a 100 nm SiO_2 layer (bottom electrodes) or on a top of RuO_2/TiO_2 bilayer (top electrodes).

The TiO_2 films were grown in a flow-type ALD reactor [8] at a temperature 400° C on 10×10 mm^2 Si substrates covered by polycrystalline RuO_2 films serving as bottom electrodes and seed layers. An ALD cycle, consisting of exposure of the substrates to the $TiCl_4$ vapour (for 2 s), purge in the flow of pure nitrogen for 2 s, exposure to the H_2O vapour for 2 s, and purge in the flow of pure nitrogen for 5 s was repeated 400 times to obtain 20 nm thick TiO_2 films.

Optical lithography and argon ion milling was used to define top RuO_2 electrodes, grown by MOCVD. The top electrodes were circle-shaped with a diameter of 100 μm. Bottom electrode was always grounded and top electrode biased. Post-deposition annealing was performed as the last step of processing.

Current density-voltage (*J-V*) characteristics were measured using an Agilent 4284A LCR meter and Keithley 2400 Source Meter, respectively.

3. Results and discussion

Annealing in oxygen and nitrogen ambient was examined in earlier studies in order to evaluate the most appropriate annealing conditions with respect to the ability of the stacks to perform resistive switching. Many successful experiments (e.g. [2], [9]) on TiO_2 incorporate oxygen deficient TiO_{2-x} layers, which seems to be necessary for RS as a source of oxygen vacancies. Test performed by Kim et. al. [9] on single TiO_2 layer, that did not reveal any RS behaviour, confirm the contribution of oxygen vacancies to RS. Therefore in this work, as expected, better results were achieved by annealing in nitrogen ambient than in oxygen. Annealing in nitrogen for 30 minutes at 400°C was found to be the most appropriate treatment to induce the switching.

Symmetrical $RuO_2/TiO_2/RuO_2$ MIM structures behave like capacitors with certain capacitance and leakage current densities. Annealing in nitrogen reduces TiO_2, forming a sub-oxide that is a n-type wide band gap semiconductor rather than an insulator [10] and has higher leakage current levels compared to stoichiometric TiO_2. To induce the switching in these "leaky capacitors" a forming operation has to be performed. This is done by sweeping the voltage to either positive or negative values, unless there is a SET (sudden voltage rise, switch from HRS -> LRS) or, in the case where the sample is intrinsically in LRS, RESET (sudden voltage drop, switch from LRS -> HRS). It is noteworthy that in contrary to other works on TiO_2 [11], we were able to give rise only to bipolar resistive switching in these stacks. As one can see in Fig. 1, in our case forming operation could be performed at both polarities, while SET was always obtained at the positive bias and RESET at the negative bias. This is probably caused by uneven distribution of defects (mainly oxygen vacancies) in the dielectric layer in its growth direction, as confirmed in our previous work on capacitors [5].

978-1-4244-8574-1/10 $26.00 © 2010 IEEE

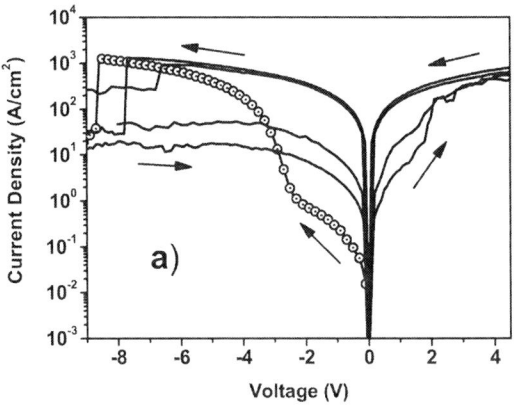 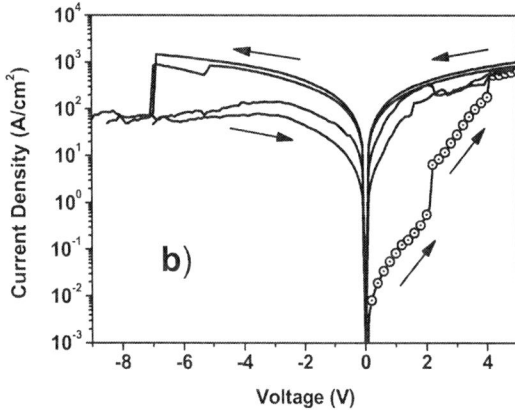

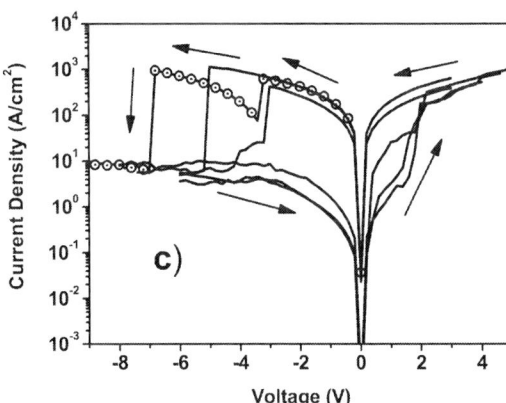

Fig. 1: Forming operations (circle symbols) at different polarities and subsequent resistive switching voltage sweeps (without symbols). a) Negative sweep is applied to the sample in virgin state (showing capacitance), the sweep continues directly to LRS and RESET appears at -8.5 V. b) Positive sweep is applied to the sample in virgin state, sample switches to LRS in two distinct steps at 2.2 V and 4.1 V. Smaller resistive ratio in the case of positive forming was not a rule. c) This sample was intrinsically in the LRS state and negative forming took place at two distinct steps: 1. switch to a virgin state, where capacitance of the MIM stack could be measured and 2. RESET. We were unable to change anti-clockwise direction of the switching loop. Arrows represent the direction in which voltage has been sweeped.

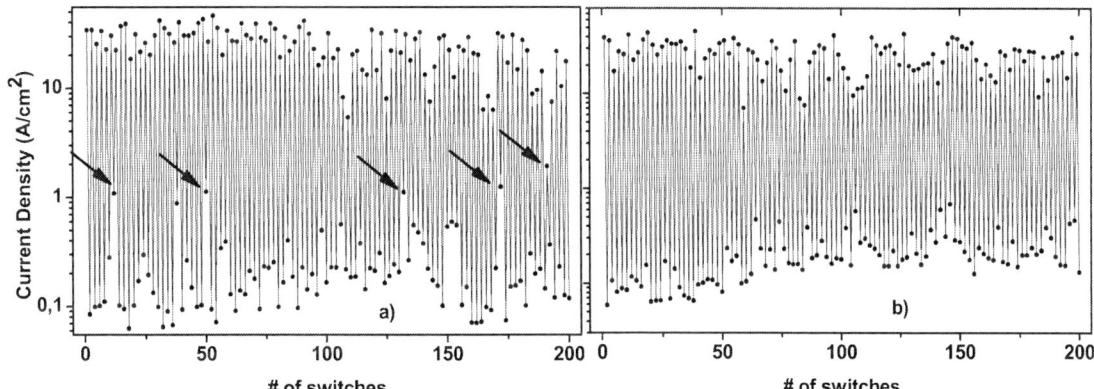

Fig. 2: Retention characteristics of the same sample. Plotted current density through an electrode at read-out voltage -0,3V a) SET at 3V and RESET at -6V. Note overlapping states, marked by arrow. It would be difficult to tell, whether the sample is in LRS or HRS. b) Same sample with higher switching voltages: SET at 4V and RESET at -7V. Overlapping states were reduced.

After forming, subsequent voltage sweeps were performed. The direction of the sweeps remained anti-clockwise (RESET at negative bias and SET at positive bias) regardless of the forming sweep polarity. In Fig. 2 retention characteristics are shown. Already electroformed samples were SET by a voltage pulse to LRS; the current was measured at the read-out voltage of -0.3 V, as this value is low enough to minimize the power consumption of the read operation and higher LRS/HRS resistance ratio was found at the negative bias (see Fig. 1); then the sample was RESET by a voltage pulse to HRS and the current was measured again at the same read-out voltage of -0.3 V. This was done in a loop 100 times at two distinct levels of SET and RESET voltages. First, values +3 and -6 V (Fig. 2a) were used, but overlapping states of LRS and HRS were found, then values +4 and -7 V (Fig. 2b) were used, with no overlapping states found.

3. Summary

Resistive switching MIM structures with MOCVD-grown RuO_2 electrodes and ALD-grown rutile-TiO_2 switching layer were prepared and investigated. To induce the switching in these stacks, nitrogen annealing at 400 °C for 30 minutes followed by electroforming voltage sweep with either positive or negative polarity was necessary. Regardless of the forming sweep polarity, the switching was purely bipolar with SET always at the positive bias and RESET at the negative bias, when the top electrode was biased and the bottom electrode was grounded in every measurement. Measurements of retention characteristics revealed that with voltages +4 V for SET and -7 V for RESET there were no overlapping states of LRS and HRS read at -0.3 V in 100 switching cycles. The HRS/LRS resistance ratio was found to be 10^3 at the best, which is definitely enough for the device operation. However, SET and RESET currents were too high for the practical device operation. More study on this subject should be performed in order to lower the switching currents and to clarify the switching mechanisms.

Acknowledgement

This work was financially supported by the Slovak Research and Development Agency project APVV-0133-07 and VEGA project 2/0031/08 and Estonian Science Foundation (Grant No. 7845).

References

[1]   ITRS 2009 Edition, http://www.itrs.net.
[2]   Yang, J. et. al.: Nature nanotechnology, Vol. 3, p. 429-433, (2008).
[3]   Lee, M. H. et al.: Appl. Phys. Lett. 96, 152909 (2007).
[4]   Szot,K.et.al.: Phys.Rev.Lett. 88, (2002).
[5]   Fröhlich, K. et. al.: J. Vac. Sci. Technol. B 27, 266 (2009).
[6]   Choi, G.-J. et. al.: J. Electrochem. Soc. 156, G71 (2009).
[7]   Cheng, C. H. et. al.: IEEE Elect. Dev. Lett. 29, 845 (2008).
[8]   J. Aarik, A. Aidla, A.-A. Kiisler, T. Uustare, V. Sammelselg: Thin Solid Films 340, 110 1999.
[9]   Kim, S., Choi, Y.: IEEE Transactions on Electronic Devices Vol. 56, No. 12 (2009).
[10] P. Kofstad, J. Less Common Metals 13 (1967) 635.
[11] D. S. Jeong, H. Schroeder, R. Waser: Electrochem. Sol. St. Lett. 10, G51 (2007).

Micro-power converters for energy harvesting devices

Enrico Sangiorgi, Aldo Romani and Marco Tartagni

ARCES, II School of Engineering, University of Bologna
Via Venezia 52, 40521 Cesena, Italy
e-mail: {esangiorgi, aromani, mtartagni}@arces.unibo.it

This paper demonstrate the feasibility of micro-power converters based on active control for harvesting power from multiple tiny piezoelectric transducers with higher efficiency than conventional passive approaches, especially in case of irregular vibrations and heterogeneous transducers. A prototype system was developed and validated in realistic conditions: three $0.5 \times 12.7 \times 31.8$ mm^3 piezoelectric cantilevers with 18g tip masses were excited with the vibrations of a train passenger car. The converter harvested up to $142\mu W@0.22g\text{-}RMS$ and up to $90\mu W@0.11g\text{-}RMS$. An implementation based on discrete components including passive start-up and implementing basic power management techniques consumes a fraction of the extra harvested power during operation.

1. Introduction

Piezoelectric transducers are a suitable solution for harvesting energy from environmental vibrations [1]. Passive rectifiers can be used as basic electronic interfaces for transferring the electrical charge originating from multiple piezoelectric transducers [2]. Unfortunately, the efficiency of energy transfers is limited and strongly depends on the operating point. This is a critical aspect in multi-source configurations: (a) it is difficult to optimally bias all transducers in case of irregular vibrations; (b) transducers producing low voltages operate inefficiently. In [3][4] it is shown that the above limitations can be overcome by switching converters synchronized with vibrations at the expenses of an additional power consumption for control.

2. Principle of operation

This paper presents a switching synchronized converter for handling in a scalable and efficient way multiple piezoelectric transducers. The scheme and the principle of operation are shown in Fig. 1. The transducers are normally disconnected. When a local voltage maximum [minimum] is detected on the i-th transducer, V_{CMAXi} [V_{CMINi}] is enabled: C_{Pi} is discharged by an increasing current flowing through L. When $V_{Pi} = 0V$ (i_L is maximum), V_{CMAXi} [V_{CMINi}] is disabled and V_{CST} is activated: i_L flows onto C_O. When $i_L = 0A$, all switches are opened again. Simultaneous maxima/minima are served sequentially.

3. Experimental Results and Circuit Implementation

This circuit was implemented and tested with three Q220-A4-303YB transducers from Piezo Systems. A shaker reproduced the vibrations previously recorded on a train passenger car. Fig. 2 shows the experimental results. The proposed scheme is more efficient than an interface based on rectifiers and harvests more power, at least +41% and up to +208% in case of weak and irregular vibrations. Moreover, the efficiency is less sensitive to the output bias than the passive interface. Another advantage is that this scheme can scalably handle multiple sources at the expenses of few additional components per source.

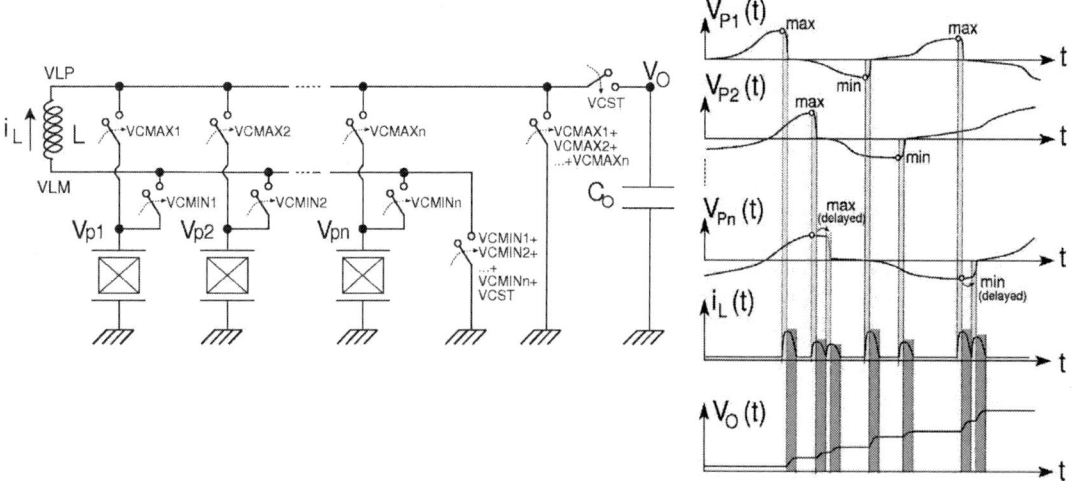

Figure 1. Scheme of the proposed multi-source switching converter and Waveforms depicting the energy conversion process..

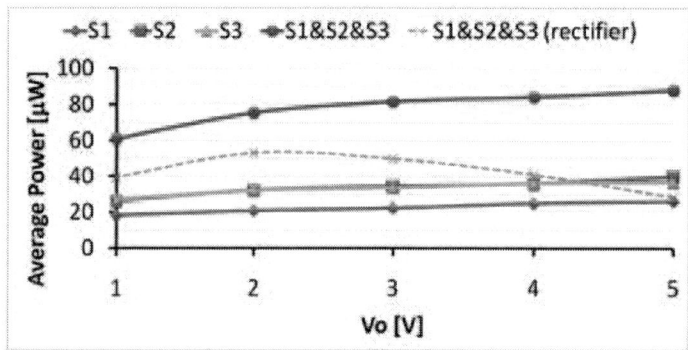

Figure 2. Measured output power as a function of the output bias with train vibrations with $a_{RMS}=0.11g$. The multi-source switching converter is also compared to a full-wave rectifier (dashed curve).

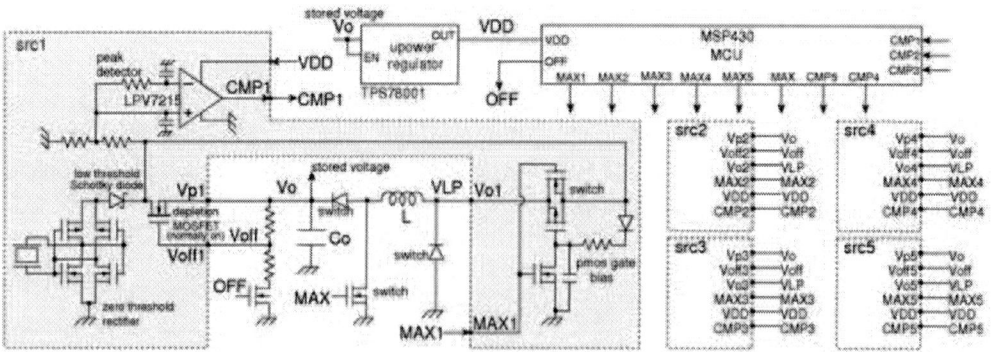

Figure 3. A circuital implementation of the converter requiring no external supply. The highlighted part of the circuit is replicated for each piezoelectric source.

A simplified schematic of the multi-source power converter is shown in Fig. 3. The behavior and the power consumption of the converter were carefully estimated with circuital simulations. Deep sleep modes of the microcontroller are fully exploited, so that the active duty cycle of the microcontroller was estimated to be < 0.4% with the reference vibrations and the average current consumption is about 1.5µA. Few additional µA are required for supplying the analog maximum voltage detectors. The dynamic power consumption due to the switching activity of the gates of the power MOSFETs can be neglected. However, the linear regulator might operate inefficiently with high input voltages. The extra harvested power is always higher than the power consumption of the active switching converter. The proposed circuit consumes just a few µW per source during operation and demonstrates the feasibility and the efficiency of energy autonomous actively controlled power harvesters.

Acknowledgement

This work was supported by Eurotech Group as part of the research project "Self-Powered Portable and Wireless Electronic Systems" and by "Fondazione Cassa dei Risparmi di Forlì".

References

[1] S. Roundy, P. K. Wright, J. Rabaey, "A study of low-level vibrations as a power source for wireless sensor nodes", Computer Communications, 26(11) (2003), pp. 1131-1144

[2] M. Ferrari, V. Ferrari, M. Guizzetti, D. Marioli, A. Taroni, "Piezoelectric multifrequency energy converter for power harvesting in autonomous Microsystems", Sens. Actuators A, 142(1) (2008), pp. 329-335

[3] E. Lefeuvre, A. Badel, C. Richard, L. Petit, D. Guyomar, "A comparison between several vibration-powered piezoelectric generators for standalone systems", Sensors and Actuators A, 126 (2006), pp. 405–416

[4] A. Romani, C. Tamburini, R. Paganelli, A. Golfarelli, R. Codeluppi, E. Sangiorgi, M. Tartagni, "Dynamic switching conversion for piezoelectric energy harvesting systems", Proc. of IEEE Sensors (2008), pp. 689-692

978-1-4244-8574-1/10 $26.00 © 2010 IEEE

A monolithic micro fuel cell based on a functionalized porous silicon membrane

N. Torres-Herrero[1], J. Santander[1], N. Sabaté[1], C. Cané[1], T. Trifonov[2], A. Rodriguez[2], R. Alcubilla[2]

[1] Instituto de Microelectrónica de Barcelona, IMB-CNM, (CSIC)
Campus UAB s/n 08193 Bellaterra (Barcelona, Spain)
[2] Departamento de Ingeniería Electrónica, Universidad Politécnica de Cataluña – UPC
Campus Nord Edifici C4, c/ Jordi Girona 1-3, 08034 Barcelona, Spain

e-mail: nuria.torres@imb-cnm.csic.es

Due to the recent advances in the microsystems field a new group of applications based in wireless applications have arisen. In order to provide these devices with the working autonomy that they need, micro direct methanol fuel cells emerge as a suitable solution because of the high energy density of the used fuel. The main challenge regarding micro fabrication technology is the fabrication of the different parts of the fuel cell as a single device. In this work different approaches of silicon-based porous membranes for micro fuel cell applications are presented. Fabrication and characterization as a proton exchange membrane of all of them has been carried out in order to drive towards a final monolithic device.

1. Introduction

As new advances and developments in the microsystems field open new applications for this kind of devices, it becomes a must to guarantee a certain degree of working autonomy to the system. In this context, micro fuel cells arise as a competitive proposal as PowerMEMS, providing advantages such as the higher energy density of the fuel, in the case of direct methanol fuel cells, in comparison with batteries [1, 2].

The main issue concerning the fabrication of fuel cells by using micro technologies is the proton exchange membrane, as the polymeric materials commonly used for this purpose are not compatible with microfabrication processes. Previous works started the development of passive direct methanol micro fuel cells by developing microfabricated current collectors combined with Nafion membranes [3, 4] . Further developments are towards a monolithic architecture, including functionalized porous silicon as a suitable solution for the membrane.

This work focuses in comparing different approaches for the functionalized porous silicon membrane. Functionalization is fulfilled by introducing Nafion inside the porous matrix, providing the proton conduction capability and at the same time avoiding the volumetric changes induced by the hydration of the polymer.

2. Fabrication

Fig. 1 shows a scheme of the proposed architecture for the monolithic micro fuel cell. The core of the device is a porous silicon membrane, obtained via an anisotropic etch process followed by an electrochemical anodization process. Afterwards, catalysts and current collectors will be added.

978-1-4244-8574-1/10 $26.00 © 2010 IEEE

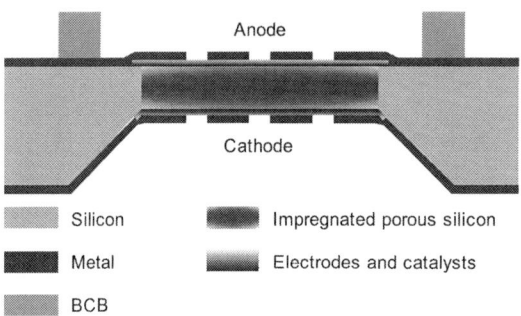

Fig 1. Scheme of the final device

Several options for the structure of the porous silicon are being considered, attending both to the porosification and to the functionalization processes. The first approach is based on a square shaped vertical channel matrix of 80x80 μm in size fabricated by means of a photolithography process which defines the channel matrix area, followed by a Deep Reactive Ion etching process (DRIE) to open the channels through the membrane. The second type of membrane consists of an ordered macroporous silicon membrane that has been fabricated as reported in detail in [5]. Basically, the process is based on a photolithography step to mark ordered pore tips at the surface of the substrate. Besides, another type of macroporous membrane has been fabricated based on a random vertically oriented macropore array fabricated as the previous one but without the photolitography step. A third type of membrane, based on mesoporous silicon and which fabrication process is illustrated in fig. 2, will be also considered, being expected to be more suitable to be integrated to obtain a fully monolithic device [6]. This fabrication process is based on a photolithography step in each side of the wafer defining a 5x5 mm membrane area and a DRIE process to micromachine the membrane. Finally, regarding the porosification process, an electrochemical anodization is carried out as reported elsewhere [6]. Afterwards, the porous area was impregnated with a certain quantity of Nafion 5% in order to provide the membranes with proton exchange properties. In fig. 3 an entire wafer of porous silicon membrane chips is shown.

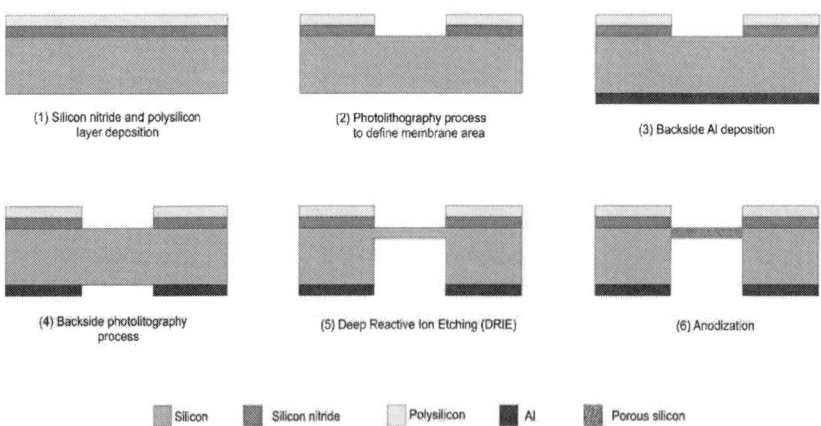

Fig. 2. Fabrication steps of a porous membrane

Figure 3. Microfabricated wafer of porous silicon membranes

3. Results

Proton conductivity data have been calculated from electrochemical impedance spectroscopy (EIS) measurements in order to demonstrate the proton exchange properties of these different approaches of porous membranes.

The electrolyte proton conductivity is related with the high frequency response. In order to fit these contributions a particular electric circuit has been chosen (depicted in the inlet in fig. 4a). The most relevant component of this model is the active electrolyte resistance R_S, which allows the proton conductivity calculation.

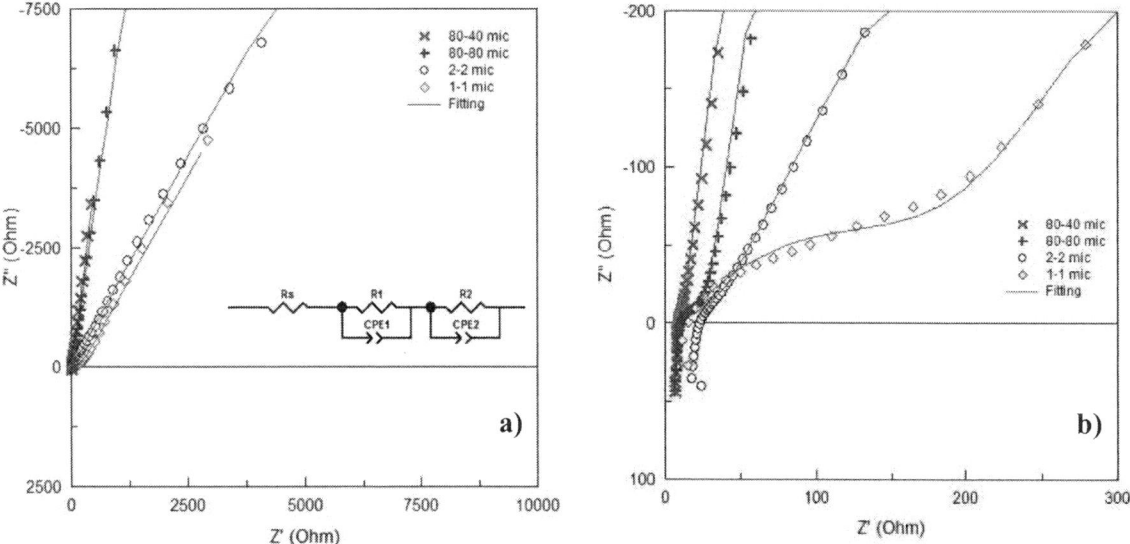

Fig 4. a) Nyquist plot of four different samples of porous silicon membranes. b) Zoom of the high frecuency range.

Fig. 4 and Table 1 show the obtained results for the macroporous silicon samples (1 and 2µm in diameter pore sizes) and for two versions of the 80µm x 80µm matrix, with the squares separated 40µm or 80µm.

The conductivity values σ_T are obtained using the total area of the measured samples. The values σ_e are obtained by using the area filled with Nafion, so corresponding to an effective conductivity to be compared directly with that of the Nafion. The presented results are in the same range of those reported in commercial Nafion film so demonstrating the validity of these silicon Nafion-filled membranes as a proton exchange element.

Sample	R_s (Ω)	R_1 (Ω)	σ_T (mScm^{-1})	σ_e (mScm^{-1})
1 μm random	13.13	172.80	2.0	4.0
2-2 μm	23.74	40.32	2.2	8.4
80_40 μm	8.78	2.74	13.6	31.0
80_80 μm	11.20	14.71	10.7	44.7

Table 1. Fitting parameters of the EIS spectra and obtained proton conductivity.

4. Conclusions and further work

Different types of silicon-based membranes able to act as the proton exchange element in a monolithic micro fuel cell have been presented and characterized, demonstrating the feasibility of such type of membrane. Future work is oriented to mesoporous membranes (pores in the range of 100 nm in diameter), in order to improve the integration of the final device. The whole device architecture and its related microfabrication process are under development.

Acknowledgement

This work was supported by the Spanish Government project TEC2009-14660-C02-01 (MICAELA).

References

[1] S.K. Kamarudin, W.R.W. Daud, S.L. Ho and U.A. Hasran, "Overview on the challenges and developments of micro-direct methanol fuel cells (DMFC)". *Journal of Power Sources, 2007. 163(2): p. 743-754.*

[2] T.S. Zhao, R. Chen, W.W. Yang and C. Xu, "Small direct methanol fuel cells with passive supply of reactants". *Journal of Power Sources, 2009. 191(2): p. 185-202.*

[3] N. Torres, J. Santander, J.P. Esquivel, N. Sabaté, E. Figueras, P. Ivanov, L. Fonseca, I. Gràcia and C. Cané, "Performance optimization of a passive silicon-based micro-direct methanol fuel cell". *Sensors and Actuators B: Chemical, 2008. 132(2): p. 540-544.*

[4] J.P. Esquivel, N. Sabaté, J. Santander, N. Torres-Herrero, I. Gràcia, P. Ivanov, L. Fonseca and C. Cané, "Influence of current collectors design on the performance of a silicon-based passive micro direct methanol fuel cell". *Journal of Power Sources, 2009. 194(1): p. 391-396.*

[5] S. Cheylan, T. Trifonov, A. Rodriguez, L.F. Marsal, J. Pallares, R. Alcubilla and G. Badenes, "Influence of the fabrication process on the light emission of macroporous silicon". in *Conference on Photonic Materials, Devices and Applications,* 2005. Seville, SPAIN.

[6] N. Torres, M. Duch, J. Santander, N. Sabate, J.P. Esquivel, A. Tarancon and C. Cane, "Porous Silicon Membrane for Micro Fuel Cell Applications". *Journal of New Materials for Electrochemical Systems, 2009. 12(2-3): p. 93-96.*

Experimental Analysis and Modeling of the Mechanical Impact during the Dynamic Pull-In of RF-MEMS Switches

M. Niessner[1], J. Iannacci[2], G. Schrag[1] and G. Wachutka[1]

[1]Institute for Physics of Electrotechnology,
Munich University of Technology,
Arcisstrasse 21, 80225 Munich, Germany,
email: niessner@tep.ei.tum.de

[2]MEMS research unit,
Fondazione Bruno Kessler,
Povo - Via Sommarive 18, I-38123 Trento, Italy,
email: Iannacci@fbk.eu

The dynamic pull-in and pull-out of an electrostatically actuated and viscously damped ohmic contact RF-MEMS switch is both measured and simulated. Three different models are used for the simulations and evaluated w.r.t. measurements performed with a white light interferometer and a laser vibrometer. The evaluation shows that all models fail in predicting the initial contact phase of the membrane correctly. Further analysis reveals that this is due to the presence of a higher eigenmode that is activated during the first impact of the membrane.

1. Introduction

The simulation of pull-in and pull-out transients is essential for predicting the switching times of new radio frequency micromechanical system (RF-MEMS) switch designs. Moreover, such simulations can help to reduce the impact energy and, thus, increase the reliability and lifetime of the components [1]. At the same time, simulations of this kind are very demanding from a modeling point of view: not only the mechanical, electrostatic and fluidic domains, as well as their nonlinear interactions, but also the mechanical contact has to be considered. In this case, simulations based on finite elements (FE) become computationally extremely expensive and difficult to handle. This originates from the nonlinear coupling between the different energy domains, the large mesh deformations and the numerically challenging contact problem. Due to this, the pull-in/-out transients are commonly simulated using macromodels with a reduced number of degrees of freedom and, thus, acceptable computational expense. The challenge of the macromodel-based simulation of pull-in transients is, however, to find a physics-based description of the impact when the moving mechanical components hit the stop pads during the closure of the switch.

The RF-MEMS device investigated in this work is fabricated at Fondazione Bruno Kessler (FBK) [2]. It consists of a perforated movable gold membrane suspended by four beams above a fixed ground polysilicon electrode (cp. fig 1). By applying a voltage beyond the so called pull-in voltage, which is 30.5V for the presented device, the membrane collapses onto elevated contact pads (cp. fig. 2) and closes an ohmic contact. The two switch designs that are analyzed in this work have a gap height g_0 between the membrane and the contact pads of 1.4 and 1.7µm, respectively. The frequency of the first fundamental mechanical eigenmode of both designs is 14.7 kHz.

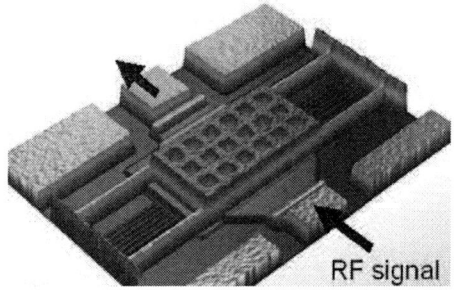 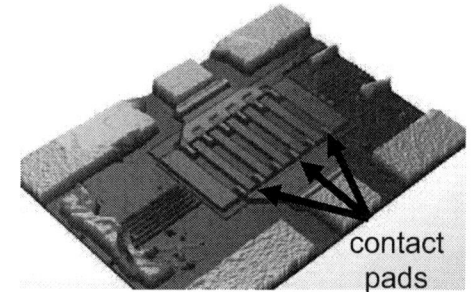

Fig. 1. Measured 3D profile of the switch without bias. A white light interferometer (WLI) was used to perform the measurement.

Fig. 2. Measured (WLI) profile of the electrodes and the 12 elevated contact pads. The membrane was removed for this measurement.

2. Modeling

Three macromodels, which were derived by flux-conserving reduced-order and/or compact modeling techniques, are used to simulate the switch. All macromodels are formulated in terms of "across"- and "through"-quantities and use generalized Kirchhoffian network theory as a theoretical framework for the coupling of multiple energy domains. Each macromodel consists of four submodels: a mechanical submodel of the suspended membrane, an electrostatic one that describes the attracting forces between the membrane and the electrode, a model comprising the viscous damping forces and a submodel of the mechanical contact between the membrane and the elevated contact pads. In order to account for residual stresses originating from the manufacturing process, all mechanical submodels were calibrated to have their first mechanical resonance frequency at 14.7 kHz.

The first macromodel (abbr. "MLM") is derived applying the approach presented by Schrag and Niessner [3, 4], which uses various techniques to generate macromodels of the different domains. The mechanical submodel is based on a superposition of eigenmodes and the electrostatic forces are modeled through a Lagrangian energy functional formulated in terms of the modal amplitudes. The mixed-level approach presented in [3], i.e. the Reynolds equation evaluated in a distributed finite network model in combination with physics-based compact models accounting for perforations and the outer boundary, is used to include the viscous damping forces, and, finally, the contact force is modeled by a dissipative term that accounts for losses and a linear force term with a contact stiffness that accounts for the deformation of the pads.

The second macromodel (abbr. "Iannacci") presented by Iannacci [2] uses lumped models only: lumped linear beam models for the four suspensions, a rigid plate model for the membrane, and a parallel-plate capacitor for the calculation of the electrostatic forces. A modified Reynolds equation that accounts for the perforations of the membrane is used to compute the viscous damping forces. The contact model is given in form of a linear force term with a contact stiffness similar to the MLM macromodel.

The third macromodel (abbr. "Architect3D") is assembled using the commercial software COVENTORWARE Architect3D^{TM} [5]. Here, the suspensions are modeled by high-order non-linear beams, the membrane by a mixed interpolation of the tensorial components approach (MITC) and the electrostatic forces by a lumped capacitor element. A generic lumped submodel based on the Reynolds equation in its original form is used for including viscous damping. For the mechanical contact, a readily available standard model is employed, which is, however, not described in detail within the manual.

4. Results and discussion

A two-stage procedure was pursued for the evaluation of the three macromodels w.r.t. white light interferometer (WLI) and laser vibrometer (LV) measurements. First, the quasi-static pull-in/-out characteristics were measured to evaluate the electromechanical coupling of the macromodels. Second, the dynamic pull-in/-out transients were measured to evaluate the fluidic-electromechanically coupled model including the contact submodel.

In order to minimize the effect of dielectric charging [6] during the measurement of the quasi-static pull-in/-out characteristics, the switch was actuated with a 70V peak-to-peak (pp) triangular waveform at 20Hz with zero mean value. The pull-in voltage of the membrane was measured to be 30.5V and the release voltage to be 24V (cp. fig. 3). All three models reproduced the pull-in characteristic correctly, but predicted a too early release. This is due to the fact that no further contact-related phenomena like dielectric charging or adhesion forces were considered in the three different contact submodels, yet. In order to measure the dynamic pull-in/-out transients, the switch was actuated with a 36V pp rectangular voltage waveform, so that the membrane first collapses onto the contact pads and is released when the voltage drops to zero. The Iannacci and MLM macromodels accurately predicted the damped oscillation after release whereas the Architect3D model failed regarding this point (cp. fig. 4, Architect3D model is omitted for the sake of clarity). This is due to the tailored and physics-based models used for the viscous damping effects in the Iannacci and MLM macromodels.

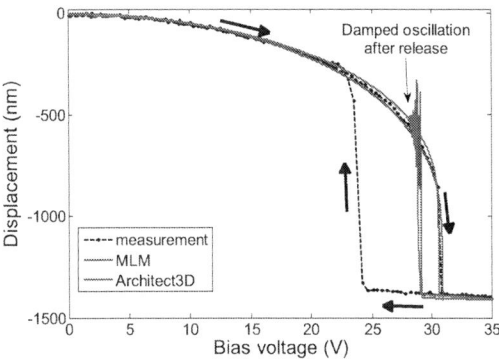

Fig. 3: Quasi-static measured (WLI) and simulated pull-in/-out of the membrane (g_0=1.4μm). The damped oscillation after release was not recorded because of the low sampling rate used in this measurement. The Iannacci macromodel is omitted for the sake of clarity.

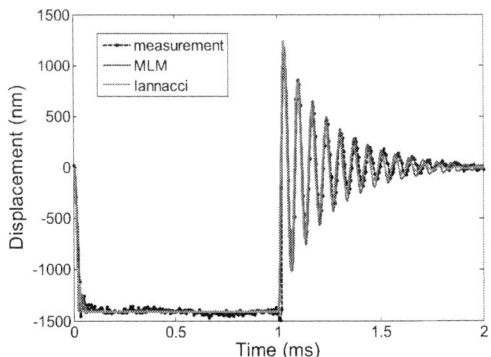

Fig. 4: Measured (WLI) and simulated pull-in/-out of the membrane (g_0=1.4μm) for a rectangular waveform at 500Hz: amplitudes 36V (on) and 0V (off). 50% duty-cycle. The Architect3D macromodel is omitted.

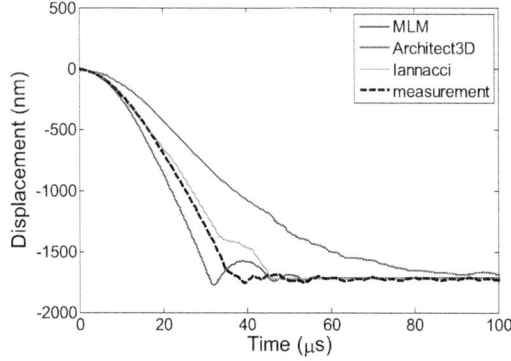

Fig. 6: Measured (LV) and simulated landing of the membrane (g_0=1.7μm) for a rectangular waveform at 250Hz: amplitudes 35V (on) and 0V (off). 50% duty-cycle. Only the first 100μs are shown.

In order to enable a detailed analysis of the initial contact phase of the membrane, we used the LV and a sampling rate of 2.56MHz. The measured data (cp. fig. 6) showed that all three simulations differ from the actual behavior of the membrane. A further analysis revealed the presence of two eigenmodes during the landing phase (cp. fig. 7). These eigenmodes can also be observed during the release phase (cp. fig. 8). FE analysis shows that the first resonance at 14.7 kHz corresponds to the fundamental eigenmode of the structure and the resonance at 136 kHz to the next higher symmetric mode. The resonance frequencies during the landing phase are much higher than during the release phase. This is due to the contact stiffness that couples with the movable structure of the switch. However, when the switch is actuated with a lower voltage, so that no pull-in occurs, only one eigenmode is observed (cp. fig. 8). These findings substantiate the assumption that a second eigenmode is activated whilst the first impact of the membrane and that it absorbs most of the kinetic energy during impact.

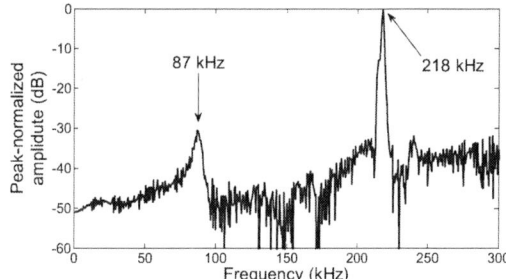
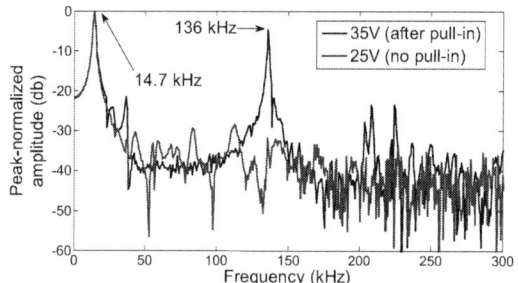

Fig. 7: Frequency spectrum of the measured (LV) landing phase (0..2ms) of the membrane (g_0=1.7µm). Voltage 35V, frequency 250Hz. A Hanning window was used for the FFT in MATLAB^{TM}.

Fig. 8: Frequency spectrum of the measured (LV) release phase (2..4ms) of the membrane (g_0=1.7µm). Voltages 35V and 25V, frequency 250Hz. A rectangular window was used for the FFT in MATLAB^{TM}.

5. Conclusion

Three macromodels were used to simulate the dynamic pull-in/-out transients of an electrostatically actuated and viscously damped ohmic contact RF-MEMS switch and evaluated w.r.t. measurements. The evaluation showed that only physics-based and tailored macromodels of viscous damping, e.g. the ones used in the MLM and Iannacci macromodels, are able to accurately predict the damped oscillation after release. Furthermore, the evaluation showed that no macromodel is, at its current state of implementation, able to capture the correct physics during the initial contact between the membrane and the contact pads, i.e. the activation of higher eigenmodes during the first impact. Consequently, future work on the MLM macromodel will focus on an extension towards a multi-modal macromodel with multiple and contact-related eigenmodes as well as a physics-based submodel of the transfer of kinetic energy to these contact-related eigenmodes during impact.

References

[1] G. Rebeiz, *RF MEMS*, John Wiley & Sons, Hoboken, New Jersey, 2003.
[2] J. Iannacci, R. Gaddi, and A. Gnudi, Journal of Microelectromechanical Systems, **19**, 3, 2010, pp. 526-537.
[3] G. Schrag, and G. Wachutka G., Sensors and Actuators A **97-98**, 2002, p. 193–200.
[4] M. Niessner, G. Schrag, G. Wachutka, J. Iannacci, T. Reutter, and H. Mulatz, in *Proceedings Eurosensors XXIII*, Lausanne, Switzerland, 2009, pp. 618-621.
[5] Coventor, Inc., *Coventorware Architect Version 2008.10 Reference*, 2008.
[6] R. Marcelli, et al., *J. Microsystem Technologies*, **16**, 7, 2010, pp. 1111-1118.

Hybrid photonic/plasmonic ZnO/Au composites for sensing applications

J. A. Zapien [a], Liu Yu [a], Chap-Hang To [a], Chen Limiao [a], J. Kováč jr. [a,b], I. Bello [a], and S. T. Lee [a]

a) Center Of Super-Diamond and Advanced Films (COSDAF), and Department of Physics and Materials Science, City University of Hong Kong, Hong Kong SAR, PRC

b) Faculty of Electrical Engineering and Information Technology, Slovak Technical University, Ilkovičova 3, 81219 Bratislava

e-mail: apjazs@cityu.edu.h

We have studied the enhancement factors (EF) in Surface Enhanced Raman Scattering (SERS) substrates prepared with self assembled ZnO nanorod arrays and Au nanocrystals. The ZnO/Au composite nanoarrays present experimental EF values as high as ~10^7 for detection of Rhodamine 6G (R6G) using an excitation wavelength of 633 nm. We have study the EF via numerical simulations of a single ZnO/Au composite nanorod using a FDTD algorithm. Our results indicate that the coupling between the photonic wave guiding in the ZnO nanorod and the plasmonic resonance in the Au nanocrystals plays a significant role in the observed EF and need to be considered to fully understand and optimize the development of high efficiency, reproducible, and stable SERS substrates.

1. Introduction

Self-assembled nanostructures are able to form highly efficient photonic cavities capable of waveguiding with low losses and high optical gain [1]. Furthermore, nanotechnology approaches enable the growth of very high quality nanostructures without the need for lattice matching substrates [2]. Indeed, composition tuning enables the growth of nanostructures capable of supporting lasing action over broad spectral ranges [3]. On the other hand, localized surface plasmons (LSP) and surface plasmon polaritons (SPP) can efficiently store optical energy in the form of electron oscillations at the interface of metals and dielectrics [4]. In fact, LSP are the main contributors of the large enhancement factor (EF) observed in Surface Enhanced Raman Scattering (SERS) that enable single molecule detection [5]. Clearly, hybrid photonic-plasmonic nanostructures provide an attractive alternative to develop ultra-small devices where light concentration or manipulation in subwavelength dimensions is advantageous [6].

2. Surface Enhanced Raman Scattering

Raman spectroscopy is widely used for chemical measurements of molecular species of organic, inorganic, and biological materials and it can be used to determine the chemical structure of complex samples in gas, vapor, aerosol, liquid, or solid form [7]. However, the Raman effect is ~ 10^{-3} times the intensity of Rayleigh (elastic) scattering, which in turn is ~ 10^{-3} times the intensity of the light source and severely limited its use until the development of the laser in 1960's; a second major breakthrough took place in 1974 with the discovery of unprecedented sensitivity enhancements on roughened silver electrodes [8] that was later recognized to be associated with an increase in the Raman cross section by 5-6 orders of magnitude [9]. To date, "Surface-Enhanced Raman Scattering (SERS)" with up to 10-11 orders of magnitude EF have been reported enabling single-molecule spectroscopy [8,10]; although even modest EF ~ 10^6 can enable single molecule detection in special conditions [5].

It is generally accepted that the enhancement factor (EF) in SERS is due primarily to (1) electromagnetic (EM) enhancement, and (2) chemical enhancement. The EM enhancement is considered to be the largest and is attributed to large local fields caused by plasmon resonances due to the collective behavior of electrons in confined metal clusters. The chemical enhancement is less well understood although it is recognized that contact between the molecule and the metal nanocluster is a requirement for effective chemical enhancement [11]. The simpler experiments, and often larger EF in SERS, are found in aggregated Ag and Au colloids although several problems hinder its widespread application [12]. Alternative substrates for SERS are surface-roughened metal electrodes, metal films evaporated on solid substrates or through shadow masks, and self-assembled metal nanostructures.

3. Experimental Section

The fabrication and Raman enhancement factors in the prepared ZnO/Au Composite Arrays have been presented elsewhere [13]. The three step process is briefly summarized as follows. *1 - Buffer layer:* aluminum-doped ZnO thin films (AZO) were growth on (100) Si wafer at 500 °C by r. f. magnetron sputtering using a 2 wt % Al_2O_3:ZnO target. *2 - Arrays:* The ZnO arrays were then synthesized in a fused quartz tube system [14] using ZnO and graphite powders, (1:1 by weight) that were backfilled with Ar (99.995% pure) to maintain a pressure of 300 mbar. The precursors and substrate were then heated to 950 and 700 °C, respectively, and a small amount of oxygen (99.99%) was admitted after reaching the target temperature; growth proceeded for 10-60 min. *3 - Composite arrays:* First, a solution with 0.1-0.5 mM of $HAuCl_4$ was prepared by adding 0.3-1.5 mL of $HAuCl_4$, 1 mL of methanol, and 20 mL of deionized water; the pH of the solution was adjusted to 7-8 using 0.01 M sodium hydroxide (NaOH). The ZnO arrays were immersed in this solution inside a stainless steel autoclave (40 mL in volume) and slowly stirring for ~1 h; hydrothermal synthesis was then carried at 120 °C for 1 h. The samples were cooled down to room temperature naturally, washed with deionized water several times and finally dried at 80 °C under vacuum.

Low concentrations solutions of Rhodamine 6G (R6G) (with concentrations as low as 10^{-10}) and a mixture of melamine (10^{-5} M) in egg white (~6 g/L) and deionized water was prepared to test the enhancement factor of the prepared composite arrays. Raman measurements were conducted with a Renishaw in-via 2000 laser Raman microscope equipped with a 633 nm argon ion laser. The laser power at the sample is estimated to be ~ 8 mW and the spot size for excitation ~ 1 μm. Numerical simulation results were calculated using the finite difference time domain (FDTD) algorithm [15]. For these calculations, the dielectric properties (ε_1, ε_2) of the ZnO nanorods were modeled after the retrieved optical properties of ZnO films [16] from spectroscopic ellipsometry measurements [17] and those of gold were directly taken from the literature [18].

4. Results and Discussion

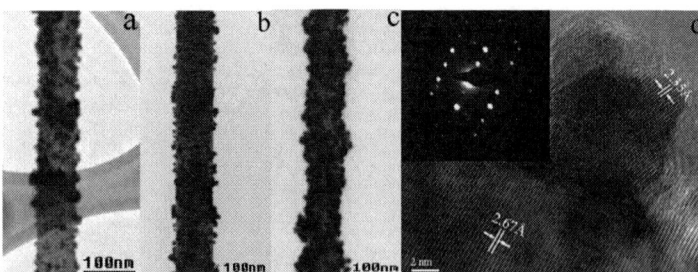

Fig. 1 TEM images of ZnO/Au nanoneedles prepared at different concentrations of HAuCl4: (a) 0.1, (b) 0.3, and (c) 0.5 mM. (d) HRTEM image of a ZnO/Au nanoneedle. (Reprinted with permission from[13]).

The morphology of the ZnO nanoarrays could be tuned between nanoneedles (with diameters of ~100-300 nm and ~50-120 nm at the root and tip, respectively) and nanorods (uniform diameters for each rod and diameters in the range ~100-450 nm) by simply adjusting the ramping temperature between 50 °C and 20 °C, respectively. The single crystal structure of both morphologies was confirmed by SAED patterns. The Au nanoparticle density on the ZnO/Au arrays can be increased by increasing the concentration of HAuCl$_4$, (Fig. 1a-1c) while the high resolution TEM (HRTEM) image (Fig. 1d) reveals two sets of lattice fringes with interlayer spacing of 0.267 nm and 0.235 nm corresponding to the d spacing of the (0001) lattice planes of hexagonal wurtzite ZnO crystal and the (001) planes of the fcc Au. These conclusions are further verified by electron diffraction (ED) patterns (inset in Fig. 4d).

The experimentally determined SERS (EF$_{exp}$) are usually calculated according to the equation $EF_{exp} = (I_{SERS}/I_{bulk})(N_{bulk}/N_{surface})$ [19], where I_{SERS} and I_{bulk} denote the integrated intensities for the Raman peak of interest adsorbed on the SERS substrate and on glass, respectively, whereas N$_{SERS}$ and N$_{bulk}$ represent the corresponding number of molecules excited by the laser beam. The EF$_{exp}$ for the highest density Au nanocrystals on nanoneedle and nanorod arrays were found to be of the order of ~10^7 and ~10^6, respectively [13].

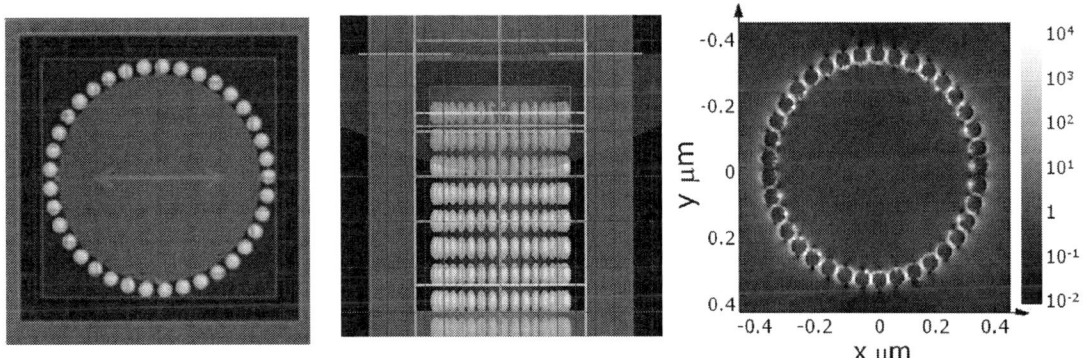

Fig. 2 Geometry used in the numerical simulations showing the (a) top view (*x,y*) and (b) side view (*x,z*) a Gaussian beam (633 nm) is incident from the top and polarized in the *x* direction (blue arrow). (e) Calculated elastic scattering enhancement for a 650 nm in diameter ZnO nanorod surrounded with Au nanocrystals with 50 nm diameter; see text for details.

Following reference [20], we estimate the Raman scattering enhancement factor (EF$_{sim}$) for the ZnO/Au composites by calculating the elastic scattering of a linearly-polarized Gaussian wave with wavelength corresponding to that of the excitation laser used (633 nm). For simplicity, the geometry used in the FDTD involves Au spheres of 50 nm diameter and 10 nm gap between them were symmetrically arranged around a ZnO nanorod of 650 nm diameter and 3 um length (Figs. 2a-2b). Light was incident from the top and coaxially with the nanorod axis. The calculated electric field from the numerical simulation has been used to estimate the elastic scattering *irradiance* $I \propto E^2$ and further normalized to that obtained in the absence of Au nanocrystals. The results (Fig. 2c) show that Au nanocrystals provide regions of intense light enhancement up to EF$_{sim}$ ~10^4. Such values are lower than EF$_{exp}$ ~ 10^7 but in reasonable agreement with reported values Au dimers with gap separation considered here [5].

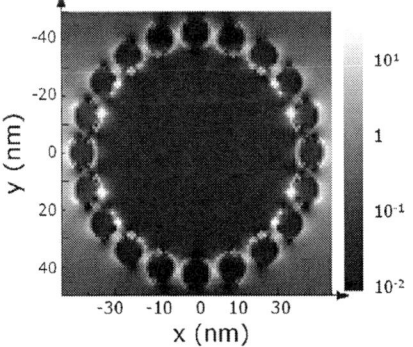

Fig. 3 As Fig. 2c with 75 nm in diameter ZnO nanorod, and 10 nm in diameter Au nanocrystals.

The normalization presented here (i.e., with respect to the case with ZnO nanorod *without* Au nanocrystal) was used to highlight and differentiate between the effects of light waveguiding due to the ZnO nanorod and the LSP resonance due to the Au nanocrystal. This is because ZnO nanorods are capable of efficient waveguiding [1]; however it is noted that the coupling of light into the ZnO optical cavity depend on the ZnO diameter and thus large losses are found for smaller diameter ZnO (75 nm) as shown in Fig. 3. This results highlight the need to further study the coupling effect between waveguiding by the supporting semiconductor nanostructure (ZnO nanorods in this case) and the resonance effects due to LSP in Au nanocrystals; such coupling will be the subject of further studies.

5. Conclusions

Using a FDTD algorithm and a simple geometrical model we have simulated the enhancement factor (EF_{sim}) in ZnO/Au composites. The simulation results show that irradiance enhancements as high as 10^4 are easily achieved in the vicinity of the Au nanocrystals at selected positions that depend on the polarization and geometry of the composite. The calculated values, however, fall short from the experimental results ($EF_{exp} \sim 10^7$) which indicate the need to consider further effects such as the gap separation between the Au nanocrystals used in the modeling and the coupling effects between the semiconductor waveguides and the plasmonic resonances in metallic Au nanocrystals.

Acknowledgement

The work has been supported by a General Research Fund (CityU 103208) of Research Grants Council of Hong Kong.

References

[1] J. C. Johnson, H. Yan, R .D. Schaller, L. H. Haber, R. J. Saykally, P. Yang, *J. Phys. Chem. B.* **105**, 11387, 2001.
[2] R. Q. Zhang, Y. Lifshitz, and S. T. Lee *Adv. Mat.* **15**, 635, 2003.
[3] (a) H. Y. Yang, S. P. Lau, and S. F. Yu, *Appl. Phys. Lett.* **89**, 081107, 2006; (b) J. A. Zapien, Y. K. Liu, Y. Y. Shan, H. Tang, C. S. Lee, and S. T. Lee, *Appl. Phys. Lett.* **90**, 213114, 2007; (c) A. L. Pan, W. Zhou, E. S. P. Leong, R. Liu, A. H. Chin, B. Zou, and C. Z. Ning, Nano Lett. 9, 784 2009).
[4] (a) W. L. Barnes, A. Dereux, T. W. Ebbesen, Nature 424, 824, 2003; (b) M. Quinten, Appl. Phys. B: Lasers Opt. 73, 245, 2001.
[5] P. G. Etchegoin and E. C. Le Ru, Phys. Chem. Chem. Phys. **10**, 6079, 2008.
[6] (a) R. F. Oulton, V. J. Sorger, T. Zentgraf, R.-M. Ma, C. Gladden, L. Dai, G. Bartal, and X. Zhang, *Nature* **461**, 629 (2009); (b) H. A. Atwater, A. Polman, *Nature Materials*, 9, 205, 2010.
[7] *Modern Techniques in Raman Spectroscopy* J. J. Laserna (Ed.) John Wiley and Sons, England, 1996.
[8] M. Fleischm, P. J. Hendra, A. J. Mcquilla, *Chem. Phys. Lett.* **26**, 163, 1974.
[9] (a) D. L. Jeanmaire, R. P.Vanduyne, *J. Electroanal. Chem.* **84**, 1, 1977; (b) M. G. Albrecht, J. A. Creighton, J. Am. Chem. Soc. **99**, 5215, 1977.
[10] (a) S. Nie, S. R. Emory, *Science* **275**, 1102, 1997; (b) K. Kneipp, Y. Wang, H. Kneipp, L. T. Perelman, I. Itzkan, R. R. Dasari, M. S. Feld, *Phys. Rev. Lett.* **78**, 1667, 1997; (c) Michaels, M.; Nirmal, M.; Brus, L. E. *J. Am. Chem. Soc.* **121**, 9932, 1999.
[11] K. Kneipp, H. Kneipp, I. Itzkan, R. R. Dasari, M. S. Feld *J. Phys.: Condens. Matter* **14**, R597, 2002.
[12] (a) H. Lin, J. Mock, D. Smith, T. Gao, M. J. Sailor, *J. Phys. Chem. B* **108**, 11654, 2004; (b) S. Chan, S. Kwon, T. W. Koo, L. P. Lee, A. A. Berlin, *Adv. Mater.* **15**, 1595, 2003.
[13] L. Chen, L. B. Luo, Z. H. Chen, M. L. Zhang, J. A. Zapien, C. S. Lee, S. T. Lee, *J. Phys. Chem. C* **114**, 93, 2010.
[14] C. Geng, Y. Jiang, Y. Yao, X. Meng, J.A. Zapien, C. S. Lee, Y. Lifshitz, S. T. Lee, *Adv. Funct. Mater.* **14**, 589, 2004.
[15] FDTD Solution, commercial professional software coded by Lumerical Inc.: *http://www.lumerical.com.*
[16] S. Flickyngerova,, K. Shtereva, V. Stenova, D. Hasko, I. Novotny, V.Tvarozek, P. Sutta, E. Vavrinsky, *Appl. Surf. Sci.* **254**, 3643, 2008.
[17] C. H. To, unpublished.
[18] E. D. Palik (Ed.) *Handbook Of Optical Constants Of Solids*, Academic Press 1985.
[19] Chaney, S. B.; Shanmukh, S.; Zhao, Y. P.; Dluhy, R. A. *Appl. Phys.Lett.*, **87**, 31908, 2005.
[20] Cao L, Nabet B, Spanier JE. *Phys. Rev. Lett.* **96**, 157402, 2006.

Constitutive Equation of the Dipole Layer in Hydrogen-sensing Metal-Oxide-Semiconductor Structures

F. Šrobár and Olga Procházková

Institute of Photonics and Electronics, Academy of Sciences of the Czech Republic,
Chaberská 57, 182 51 Praha 8, Czech Republic
e-mail: srobar@ufe.cz and olgap@ufe.cz

*Essential part of MOS-based hydrogen sensors is constituted by a dipole layer of polarized H atoms at the metal-oxide interface. This layer decreases barrier height of the Schottky diodes or causes a shift in the threshold voltage of the FET devices. Constitutive P – E (dielectric polarization versus external electric field intensity) equation of the dipole ensemble is derived supposing Langmuir-type hydrogen absorption and a two-valley model of the elementary dipole potential energy. Depending on the energy difference of the minima, the P – E isotherms suggest a rather sudden flip of the dipole orientation from antiparallel to parallel with respect to the **E** vector. Rabi resonance experiments could be based on this phenomenon, providing information about the adsorbed hydrogen atoms. Dependences of polarization on parameters of the potential profile and on temperature at selected E values are also presented.*

1.   Introduction

An important category of hydrogen-sensing devices is based on dissociative adsorption of H_2 molecules on the surface of a catalytic metal (usually Pd or Pt) with subsequent diffusion of H atoms to the metal–insulator interface where they form a dipole layer. The latter causes a decrease of the barrier height with a consequent increment in current. Both Schottky diode and MOSFET structures are used, Si, InP, and GaN being the most commonly employed semiconductors.

The research in this area, initiated by the seminal work of Lundström et al. [1], has covered mainly the kinetics and thermodynamics of hydrogen adsorption and desorption, with the possible participation of oxygen or water [2, 3], transient responses to introduction and removal of the analyte [4], and various microelectronic configurations of the sensor, often incorporating nanoparticle or nanowire components [5 - 7].

The subject of this study are the polarization properties of the atomic-hydrogen based dipole layer that lies at the heart of the sensing function of the device. The layer is extremely thin, given the fact that H atoms are not ionized, only polarized due to difference in electron affinity of materials forming the interface: the catalytic metal and the oxide. The elementary atomic dipole **p** points from metal to insulator, its vector is perpendicular to interface and its magnitude for $Pd/SiO_2/Si$ MOS structure is consensually taken as 2 Debye = 6.68×10^{-30} C.m. This yields the thickness value of the dipole layer $d \approx p/e = 0.042$ nm (e is the elementary charge). The quantity p is considered to be largely independent on the relative surface coverage of the M/O interface by the adsorbed H atoms, the quantity $\Theta = n_i / N_i$ (occupied / total available adsorption sites at the interface) [8]. We also assume that mutual dipole interaction can be neglected, which corresponds to the in this context often evoked Langmuir model of adsorption processes [9].

978-1-4244-8574-1/10 $26.00 © 2010 IEEE   275

2. The model

For an individual dipole, we have, as shown in Fig. 1, two possible states with energies $U_l = V_1 + pE$ and $U_h = V_h - pE$, respectively. Corresponding partition function is $Z = \exp(-\beta U_l) + \exp(-\beta U_h)$, with $\beta = 1/(k_B T)$. The Helmholtz free energy of the canonical ensemble of N dipoles is then $F(E, T) = - N \ln Z / \beta$. Using the relation $P = - (\partial F/\partial E)_T / \Delta V$ (with ΔV the volume element), we finally get the constitutive equation

$$P = P_m \frac{\exp[-\beta(V_h - pE)] - \exp[-\beta(V_l + pE)]}{\exp[-\beta(V_h - pE)] + \exp[-\beta(V_l + pE)]} \cdot \tag{1}$$

Here $P_m = Np / \Delta V$ is the maximal value of P corresponding to perfect alignment of all dipoles along the direction of $\boldsymbol{E}$, and $N / \Delta V$ is the dipole volume density.

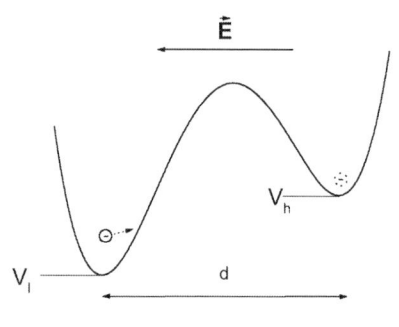

Fig. 1. Energetic model of the absorbed hydrogen-atom dipole.

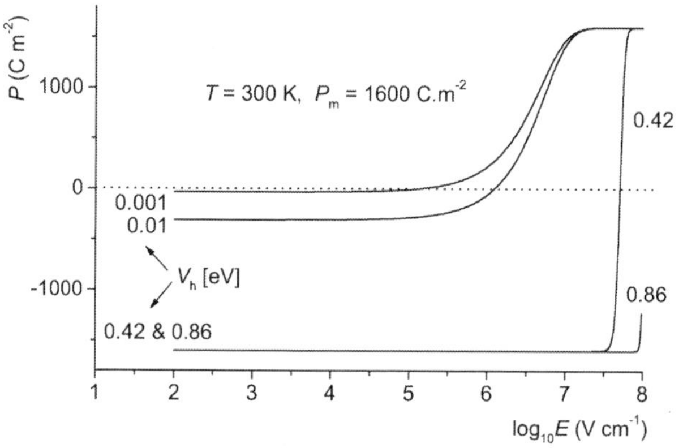

Fig.2. Polarization isotherms $P - E$ for selected values of the higher energy minimum position V_h.

3. Numerical evaluations and discussion

The room-temperature P - E isotherms computed for several values of V_h are presented in Fig. 2, choosing $V_1 = 0$. We have taken the surface dipole density value of 1×10^{14} cm^{-2}, found for the Pd/SiO$_2$ interface [10] and volume element $\Delta V = d .1\text{cm}^2 = 4.17 \times 10^{-9}$ cm^3. In all evaluations, dipole moment magnitude $|p| = 2$ D was assumed. One can see that for small values of the energy V_h the two minima of the potential profile in Fig. 1 are almost equally populated, hence the average polarization remains small, with a slight preference for the lower valley. At fields in the vicinity of 10^6 Vcm^{-1} there occurs a relatively slow reversal of $\boldsymbol{P}$ and for E above 10^7 Vcm^{-1} the saturation level P_m is achieved. The two curves labeled 0.86 eV and 0.42 eV represent upper limits of the quantity V_h for the Pd/SiO$_2$ [11] and Pd/InP-related oxide, respectively, namely the maximal values of the hydrogen adsorption energy at the interface, ΔH_{i0}. For these large values of the minima difference the behaviour of the dipole system is quite rigid: the original dipole orientation is preserved up to fields above 10^7 Vcm^{-1}, then there is an abrupt flip to conformity with the external field. A note is in place here: the observed changes in the metal work function due to adsorbed hydrogen are on the order of tenths of eV which translates to numerically equal changes in threshold voltages of FET transistors or Schottky barrier heights. Given the small value of d, this leads to fields on the order of 10^8 V cm^{-1} that have been envisaged in our computations.

Figure 3 shows dependences of polarization P on the minima difference V_h for several values of the applied external field E. At feeble fields (10^2 Vcm^{-1}) and small V_h values, thermal chaos predominates, leading to $P \approx 0$. At higher fields ($E > 10^6$ Vcm^{-1}) small V_h values afford easy flopping of the P vector in the direction of applied field, a process that becomes increasingly difficult for higher V_h values ($V_h > 100$ meV), where fields in excess of 10^7 Vcm^{-1} are required to align polarization with external field. The flopping event is then very sensitive to small variations of V_h.

Influence of varying temperature on polarization for selected values of the field intensity is apparent from Fig. 4. (The chosen temperature interval 0 – 200 ^{0}C covers the commonly used operating regimes of the hydrogen sensors. The fixed value $V_h = 100$ meV represents plausibly the potential profile in real structures, which is not known exactly.) At fields not high enough to effect the polarization flip ($E < 2\times10^7$ Vcm^{-1}), the absolute value of P diminishes with rising temperature due to increasing thermal disorientation of the dipole ensemble. When the alignment finally occurs for E close to 10^8 Vcm^{-1}, the dipoles are largely resilient to thermal randomization; hence P is virtually temperature independent.

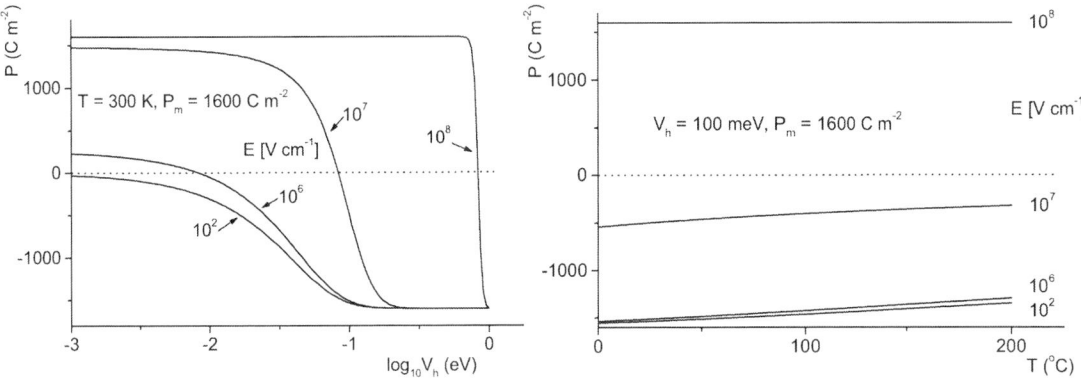

Fig. 3. Dependences of the polarization P on the energetic minima difference V_h for several values of the external field E.

Fig. 4. Temperature variation of polarization for various values of the electric field intensity.

4. The Rabi resonance

A two-level electronic system like that in Fig. 1 can, under suitable conditions, exhibit Rabi oscillations [12] near the frequency $\chi = p.E / h$, with h = 6.6261×10^{-34} J.s the Planck constant. The resonant transition frequency in our case is $f_{tr} = V_h / h$, which for $V_h = 100$ meV = 1.602×10^{-20} J yields 2.418×10^{13} Hz, a value corresponding to the long-wavelength infrared region ($\lambda = 12.37$ μm). Taking $p = 2$ Debye, one finds that field intensity value of 2.38×10^7 Vcm^{-1} is required to observe the Rabi resonance in the polarized hydrogen layers adsorbed at the metal-oxide interface.

This value falls well within the range of field intensities used in the above calculations. One can therefore envisage Rabi resonance experiments, apparently involving picosecond laser pulses of high intensity (10^{10} Wcm^{-2} is now available) to probe the properties of hydrogen sensor structures.

5. Conclusions

Constitutive P - E equation of the dipole ensemble at the core of an important category of solid-state hydrogen sensors has been derived within the canonical ensemble framework using a two potential minima model. The resulting formula affords numerical evaluation of the $P - E$ isotherms and of the influence on polarization of varying the parameters of potential profile and temperature. One can observe qualitatively differing pictures for low and high applied electric field intensities, as the balance of Lorentz forces and thermal randomizing forces changes. Occurrence of Rabi-type resonances is suggested for this kind of sensors.

Acknowledgement

Financial support of the Czech Science Foundation (project # 102/09/1037) is gratefully acknowledged.

References

[1]   I. Lundström, S. Shivaraman, C. Svensson, and L. Lundkvist, *Appl. Phys. Lett.* **26**, 55, 1975.

[2]   J. Fogelberg and L.-G. Petersson, *Surf. Sci.* **350**, 91, 1996.

[3]   T.- H. Tsai et al., *Sens. Actuators* **B 129**, 292, 2008.

[4]   Y.-Y. Tsai et al., *Sens. Actuators* **B 134**, 750, 2008.

[5]   Y.– I. Chou, C.– M. Chen, and W.– C. Liu, *IEEE Electron Dev. Lett.* **62**, 62, 2005.

[6]   S.-Y. Chiu et al., *IEEE Electron Dev. Lett.* **30**, 898, 2009.

[7]   K. Skucha, Z. Fan, K. Jeon, A. Javey, and B. Boser, *Sens. Actuators* **B 145**, 232, 2010.

[8]   I. Lundström and L.-G. Petersson, *J. Vac. Sci. Technol.* **A 14**, 1539, 1996.

[9]   L.-G. Petersson, H.M. Dannetun, J. Fogelberg, and I. Lundstöm, *J. Appl. Phys.* **58**, 404, 1985.

[10]   M. Eriksson, I. Lundstöm, and L.-G. Ekedah, *J. Appl. Phys.* **82**, 3143, 1997.

[11]   W.-C. Liu et al., *IEEE Trans. Electron Dev.* **48**, 1938, 2001.

[12]   I.I. Rabi, *Phys. Rev.* **51**, 652, 1937.

Radiation Effects on CMOS Image Sensors due to X-Rays

Jiaming Tan[1], Bernhard Büttgen[1] and Albert J. P. Theuwissen[1,2]

[1]Electronic Instrumentation Lab, Delft University of Technology, Delft, the Netherlands
[2]Harvest Imaging, Bree, Belgium
e-mail: j.tan@tudelft.nl

This work presents a study on X-Ray radiation induced degradation mechanism on both CMOS Image Sensors (CIS) with 4-Transistor (4T) pixels and its elementary test structures. The major degradation shows an increase of dark random noise and leakage current for both the sensor and the test structures. Moreover, the quantum efficiency of the pinned-photodiode (PPD) shows a post-irradiation variation at the short wavelength region. It is found that the $Si-SiO_2$ interface trap generation and charge trapping in the shallow trench isolation oxide are the main failure mechanisms.

1. Introduction

Nowadays, CMOS image sensors are getting popular to be used for the medical/space application thanks to their advantage of low power, low cost and high integration capability compared with CCD sensors. However, the radiation tolerance then becomes a great concern since the sensors will be degraded by the radiation environment during the application. Based on a wide study about the radiation performance of 3-Transistor Active Pixel Sensors (3T APS), it was found that oxide trapped charges will induce a total dose effect on the dark current increase[1]. However, very few papers have been published on the radiation effects on 4T APS and its in-pixel elementary devices so far. Since the 4T APS has different pixel architectures and a different device fabrication process compared with the 3T ones due to the pinned-photodiode, the transfer gate (TG) and the reset transistor, the previous knowledge about the radiation effect on 3T APS cannot be directly applied to 4T pixels[2]. This work aims to get an initial insight of the X-Ray radiation induced degradation mechanism of a 4T APS sensor and its test structures fabricated in a commercially available 0.18 μm technology.

2. Radiation Degradation of CMOS Image Sensors and Elementary Test Structures

In order to characterize the post-irradiation performance of a 4T APS CIS due to X-Rays, some elementary in-pixel test structures and an image sensor are designed and fabricated. These devices were irradiated by an X-ray source at Philips Health Care at room temperature with a total ionizing dose (TID) level of 31krad, 86krad, 106krad, 109krad and 137krad after 3-turn radiation, with an average energy of 46.2keV. Meanwhile, some simulations have been implemented as well in order to get some supplementary insights into the device performance.

2.1 Test Structures and its Simulation

In the test structures several single MOSFETs with different implants, sizes and gate-shapes are designed to be compared with each other. Moreover, several in-pixel devices consisting of pinned-photodiodes and transfer-gate transistors are also included. Fig.1 (a) shows the schematic of a 4T APS pixel, which consists of a PPD, a TG transistor, a reset transistor and a source follower, while Fig.1 (b) demonstrates a cross section of a PPD, a TG transistor and a reset transistor from the simulator[3].

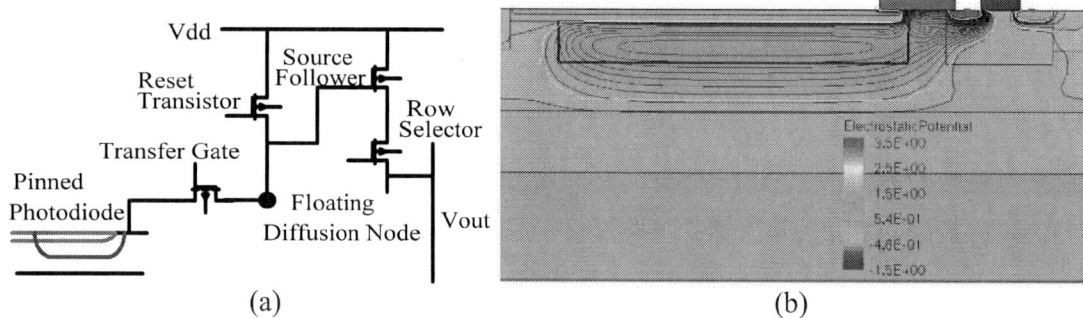

(a)   (b)

Figure.1 A 4T APS pixel architecture (a) the schematic, (b) the simulated cross section

The transfer-gate transistor takes the pinned photodiode as its source region and is overlapping with the pinning layer of the PPD[2]. It can be seen in Fig. 1 (b) that there is a high electric field distribution in that overlapping region. Together with the gate oxide layer, these kinds of high electric field regions inside the pixel will be very sensitive to the radiation degradation.

2.2 Degradation of MOSFETs

Since the MOSFETs are the elementary devices inside the pixel, it is always necessary to assess their post-irradiation performance to study the basic degradation mechanism within the pixel.

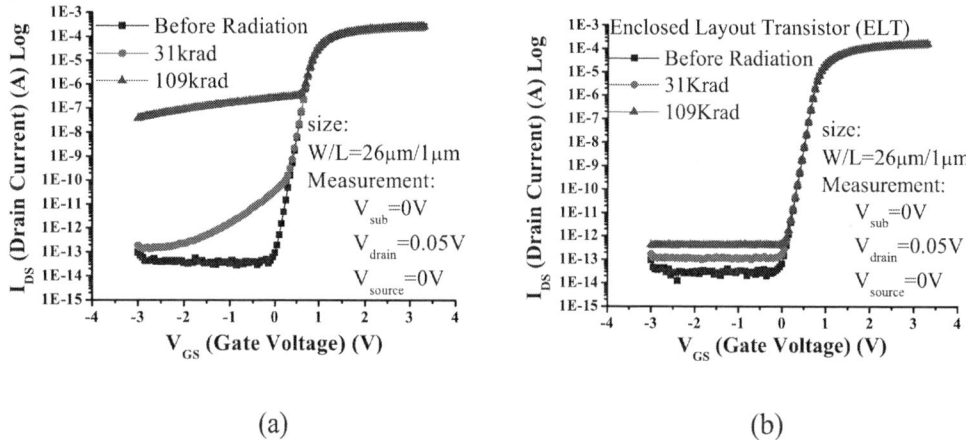

(a)   (b)

Figure. 2   N-type MOSFET characteristic after different radiation doses (a) standard layout transistor, (b) enclosed layout transistor

Fig.2 (a) shows the ionizing radiation effects on a transistor with a standard layout. It shows only a large drain leakage current increase, while the threshold voltage (V_{th}) does not shift due to the thin gate oxide thickness of this technology node. The shallow trench isolation oxide used to isolate the devices in this technology node can trap some holes generated from radiation. Due to these trapped charges, a lateral leakage path forms between the transistor's source and drain nodes by a parasitic field oxide transistor[4], which is the reason for a large increase of drain leakage current in Fig.2 (a). Compared with Fig.2 (a), the result from an enclosed layout transistor in Fig.2 (b) shows a much lower drain leakage current increase and

978-1-4244-8574-1/10 $26.00 © 2010 IEEE   280

meanwhile there is neither a shift of V_{th}. That is because the enclosed layout transistor (ELT) has an edgeless drain/source node which can suppress the parasitic transistor formation. Therefore the drain leakage current increase is much lower compared to the standard layout after a certain radiation dose level[4]. But there is still a tiny post-irradiation drain leakage current increase of the ELT transistor, which can be originated from some interface trap generation at the Si-SiO$_2$ interface. These donor-like interface traps are mostly located in the lower half of the band gap, which mainly make the drain leakage current up while it has no effect on the sub-threshold slope and threshold voltage shift[5].

2.3 Post-irradiation Dark Random Noise Degradation of CMOS Image Sensors

The device noise performance will be closely correlated with the interface trap generation. Therefore, the dark random noise of the CIS is measured before and after radiation by taking an average of the pixel output of several continuous frames. The sample is irradiated up to 31krad and 109krad. In the meanwhile, the transfer gate transistor is turned off in order to exclude the noise source from the PPD and the TG transistor.

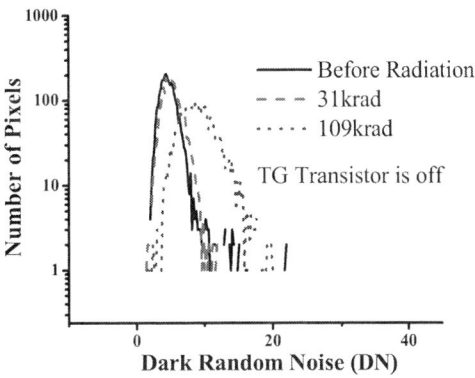

Figure. 3 Dark random noise histogram of a 4T APS CIS before and after radiation

As seen in Fig. 3, after the radiation the sensor presents a right shift of the dark random noise histogram with a larger digital number (DN) and meanwhile the width or the distribution of the histogram is getting wider. During the radiation, the passivated Si-SiO$_2$ surface beneath the transistor gate oxide is damaged and is generating some interface traps again after the incidence of the X-Rays, which will leave some holes transporting in the gate oxide to the Si-SiO$_2$ interface. Taking the noise histogram after 109krad radiation in Fig.3 as an example, a larger tail is present, compared to the original one. This presents a higher random telegraph signal noise due to the increasing number of interface traps[6]. Therefore, it can be concluded that the post-irradiation noise performance of the sensor is becoming worse due to the interface trap generation induced by the X-Ray radiation. Since during the measurement, the TG transistor is off, the noise performance shown in Fig. 3 mainly comes from the reset transistor and the source follower. Thus, it confirms that the X-Ray radiation does increase the interface trap generation of the elementary MOSFETs inside the pixel.

2.4 Quantum Efficiency Variation due to Radiation

The quantum efficiency is defined as the pixel output signal (expressed in electrons) over input photons on a pixel. It is measured before and after radiation for the dose level of 86krad and 106krad. As shown in Fig. 4, there is no significant degradation of the quantum efficiency for most of the wavelengths after the radiation, while there is a small reduction at the short wavelength region between 400nm and 550nm.

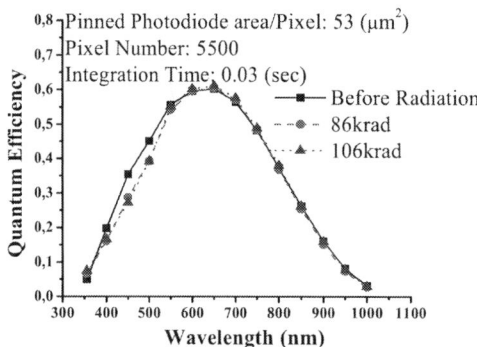

Figure 4. Quantum efficiency of a pinned photodiode of 4T CIS

The pinned-photodiode has a shallow p^+ pinning layer used to reduce the generation of dark current. But the passivated interface states will be damaged after the radiation. Then, some generated interface traps can attenuate the sensor's output and reduce the quantum efficiency as shown in Fig. 4.

3. Conclusion

Radiation effects on 4T APS are evaluated based on the measurements implemented on elementary MOSFETs in the pixel and complete CMOS image sensors. STI charge trapping induces a parasitic leakage path formation inside the MOSFETs and between devices, which raises the post-irradiation leakage current. Furthermore, the interface trap generation after X-Ray radiation can also increase the leakage current of a MOSFET. The shift of the dark random noise histogram and the reduction of quantum efficiency at the short wavelength region measured from a sensor confirm the radiation-induced interface trap generation.

Acknowledgement

The authors appreciate the involvement of Hans Stouten and Tim Poorter from Philips Health Care to help with the irradiation work on the samples.

References

[1] M. Cohen and J. P. David, "Radiation-induced dark current in CMOS Active Pixel Sensors," *IEEE Trans. Nucl. Sci.,* vol. 47, no. 6, pp. 2485-2491, Dec. 2000.

[2] X. Wang, P. R. Rao and A. J. P. Theuwissen, "Fixed-pattern noise induced by transimission gate in pinned 4T CMOS image sensor pixels," *ESSDERC 2007-Proceedings of the 37th European Solid-State Device Research Conference,* art. no. 4430955, pp. 370-373, 2007.

[3] Sentaurus device user guide, 2006.

[4] V. Goiffon, et al., "Ionizing Radiation Effects on CMOS imagers manufactured in deep submicron process," *Proceedings of SPIE - The International Society for Optical Engineering, 6816,* art. no. 681607.

[5] A. Baiano, J. Tan, R. Ishihara, K. Beenakker, "Reliability analysis of single grain si TFTs using 2d simulation," *ECS Trans.,* vol. 16, pp. 109-114, 2008.

[6] X. Wang, "Noise in Sub-Micron CMOS Image Sensors," pp. 73-108, 2008, ISBN: 978-90-813316-4-7

First Measurement on the DEPFET Mini-Matrix Particle Detector System

J. Scheirich[1,2], C. Oswald[2] and P. Kodyš[2]

[1] Department of Microelectronics, Czech Technical University in Prague, Faculty of
Electrical Engineering,
Technická 2, 166 27 Prague, Czech Republic
e-mail: jan.scheirich@cern.ch

[2] Institute of Particle and Nuclear Physics, Charles University in Prague, Faculty
of Mathematics and Physics,
V Holešovičkách 2, 180 00 Prague, Czech Republic

The DEPFET is a new type of active pixel particle detector. A MOSFET is integrated in each pixel providing the first amplification stage of the readout electronics. Excellent noise parameters are obtained with this layout. The DEPFET sensor will be integrated as an inner detector in the BELLE II experiment. A flexible measuring system with a wide control cycle range and minimal noise was designed for testing small detector prototypes. Noise of 19 electrons of the equivalent input charge was achieved during the first measurements on the system.

1. Introduction

Particle physicist use linear or cyclic accelerators to accelerate charged particles. The particles are collided at an interaction point surrounded by various types of detectors measuring type, track, momentum and energy of the newly-originated particles. The detector closest to the interaction point is called vertex detector and usually built of semiconductor strip and/or pixel sensors. In 1987 Kemmer and Lutz proposed a new type of pixel detectors: the DEPFET (= DEPleted Field Effect Transistor) [1]. Today, DEPFET based vertex pixel detectors are developed by an international collaboration for the BELLE II experiment in Japan and the future International Linear Collider. A new system has been developed for measuring and characterizing small samples of the DEPFET detector. This system enables precise and low-noise charge measurements and a flexible high resolution configuration of control signals.

2. The DEPFET Detector

The DEPFET detector itself consists of a high-resistivity depleted n-substrate and two p-regions, creating a pnp-sandwich structure (p-frontside-implantation, n-substrate, p-rearside). Figure 1 left illustrates the pixel cross-section. The n-substrate is depleted sidewards [2, 4]. It means the n-substrate (bulk) is depleted from both sides by applying negative voltages to both p-implantations with respect to the bulk. The minimum of the electron potential is in a plane parallel to the front surface. Asymmetric depletion voltages are applied in the DEPFET pixel to shift the electron potential minimum close to the front surface, where the MOSFET is located. Additional n-implants hinder electron lateral diffusion and the electrons are concentrated in a small region under the MOSFET channel. This region is called an internal gate. An impinging ionizing radiation generates electron-hole pairs in the depleted n-substrate (bulk), the holes drift to the rear-side contact, and the electrons are trapped in the internal gate. The internal gate is located directly under the MOSFET channel

---

978-1-4244-8574-1/10 $26.00 © 2010 IEEE   283

below the external gate contact, so the charge stored in the internal gate affects the MOSFET channel.

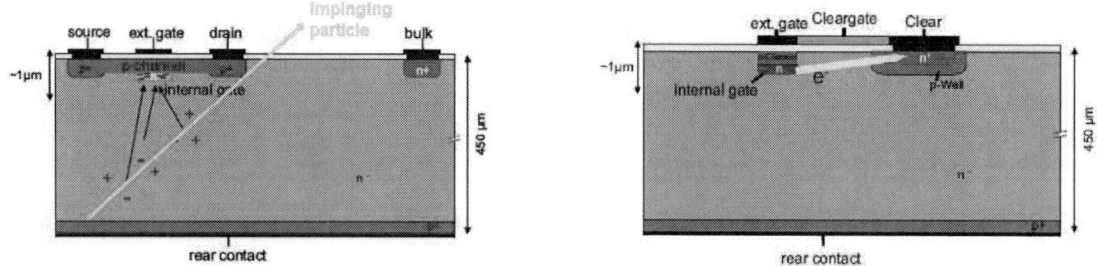

.Figure 1 - The DEPFET Pixel Cross-section (Left: MOSFET, Right: Clear Structure)

For a fixed drain to source voltage V_{DS} and a constant external gate voltage V_{GS}, the drain current I_D is proportional to the stored charge in the internal gate. The amplification g_q is given by the change of the transistor current δI_D due to the collected charge δQ [2]

$$g_q = \frac{\delta I_D}{\delta Q}\bigg|_{V_{GS},V_{DS}} . \tag{1}$$

Amplification 400 – 800 pA/e- can be obtained for mini-matrices with channel effective length 3 µm.

When new charge collection is needed, it is necessary to empty the internal gate. For clearing out the internal gate, there is a clear contact next to the MOSFET transistor. .Figure 1 right shows a cleargate cross-section [5]. The electrons are extracted from the internal gate by applying a high positive voltage to the clear contact. This causes the electrons drift to the clear contact, where they are taken away. To prevent losses during charge accumulation, the n^+-region below the clear contact is surrounded by the p-well. The n^+-region provides an ohmic contact to the clear electrode and with the p-well it provides a reverse biased PN junction that represents a potential barrier for the electrons in the internal gate. When the voltage applied to the clear electrode is high enough, the depleted region in the p-well is able to pass through the p-well and touches the p-well boundary [3]. Then there is no barrier for electrons in the internal gate and they are extracted. In order to control the potential barrier between the internal gate and the clear contact, an additional MOS structure called cleargate is added. If the cleargate is on a positive potential during the clear process, it helps to form an n-channel in the p-well.

3. The Measuring System

The Mini-matrix Measuring System is able to measure and characterize small prototypes of a DEPFET. The small sensor has 4 x 12 (or 2 x 12) active pixels, allowing studies of the DEPFET structure behaviour and processes during operation of the sensor. The Mini-matrix readout setup allows us to make a precise collected charge measurement in each pixel with low noise, charge sharing among multiple pixels, clustering, charge-loss measurement, trimming steering voltage values and timing of driving signals. The system is made of commercial and custom-made blocks as a PC with 14-bit PCI data acquisition card, an FPGA control card, a current readout and a switching circuit. The custom-made current readout circuit is made of 8 low noise trans-impedance amplifiers and the switching circuit with 12 individual analogue pulse generators that are necessary for the DEPFET matrix gate and clear electrodes control. The switching circuit can perform gate voltage timing with resolution of 7.5 ns, a voltage for a pedestal current subtraction is reconfigurable and the measuring system is controlled and configured by the PC.

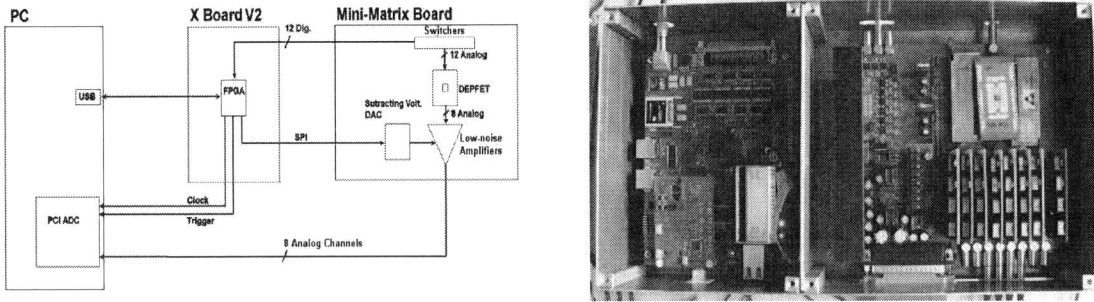

Figure 2 – The Measuring System (Left: The Conception, Right: Experiment Setup)

The readout signals from the DEPFET detector are the current base. The source of the MOSFET pixel is kept at the ground potential and the MOSFET's drain is at -5 V. The output signal has two components, constant pedestal current and signal current proportional to the charge in the internal gate. The pedestal current is subtracted at the analogue level at the inputs of the low noise readout amplifiers. The drain signal currents are read out by 8 amplifiers in parallel and digitized by a GaGe Octopus data acquisition card. The signal stream is recorded and acquired by the data acquisition system (DAQ).

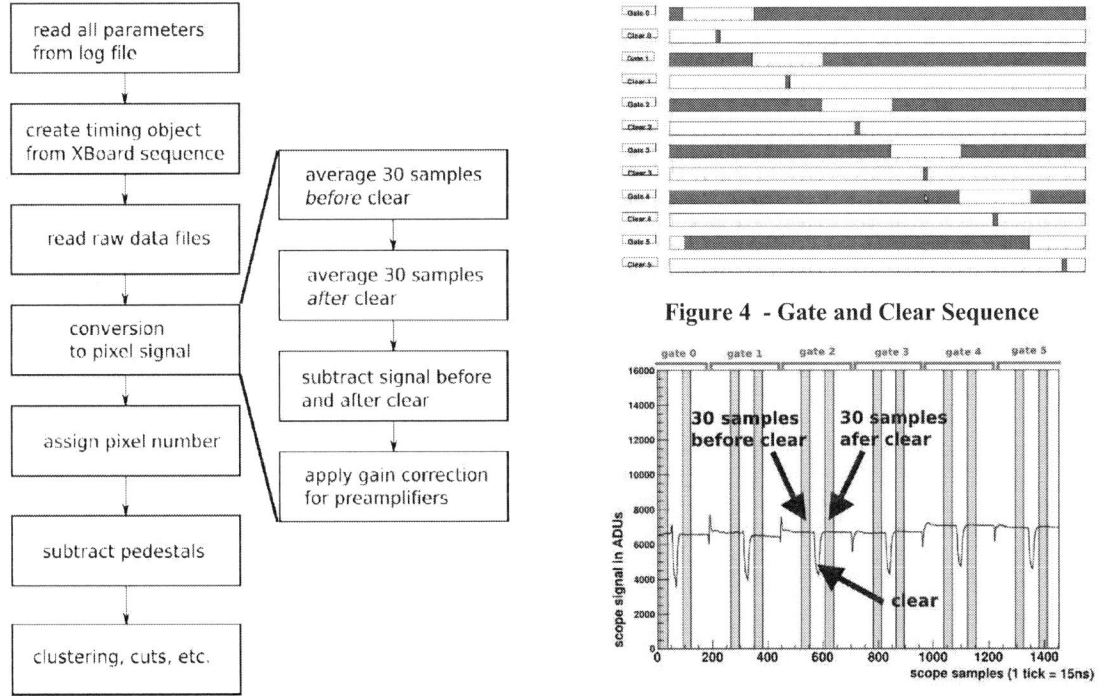

Figure 4 - Gate and Clear Sequence

Figure 5 - Signal Acquisition

Figure 3 – DAQ Structure

Figure 3 illustrates the structure of the DAQ system. Figure 4 shows the configuration of the gate and clear pulses. The final pixel signal is obtained by the calculation of difference between measurement taken before and after clearing pulse what is illustrated on one channel in Figure 5.

978-1-4244-8574-1/10 $26.00 © 2010 IEEE

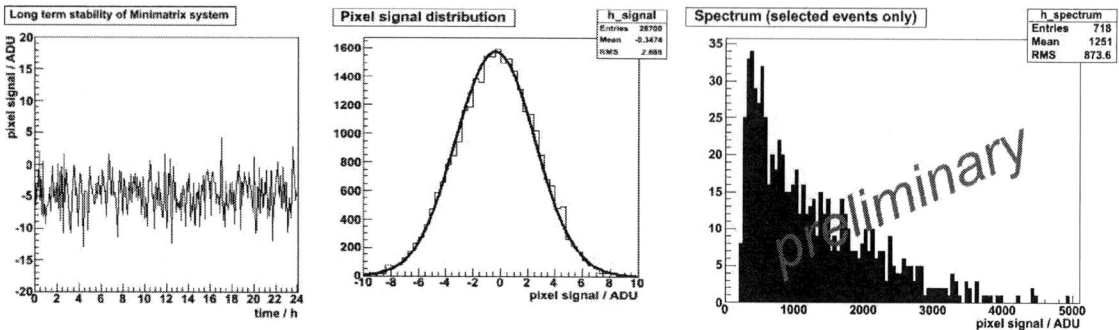

Figure 6 – First Measurements (Left: Long Term Stability, Middle: Pixel Signal Distribution, Right: Source Spectrum

First measurements results are shown in Figure 6. The left plot shows a long term stability measurement. Middle shows the pixel distribution, when the sensor is in the dark without any radioactive source. 1 ADU corresponds to 4 nA of input current. The RMS noise is than 12 nA (approximately 23 electrons of charge in DEPFET sensor with amplification of 500 pA/electron). The right plot shows an example of a spectrum of a radioactive source Sr90.

4.   Conclusions

A measuring system has been designed for the DEPFET Mini-matrix particle detector. The matrix of the detector with maximum 4 x 12 pixels can be repeatedly read out with noise lower than 20 electrons. The steering pulses can by configured with high resolution of 7.5 ns. These parameters are efficient enough for testing and characterizing the prototypes of the DEPFET particle detector.

Acknowledgement

This project was supported by the Czech Science Foundation Grant No. 203/10/0777, by the Ministry of Education, Youth and Sports of the Czech Republic under contract No. LA10033 and MSM0021620859. and by the Grant Agency of the Czech Technical University in Prague, grant No. SGS10/075/OHK3/1T/13 "Testing and Characterization of a Mini-matrix DEPFET Particle Detector". The DEPFET sensors and the FPGA X Board card was developed and supplied by the Max-Planck Institute for Physics in Munich in the frame of the DEPFET Collaboration.

References

[1]   KEMMER, J., LUTZ, G. New detector concepts. Nuclear Instruments and Methods in Physics Research A253 (1987) 365-377

[2]   TRIMPL, M. Design of a current based readout chip and development of DEPFET pixel prototype system for the ILC vertex detector. *PhD Thesis,* Bonn University, 2005.

[3]   CHU, J. L. Thermionic injection and space-charge-limited current in reach-through p+np+structures. *Journal of Applied Physics*, 1972.

[4]   NICULAE, A. S. Development of a low noise analog readout for a DEPFET pixel detector. *PhD Thesis,* Siegen University, 2003.

[5]   ANDRICEK, L., FISCHER, P., HEINZINGER, K., et. al. The MOS-Type DEPFET Pixel Sensor for the ILC Environment. *Nuclear Instruments and Methods in Physics Research.* Elsevier Science, 2003

Due to formatting issues there is a gap in pagination.

Pages 287 - 294

SEM techniques for characterization of GaN nanostructures and devices

A. Šatka[1,2], J. Kováč[1,2], J. Priesol[1], A. Vincze[2], F. Uherek[1,2] and M. Michalka[2]

[1] Slovak University of Technology, Faculty of Electrical Engineering and Information
Technology, Ilkovičova 3, 812 19 Bratislava, Slovakia
[2] International Laser Center, Ilkovičova 3, 841 04 Bratislava, Slovakia
e-mail: alexander.satka@stuba.sk

The scanning electron microscope (SEM) techniques play a key role in the characterization of various inorganic and/or organic semiconducting materials, micro-/nanostructures and devices. The power of the SEM methods is mainly in imaging, characterization and diagnostics of local near surface properties. Among a variety of the SEM methods, Cathodoluminescence (CL) and Electron Beam Induced Current (EBIC) methods have been extensively used for characterization of generation/recombination phenomena and electrical properties of bulk semiconductors and semiconductor structures [1]. Exploitation of these methods allow to visualize the electrically active defects such as dislocations, grain boundaries and inhomogeneities and to estimate the diffusion length, lifetime and surface recombination velocity of minority carriers. In addition, EBIC can be used for investigation and diagnostics of SCR of Schottky and p-n junctions, which extends its applications, in combination with other techniques such as Voltage Contrast (VC), to functional diagnostics of electronic devices biased at operation point. Currently, various modifications and improvements of "standard" SEM methods are developed to improve their resolution and sensitivity for investigations of advanced structures and devices structural, optical and electrical properties. Various sample preparation and micromanipulation / microprobe techniques are examined to localize appropriate signals with the aim to improve the spatial resolution while for extreme depth resolution low acceleration voltages of SEM techniques are utilized [2]. In addition, complex signal acquisition and processing techniques are implemented to achieve time and frequency resolution and to gain required information about the sample properties.

Cathodoluminescence system for measurement of luminescence spectra and panchromatic and spectrally resolved CL mapping in 300-1550 nm wavelength region was implemented in FEG SEM LEO-1550. An internal parabolic mirror and an external collimator were used to collect light to an optical fibre spectrophotometer. For monochromatic CL maps the filters of 8 nm FWHM in front of a thermoelectric-cooled photomultiplier with 3 ns rise time were employed. In conjunction with digital signal processing unit, the luminescence noise and mechanical instabilities of the nanostructures were investigated in frequency range from 1 Hz to 25 kHz. As a result, luminescence properties of various semiconductor structures can be investigated with detail spatial resolution determined mainly by the acceleration voltage for which the generation volume is below the diffusion length of investigated semiconductors. The optical emission properties of InGaN/GaN single quantum well (QW) light-emitting diode structures grown on the facets of GaN nanopyramids forming moth-eye arrays were studied by using depth and spectrally resolved CL mapping at room temperature. To distinguish between luminescence from the GaN buffer layer, GaN nanopyramids and InGaN quantum wells the dependence of the spatially and spectrally resolved CL on the electron beam acceleration voltage in the range between 2 to 20 keV has been used. Three main typical peaks were detected in CL spectra taken from the sample at low magnification. It has been found that a high-intensity CL signal from the nanopyramids facets is observed for wavelength between 400 and 450 nm, with maxima at ~405-425 nm depending on InGaN QW growth conditions, InGaN QW thickness and In content in particular. The peak at ~365 nm corresponds to band-to-band luminescence in GaN, whereas relatively broad peak with maxima at ~550 nm corresponds to recombination via defect states in GaN [3]. The generation and diffusion of the carriers in the structure has been simulated using Monte Carlo method [4] for the interpretation of the measured CL spectra and their dependence on electron beam acceleration voltage and sample parameters. Currently, the gained knowledge and methodology is extensively applied for investigation of various advanced GaN structures prepared e.g. using a two-step nano-pendeo coalescence overgrowth on an array of GaN nanopillars [5].

Various modifications of EBIC technique have been developed and implemented in FEG SEM with a beam-blanking system as stationary and quasi-stationary EBIC, time-resolved EBIC (TREBIC) [6] combining box-car integration technique and digital signal oscilloscope, and frequency-resolved EBIC (FREBIC) [7] techniques. These techniques will be presented as effective tools to study of GaN-based heterostructures and device properties such as Metal-Semiconductor-Metal (MSM) and High Electron Mobility Transistor (HEMT) [8], [9] under various bias conditions. Utilizing EBIC method the lateral distribution of the charge collection between drain and gate electrodes was determined in dependence on V_{DS} and V_{GS}. We found out that the EBIC signal is strongly inhomogeneous in response to the beam current switched on and off. This effect reveals inhomogeneous distribution of electric field in G-D region owing either to the inhomogeneous doping, domains formed during the epitaxial growth or inhomogeneous lateral distribution of traps in HEMT structure. The local dynamic behaviour of the HEMT structures has been investigated by TREBIC point measurements, and in conjunction with TREBIC mapping has revealed inhomogeneous lateral distribution of the induced current, demonstrating the role of the traps on the time-dependent behaviour of HEMT. Likewise, the FREBIC method has been utilized for identification of the origin of structure breakdown, time instabilities and low-frequency noise sources.

Acknowledgement

The authors gratefully acknowledge the financial support of the Slovak Grant Agency projects VEGA 1/0716/09 and 1/0689/09. Part of the work has been done in Centre of Excellence CENAMOST (VVCE-0049-07) with support of the NanoNetII project (ITMS code 26240120018).

References

[1] L. Reimer, *Scanning Electron Microscopy*. Springer-Verlag Berlin Heidelberg, 1985
[2] M. Ledra, N. Tabet, *Superlattices and Microstructures* **45**, 444-450, 2009
[3] C. Liu et al., *Phys. Status Solidi* C 7, No. 1, 32–35 (2010)
[4] D. C. Joy, *Monte Carlo Modeling for Electron Microscopy and Microanalysis*. Oxford University Press Inc., New York, 1995, 216 pp.
[5] P. Shields et al., *Phys. Status Solidi,* in press
[6] T. Geinzer, R. Heiderhoff, L. J. Balk, In: "IEEE 47th Annual Int. Reliability Physics Symp. IRPS2009", Montreal, Canada, 2009, p.796-800
[7] A. Šatka, D. Donoval, F. Mika, *Mater. Sci. Engineering B* **91-92C**, 239-243, 2002
[8] C. Moreau et al., Microelectron. Reliab., (2010), doi:10.1016/j.microrel.2010.07.093
[9] J. Kuzmík et al. *IEEE Trans. Electron Devices* **53** (3), 422-426, 2006

Study of optical and electrical properties of sputtered indium oxide films

M. Predanocy[1], I. Fasaki[2], M. Wilke[3], I. Hotovy[1], I. Kosc[1], L. Spiess[3]

[1]Department of Microelectronics, Slovak University of Technology, Ilkovicova 3, 812 19 Bratislava, Slovakia
[2]National Hellenic Research Foundation, Theoretical and Physical Chemistry Institute, Vasileos Konstantinou Ave. 48, 11635 Athens, Greece
[3]FG Werkstoffe der Elektrotechnik, Institut für Werkstofftechnik, TU Ilmeau, Postfach 100565, 98684 Ilmenau, Germany
e-mail: martin.predanocy@stuba.sk

The indium oxide films were deposited by dc reactive magnetron sputtering from In target on unheated Si substrate with oxygen flow in the ranging from 40 to 80 sccm. The deposited films were annealed in a conventional tube at $T=400^{\circ}C$ for 1 hour in N_2 atmosphere. Measured absorption coefficients of all indium oxide films were in the range the values of $2.6 \div 12.5 \times 10^6$ m^{-1}. It was found that calculated values of the direct band gap and indirect band gap depend on oxygen content in the sputtering gas mixture. Electrical resistivity increased from 6.8×10^3 to 28.5×10^3 Ωcm with increasing oxygen flow $40 - 80$ sccm. Finally, the correlation between optical properties and surface roughness of examined samples was identified.

1. Introduction

Indium oxide (In_2O_3) is a wide band gap ($E_g \sim 3.7$ eV) semiconductor with many applications in microelectronics and optoelectronic devices. In its stochiometric form, it behaves as an insulator, while in its non stochiometric form (In_2O_x), it appears to have semiconducting properties, providing high transparency in visible spectrum and high reflectivity in the infrared region of the frequency spectrum and hence there has been great deal of work on investigating their preparation processes and optimizing their properties. This material can be used as transparent electrode for solar cells and flat panel displays and as grating material [1]. Highly resistive In_2O_3 thin films are utilized as active layers of gas sensors, especially in ozone sensors. Polycrystalline In_2O_3 films can be prepared by various methods such as evaporation, sputtering, sol-gel process and chemical pyrolysis. In_2O_3 nanoparticles can be obtained by metal organic chemical vapor deposition [2].

In the present work we report the experimental results of effect of oxygen flow on the electrical and optical properties of In_2O_3 thin films. The correlation between optical and electrical properties will be determined.

2 Experiment details

The In_2O_3 films were deposited by dc reactive magnetron sputtering from an In target (3" in diameter, 99, 99% pure) in a mixture of oxygen and argon onto unheated Si and glass substrates. A sputtering power of 75W was used. The flow oxygen was changed in the range of $40 - 80$ sccm. Some of the deposited films were annealed in a conventional tube at temperature $400^{\circ}C$ for 1 hour, in N_2 atmosphere.

The study of the optical properties allowed us to get some important parameters such as transmittance, absorption coefficient α, energy band gap E_g of material and others. The values optical transmittance of as-deposited in comparison with annealed In_2O_3 films were measured in the wavelength range 300–1000 nm using spectrometer OceanOptics.

The electrical resistance of the deposited films was determined using standard Van der Pauw method. Hall voltage developed in the films was measured by applying a magnetic field of 0.33 T and used for the determination of Hall mobility and carrier concentration.

3   Results and discussions

3.1   Optical properties of In_2O_3

All examined films were transparent in the visible range at ~ 80 % and higher. It was observed that the films deposited at oxygen flows of 40, 50 and 60 sccm have approximately the same transmittance. But at higher oxygen flows the transparency increases and the sample deposited at 80 sccm achieved maximum transmittance. In general, we can say the value of transmittance decreases due to the following factors: presence of mixed phases, increase in thickness, presence of defects and oxygen vacancies, large rms surface roughness, porous nature of the films and grain boundary scattering [2]. The measured transmittance T (550 nm) was converted into absorption coefficient using the relationship (1)

$$\alpha = \frac{\ln(1/T)}{d} \tag{1}$$

where d is the thickness of the films. The direct band gap values were estimated from the dependence $(\alpha h v)^2 \sim E\text{-}Eg$ and the indirect band gap from $(\alpha h v)^{1/2} \sim E\text{-}Eg$ at the interception of the tangents to the $(\alpha h v)^2$ or $(\alpha h v)^{1/2}$ plots with the photon energy axis. The calculated values of the direct band gap are in the range of 3.35-3.51 eV. The values in the literature of direct band gap are 3.32-3.81 eV [2], 3.59-4.17 eV [3]. The calculated values of the indirect band gap are in the range of 2.18-2.45 eV in the literature the values are 2.1–2.7 eV [2]. The calculated values for both direct and indirect band gap are close to the ones in the literature [2, 3]. The oxygen content during deposition seems to affect slightly these values in most cases. The sample deposited at the lower oxygen flow has a significant lower value for both direct and indirect band gap regardless of layer thickness. This might be due to oxygen vacancies (lower oxygen content) into the lattice of the In_2O_3.

All In_2O_3 films after annealing are transparent in the visible range at ~ 85 %. There is a slightly increase of the transmittance after annealing, mostly for the samples deposited at lower oxygen flows. Again the sample deposited at 80 sccm is the most transparent. It can be suggested that the lower transmittance of as-deposited films were due to an amorphous structure as reported [2]. The band gap was estimated as it was described above. The annealing in all cases caused an increase of the band gap.

We suppose that the presence of defects and oxygen vacancies affected the transparency of indium oxide thin films. This indicates decrease of the density of free electrons in the films with increasing O_2 content [2]. The measured results of roughness surface and calculated absorption coefficients are shown in Fig. 1a. It can be seen a decreasing tendency of absorption coefficient values with increasing the RMS values, especially at higher oxygen flows. This phenomenon can indicate the evolution of optical properties and surface roughness of indium oxide thin films observed in our previously work [6]. We suppose that the decreasing of absorption coefficients in this range oxygen flow (40 – 80 sccm) is chiefly due to presence of defects and oxygen vacancies in films, crystallinity, increasing roughness surface, the coalescence and the diffusion among the grains. Annealed samples revealed the same tendency as as-deposited samples, but dispersion of values was different.

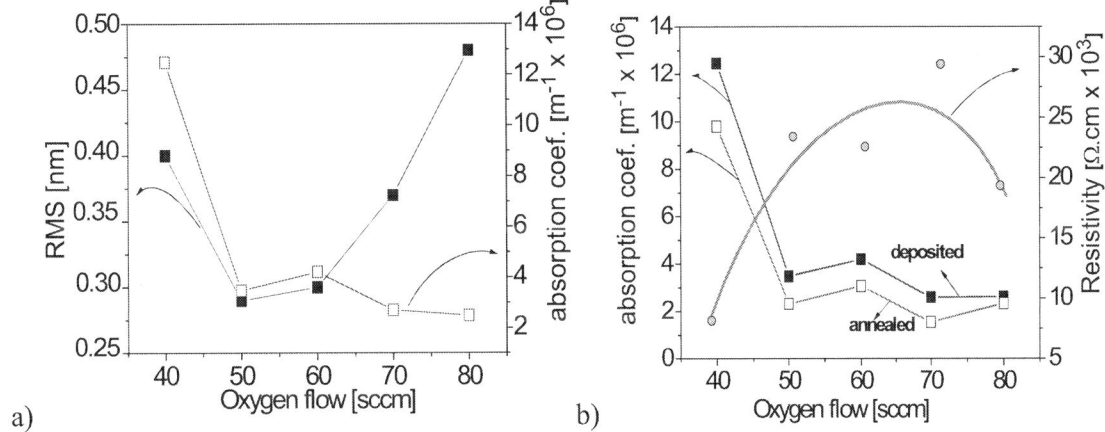

a)   b)

Fig. 1 Absorption coefficients, RMS (a) and resistivity (b) in dependence on the oxygen content in the working gas

3.2 Electrical properties

All examined indium oxide thin films were high resistance and their values were about 10^3 Ωcm at the room temperature. It has tendency to increase from 6.9×10^3 to 28.5×10^3 Ωcm with oxygen flow $40 - 80$ sccm. The values of absorption coefficients were compared with electrical measurements (Fig. 1b). One can see that with increasing oxygen flow ($40 - 80$ sccm) decreased absorption coefficient in the range of $12.47 \times 10^6 \div 2.6 \times 10^6$ m^{-1} in In_2O_3 films. We can suppose that reason of increase resistivity of our films from 6.9×10^3 to 28.5×10^3 Ωcm related to the increasing oxygen incorporation during deposition. Beena [5] presented that lack of stochiometry in the indium oxide films due to the presence of oxygen-array vacancies and the oxygen vacancies can act as doubly ionized donors which can contribute two electrons for electric conduction. Hence the presence of oxygen deficiency plays a major role on the electric conductivity of indium oxide films. In general, incorporated oxygen atoms into the lattice materials may increase the resistance of material. Resulting is reduced flow of charge carriers by material. Our Hall measurements were not successful. We assume that we were unable to create a suitable ohmic contact between measuring tips and In_2O_3 thin layer.

The temperature variation of conductance has been measured between 50 and 200°C. Figure 2 shows a representative temperature dependence of conductance for investigated In_2O_3 films. The calculated values of the activation energy for sample with 80 sccm oxygen flow are E_a=0.37eV (for temperature range between 50 and 85°C) and E_a=0.43eV (for temperature range between 120 and 200°C). Activation energy is similar to published in [4]. Very low activation energy for electrical conduction indicates the presence of large number of donors to form a donor band which approach or overlap the conduction band [4].

978-1-4244-8574-1/10 $26.00 © 2010 IEEE   299

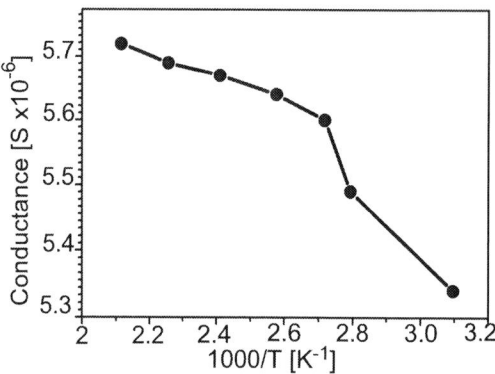

Fig.2 The conductance of In_2O_3 thin films in air as a function of inverse temperature

4.   Conclusion

In_2O_3 thin films were prepared by dc reactive magnetron sputtering at different oxygen flow (from 40 to 80 sccm). All investigated In_2O_3 films were transparent in the visible range at T ~ 80% and 85%, respectively for as-deposited and annealed samples, and higher. In_2O_3 film prepared at 80 sccm oxygen flow is evaluated as the most transparent. The calculated values of the direct band gap were in the range of 3.35÷3.51 eV and indirect band gap were in the range of 2.18÷2.45 eV. It was found a decreasing tendency of absorption coefficient values with increasing the RMS values. We can suppose that reason of increase resistivity of our films from 6.9×10^3 to 28.5×10^3 Ωcm related to the increasing oxygen incorporation during deposition.

Finally, we can conclude that oxygen content in sputtering gas plays important role in optical and electrical properties of sputtered indium oxide films.

Acknowledgement

The work has been supported by Scientific Grant Agency of Ministry of Education of Slovak Republic and Slovak Academy of Sciences, No. 1/0553/09, by Science and Technology Assistance Agency under contract No. VVCE-0049-07, APVV-0655-07 and PPP Programme project DAAD No. D/08/07742.

References

[1]   M. Suchea, N. Katsarakis, S. Christoulakis, S. Nikolopoulou and G. Kiriakidis, *Sensors and Actuators* **B 118**, 135, 2006

[2]   L. Kerkache, A. Layadi and A. Mosser, *Journal of Alloys and Compounds* **479**, 156, 2009

[3]   P. Prathap, G. Gowri Devi, Y.P.V. Subbaiah, K Ramakrishna, *Current Applied Physics* **8**, 120

[4]   A.M Orlov, B. M. Kostishko, and L. I. Gonchar, *Technical Physics Letters* **24**, 81, 1998

[5]   D. Beena, K.J. Lethy, R. Vinodkumar, V. Ganesan , D.M. Phase, *Applied Surface Science* **255**, 8334, 2009

[6]   M. Predanocy, I. Fasaki, M. Wilke, I. Hotovy, I. Kosc, in *Proceedings of the 12th Conference of Doctoral Students ELITECH '10*, Slovakia, 2010, p. 6

Characterization and optical properties of TiO_2 prepared by pulsed laser deposition

O. Kádár[1], F. Uherek[1,2], J. Chlpík[3], J. Remsa[4], J. Bruncko[2], A. Vincze[2], M. Jelínek[4]

[1] Department of Microelectronics FEI STU, Ilkovičova 3, 812 19 Bratislava, Slovak Republic
[2] International Laser Centre, Ilkovičova 3, 841 04 Bratislava Slovak Republic
[3] Department of Physics, FEI STU, Ilkovičova 3, 812 19 Bratislava, Slovak Republic
[4] Academy of Science CR, Na Slovance 2, Praha, The Czech Republic
E-mail: ondrej.kadar@stuba.sk

The contribution reports on fabrication and characterization of TiO_2 thin films on Si substrate. The TiO_2 thin films are used for photonic devices in optical communication systems. The most important parameter for the waveguide structures is the refractive index. The crystalline TiO_2 polymorphs can have different refractive index depending on the structure phase, which is technology dependent. The waveguide structures was prepared by pulsed laser deposition using 248 nm KrF laser source and targets of Ti or rutile in oxygen atmosphere. The optical characterization were done using spectroscopic ellipsometry providing the information about electron band gap values and determining the optical parameters (thickness and refractive index) of prepared TiO_2 thin films. Secondary ion mass spectroscopy (SIMS) and secondary electron microscopy (SEM) was done to verify the vertical and chemical structure.

1. Introduction

This article is aimed at investigation of optical properties of titanium dioxide (TiO_2) waveguide structure prepared by pulsed laser deposition. Titanium dioxide forms three distinct crystalline polymorphs with corresponding refractive indices – 2,488 (anatase), 2,583 (brookite) and 2,609 (rutile). Rutile has the simplest and best known structure. Anatase and rutile are both tetragonal, containing six and 12 atoms per unit cell. Each Ti atom is coordinated to six O atoms and each O atom is coordinated to three Ti atoms. The distortion is greater in anatase than in rutile. The third form of TiO_2 – brookite shown Fig. 1 (c) has a more complicated structure. It has eight formula units in the orthorhombic cell. [1]

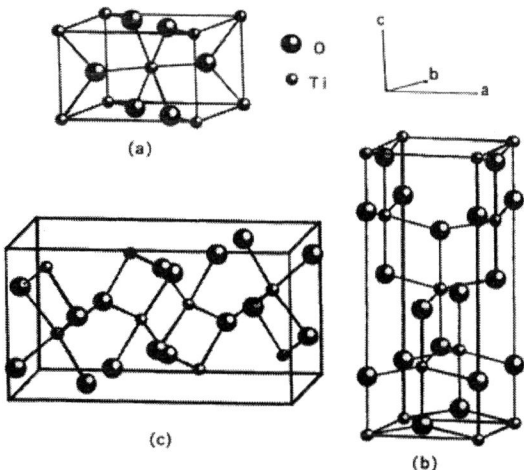

Fig.1 Crystal structures of TiO_2 (a), Rutile (b), Anastase (c), Brookite [1]

2. Thin films description and their preparation

The TiO_2 thin films were grown using PLD with the Compex Pro 205 F KrF$^+$ excimer laser ($\lambda = 248$ nm, $\tau_{FWHM} = 20$ ns, repetition rate 1-50 Hz), a stainless steel vacuum chamber and a target holder with a resistive heating. Preliminary parametric studies were done for optimizing the TiO_2 thin film deposition conditions. The laser beam was focused by a fused silica lens (f = 250 mm) on the target, which was a disc with the diameter of 50 mm and thickness 5 mm made from titanium (99,95%) or rutile (99,99%, pressed pellets). Silicon wafers (111) with dimensions of 10×10 mm^2 were used as substrates. The chamber was evacuated to 10^{-4} Pa before the deposition, and then filled with oxygen. One sample was prepared utilizing radio-frequency discharge (RF) (13,56MHz) with 50W power. During PLD+RF deposition no substrate heating was applied, but the substrate was heated by process to 90°C (measured by thermocouple and IR camera).

Detailed information on deposition conditions can be found in Table 1.

Sample No.	Target	Focus spot [mm^2]	Fluency [J/cm^2]	No of pulses [x10^3]	O$_2$ pressure [Pa]	T$_S$ [°C]	d$_{TS}$ [m]	Laser freq. [Hz]
175	Rutile	1.9x3.2	3.2	9	3	RF	40	5.1
178	Rutile	1.9x3.2	2.0	10	10	100	40	2.1
181	Rutile	1.9x3.2	4.0	20	3	400	60	10
182	Ti	1.9x3.2	4.0	20	3	300	60	10.1

Table 1: Deposition conditions of prepared thin films

3. Thin films characterization

Thin films were measured using spectroscopic ellipsometer Angstrom Advanced PHE 102 at Microelectronics Department. The angle of incidence was 70° with measuring steps 2, 5 and 10 nm. The measurements were done in spectral range 250 nm to 1100 nm. [2] We find the model Amorphous as a best describing model for approximation/fitting of our measured data. This model is valid in the range of wavelengths from 310 nm to 1650 nm. The results of the fitting procedure are the parameters of the thin films; among others parameters (f_j, Γ_j, ω_j, ω_g) and the TiO_2 layer thickness. [3].

The ellipsometer measures the change in polarization state of light reflected from (or transmitted through) the surface of a thin film. The measured values are expressed as psi (ψ) and delta (Δ), which are related to the ratio of Fresnel reflection coefficients R_p and R_s for p- and s- polarized light respectively [4]:

$$\rho = \frac{R_p}{R_s} = \tan(\psi)e^{i\Delta} \qquad (1)$$

Secondary Ion Mass Spectrometry (SIMS) profiling as well as Scanning Electron Microscopy (SEM) topography were employed for TiO_2 thin film structure characterisation and composition determination. The time of flight based SIMS instrument (Ion-TOF) with high energy Bi$^+$ primary source was employed for TiO_2/Si waveguide structure analysis. For depth profiling of the structure high energy pulsed primary source (25 keV) was combined with low energy sputter gun at 2 keV (Cs^+) in 45° to sample surface. Sputtering ion beam is rastered over 300x300 μm^2 area while the primary beam within 80x80 μm^2 area in the centre of sputtered area [5].

4.   Results

Dependence of refractive index *n* on wavelength for TiO_2 thin film 182 is shown in Fig.2. The refractive index is very high (~2.9 for wavelength around 400 nm) and its value is decreasing with increasing wavelengths. For the waveguide structures we are interested for the refractive index at two different wavelengths used in optical communication systems 850 and 1550 nm, respectively.

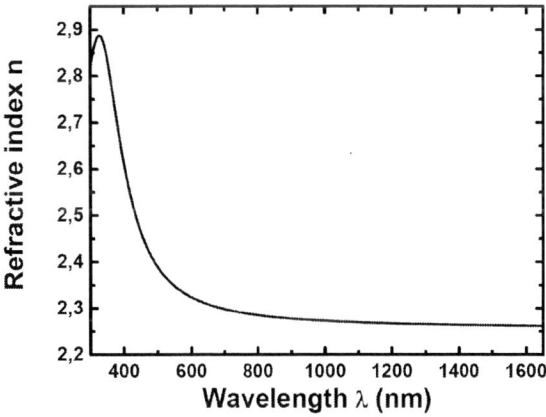

Fig.2 Dependence of refractive index n on wavelength for thin film 182

The thicknesses of TiO_2 thin films were obtained by spectroscopic ellipsometry measurements. Their comparison are shown in the Table 2 also with the corresponding refractive indices of thin films.

Number of thin film	175	178	181	182
Thickness PHE 102 (nm)	535	560	495	540
n at 1550 nm	2.279	2,368	2,658	2,285
n at 850 nm	2.304	2,149	2,652	2,306

Table 2. Comparison of TiO₂ thin film thicknesses of and refractive indices at 850 and 1550 nm

The TiO_2 thin films are homogenous from the SIMS profiling point of view. The SIMS profile of TiO_2 thin film No.181 is shown in the Fig. 3. In comparison of the SIMS profiles of all thin films only small differences are possible to observe. The differences can be expressed as a consequence of incorporation of the impurities like C and H during the deposition process. The elevated growth temperature is the implication of the related to SiO_2 in the TiO_2/Si interface. The surface properties of thin films can be correlated to ellipsometry measurements and their resulting optical properties.

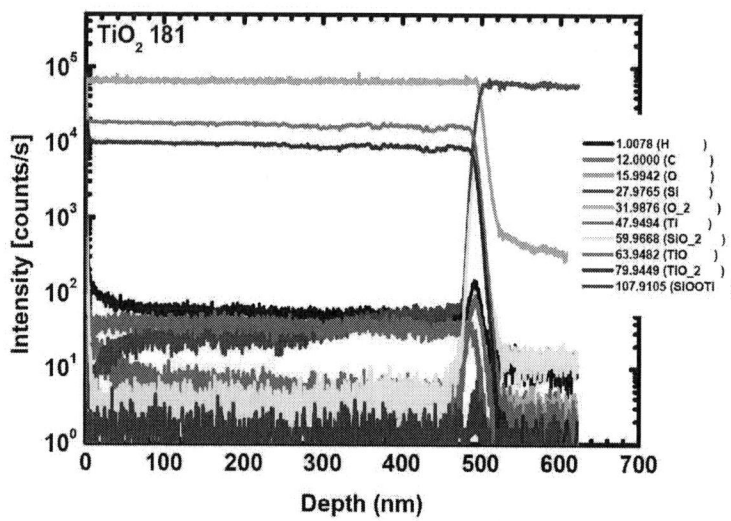

Fig. 3. SIMS depth profile of TiO₂ thin film No 181

5. Conclusion

We deposited and characterised TiO$_2$/Si structures using various methods. The TiO$_2$ thin films are used in optical communication systems as waveguide structures. Using spectroscopic ellipsometry is the refractive index and thin film thicknesses were evaluated and extracted from the measured data of (psi (ψ) and delta (Δ)). The highest refractive index is around 2.65 considered be rutile for TiO$_2$ thin film No. 181 for both optical communication wavelengths. The dependence of the refractive index on wavelength was calculated using an approximation/fitting model in the range of 310 nm to 1650 nm. Also the thin film thicknesses were obtained by ellipsometry. The SIMS depth profiling of TiO$_2$ thin films shows homogenous profiles. Some small differences can be expressed as a consequence of incorporation of the impurities like C and H during the deposition process. The surface properties and characterisation of TiO$_2$ thin films can be correlated to ellipsometry measurements and their resulting optical properties.

Acknowledgement

This work was supported by the Slovak Grant Agency contract VEGA 1/0787/09, under the frame of Centre of Excellence CENAMOST (VVCE 0049-07), APVV project SK – CZ – 0174 – 09 and Institute Research Plan AVOZ 10100522.

References

[1] Shang – Di Mo and W.Y.Ching: Electronic and optical properties of three phases of titanium dioxide: Rutile, anatase, and brookite.

[2] Manual for spectroscopic ellipsometer Angstron Advanced PHE 102 http://www.angstromadvanced.com/Products/PHE102.asp

[3] Horiba Jobin Yvon, Spectroscopic Ellipsometry, User Guide, 2008

[4] J.A.Woollam., Inc, A short Course in Ellipsometry, Revised January 1, 2001 (version 3.335)

[5] A. Vincze, A. Šatka, L. Peternai, J. Kováč, S. Hasenöhrl, M. Veselý: SIMS and SEM Analysis of In$_{1-x-y}$Al$_x$Ga$_y$P LED Structure grown on In$_x$Ga$_{1-x}$P Graded Buffer, Applied Surface Science 252, 2006, pp. 7279-7282

978-1-4244-8574-1/10 $26.00 © 2010 IEEE

New InP Based pHEMT Double Stage Differential to Single-ended MMIC Low Noise Amplifiers for SKA

N.Ahmad[1], S.Arshad[2] and M. Missous[1]

[1]Microelectronics And Nanostructures (M&N) Group, School of E&EE,
The University of Manchester, UK, M60 1QD.
[2] CAE, National University of Sciences and Technology, Risalpur, Pakistan.
E-mail: norhawati.ahmad@postgrad.manchester.ac.uk

A series of room temperature operating double stage differential to single-ended MMIC low noise amplifiers (LNA) design are presented in this work and are based on novel high breakdown InGaAs/InAlAs/InP pHEMTs that have been developed and fabricated at the University of Manchester [1]. All designs are optimised for the frequency range of 0.3 to 1.4GHz in line with the Square Kilometre Array (SKA) requirements [2]. A noise figure of less than 0.5dB with unconditional stability over the entire frequency band of interest is achieved for large periphery gate transistors. Very low power dissipation of 190mW are expected for these design. LNA designs as well as the test mechanisms have been verified by performing simulations in Agilent ADS. Both the double stage differential to single-ended LNAs are being fabricated.

1. Introduction

SKA is a future radio telescope that will have one square kilometre size of effective collecting area. At frequencies up to 1.4GHz, this area will contain in the order of $\sim 10^7$ receivers [2]. Concomitant with this demand, the Low Noise Amplifiers (LNA) which are located at the front end of the receiver chain need to be extremely low noise. Therefore, InP based pHEMT are chosen as an improvement over the GaAs based HEMT device largely due the higher electron mobility and charge density in the channel. This technology has previously demonstrated superior noise performances at millimetre-wave frequencies though at cryogenic temperatures [3].

This work reports the design of fully MMIC differential to single-ended LNA that give promising low noise characteristics and that might be suitable for implementation in the aperture array concept of the SKA project. A differential input amplifier is chosen as the LNA will be interfaced with a differential output receiving antenna while the single-ended output is preferable for the subsequent stages in the analogue chain. The linear and nonlinear models for the InP based pHEMTs used in this work are based on the same technique that has been presented in [4].

2. Epitaxial Layer: Growth and Fabrication

In this work, the whole epitaxial layer structure is grown on an InP semi-insulating substrate with a strained InGaAs channel with indium composition of $\sim$ 70%. The fabricated devices use different gate widths (from 100 to 1200μm), with the gate length optimised to 1μm for both RF and noise performance. The source-gate and gate-drain separation are both 2μm. **Error! Reference source not found.**The epitaxial structure for the device used throughout this work is the same as that reported in [3].

978-1-4244-8574-1/10 $26.00 © 2010 IEEE

3. DC and RF Characteristics

Figure 2 and Figure 3 shows the comparison of modelled and measured DC and RF characteristic of a 4x200μm device fabricated on InP based pHEMTs. The graphs show good agreement between the measured and modelled parameters. The device showed threshold voltage of -1.1V with maximum transconductance of approximately 250mS at V_{GS} of -0.5V. These are the key starting point models for the Monolithic Microwave Integrated Circuits (MMIC), which will be presented later on.

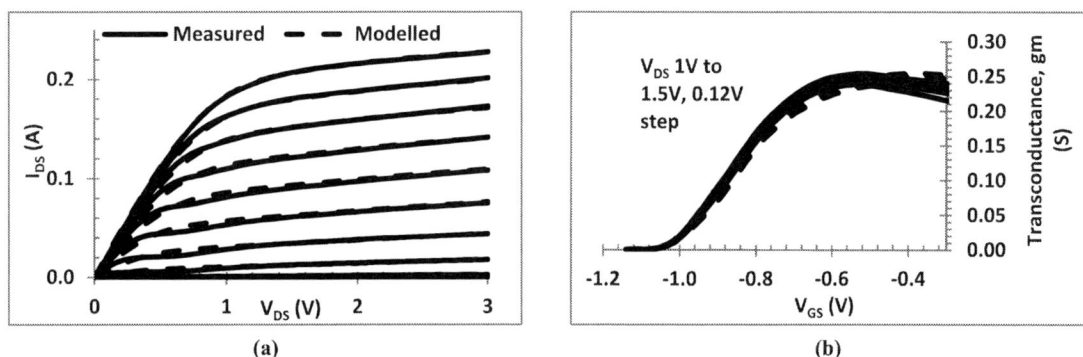

(a)   (b)

Figure 1 Comparison of modelled and measured DC Characteristics of 4x200μm device fabricated on InP based pHEMT: (a) IV curve and (b) Transconductance

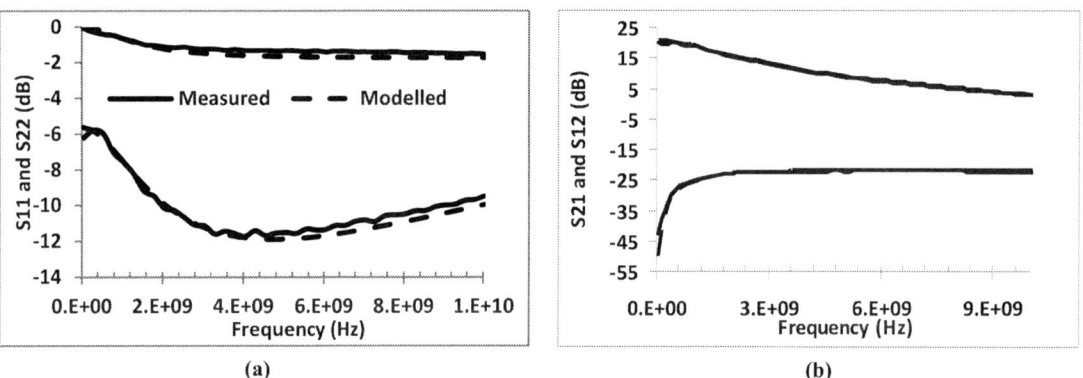

(a)   (b)

Figure 2 Comparison of modelled and measured RF Characteristics of 4x200μm device fabricated on InP pHEMT at V_{DS}= 1V, 10 % I_{DSS} : (a) Input Reflection Coefficient (S11) and Output Reflection Coefficient (S22); (b) Forward Transmission Coefficient (S21) and Reverse Transmission Coefficient (S12)

4. Double stage Differential to Single-Ended MMIC LNA Designs

Two designs for differential to single-ended MMIC LNA have been investigated in this work. Both designs have the same topologies; the first stage is fully differential while the second stage is differential to single-ended. The only difference between the two stages is the transistor size used in the design. The purpose of conducting these experiments was to study the effect of using different transistor sizes on the noise figure and gain. Figure 4 shows the LNA schematic for the first and second stages while Table 1 lists the different of transistor size used for each design.

Transistors M1 and M2 are used as the differential RF input. These transistors are biased at their optimum noise performance i.e. at 10%I_{dss} (~20mA), V_{ds} = 1V and V_{gs} = -0.7V [5]. It is very important to use larger transistor size at the input as it provide low noise resistance (R_n), and hence improved noise figure (*NF*) and input matching for the LNA design. Refering to Eqn. (1) for noise factor (F) and Eqn. (2) for the noise figure (NF), R_n is seen to be directly proportional to F. Therefore, low R_n will give low F and hence low NF.

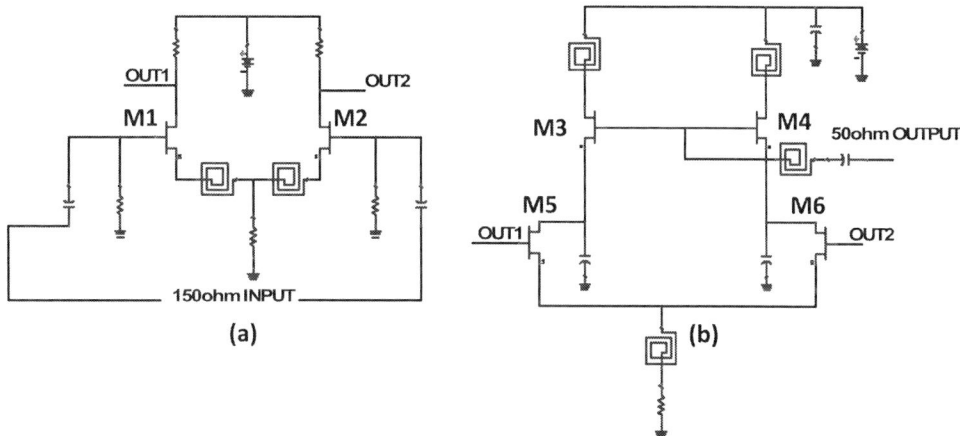

Figure 3 InP based pHEMT Double Stage Differential to Single-ended MMIC LNA: (a) First Stage: Fully differential; (b) Second Stage: Differential to Single-ended

Table 1 Different transistor sizes used in LNA Design1 and LNA Design2

Design	Transistor Size (μm)					
	M1	**M2**	**M3**	**M4**	**M5**	**M6**
Design1	4x200	4x200	4x50	4x50	4x200	4x200
Design2	6x200	6x200	2x50	2x50	4x200	4x200

$$F = F_{\min} + 4R_n \frac{\left|\Gamma_s - \Gamma_{opt}\right|^2}{\left|1 + \Gamma_{opt}\right|^2 \left(1 - \left|\Gamma_s^2\right|\right)} \quad \text{----- (1)} \qquad NF = 10\log_{10}(F) \quad \text{----- (2)}$$

The differential to single-ended stage utilises transistors M3 to M6. Transistor M3 and M4 act as the active load, while the large gate width pHEMT (4x200μm) are used for transistors M5 and M6. Transistor M3 and M4 are chosen to be smaller in order to achieve minimum noise figure ($NFmin$). It is quite a challenge to control the bias conditions for this second stage. This is due to the sensitivity of the output voltage from fully differential stage to the gate arm of both transistors M5 and M6. A 16Ω resistor is used as a current source to drain, with approximately 40mA of current. Therefore, each leg of the differential pair will share half of the current flowing through the resistor. The supply voltage for the first stage is 2V, while the second stage is 3V.

The s-parameters for the second design are illustrated in Figure 5. This LNA shows excellent gain flatness 22±1dB over a wide bandwidth (0.3-1.4GHz). The NF is less 0.5dB with unconditional stability over the entire frequency band of interest. The designed LNA demonstrate good input and output reflection coefficient especially at higher frequency. The power dissipation for this design is 190mW. Comparison of both s-parameter designs are shown in Table 1.

5. Conclusions

New InP based pHEMT Double Stage Differential to Single-ended MMIC low noise amplifiers for SKA have been successfully designed to yield noise figures compatible with the stringent requirements of the SKA. Designs with larger transistors in the first stage show improvement for noise figure and gain flatness with unconditional stability over 10GHz. Power dissipation for both designs are considerably low at ~190mW.These MMIC are being fabricated.

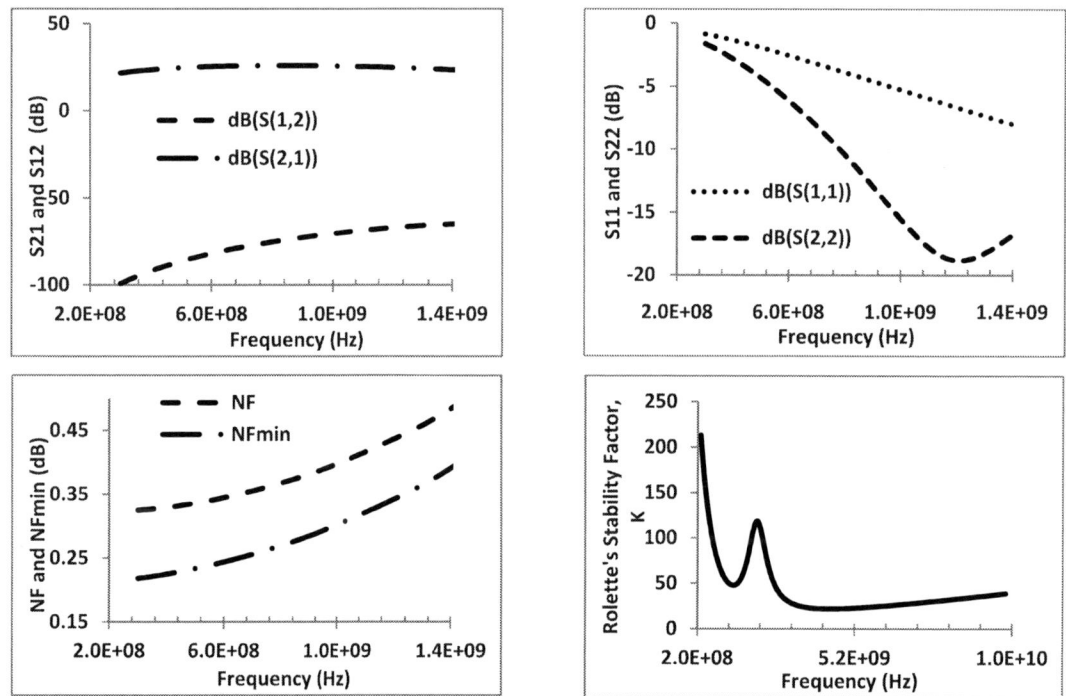

Figure 4 Simulated response of InP based pHEMT Double Stage Differential to Single-ended MMIC LNA (Design2)

Table 2 Comparison of S-parameter for InP based pHEMT Double Stage Differential to Single-ended MMIC LNA (Design1 and Design2)

Design	S21 (dB)		NF (dB)		S11 (dB)		Pdiss
	0.3GHz	1.4GHz	0.3GHz	1.4GHz	0.3GHz	1.4GHz	(mW)
Design1	18.5	23.9	0.52	0.66	-0.5	-4.9	186
Design2	21.8	23.6	0.33	0.48	-0.9	-8	189

Acknowledgement
This work was partially supported by UK's STFC as part of the SKADS programme.

References
[1]   A. S. A. Bouloukou, D. Kettle, J. Sly, M. Missous "Novel High Breakdown InGaAs/InAlAs pHEMTs for radio astronomy application" in *Proceedings of the 4th ESA Workshop on Milimeter Wave Technology and Applications (7th MINT Milimeter-wave International Symposium)*, Finland, 2006, pp. 221-226.
[2]   www.skatelescop.org.
[3]   A. Bouloukou, B. Boudjelida, A. Sobih, S. Boulay, J. Sly, and M. Missous, "Very low leakage InGaAs/InAlAs pHEMTs for broadband (300 MHz to 2 GHz) low-noise applications," *Materials Science in Semiconductor Processing,* vol. 11, pp. 390-393, 2008.
[4]   B. Boudjelida, A. Sobih, A. Bouloukou, S. Boulay, S. Arshad, J. Sly, and M. Missous, "Modelling and simulation of low-frequency broadband LNA using InGaAs/InAlAs structures: A new approach," *Materials Science in Semiconductor Processing,* vol. Volume 11, pp. 398-401, 2008., October 2008.
[5]   H. T. Friis, "Noise Figure of Radio Receivers," in *Proceedings of the IRE*, 1944, pp. 419-422.

SIMS depth profile characterisation of InAlN/GaN structures

A. Vincze[1], J. Kovac[1,2], H. Behmenburg[3,4], R. Srnanek[2], F. Uherek[1,2], D. Donoval[2], M. Heuken[3,4]

[1] International Laser Centre, Ilkovicova 3, 841 04 Bratislava Slovak Republic
[2] Department of Microelectronics FEI STU, Ilkovicova 3, 812 19 Bratislava, Slovak Republic
[3] Institut für Theoretische Elektrotechnik, RWTH Aachen University, Kackertstrasse 15-17, 52072 Aachen, Germany
[4] AIXTRON AG, Kaiserstrasse 98, 52134 Herzogenrath, Germany
e-mail: vincze@ilc.sk

This contribution reports on properties and characterization of InAlN/GaN structures prepared by metal organic chemical vapour deposition (MOCVD) using secondary ion mass spectroscopy (SIMS). The SIMS revealed the vertical cross section of the InAlN/GaN sample structures on SiC substrate and also visualizes the different growth procedure results. The SIMS comparison of the structures shows the Al, In and Si incorporation to the neighbouring layers and has influence on the in-plane stress in AlN layer and measured electrical properties of the fabricated HEMT devices.

1. Introduction

The microelectronic devices fabrication usually starts with the growth on the appropriate substrate. The using of the nitride groups showed us in past years many challenges in the growth and device fabrication as well as in the characterisation methods evolution. Recently, the research and development of GaN-based high-electron-mobility transistors (HEMTs) for high-frequency and high-power operation have rapidly advanced. $In_{1-x}Al_xN$ ternary alloy is an attractive alternative to AlGaN as the barrier layer. $In_{1-x}Al_xN$ with high Al composition is lattice-matched to GaN results higher Schottky barrier height and higher current densities than AlGaN [1, 2]. To prepare specific devices like HEMT transistors the growth on the GaN buffer layers is quite common. This approach is convenient because of relatively simple bandgap engineering by varying the material contents. To improve the material quality pulsed growth of AlN, alternate the V/III ratio, use of AlN interlayers grown at low temperature or epitaxial lateral overgrowth is possible [3]. The GaN buffer layer needs to be prepared on silicon carbide (SiC), which is possible with AlN buffer layer. The technology for this is already available using metal organic chemical vapour deposition (MOCVD) also in commercial field. SiC is at present the reference substrate material for growth of GaN-based HEMTs. As discussed before ther advantages are due to its relatively small lattice mismatch to GaN and a high thermal conductivity. To reduce the lattice mismatch AlN can be employed AlN as a buffer layer on SiC. However, the AlN layer thickness, surface morphology and crystal quality play an important role since they strongly influence the quality and insulating properties of the subsequent GaN buffer.

2. Material and methods

The series of four samples was grown on semi-insulating 6H-SiC substrates using AIXTRON metal-organic vapour phase epitaxy reactor with standard precursor configuration. The sample structure consist on a top of SIC of an AlN buffer layer 270–350 nm thick, followed by a 2.5 µm GaN layer, 1 nm thick AlN and 7 nm $In_{0.14}Al_{0.86}N$ barrier layer as shown in Fig 1. The AlN buffer layer was grown by changing the NH_3 flux resulting in different V/III flow ratios (Tab. I.). The different NH3 flux changes the properties and enables to investigate the impact on growth mode, coalescence and crystal quality. The GaN/AlN/InAlN layers were grown at identical conditions for all samples [4]. For detailed investigations of structure and interface layers secondary ion mass spectroscopy (SIMS) and micro-Raman spectroscopy were used.

To get the full profile information about the structures they were analysed using time of flight based SIMS instrument (Ion-TOF, SIMS IV) with high-energy Bi^+ primary source with combination of low energy Cs^+ sputter gun. A Horiba Jobin Yvon HR800 monochromator equipped with a CCD detector was used to record the micro-Raman spectra at room temperature in backscattering geometry. The spectra were obtained with excitation of He–Ne laser using a laser line of 633 nm wavelength.

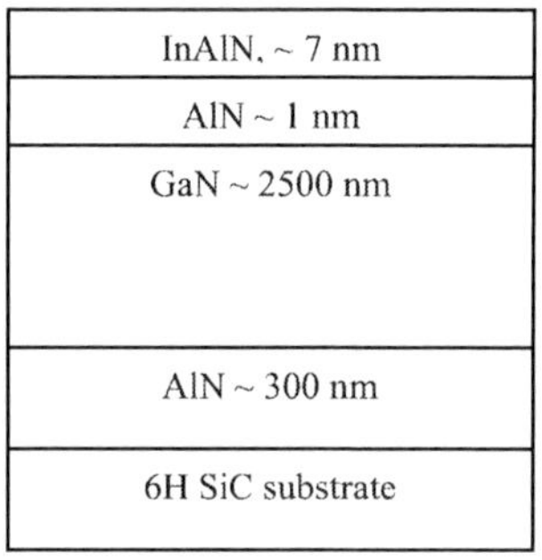

InAlN, ~ 7 nm
AlN ~ 1 nm
GaN ~ 2500 nm
AlN ~ 300 nm
6H SiC substrate

Fig. 1. Schematic structure of investigated samples

3. Results

All structures were investigated by micro-Raman spectroscopy mapping. By this way the in-plane stress in the AlN layer was observed. The stress was calculated from the frequency shift (645 - 660 cm^{-1}) of the peak E_2^H of AlN layer in respect to unstrained AlN compound (656 cm^{-1}). The obtained value of the stress depends on the technological conditions of preparation in different structures, especially on the V/III molar flow ratio during AlN layer growth.

Tab.1 V/III molar flow ratio and evaluated strain of investigated InAlN/GaN structures

Sample	V/III ratio	Strain Raman/XRD [GPa]
42	240	2.0/1.63
40	1200	0.44/-0.35
43	4700	0.14/-1.8
41	8200	-0.8/-2.24

Value of the stress ranged from 2 GPa of tensile to -0.8 GPa of compressive stress measured by Raman spectroscopy and 1.63 GPa to – 2.24 GPa measured by XRD as shown in Tab.1. This results in different residual strain in the AlN layer which was evaluated by micro-Raman spectroscopy and X-ray diffraction measurements [4].

The SIMS investigations started with a survey depth profile measurement as is shown in the Fig. 2. The profile shows the most important elements in the depth profile trough the Au contact on the top of the structure. The detailed view reveals the different distribution of oxygen in the structure as well as homogeneity of AlN layer for different structures. Fig. 3a shows detail of InAlN layer below the surface of Au. Due to the surface roughness of the top layers the sharp interfaces of InAlN/AlN/GaN structure disappear.

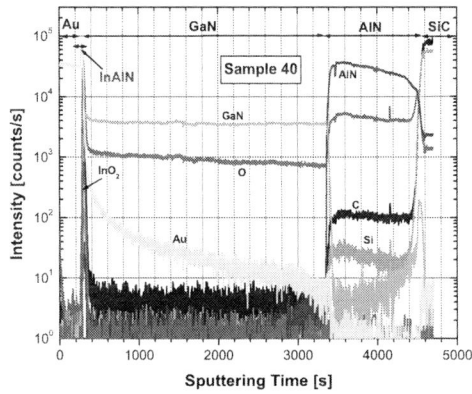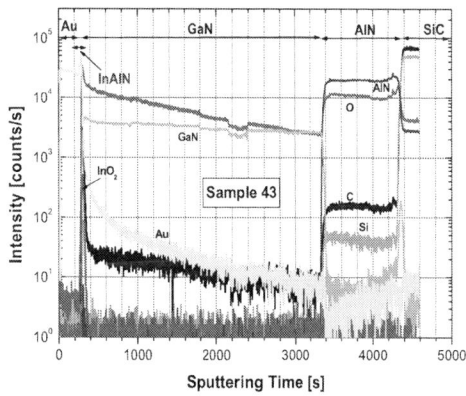

Fig. 2. a) Survey depth profile of sample 40 and b) sample 43

The detail SIMS profile of AlN, GaN interface for all structures is compared in Fig 3b. There are differences in the AlN layer profile while homogenous layer composition and sharp interface exhibit the structure with lowest strain (sample 43). The layers with highest strain exhibit the composition changes in AlN layer with simultaneous increased inter-diffusion of the GaN (Fig. 3b) and C (Fig. 4a) in AlN layer.

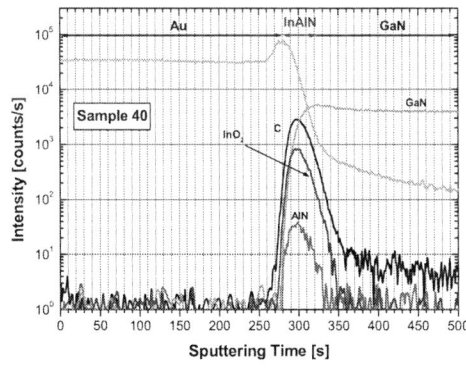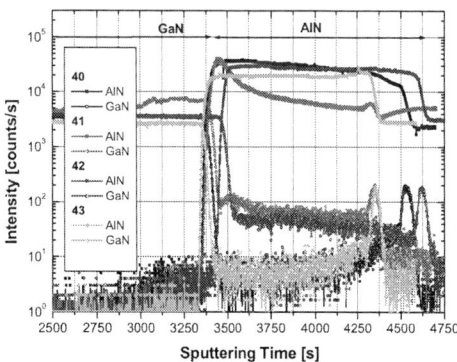

Fig. 3. a) Detail to the InAlN/AlN of structure 40 and b) detail GaN/AlN profile comparison

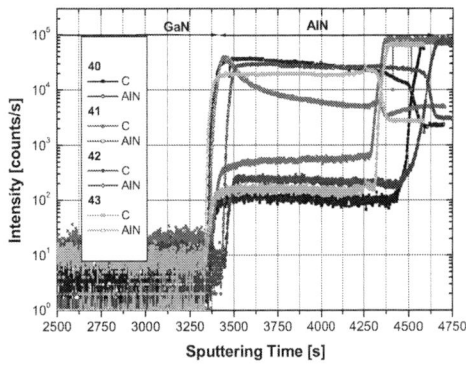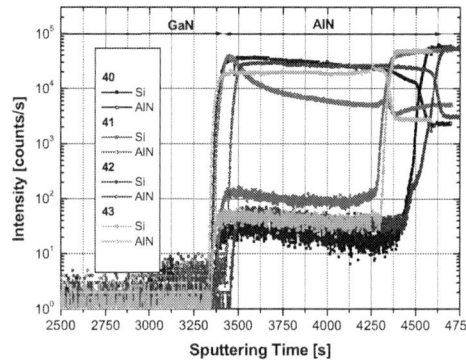

Fig. 4. Detail to a) AlN/C and b) AlN/Si depth profile comparison

Due to differences during the growth of the AlN buffer layer on the SiC substrate we observed different distribution of Si, which suggests the diffusion process of Si into to the AlN and GaN buffer layer as shown in Fig. 4b.

4.   Conclusions

The InAlN/GaN structures are important for the HEMT fabrication. Due to the modified process of growth method the resulting properties of the SIMS profiles are modified. The most important changes are in the profiles between structure 40 and 41, where the Si is diffused to the upper AlN layer. The SIMS depth profiling of selected elements and compounds revealed the differences during the structure growth. Investigations by XRD and micro-Raman spectroscopy revealed that the AlN lattice constants are strongly affected by the changing growth mode accompanied by creating in-plane strain in the buffer layer. This was confirmed by changes in SIMS depth profile of AlN layers, which shows different composition profile and inter-diffusion of Ga, Si and C at the GaN/AlN and AlN/SiC interfaces.

Acknowledgements

The authors acknowledge financial support from the MORGAN EU FP7 program and also the support from Center of Excellence CENAMOST (Slovak Research and Development Agency Contract. No. VVCE-0049-07) with support of projects APVV-0290-06, VEGA-1/0689/09, VEGA-1/0787/09.

References

[1]   M. Hiroki, H. Yokoyama, N. Watanabe, T. Kobayashi, *Superlattices and Microstructures* 40 (2006) 214–218

[2]   P. Tasli, B. Sarikavak, G. Atmaca, K. Elibol, A.F. Kuloglu, S.B. Lisesivdin, *Physica B* 405 (2010) 4020–4026

[3]   O. Reentilä, F. Brunner, A. Knauer, A. Mogilatenko, W. Neumann, H. Protzmann, M. Heuken, M. Kneissl, M.Weyers, G. Tränkle, *Journal of Crystal Growth* 310 (2008) 4932–4934

[4]   H. Behmenburg, C. Giessen, R. Srnanek, J. Kovac, H. Kalisch, M. Heuken, R. H. Jansen – IC MOVPE XV, *submitted to Journal of Crystal Growth*

Physics-Based Modeling of Electromagnetic Parasitic Effects in Interconnects and Busbars

Gerhard Wachutka and Peter Böhm

Inst. for Physics of Electrotechnology, Munich University of Technology,
Arcisstrasse 21, 80290 Munich, Germany
e-mail: wachutka@tep.ei.tum.de

The 3D-simulation of electromagnetic fields and current flow in real-life interconnect structures enables the detailed analysis of parasitic inductive effects and, thus, provides the basis for the optimization of bus bars in high power modules.

1. Motivation

The optimization of high power components based on fast switching power semiconductor devices requires an integral approach analyzing the overall performance of a module including the system-specific parasitic effects. With increasing switching frequencies and ever shorter pulse width, the eigendynamics inherent in the system becomes relevant and, hence, has to be included in the system analysis. Along with electrothermal aspects, the transient electromagnetic behavior of the interconnects can no longer be neglected.

It turns out that the characterization of a complex, multiply contacted bus bar, or a system of bus bars, respectively, has to include the concept of time-dependent inductance. Induced overvoltage peaks, not only resulting from the time-derivative of the terminal current, but alsoreflecting the time-dependence of the inductance, have to be considered as they can damage the attached semiconductor devices. As already demonstrated [1], inhomogeneous current distributions in the interconnects do not only cause significant and, hence, non-negligible electrothermal heating, but they can also lead to a malfunction of large-scaled semiconductor switches, because their function relies on a nearly homogeneous current distribution along the contacts. Moreover, the switching behavior may no longer be governed by the turn-on time of the semiconductor switches in use, but rather by the turn-on delay caused by the "electromagnetic inertia" of the bus bars.

2. Conventional approach

The traditional way of describing and characterizing the electromagnetic behavior of high power bus bar structures is confined to the extraction of the static inductance matrix of self- and mutual inductance coefficients. Apart from the commonly employed heuristic approaches [2,] there exist well-established numerical methods for inductance extraction; in particular, the "Partial-Element Equivalent-Circuit Method" (PEEC) [3] is widely used. This method exploits Neumann's formula to calculate the partial inductances. To this end, the interconnects are partitioned into brick-shaped parts, where each of them is associated with an elementary current filament [4]. Different alternative integration techniques are employed to evaluate this formula, among others multipole expansion and Monte Carlo integration.

3. Extended FEM-based approach

However, all these methods are restricted to the quasi-magnetostatic or the time-harmonic case and require a qualitative a-priori knowledge of the spatial distribution of current flow in the bus bars. But for the analysis of the fast transient behavior of interconnects, a detailed understanding of the electromagnetic fields inside and outside the interconnects is essential.

978-1-4244-8574-1/10 $26.00 © 2010 IEEE

In that case it is more appropriate to stay in the time domain and not to switch over to the frequency domain by Fourier analysis.

The transient skin effect is one of the topics arising in this context. Actually it is a special case of an eddy current problem: Time varying pulse-shaped current is flowing in one or more conductors with unknown distribution of the current density, while either the voltages at the terminals of the conductors (voltage-driven case) or the terminal currents (current-driven case) are given as boundary data. The basic configuration is depicted in Fig. 1. Maxwell's equations have to be solved in the so-called quasistationary approximation, where the propagation of electromagnetic waves is neglected. If the terminal voltages $u_k(t)$ are given data, we make use of a magnetic vector potential A and an electric scalar potential ϕ:

$$\boldsymbol{B} = \boldsymbol{curl\,A} \quad \text{everywhere}, \quad \boldsymbol{E} = -\frac{\partial \boldsymbol{A}}{\partial t} - \boldsymbol{grad}\,\phi \quad \text{in } \Omega_\text{c}$$

In this case, the boundary data $u_k(t)$ control the scalar potential at the terminals, while we have homogeneous boundary conditions elsewhere. If the terminal currents $i_k(t)$ are given, the combination of a current vector potential T and a magnetic scalar potential Φ is expedient [5]:

$$\boldsymbol{J} = \boldsymbol{curl(T_0 + T)} \quad \text{in } \Omega_\text{c}, \quad \boldsymbol{H} = \boldsymbol{T_0 + T - grad}\,\Phi \quad \text{in } \Omega_\text{c}, \quad \boldsymbol{H} = \boldsymbol{T_0 - grad}\,\Phi \quad \text{in } \Omega_\text{n}$$

These two different and complementary formulations allow us to consider a large variety of real-life operating conditions and to extract all quantities required to describe the distributed parasitic behavior of realistic interconnect structures.

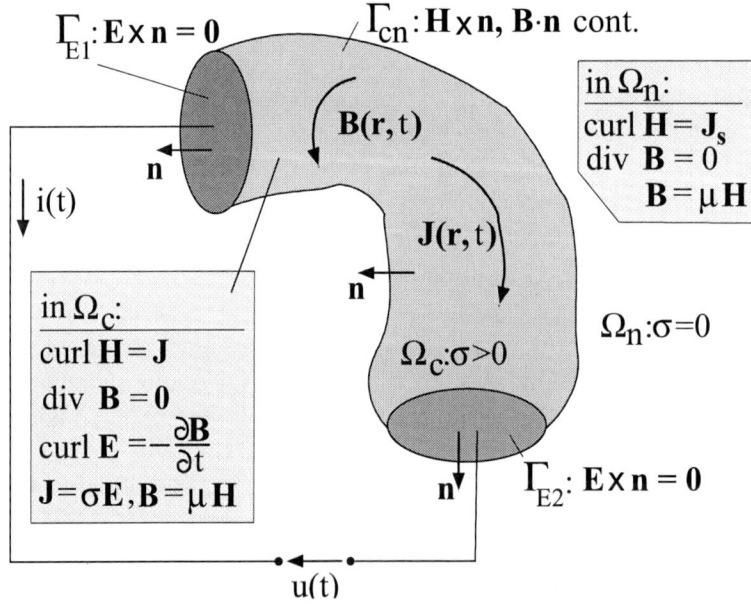

Fig. 1: Schematic representation of a quasistationary electromagnetic field problem (skin-effect problem). The equations to be solved in the conducting region(s) Ω_c and in the non-conducting exterior region Ω_n are also shown.

To the authors knowledge, there exist no commercial finite element programs to simulate the three-dimensional transient current-driven skin effect. Therefore a new simulator has been developed based on the C++ library DIFFPACK [6], an object-oriented platform for the solution of partial differential equation systems. Extending the already available functionality of this software package to our particular needs, we developed two variants of a numerical simulator for the voltage- and the current-driven inductance analysis for short-pulsed current flow. The spatial discretization of the simulation domain employs classical node elements as well as special edge finite elements (necessary for current-driven operation [7]).

4. An Illustrative Example

A typical high power bus bar structure, as it may be encountered in industrial applications, has been analyzed. The bus bar is made of thin copper plates with a thickness of 2 mm. The total length of the structure is in the range of 10 cm. Two cases have been considered: First, in the voltage-driven case, a voltage of 100 V is applied as bias with a ramping time of 100 μs. Second, in the current driven case, a current of 1 kA is switched on at t = 0 s with a rising time of again 100 μs. Fig. 2 shows the current density in the voltage-driven case at t = 10 μs. Because of self-induced eddy currents, the current density is concentrated at the outer edges of the structure. This phenomenon occurs for a certain period of time (t < 1 ms) and is clearly represented in the simulation. In the quasi-static state (t > 10 ms) the current distribution looks like a laminar, potential-driven current flow, not disturbed by eddy currents (Fig 3). The current flowlines in the current-driven case show almost the same pattern as in the voltage-driven case.

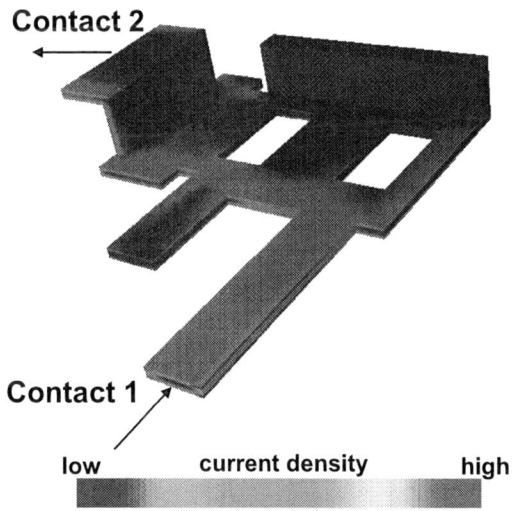

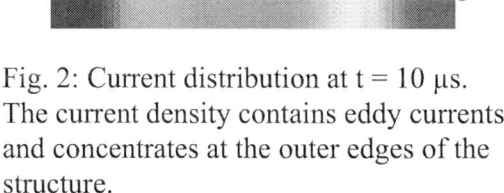

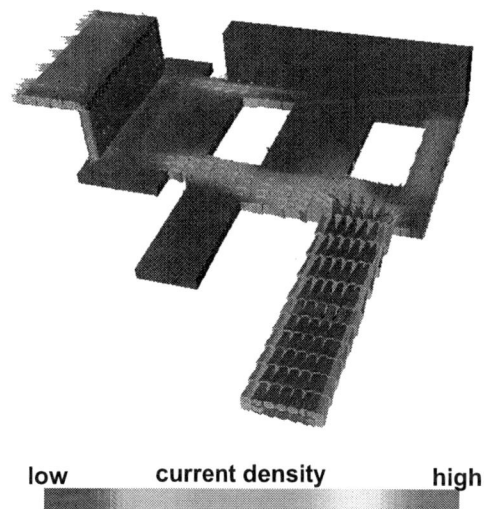

Fig. 2: Current distribution at t = 10 μs. The current density contains eddy currents and concentrates at the outer edges of the structure.

Fig. 3: Current distribution at t = 0.5 s. The current density reflects the solution of a Poisson problem (potential-driven current flow).

The terminal behavior of the voltage- and the current-driven case is quite different (Fig. 4). In the voltage-driven case, the voltage controls a specific, time-varying amount of energy. This energy is split up into one part building up the magnetic field and another part associated with the current flow. As a consequence, we observe a delay in the turn-on time of the terminal current. For the current-driven case, the terminal behavior is dominated by an overvoltage peak at turn-on time. Also the total self-inductance can be decomposed in an "internal" and an "external" contribution, which show different transient behavior. The internal self-inductance of the copper bar is mostly determined by the redistribution of the current density, whereas

978-1-4244-8574-1/10 $26.00 © 2010 IEEE   315

the external self-inductance is nearly constant over time. However, as most of the magnetic energy is stored in the field outside the conducting domains, the internal self-inductance is about one order of magnitude lower than the external selfinductance.

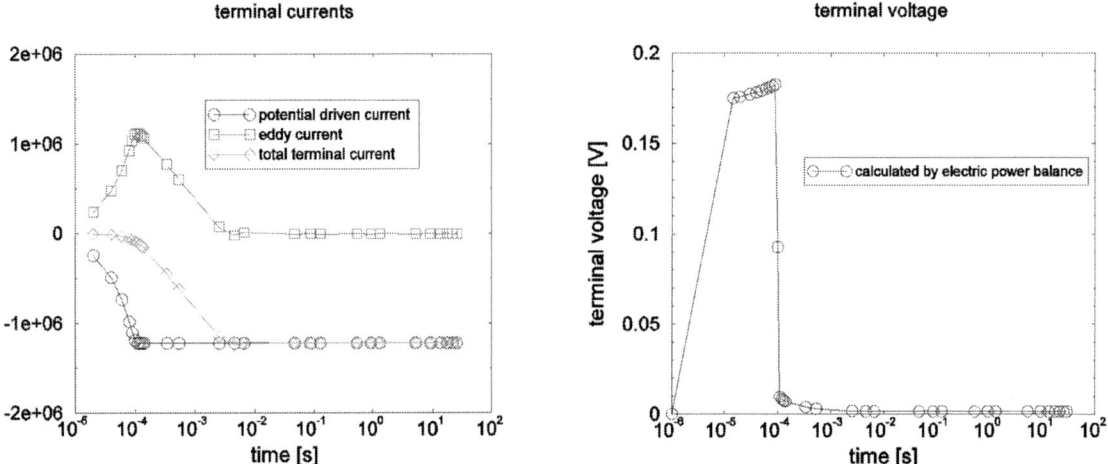

Fig. 4: Terminal currents in the voltage-driven case (left). Terminal overvoltage peak in the current-driven case (right).

5. Conclusion

The three-dimensional transient simulation of electromagnetic fields and current flow in real-life interconnect structures using a special variant of the finite element method provides a detailed insight into the electromagnetic behavior of such structures. It also enables the accurate extraction of integral quantities serving as target criteria in the optimization of bus bars in high power modules. Complementary formulations of the basic equations in terms of alternative potentials are the key for the proper treatment of current- and voltage-driven skin effect problems. The capabilities of the new simulation method have been illustrated by the numerical analysis of a realistic bus bar structure adopted from industrial applications.

References

[1] Böhm, P., Falck, E., Sigg, J., Wachutka, G.: Continuous Field Analysis of Distributed Parasitic Effects by Interconnects in High Power Semiconductor Modules. Proc. of SISPAD, Leuven, Belgium (1998) 340-343.

[2] Skibinski, G.L., Divan, D.M.: Design Methodology & Modeling of Low Inductance Planar Bus Structures. Proc. of EPE, Brigthon, U.K. (1993) 98-105.

[3] Ruehli, A.E.: Inductance Calculation in a Complex Integrated Circuit Environment. IBM Journal of Research and Development (1972) 470-481.

[4] Clavel, E., Schanen, J.L., Roudet, J.: Electromagnetic Modeling of a Power Module Case. IEEE EUROEM (1994).

[5] Bíró, O., Preis, K., Renhart, W., Vrisk, G., Richter, K.R.: Computation of 3D Current Driven Skin Effect Problems Using a Current Vector Potential. IEEE Transaction on Magnetics **29**, No. 2 (1993) 1325-1328.

[6] Langtangen, H. P., Tveito, A.: Advanced Topics in Computational Partial Differential Equations, Numerical Methods and Diffpack Programming. Lecture Notes in Computational Science and Engineering, vol. **33**, Springer, Berlin (2003).

[7] Bíró, O., Preis, K., Böhm, P., Wachutka, G.: Edge Finite Element Analysis of Transient Skin Effect Problems. IEEE Transactions on Magnetics **36**, No. 4 (2000) 835-839.

Compact Model Extraction from Quantum Corrected Statistical Monte Carlo Simulation of Random Dopant Induced Drain Current Variability

Urban Kovac, Craig Alexander, Gareth Roy, Binjie Cheng, and A. Asenov

Dept. Electronics & Electrical Engineering, University of Glasgow,
Glasgow, G12 8LT, UK
e-mail address : kovac@elec.gla.ac.uk

An efficient method to accurately capture quantum confinement effects within Monte Carlo (MC) simulation while simultaneously resolving 'ab initio' ionized impurity scattering via the density gradient (DG) formalism is presented. The model is applied to study the impact of transport variability due to scattering from random discrete dopants on the on-current variability in realistic nano CMOS transistors. Such simulations result in an increase in drain current variability when compared with similarly quantum corrected drift diffusion (DD) simulation. Following this, an efficient three-stage hierarchical strategy is presented that propagates the increased on-current variability captured in 3D quantum corrected 'ab initio' MC into efficient 3D DD simulations that are in turn used to obtain target I_D-V_G characteristics for the extraction of statistical compact models.

I INTRODUCTION

With transistors approaching nanometer dimensions, statistical variability has become a major concern to future scaling and integration [1]. At the 45nm technology generation, the magnitude of the variation introduced by statistical variability can be comparable to the global process variation. In the 32nm technology generation and beyond, statistical variability will become the major source of variation. The ever-increasing statistical variability calls for a change in digital circuit design practices in order to create variability aware design technologies, the prerequisite for which is reliable statistical compact models. Intrinsic parameter fluctuations introduced by the discreteness of matter and charge, which cannot be eliminated by tightening of process controls, are the major source of statistical variability [2]. The source of greatest variability in conventional (bulk) MOSFETs is the random discrete dopant distribution (RDD) within the channel and the source/drain extensions [3]. A significant amount of work has been done in examining variability both experimentally and numerically in idealized [2,4] and realistic devices [5].

Because a 3D treatment is necessary for statistical variability simulation studies, DD simulations have been routinely used [6]. These capture the electrostatic impact of the variability in subthreshold and provide an accurate estimate of threshold voltage variability [7]. Unfortunately, DD simulations underestimate the current variability associated with transport variations due to scattering from discrete dopants [8,9], resulting in a deleterious effect on any subsequent circuit analysis.

In this paper we present a three-stage hierarchical simulation strategy that propagates transport variability captured in statistical 3D MC simulations into statistical 3D DD. The DD simulations in turn are used to generate a set of enhanced I_D-V_G characteristics for the extraction of statistical compact models. To the best of our knowledge this is the most comprehensive, efficient and accurate approach for predictive simulation of RDD variability and its incorporation within industry standard compact models for use in statistical design. The inclusion of quantum corrections in MC simulations is discussed in the second section. The third section describes the hierarchical simulation strategy that is further illustrated in the fourth section using a statistical ensemble of 50 devices, the results of which are presented with discussion before conclusions are drawn.

II QUANTUM CORRECTED MONTE CARLO

Both 3D DD and self-consistent MC simulations are performed with quantum corrections via the density gradient (DG) effective quantum potential [10]. This is essential for resolving confinement effects in the accurate simulation of nanoscale MOSFETs and in MC this is applied using a previously developed methodology [9]. RDD induced transport variations are included directly in MC via the '*ab initio*' treatment of impurity scattering [8]. However, in place of the short-range correction to the classical mesh-resolved Coulomb interaction [8,11,12], the carrier-ion interaction is here uniquely captured through the quantum corrected potential alone. The quantum potential obtained from the DG solution surrounding a single attractive impurity potential is shown in Fig.1. Solutions corresponding to cubic discretization meshes with different mesh spacings are compared to a strong mesh spacing dependence of the classical Coulomb potential [6]. The effective DG quantum potential has only a weak mesh spacing dependence and shows convergence of the potential solution for mesh spacings of 1 nm and smaller.

Also shown in Fig.1 is the analytic Coulomb potential and short-range correction model applied previously [8]. It is clear that the analytic short-range model closely agrees with the converged DG effective quantum potential. The similarity hints at the suitability to simultaneously include quantum corrections and short-range electron-impurity interactions from the mesh resolved quantum potential in 'ab initio' MC. This would provide an efficient and consistent mesh based treatment that would automatically include confinement effects as well as 'ab initio' Coulomb interactions and would accurately define boundary and image-charge effects. The prior short-range correction approach to impurity scattering was originally validated by the reproduction of concentration dependent mobility [9]. The quantum potential approach is similarly validated and bulk mobility results are compared in Fig. 2. The new model agrees well with experimental mobility and previous short-range corrected MC results.

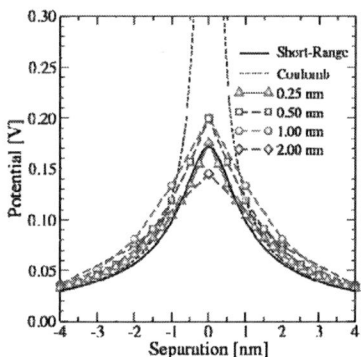

Fig.1 Comparison of analytic short-range correction applied in [8] to density gradient solution around point charge.

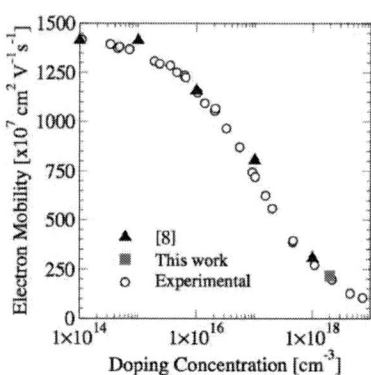

Fig.2 The verification of 'bulk' concentration dependent mobility from quantum corrected simulations in comparison with classical results obtained in [8].

III STATISTICAL COMPACT MODEL SIMULATION METHODOLOGY

In the first stage of the hierarchical approach, which is illustrated in its entirety in Fig. 3, statistical DD simulation of drain current variability is performed at a single, high, gate bias and at low and high drain biases. Identical self-consistent quantum corrected statistical MC simulation follows, initialized from the prior DD solution, and results in an improved estimate of the drain current variation.

The second stage incorporates the more accurate, increased, variation from MC into DD. This is achieved through a two-step self-consistent calibration of the mobility models within DD, such that the relative drain current variation agrees with MC at both low and high drain bias. In the first step, concentration and vertical field dependent low field mobility models are altered to match MC variation at low drain. In the second step, the saturation velocity within the lateral field dependent mobility model is altered to match MC at high drain. The two-step process is iterated until a single solution satisfies global agreement. After mobility calibration, a statistical set of complete I_D-V_G characteristics may be efficiently generated from DD That now includes the additional variation seen in MC.

A final, third, stage extracts compact model parameters from the enhanced I_D-V_G curves following the methodology reported in [13,14]. In the first step, the complete set of BSIM4v4 [15] model parameters using Synopsis tool Aurora [16] are extracted from the simulated characteristics of transistors with continuous doping profiles and different channel lengths and widths. These account for the long channel behaviour, threshold voltage in the short channel regime and the behaviour of the drain current at high field. The second step is carried out in two sub-extraction loops and is based on the enhanced I_D-V_G characteristics of each device from the simulated statistical ensemble. In this step, seven key BSIM4v4 model parameters are chosen and extracted that, from a device operation point of view, account for the effects of the unique distribution of dopants.

IV RESULTS AND DISCUSSION

Simulations of channel dopant induced drain current variability within a 35 nm test bed MOSFET, described elsewhere [17], were performed for an ensemble of 50 randomly generated devices at high gate bias of 1.0 V and low and high drain biases of 0.05 V and 1.0 V respectively. A typical potential distribution from self-consistent quantum corrected MC simulation is presented in Fig. 4. Individual acceptors are visible and their impact on the inversion layer electron concentration and current density is shown in Fig. 5 for devices with the highest and the lowest current. It is clear that in the low current case a crowding of dopants near the source considerably reduces the carrier and current densities. This line of dopants across the width of the device efficiently impedes current flow and is responsible for the greatly reduced current. The effect however is extended, reaching out with the screened impurity potential. In particular, the acceptors near the source and lying close to the inversion layer backscatter electrons into the source region and impact the current more than through the localised carrier reduction alone. In this case, the positioning of the dopants as well as their number effects the current in MC.

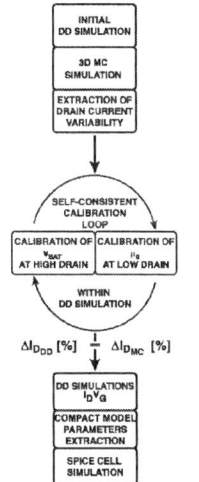

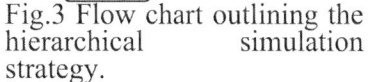

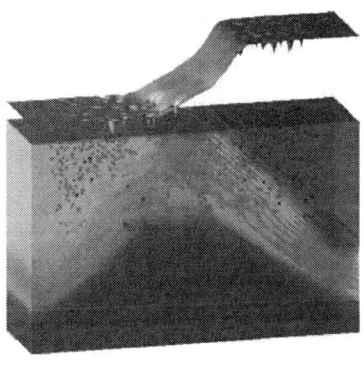

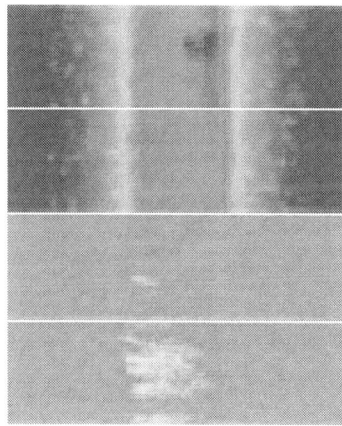

Fig.3 Flow chart outlining the hierarchical simulation strategy.

Fig.4 Potential distribution within the channel and substrate for 35nm atomistic nMOSFET for highest current.

Fig.5 Inversion layer (top) electron density $[\log_{10}(cm^{-3})]$ from source to drain (left-right). Inversion layer current (bottom) density $[mA\ mm^{-2}]$ for devices with lowest current and highest current.

This is contrasted with the lack of acceptors near the source within the high current device, which creates large 'un-doped' regions. The impact of this on current flow can be seen through the greatly increased current density in Fig. 5. Local reductions in current density associated with individual acceptors can still be seen.

The original I_D-V_G curves obtained from DD are compared with the MC calibrated DD results in Fig. 6. The calibration preserves the mean drain current at high gate bias while the significantly increased standard deviation of the drain current variation at high gate after calibration can be clearly seen. At high drain bias, the enhanced on-current distribution is flat topped and skewed to lower currents, while the original distribution has a sharper centralised peak. The increase in current variability due to transport variation, inferred from the difference between original and 'enhanced' standard deviations, is greatest at drive current conditions but reduces with the reduction of the gate bias. The mobility calibration does not significantly effect the results in the subthreshold region, validating the use of DD for simulation of threshold voltage variability. The reduction at low drain bias is from more than 50% at V_G=1.0 V to approximately 10% at V_G=0.4 V. It is clear that if unaltered, DD significantly underestimates the on-current variability, introducing errors in the statistical static timing analysis at the design stage.

The distribution of the RMS error in the extraction process for both original and enhanced I_D-V_G curves is presented in Fig. 7. The distribution of the RMS error from compact model extraction for both cases shows similar, minimal error. The average statistical compact model error and its standard deviation are larger when the transport variability is taken into account compared to the reference DD simulations. This is expected when identical compact model parameter sets are used to capture different magnitudes of statistical variability. Still, the relatively tight error distribution of the final statistical compact model set is sufficient not only for digital but also for analogue applications.

CONCLUSIONS

The introduction of quantum corrections within MC simulation yields an efficient carrier-ion interaction comparable to previous 'ab initio' methods. The new hierarchical extraction methodology provides DD results that maintain threshold voltage variability while also accurately incorporating the on-current variability obtained from MC. The simple approach based upon calibration of the DD mobility models works well over the whole range of transistor characteristics. Finally, the methodology is an efficient technique capable of generating sets of statistical I_D-V_G curves from DD simulations with information obtained from a restricted set of MC simulations.

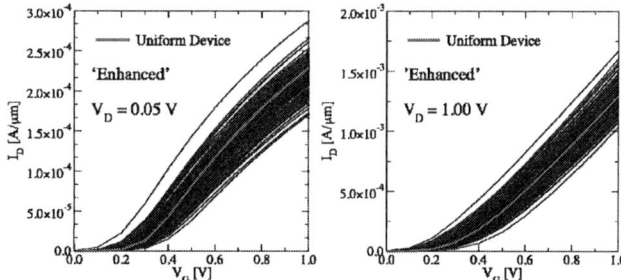 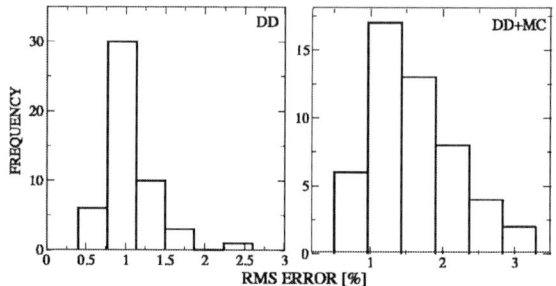

Fig.6 The original I_D-V_G curves (left) obtained from DD at high drain bias and I_D-V_G curves (right) obtained from the MC calibrated DD results.

Fig.7 The histogram of RMS error of the compact model extraction from original ID-VG curves (left) and from ID-VG curves obtained from the MC calibrated DD results (right).

REFERENCES

[1] B. Cheng, S. Roy, G. Roy, F. Adamu-Lema, and A. Asenov, *Solid-State Electronics*, vol. 49, no. 5, pp. 740–746, May 2005.

[2] A. Asenov, A. R. Brown, J. H. Davies, S. Kaya, and G. Slavcheva, IEEE Trans. Electron Devices, vol. 50, no. 9, pp. 1837-1852, Sep. 2003.

[3] T. Mizuno, J. Okamura, and A. Toriumi, *IEEE Trans. on Electron Devices*, vol. 41, no. 11, pp. 2216–2221, Nov. 1994.

[4] A. Asenov, S. Kaya, and A. R. Brown, *IEEE Trans. on Electron Devices*, vol. 50, no. 5, pp. 1254–1260, May 2003.

[5] Frank DJ, Taur Y, Ieong M, Wong H-SP, In: VLSI symp tech dig; p. 169–170.1999.

[6] G. Roy, A. R. Brown, F. Adamu-Lema, S. Roy, and A. Asenov, *IEEE Trans. on Electron Devices*, vol. 53, no. 12, pp. 3063–3070, Dec. 2006.

[7] C. Millar, D. Reid, G. Roy, S. Roy, and A. Asenov, IEEE Electron Device Letters, vol. 29, iss. 8, pp. 946–948, Aug. 2008.

[8] C. L. Alexander, G. Roy, and A. Asenov, *IEEE Trans. on Electron Devices.*, vol. 55, no.11, pp.3251-3258, 2008.

[9] C. Riddet, A. R. Brown, C. L. Alexander, J. R. Watling, S. Roy, and A. Asenov, *IEEE Transactions on Nanotechnology*, vol. 6, no. 1, pp. 48–55, 2007.

[10] M. G. Ancona and G. J. Iafrate, *Physical Review B*, vol. 39, no. 13, pp. 9536–9540, May 1989.

[11] R. W. Hockney and J. W. Eastwood, Computer Simulation Using Particles. New York: McGraw-Hill, 1981.

[12] C. J. Wordelman and U. Ravaioli, *IEEE Trans. on Electron Devices*, vol. 47, no. 2, pp. 410–416, Feb. 2000.

[13] S. Roy, B. Cheng, G. Roy, and A. Asenov, *Journal of Computational Electronics*, vol. 2, pp. 427–431, 2003.

[14] B. Cheng, S. Roy, and A. Asenov, In: *Proc. Simulation of Semiconductor Processes and Devices*, T. Graser and S. Selberherr, Eds., Vienna, Austria, pp. 301–304,Sept. 2007.

[15] BSIM4v4 manual. Available from: http://www-device.eecs.berkeley.edu/~bsim3/bsim4_get.html

[16] Taurus User Guide Version W-2004.09, Synopsis, Mountain View, CA,Sep. 2004.

[17] C. Alexander, U. Kovac, G. Roy, S. Roy and A. Asenov, In: Ultimate Integration of Silicon, pp.43-46, 2009.

Monte Carlo Simulations of Channel Scaling to Ultimate Limit in Si and In$_{0.3}$Ga$_{0.7}$As Bulk MOSFETs

Aynul Islam and Karol Kalna

School of Engineering, Swansea University, Swansea SA2 8PP, Wales, United Kingdom

Monte Carlo device simulations are carried out to analyse electron transport in scaled Si and In$_{0.3}$Ga$_{0.7}$As MOSFETs starting from a 25 nm gate length Si and In$_{0.3}$Ga$_{0.7}$As MOSFETs monitoring the electron velocity, kinetic energy and sheet density along the channel at a supply voltage of 1.0 V. We have found that while the drive current is scaled Si MOSFETs dramatically increases, the current increase in scaled In$_{0.3}$Ga$_{0.7}$As MOSFET is less pronounced. The drive current increases despite the decline of the injection velocity in Si MOSFETs from 15 nm gate length. A principal reason of the current increase is the increase in the velocity at the drain side of the device.

1.   Introduction

The roadmap (ITRS) provides guidance to future technology for semiconductor industry. ITRS predicts that the scaling of planar CMOS technology will continue till the 22 nm [1] technology node beyond which a possible extension is very appealing [2]. This desire to prolong the life of planar technology is driven by lower costs when compared to novel, non-planar technology like multi-gate architectures or nanowires [3,4]. In this work, we compare the performance of scaled channel Si MOSFETs with that of equivalently scaled InGaAs MOSFETs using our Monte Carlo (MC) device simulator MC/MOS [5]. The MC simulations of Si MOSFETs consider all phonon scattering mechanisms [6, 7], interface roughness based on Ando's model [8], and ionized impurity scattering. The simulations of InGaAs MOSFETs consider scattering with all phonons and ionized impurities, interface roughness and phonons [9]. Quantum corrections are implemented using the effective quantum potential method [7].

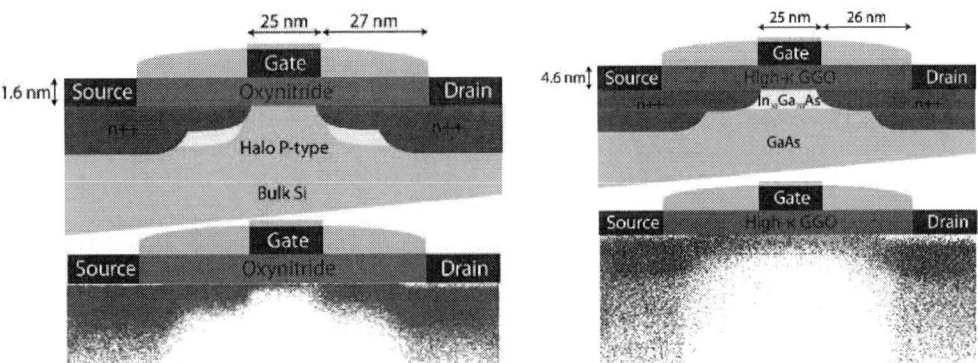

Fig. 1. Cross-section of the 25 nm gate length *n*-channel Si MOSFET is in the left upper panel. The corresponding particles distribution in X (blue) and L valleys (red) at V_G-V_T = 0.9 V and V_D = 1.0 V is in the left lower panel. The equivalent InGaAs MOSFET is in the right upper panel and particles distribution in Γ (blue), L (green) and X (red) valleys at V_G-V_T = 1.0 V and V_D = 1.0 V in the right lower panel.

978-1-4244-8574-1/10 $26.00 © 2010 IEEE   321

The verification of MC/MOS is carried out by simulating the 25 nm gate length n-channel Si device with an ON gate stack with a thickness of 1.6 nm and dielectric constant of $\varepsilon_{OX} = 7$, and a metal gate [10] shown in Fig. 1. The transistor has S/D n-type regions and extensions doped to 10^{20} cm^{-3}, a halo p-type doping of 8 x10^{18} cm^{-3}, and a p-type substrate doping concentration of 3 x10^{18} cm^{-3}. The device for the n-channel InGaAs MOSFET scaling has the same, 25 nm gate length with a spacer of 26 nm shown in Fig. 1 (right). This device consists of a 400 nm GaAs substrate, a 7 nm thick InGaAs channel with a 4.6 nm layer of high-κ Ga$_2$O$_3$ (Gd$_x$Ga$_{1-x}$) $_2$O$_3$ (GGO, κ=20) separating the channel from a metal gate. The structure has a background uniform p-type doping of 1×10^{18} cm^{-3} and 1×10^{19} cm^{-3} n-type peak doping for the S/D contacts. The particle distributions at the end of simulation in these devices are in the lower panels of Fig. 1 in the both transistors. Figure 2 (a) compares MC simulated intrinsic I_D-V_G characteristics for the 25 nm gate length Si MOSFET at low and high drain biases of 0.1 V and 1.1 V and the drive current as a function of the gate length in Fig. 2 (b). The I_D-V_G characteristics of the 25 nm gate length InGaAs MOSFET are shown in Fig. 3 (a) and the drive current as a function of the gate length in Fig. 3 (b). The drain current in the 25 nm InGaAs MOSFET is much higher than that in the equivalent Si device due to a much higher injection velocity even carrier density is smaller.

2. I_D – V_G CHARACTERISTICS

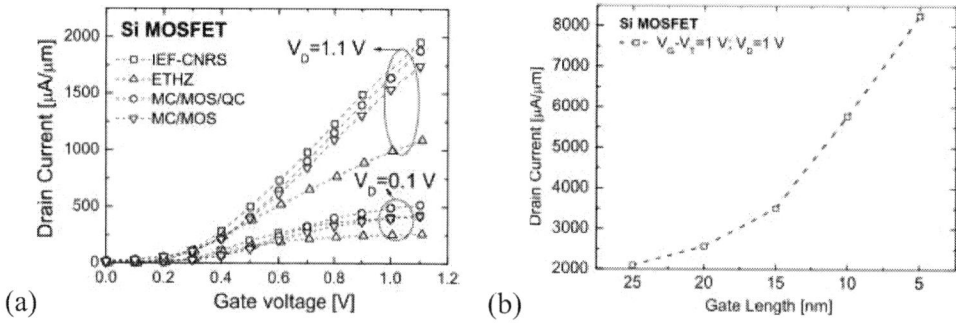

(a)   (b)

Fig. 2. (a) Simulated I_D-V_G characteristics (downward triangles) for the 25 nm gate length Si MOSFET at $V_D = 0.1$ V and $V_D = 1.1$ V, comparing our results with those from [5] (squares and upward triangles). The MC simulations using quantum corrections (open circles) are also shown for comparison, and (b) the drive current as a function of the gate length biased at V_G-$V_T = 1.0$ V and $V_D = 1.0$ V.

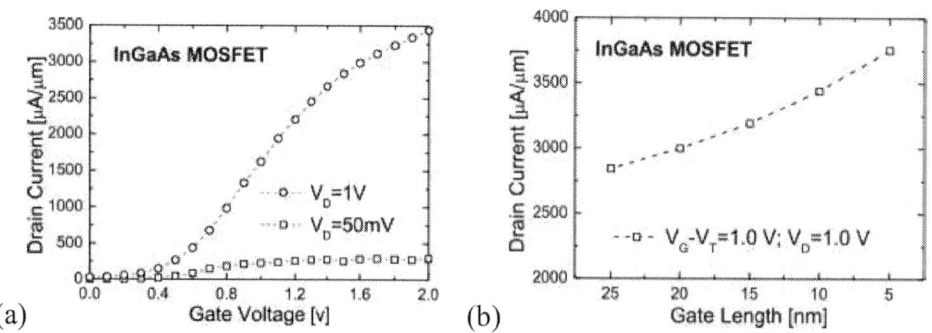

(a)   (b)

Fig. 3. (a) Simulated I_D-V_G characteristics curves for the 25 nm gate length InGaAs MOSFET at low and high drain biases of 0.05 V and 1.0 V, and (b) the drive current as a function of the gate length biased at V_G-$V_T = 1.0$ V and $V_D = 1.0$ V.

978-1-4244-8574-1/10 $26.00 © 2010 IEEE

3.   CARRIER DENSITY, AVERAGE VELOCITY AND ENERGY

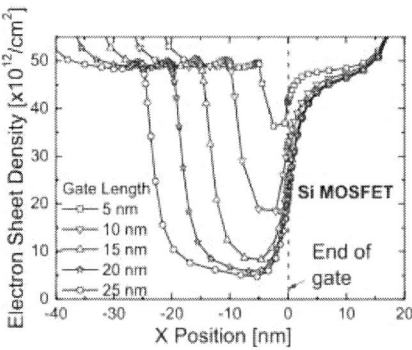 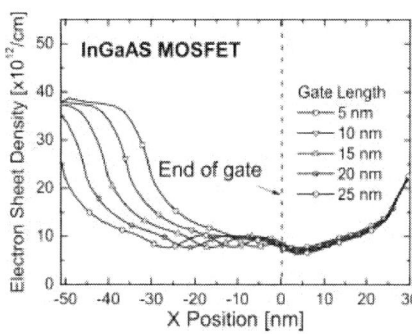

Fig. 4. Electron sheet density as a function of the distance along the channel of Si, and InGaAs MOSFETs monitored by MC simulation biased at V_G-V_T = 1.0 V and V_D = 1.0 V. Zero X position is set at the end of gate.

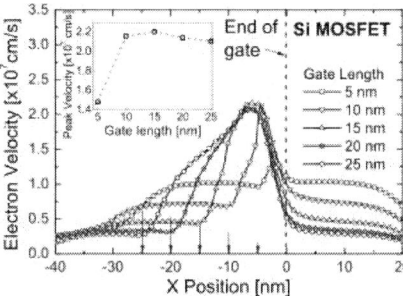 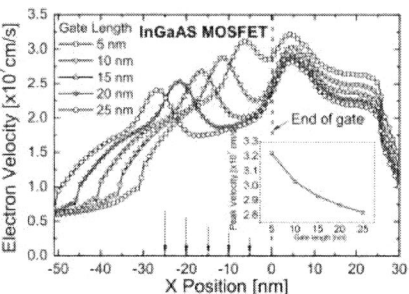

Fig. 5. Average electron velocities along the channel of Si, and InGaAs MOSFETs biased at V_G-V_T = 1.0 V and V_D = 1.0 V The inset shows the peak velocities vs. the gate length.

We have monitored density in the both MOSFETs scaled from a gate length of 25 nm to 20 nm, 15 nm, 10 nm, and 5 nm at an on-current (V_G-V_T=1.0 V, V_D=1.0 V) as presented in Fig. 4. During the channel length scaling, the depletion region is reduced and the electron sheet density in the channel gradually increases (reaching 3.5 x 10^{13} cm^{-3} at the drain side in the 5 nm Si device). On other hand, the increase of the carrier density at the source during the scaling of InGaAs MOSFETs leads to the larger increase in electron scattering thus reducing the electron density in the channel (a source starvation due to low density of states) [11].

Figure 5 represents the average velocity along the channel in the scaled Si, and InGaAs MOSFETs. The inset of the Fig. 5 shows a peak velocity of 2.2 x10^7 cm/s for the 15 nm gate length and its decline to 1.5x10^7 cm/s at the 5 nm gate length device due to the strong impact of long-range Coulomb interactions from the heavily doped S/D regions. The overall velocity along the scaled InGaAs MOSFET channel is much higher than that in the equivalent Si device thanks to a lower effective mass and reduced scattering and additionally exhibits the distinctive double peaks (right). The electron kinetic energy along the channel of scaled Si and InGaAs MOSFETs are shown in Fig. 6. The average electron kinetic energy steeply increases in the channel towards the drain region caused by the high fringing electric fields in the both transistors but will quickly decline in the Si devices in a heavily doped region.

978-1-4244-8574-1/10 $26.00 © 2010 IEEE   323

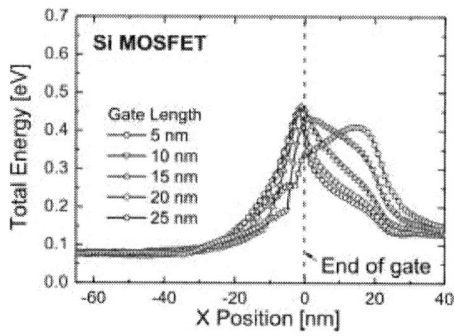 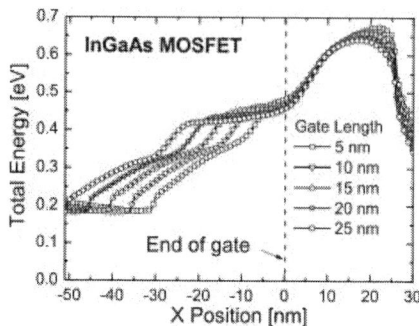

Fig. 6. Profiles of average carrier energy along the channel of Si and InGaAs MOSFETs biased at V_G-V_T = 1.0 V and V_D = 1.0 V. Zero X position is set at the end of gate.

4. CONCLUSIONS

The performance of scaled 25 nm gate length Si and InGaAs MOSFETs is carried out using ensemble MC device simulations. We studied the physical mechanisms, which determine the average electron velocity and sheet density as a function of position along the channel of MOSFETs with the decreasing gate length. The study is based on verification of our MC device simulations by comparing I_D-V_G characteristics at low and high biases for the template 25 nm gate length Si MOSFET with other state-of-the-art MC simulators [10] as well as on previous work on III-V MOSFETs [5,7,8]. As we have seen the electron velocity in scaled InGaAs MOSFETs is much higher than that in the equivalent Si devices thanks to a lower effective mass and reduced scattering and additionally exhibits the distinctive double peaks. The peak velocity is steadily increasing till the 5 nm gate length device as compared to the Si channel, promising much better switching speed with the scaling. However, we do not observe reduction of the drive current in scaled MOSFETs, which indicates that there is a missing scattering mechanism or mechanisms like the scattering with neutral impurities [4].

References:

[1] International Technology Roadmap for Semiconductors 2007, [http://public.itrs.net]

[2] G. Shahidi, *IEDM Short Course*, p 25 (2009).

[3] T. Skotnicki, J. A. Hutchby, T.-J. King, and F. Boeuf, *IEEE Circ. Dev. Mag.*, Jan-Feb, pp. 16-26 (2005).

[4] D. Fleury, G. Bidal, A. Cros, F. Boeuf, *VLSI Symp. Tech. Dig. Pap.*, pp. 16-17 (2009).

[5] K. Kalna, S. Roy, A. Asenov, K. Elgaid, and I. Thayne, *Solid-State Electron*, vol.46, no.5, pp. 631-638, 2002.

[6] C. Jacoboni, and P. Lugli, *The Monte Carlo Method for Device Simulation*, ISBN 3-211-82110-4, (1989).

[7] K. Kalna, L. Yang, A. Asenov, *in Proc. ESSDERC*, pp. 169-172 (2005); K. Kalna, N. Seoane, A. Asenov, *IEEE Trans. Electron Devices* 55, 2297-2306 (2008).

[8] K. Kalna, L. Yang and A. Asenov 2005, *in Proc. ESSDERC* 2005, ed. by G. Ghibaudo, T. Skotnicki, S. Cristoloveanu and M. Brillouet, (Grenoble, France) 169-172.

[9] B. Benbakhti, K. Kalna, A. Asenov, *in Proc. Silicon Nanoelec. W.*, pp. 147-148 (2009).

[10] C. Fiegna, M. Braccioli, C. Brugger, F. M. Bufler, P. Dollfus, V. Aubry-Fortuna, C. Jungemann, B. Meinerzhagen, P. Palestri, *in Proc. SISPAD* 2007, pp.~57-60.

[11] M. V. Fischetti, S. Jin, T-W. Tang, P. Asbeck, Y. Taur, S. E. Laux, M. Rodwell and N. Sano 2009, *J. Comput. Electron.* 8, 60-77.

Analytical Modelling of InGaP/GaAs HBTs

Gourab Dutta* And Sukla Basu**

*Research Scholar, ATDC, IIT Kharagpur,India
**Professor, ECE Dept., Kalyani Government Engineering College, Kalyani, India

Abstract—Heterojunction Bipolar Transistors with GaAs base and InGaP emitter have increasingly become important since they have great potential for numerous low- and high-frequency microwave circuit applications due to their high linearity, good reliability and nearly ideal current-voltage characteristics. Current gain and transit time are two important factors for determining the performance of these devices as amplifiers and as switches. Switching speed is mainly determined by transit time of minority carriers across a device and forward transit time is an important component of the total transit time. An analytical model is developed here to predict the variation of current gain and forward transit time with composition of InGaP emitter, as well as with emitter and base doping profile. Dependence of band gap and diffusion constants on composition of InGaP emitter is also considered in the analysis. Performance of these devices is compared with that of AlGaAs/GaAs and Si/SiGe HBTs.

Index Terms—Heterojunction Bipolar Transistor (HBT), InGaP /GaAs HBT, Transit Time

1. Introduction

Recently, InGaP-GaAs HBT technology has attracted much attention due to its uniformity[1,2] and reliability [3]. Very low power consumption of 5 GHz band receiver front end using InGaP-GaAs HBT have proved the potential of InGaP-GaAs HBT technology for future communication systems [4]. As reported in literature this device has potential for high voltage application also. In the present paper dependence of current gain and forward transit time on doping profiles of emitter and base regions and on composition of InGaP emitter is studied. Variation of band gap and minority carrier diffusion constant with composition of emitter is also taken into consideration. Comparison is done with AlGaAs/GaAs and Si/SiGe HBTs.

2. Theory

Collector [Base] current density ($J_C[J_B]$) of a BJT are given by [5]

$$J_C = \frac{q\,n_i^2}{G_B} \cdot \exp\left(\frac{q\,V_{BE}}{KT}\right) \quad \text{and} \quad J_B = \frac{q\,n_i^2}{G_E} \cdot \exp\left(\frac{q\,V_{BE}}{KT}\right) \tag{1,2}$$

where, q is charge of an electron, n_i is the intrinsic carrier concentration, V_{BE} is the base –emitter junction voltage, K is Boltzman Constant and T is temperature in Kelvin. For npn BJT with acceptor concentration profile $N_A(y)$ and donor concentration profile $N_D(y)$, emitter[base] Gummel number(G_E [G_B]) are given as [5]

$$G_B = \int_0^{W_B} \frac{N_A(y)}{D_n(y)} \cdot \frac{n_i^2(y)}{n_{iB}^2(y)}\,dy \quad \text{and} \quad G_E = \int_0^{W_E} \frac{N_A(y)}{D_n(y)} \cdot \frac{n_i^2(y)}{n_{iB}^2(y)}\,dy \tag{3,4}$$

where, $\dfrac{n_i^2(y)}{n_{iB}^2(y)} = \exp\left(\dfrac{\Delta V_{gBase}}{KT}\right)$ and $\dfrac{n_i^2(y)}{n_{iE}^2(y)} = \exp\left(\dfrac{\Delta V_{gEmitter}}{KT}\right)$

978-1-4244-8574-1/10 $26.00 © 2010 IEEE

Here $D_n(y)[D_p(y)]$ is the diffusion constant of electron[hole] in base[emitter], $n_i(y)$ is intrinsic carrier concentration, $W_B(W_E)$ is the width of base(emitter), ΔV_{gBase} ($\Delta V_{gEmitter}$) is energy gap between conduction and valance band of base(emitter), K is Boltzman Constant, T is temperature in Kelvin, $n_{iB}(y)$ ($n_{iE}(y)$) are effective intrinsic base and emitter doping concentration including band gap narrowing effect where y denotes position co-ordinate. Using equations (1),(2),(3) and (4) expression for common emitter gain(β) can be written as [5]

$$\beta = \frac{J_C}{J_B} = \frac{G_E}{G_B} = \frac{\int_0^{W_E} \frac{N_D(y)}{D_P(y)} \exp(\frac{\Delta V_{gEmitter}}{KT}) dy}{\int_0^{W_B} \frac{N_A(y)}{D_n(y)} \exp(\frac{\Delta V_{gBase}}{KT}) dy} = \frac{\int_0^{W_E} \frac{N_D(y)}{D_P(y)} dy}{\int_0^{W_B} \frac{N_A(y)}{D_n(y)} dy} \exp(\frac{\Delta E_g}{KT})$$

(5)

where, ΔE_g is the effective band-gap difference between emitter and base regions. Again base transit time (τ_B) and emitter transit time (τ_E) for $\beta \gg 1$ are given by [6]

$$\tau_B = \int_0^{W_B} \frac{n_i^2(y)}{N_A(y)} \int_y^{W_B} \frac{N_A(z)}{n_i(z)D_n(z)} dz\,dy \quad \text{and} \quad \tau_E = \frac{1}{\beta} \int_0^{W_E} \frac{n_i^2(y)}{N_A(y)} \int_y^{W_E} \frac{N_A(z)}{n_i(z)D_n(z)} dz\,dy$$

(6,7)

For uniform doping in base and emitter, $\tau_B = W_B^2/2D_n$ and $\tau_E = W_E^2/2D$ (8)

Using (6) for exponential base doping of the form $N_A(y) = N_{AB}\exp(-\alpha y/W_B)$ in the base region τ_B comes out to be

$$\tau_B = \frac{W_B^2[\alpha - 1 + \exp(-\alpha)]}{\alpha^2 D_n}$$

(9)

2.1 Current gain for uniform emitter and base doping concentration:

Considering uniform base and emitter doping concentrations $N_A(y)$ and $N_D(y)$ as N_{AB} and N_{DE} respectively equation (5) reduces to

$$\beta = \frac{N_{DE}W_E}{N_{AB}W_B} \frac{D_{nB}}{D_{pE}} \exp(\frac{\Delta E_g}{KT})$$

(10)

From Einstein relationship [5], we have at 300K, $D_p = 0.026\mu_p$ and mobility of electron (μ_n) in GaAs is 8500 cm^2/V-s [7]. Following [8] hole mobility (μ_p) will be

$$\mu_p = x\mu_p(GaP) + (1-x)\mu_p(InP)$$

(12)

μ_p for GaP is 75 cm^{2}V^{-1}S^{-1} and μ_p for InP is150 cm^{2}V^{-1}S^{-1} at 300 K [7]. Band gap difference between emitter and base regions is

$$\Delta E_g = xE_g(GaP) + (1-x)E_g(InP) - E_g(GaAs)$$

(13)

Energy band gap of GaP, InP and GaAs are 2.26 eV, 1.35 eV and 1.42eV respectively[7], ΔE_g comes out to be

$$\Delta E_g = 0.91x - 0.07$$

(14)

2.2 Current gain for uniform emitter doping and exponential base doping concentration:

Using eqn.(5) for exponential base doping with slope α, current gain (β) will be

$$\beta' = \frac{N_{DE}W_E}{\int_0^{W_B} N_{AB} * \exp(\alpha\,y/W_B)dy} \cdot \frac{D_{nb}}{D_{pe}} \cdot \exp(\frac{\Delta E_g}{kT}) \text{ , so } \beta' = \frac{\alpha}{1 - e^{-\alpha}} * \beta$$

(15)

β is the current gain with uniform emitter and base doping , y is distance along base.

2.3 *Current gain for exponential emitter doping and uniform base doping concentration:*

From (5) for exponential emitter doping of the form $N_D(z) = N_{DE} \exp(\alpha' z / W_E)$ with emitter surface doping concentration N_{DE}, current gain β can be written as

$$\beta' = \frac{\exp(\alpha) - 1}{\alpha} * \beta \tag{16}$$

2.4 *Current gain for exponential emitter doping and exponential base doping concentration:*

And when both emitter and base is exponentially doped with slope α, then

$$\beta' = \frac{\exp(\alpha) - 1}{1 - \exp(-\alpha)} . \beta \tag{17}$$

3. Results and Discussions

In this analysis temperature 300K, both emitter and base width 60nm and doping in base and emitter regions $7.5 \times 10^{16}/cm^3$ are taken unless otherwise stated. Variation of current gain with mole fraction (x) of $Ga_xIn_{1-x}P$ emitter with uniformly doped base and emitter regions is determined from (5) and shown in Fig. 1.For $Ga_xIn_{1-x}P$ as emitter and InP as base gain is of the order of hundred at mole fraction(x) =0.1 and of the order of thousand at x=0.2 where as for Si/SiGe HBT, β is nearly equal to 100 for x=0.2. But at x=0.08 $Ga_xIn_{1-x}P$/GaAs HBT, β =45(Fig.1) but for $Al_xGa_{1-x}As$ HBT, $\beta>100$(Fig.2). So at low mole fraction level current gain of $Ga_xIn_{1-x}P$/GaAs is better than Si/SiGe HBT but less than Al_xGa_{1-x} As/GaAs HBT. Fig.3 shows for uniformly doped emitter and exponentially doped base with the increase of slope of base doping current gain increases. Similarly if emitter is exponentially doped and base doping is uniform then also with the increase of slope of donor concentration current gain increases (Fig.4).

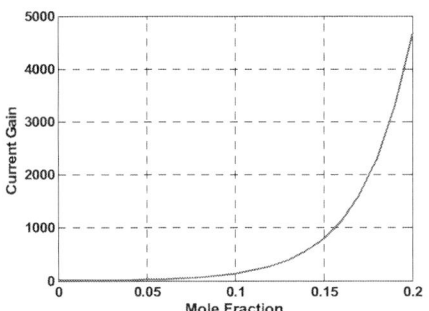

Fig.1 Variation of current gain with mole fraction for $Ga_xIn_{1-x}P$/GaAs HBT

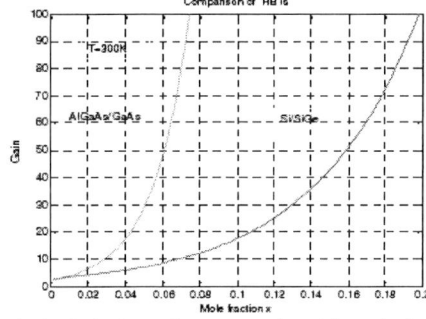

Fig.2 Variation of current gain with mole fraction for Al_xGaAs_{1-x}/GaAs HBT and $Si/Si_{1-x}Ge_x$

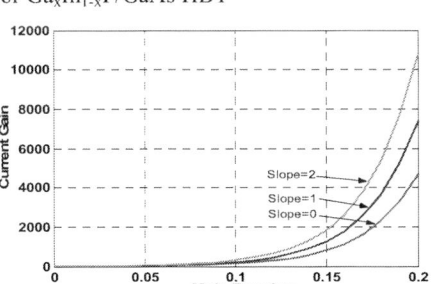

Fig.3 Variation of current gain with mole fraction at different value of Slope(α) of doping in base region for $Ga_xIn_{1-x}P$/GaAs HBT

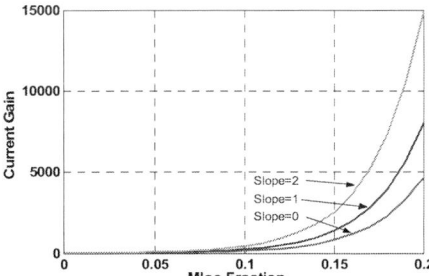

Fig.4 Variation of current gain with mole fraction at different value of Slope(α) of doping in emitter region for $Ga_xIn_{1-x}P$/GaAs HBT

When both emitter and base are doped exponentially with same slope(α) then gain increase more rapidly with x than the earlier cases (Fig. 5).In Fig. 6 variation of transit time with mole fraction is shown .Base transit time is independent of mole fraction but emitter transit time will decrease with mole fraction. At x=0.09 both of these transit time will be same. Forward transit time which is the sum of these two transit times should be as low as possible and for mole fraction x>0.09 emitter transit time has not much contribution towards forward transit time. If mole fraction(x) is changed linearly across emitter, keeping x constant at one end and varying x at other end, gain will change compared to the gain for uniform x. For x=0.2 at emitter-base(E-B) junction and varying x at surface of emitter is shown in Fig. 7(A). Gain for this case is higher than that of uniform x[Fig. 1]. But for x=0.2 at surface and variation of x occurs at E-B junction then the gain[Fig. 7(B)] will be almost same as that for uniform x [Fig. 1].

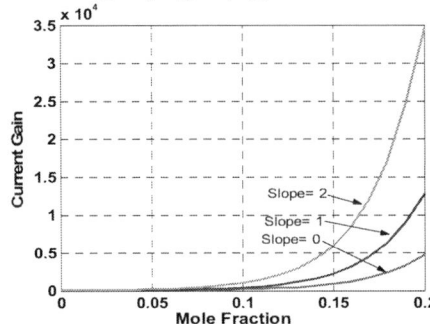

Fig.5 Variation of current gain with mole fraction at different value of slope(α) of doping in both emitter and base region for $Ga_xIn_{1-x}P/GaAs$ HBT

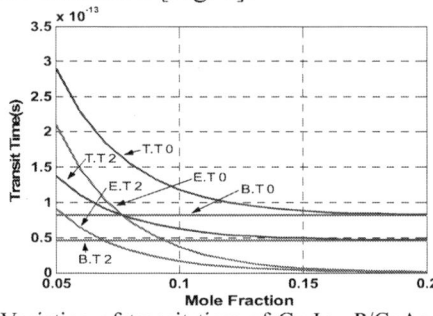

Fig.6 Variation of transit time of $Ga_xIn_{1-x}P/GaAs$ HBT with mole fraction for different slope of base doping. B.T0(2) is the base transit time,E.T0(2) is the emitter transit time and T.T0(2) is the total transit time for slope=0(2)

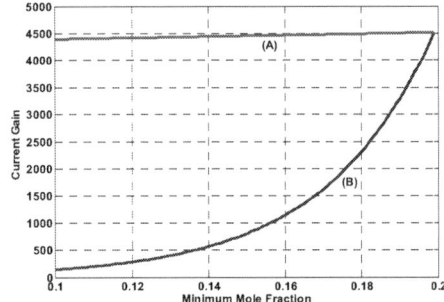

Fig. 7 Variation of current gain with minimum mole fraction at Surface(A) and at emitter-base junction(B) for linear variation of mole fraction in emitter while the other end of emitter i.e. E-B junction(in A) and Surface(in B) have mole fraction x=0.2. Here $N_{DE} = N_{AB} = 7.5 \times 10^{16} / cm^3$, $W_E = W_B = 60$nm, and T= 300° K

References

1. Lester, T., Svilans, M., Maritan, P., and Postolek, H.: 'A manufacturable process for HBT circuits', Inst. Phys.Conf. Ser., 1993, 136, pp. 449–451.

2. Lin, Y.S., Lu, S.S., and Wang, Y.J.: 'High-performance Ga0.51In0.49 P-GaAs airbridge gate MISFETs grown by gas-source MBE', IEEE Trans. Electron Devices, 1997, 44, pp. 921–929.

3. High reliability InGaP/GaAs HBT Pan, N | Elliott, J | Knowles, M | Vu, D P | Kishimoto, K | Twynam, J K | Sato, H | Fresina, M T | Stillman, G E IEEE Electron Device Letters. Vol. 19, no. 4, pp. 115-117. Apr. 1998
   4. K.-Y. Yeh, S.-S. Lu and Y.-S. Lin, 'Monolithic InGaP-GaAs HBT receiver front-end with 6mW DC power consumption for 5 GHz band WLAN applications' Electronics Letters , 25th November 2004 Vol. 40 No. 24.

5. C.C.Hu : 'Modern Semiconductor Devices for Integrated Circuits', First Edition, Pearson Education, 2010.

6. Kunihiro Suzuki ' Emitter and Base Transit Time of Polycrystalline Silicon Emitter Contact Biploar Transistors' IEEE Transactions on Electron Devices, Vol.. 38. No. 11. November, 1991.

7. S.M..Sze: 'Physics of Semiconductor Devices', Second Edition,Wiley Eastern Limited, 1991.

8. Pallab Bhattacharya, 'Semiconductor Optoelectronic Devices ,' Pearson Education

Structure and optical properties of the hydrogen diluted a-Si:H thin films prepared by PECVD with different deposition temperatures

Marie Netrvalová[1], Marinus Fischer[2], Jarmila Mullerová[3], Miro Zeman[2], Pavol Šutta[1]

[1]New Technologies – Research Centre, University of West Bohemia,
Univerzitní 8, 306 14 Pilsen, Czech Republic
e-mail: mnetrval@ntc.zcu.cz
[2]Delft University of Technology, DIMES
Feldmannweg 17, 2628 CT Delft, Netherlands
3Department of Engineering Fundamentals, University of Žilina
031 01 Liptovský Mikuláš, Slovakia

The paper deals with the hydrogenated amorphous silicon (a-Si:H) films about 300 nm in thickness prepared by using rf-PECVD with hydrogen dilution R = 10 of the silane source gas in the amorphous growth regime onto clean Corning Eagle 2000 glass substrates at different deposition temperatures ranging from 50 to 200 °C. Structural and optical properties of the films were obtained from X-ray diffraction and UV-Vis spectrophotometry. The full width at half maximum of the first scattering peak decreases with increasing of the deposition temperature up to 150 °C and then remains constant. Optical band-gaps are from 1.65 to 1.76 eV, which slightly decrease with increasing deposition temperature, whereas the refractive index increases with increasing deposition temperature. This indicates that the density of the films at higher temperature has increased.

1. Introduction

Hydrogenated amorphous silicon (a-Si:H) is a well established material for low-cost tandem solar cells, which can be obtain usually at temperatures up to 200 °C. Nevertheless, the solar cells prepared from un-diluted a-Si:H films degrade during their exploitation due to insolation known as Staebler-Wronski effect [1]. One of the possibilities how to improve the stability of the a-Si:H films is silane dilution with hydrogen ($R = H_2/SiH_4$) during the deposition. Already dilution R = 10 significantly improve the stability of the films [2,3]. Hydrogen dilution among others influences the microstructure of the films nevertheless the deposition temperature has the most significant influence on the microstructure of the films.

2. Experimental details

The a-Si:H thin films were deposited using rf-PECVD AMOR cluster tool with hydrogen dilution R = 10 and substrate temperatures from 50 to 200 °C. The thickness of all films was approximately 300 nm, which corresponds to the thickness of the absorber layer in solar cells. The individual films were simultaneously deposited on Corning Eagle 2000 type glass substrates and c-Si substrates.

The structural properties of the films were studied on a-Si:H films deposited on c-Si substrates by X-ray diffraction (XRD) analysis using an automatic powder diffractometer X´Pert Pro with a thin-film attachment (parallel beam with asymmetric ω-2□ geometry, where ω is fixed incident angle of X-rays) and a point detector Pixcel. Copper Kα characteristic radiation (λ = 0.154178 nm) was used. The XRD patterns were measured from 15 to 65 degrees in 2θ scales. The medium order of the films was evaluated analysing full width at half maximum (FWHM) of the first scattering peak of the XRD patterns. Although the a-Si:H films have two scattering peaks in the range of indicated above 2θ angles, the only FWHM of the first scattering peak is usually used as a measure of the order of the atoms in amorphous silicon structure [4,5]. It is worth mentioning that in the amorphous materials X-ray diffraction takes place on the pairs of atoms and not on the ordered atomic planes as it is usual in the case of diffraction on a typical polycrystalline material.

Optical properties were analyzed on the films deposited on Corning glass substrates using transmittance spectra recorded on a Specord 210 spectrophotometer in spectral region from 190 to 1100 nm. Spectral refractive indices and absorption coefficients of the films were extracted from the measured spectra using a Delphi-based program based on an optimization procedure using genetic algorithm. The optimization procedure minimizes differences between the experimental and theoretical transmittance in the broad spectral region. The transmittance of a thin film with parallel interfaces deposited on a thick substrate is a nonlinear function of the wavelength, the film thickness, refractive indices and absorption coefficients of the film and substrate. The theoretical transmittance was calculated using the theory in [6] and the Tauc-Lorentz dispersion model for refractive index and extinction coefficient [7].

3. Results and discussion

3.1 Structure properties

X-ray diffraction analysis indicated that the all studied a-Si:H films were amorphous. Figure 1 shows typical part of XRD patterns for amorphous silicon films deposited at different deposition temperature recorded using a thin film attachment with asymmetric ω-2θ geometry. Dependence of FWHM of the first scattering peak on deposition temperature is presented in Fig. 2. The FWHM of the first scattering peak decreases with increasing of the deposition temperature up to 150 °C and then remains constant. It follows from Fig. 2 that the structure of the film improves from 50 to 150 °C. Further temperature increasing does not change in average the FWHM of the first scattering peak.

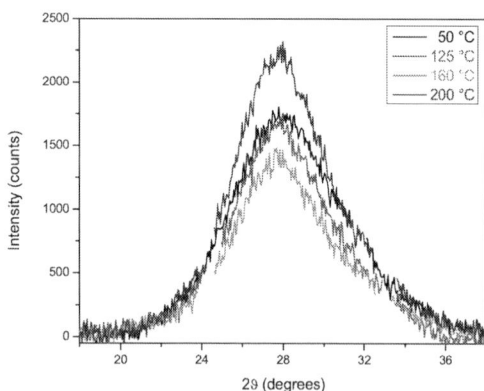

Fig. 1: XRD patterns for a-Si:H films deposited at different deposition temperatures

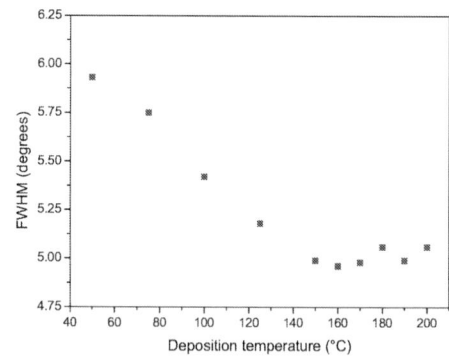

Fig. 2: Dependence of FWHM of the first scattering peak on the deposition temperature

3.2 Optical properties

Optical spectra of selected films are presented in Fig. 3 and dependence of the optical band-gaps on the deposition temperature is in Fig. 4. The optical band-gaps were calculated from the transmittance spectroscopic data near the absorption edge by using Tauc's procedure:

$$(\alpha h\upsilon)^{1/2} = B\left(h\upsilon - E_g^{opt}\right). \tag{1}$$

Values of optical band- gaps fall down only slightly with increasing deposition temperatures except for the last value deposited at 200 °C.

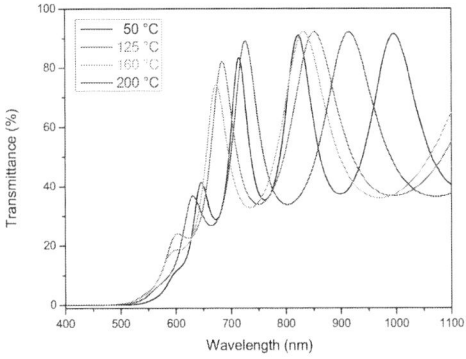

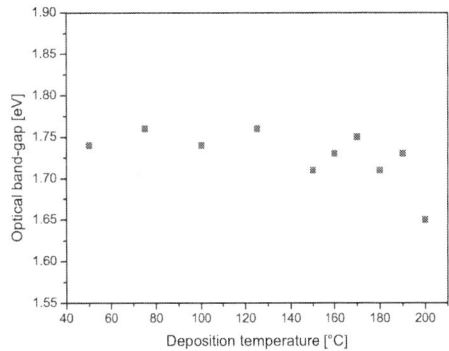

Fig. 3: UV-Vis spectra for a-Si:H films deposited at different deposition temperatures

Fig. 4: Dependence of optical band-gap for the a-Si:H films on deposition temperature

Spectral absorption coefficients and refractive indices for the investigated a-Si:H films are demonstrated in Fig. 5 and 6. While the absorption properties of the films differ only slightly with deposition temperature used without particular dependence on the temperature, the refractive indices increase (especially in the long wavelength region) with the deposition temperature. As the refractive index is an important wavelength-independent optical parameter related to the atomic structure and the mass density of the material, the results imply that samples deposited at higher temperatures have better atomic order and higher mass densities. This corresponds very well with the knowledge extracted from XRD patterns.

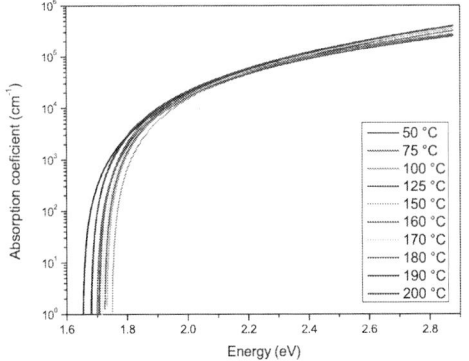

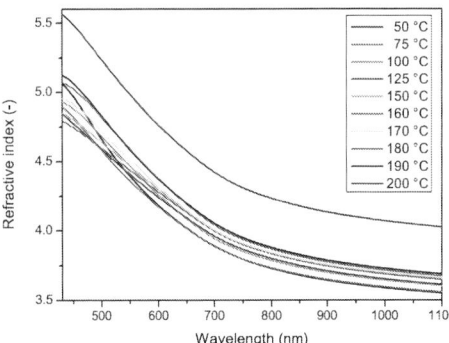

Fig. 5: Absorption coefficients for samples deposited at different deposition temperature

Fig. 6: Refractive indices for samples deposited at different deposition temperature

4. Conclusions

X-ray diffraction analysis indicated that the all investigated a-Si:H films are amorphous. Full widths at half maximum of the first scattering peaks have 5 – 6 degrees, which is typical for such amorphous materials. Nevertheless, they have slightly different micro-structure. This difference is fairly well demonstrated taking into consideration the optical properties of the films, namely the refractive index (Fig. 6). All a-Si:H films except of that deposited at 200 °C have their refractive index much lower than that usual for a typical a-Si:H material. The explanation of this observation is that these films having lower refractive index contain more voids than that deposited at 200 °C. For the better understanding of the behaviour of the structural and optical properties of the a-Si:H films prepared by PECVD at different deposition temperatures also an infrared and a Raman spectroscopy should be used. Nevertheless, from the XRD and UV VIS spectrophotometry point of view the a-Si:H film deposited at 200 °C has the best structural and optical properties.

Acknowledgements

This work was supported by the project of Ministry of Education, Sports and Youth of the Czech Republic No. 1M06031, by the Slovak Grant Agency under the project VEGA 2/0070/10 and by Delft University of Technology, financially supported by Nuon Helianthos.

References

[1] D. L. Staebler and C. R. Wronski, *Appl.Phys.Lett.* **31** (4), 292 (1977)

[2] G. Van Elzakker, P. Sutta, M. Zeman: Structural Properties of a-Si:H Films with Improved Stability against Light Induced Degradation. *MRS, 2009.* DOI: 10.1557/PROC-1153-A18-02

[3] M. Zeman, G. Van Elzakker, F. Tichelaar, P. Sutta: Structural properties of amorphous silicon prepared from hydrogen-diluted silane. *Philosophical Magazine,* 2009, 89, No. 28-30, p. 2435-2448. ISSN: 1478-6435.

[4] A.H. Mahan, D.L. Williamson, T.E. Furtak. *Mat. Res. Soc. Symp. Proc.* **467**, 657 (1997).

[5] D.L. Williamson. *Mat. Res. Soc. Symp. Proc.* **557**, 251 (1999).

[6] I., Chambouleyron, J.M., Martinez, A.C., Moretti, M., Mulato, "Retrieval of optical constants and thickness of thin films from transmission spectra", *Appl. Optics* (1997) 36, 8238-8247.

[7] Jr., G.E., Jellison, F.A., Modine, "Parameterization of the optical functions of amorphous materials in the interband region", *Appl. Phys. Lett.* (1996) 69, 371-373, Erratum Appl. Phys. Lett. (1996) 69, 2137.

Structural and chemical analysis of self-aligned titanium silicide formed by furnace annealing

Elena Barbarini, Salvatore Guastella, Fabrizio Pirri
Department of Material Science and Chemical Engineering,
Politecnico di Torino,
Corso Duca degli Abruzzi 24, 10129 Torino, Italy
e-mail: elena.barbarini@polito.it

In this paper the furnace annealing effects on the titanium silicide formation over a range of temperatures are investigated using physical and chemical measurements. In particular the formation steps and the properties of the interface between $TiSi_2$ and Si have been characterized by mean of Transmission Electron Microscopy. The experiments have been performed with the final aim to obtain Schottky barrier diodes (SBDs) whose fabrication process is suitable for a production line.

1. Introduction

Titanium silicide is commonly used in semiconductor industry to produce interconnection, barriers and contacts; moreover because of his self-aligned fabrication process and a very low electrical barrier it can be implemented in the formation of Schottky Barrier Diodes (SBDs) with a very low forward voltage drop. The control of metal-semiconductor interface of Schottky Barrier Diodes (SBDs) is of fundamental importance to minimize the power loss of such devices, because its properties affect the overall performances.

As described in literature the silicide formation technique of directly depositing Ti on a silicon surface to form the required silicide layer employs the process of direct metallurgical reaction. After the metal is deposited on the silicon, the wafer is exposed to high temperatures that promote the mass diffusion of the silicon atoms across the Ti/Si interface and chemical reactions between the metal and the silicon needed to form the silicide. In such a reaction, metal-rich silicides generally form first, and continue to grow until all the metal is consumed. In general, the silicide growth is often diffusion controlled or interface-reaction controlled and its thickness and its electrical properties are proportional to the annealing time and temperature and the silicon substrate. The presence of contaminating or doping impurities was found to influence the growth rate. Kinetic data are crucial for a basic understanding of interfacial reactions between metal thin films and silicon.

In literature are present some experiments of titanium silicide formation by rapid thermal annealing; here we present the results obtained by standard furnace annealing and standard production processes.

The samples have been analyzed with different superficial and deep characterization techniques. Four point probe sheet resistance measurement it is the first and simpler method to verify the formation of a silicide layer; while transmission electron microscopy (TEM) and depth X-ray photoemission spectroscopy (XPS) provide direct and accurate data, such as the sequence of phase formation, the morphology of phase and interface structure in the growth of silicides on silicon. To verify the correct electrical applicability the SBH of Schottky diodes has been measured.

2. Experimental

Titanium silicide samples were formed depositing, via electron-beam evaporation, Ti layers with different thicknesses, from 1000 Å up to 2000 Å, on <111> n-type Si wafers. These wafers were previously cleaned with a standard ammonium fluoride solution, in order to prepare in a optimal way the Ti/Si interface and avoid unexpected reactions during the silicide formation. Then the samples have been annealed in a standard furnace with nitrogen flow, in a temperature range from 550° up to above 750°C. The furnace recipe was customized specifically for this application in accordance with the furnace capability. The nitrogen flux into the furnace chamber is necessary to avoid metal oxidation process. Nevertheless the formation of a top layer composed of TiN and TiN_xO_y compounds has been observed. To remove these layers and the residual un-reacted Ti, after the annealing, the samples were cleaned in a selective H_2SO_4 solution.

To produce a schottky diode the TiSi layer formation process was implemented in a standard planar SBDs; a Ti film and Al were deposited and patterned on the front side of the wafer for contact metallization. Ohmic contact deposition on the wafer backside and standard commercial packaging (TO220) completed the device fabrication process.

3. Results and discussion

The experiments of titanium silicide formation presented in this paper show the formation steps of Ti silicide by standard furnace annealing.

After the evaporation, because of its high electro-negativity, the Ti reacts with oxygen present in the atmosphere and form a top layer of native oxide. At a temperature below 550°C the Si starts diffusing into the Ti, giving formation to a not ordered phase.

Increasing the annealing temperature upon around 650°C the diffusion process of Si into Ti continues, giving formation to a crystalline phase with Ti clusters and $TiSi_2$ or Ti_5Si_3 agglomerates at the Ti/Si interface. Moreover, at high temperatures, the Ti oxides react with the nitrogen fluxed into the annealing chamber leading to the formation of different nitrides and oxide precipitates. These agglomerates influence the overall formation process producing a non-uniform and unstable silicide layer; in fact the selective etch removes all the un-reacted Ti leaving islands of silicide.

In figure 1 the values of sheet resistance of a sample with a thickness of 1000 Å annealed at different temperatures before and after selective etch are presented. It is possible to see how at 550°C the value of sheet resistance doubles after the selective etch process. Moreover, the electrical measurements confirm the non-uniformity of the silicide layer: the high values of reverse leakage current of the final TiSi-based SBDs are linked to the simultaneous presence of Ti_xSi_y/Si and contact metal/Si interfaces.

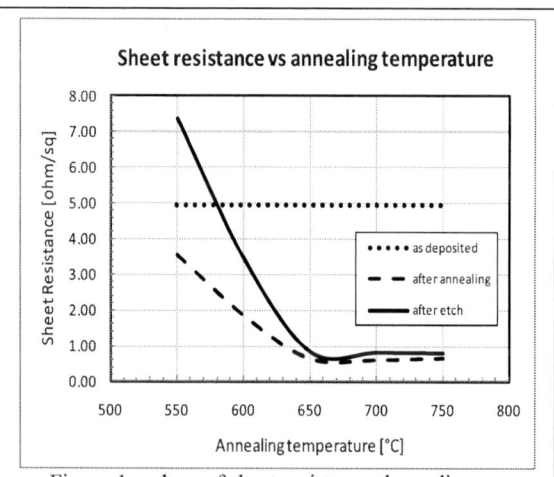

Figure 1: values of sheet resistance depending on annealing temperature, for a layer of 1000 A of Ti.

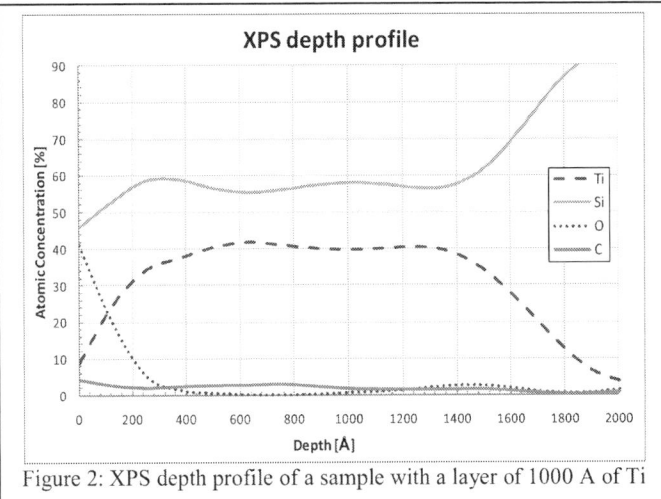

Figure 2: XPS depth profile of a sample with a layer of 1000 A of Ti

At temperatures above 650°C the formation of TiSi$_2$ is stabilized and the eventual low quantity of contaminants, subsequently etched, doesn't modify the values of sheet resistance.

In this process step the silicide thickness increases by increasing the annealing temperature and the evaporated Ti film thickness, because of the continuous diffusion process of Si in the unreacted Ti. Some top and cross-section TEM images show that, just after the annealing process, some crystalline non-uniform structures are present on the top of the structure; they are removed by the chemical wet etch, leading to the formation of a silicide layer with holes and rough surface. A deeper TEM analysis has been performed to verify the grain boundaries and the crystal orientation of the silicide. It is possible to see how the grains with different orientation and crystallographic composition are disposed in the silicide layer with a precise order. Near the Si/Ti interface there are almost no grains while they start increasing toward the upper surface. They reach their maximum quantity near the upper layer of residual titanium. The top layer of every sample is characterized by an high presence of un-reacted Ti, with its typical orientation. The TiSi layer has an uniform aspect while the grains have different orientations. Moreover, chemical analysis has shown that the Si percentage into the Ti decreases going toward the top layer, in accordance with the physics of the formation process. The stoichiometric analysis highlights that the Ti percentage increases with respect to Si percentage going towards the top of the structure until it stabilizes at values of Ti (33%) and Si (67%), confirming the complete formation of TiSi$_2$. With increasing the process temperatures above 700°C the physical and chemical properties of the silicide are subject to further modifications.

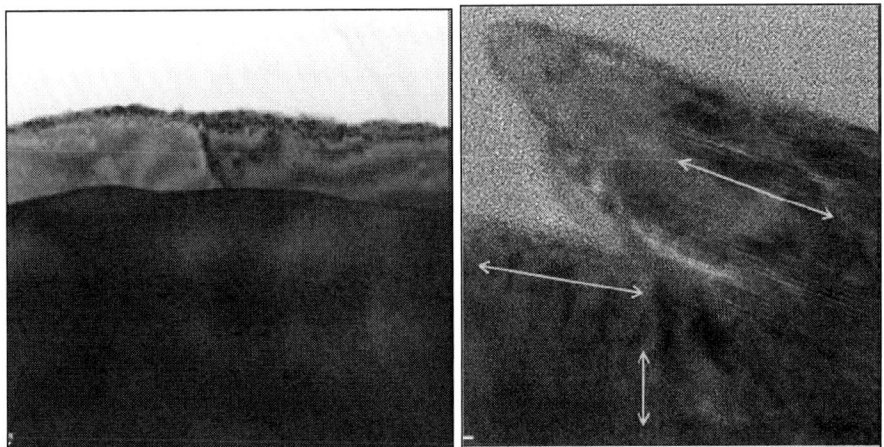

Figure 3: TEM images of the sample before self alignment etch

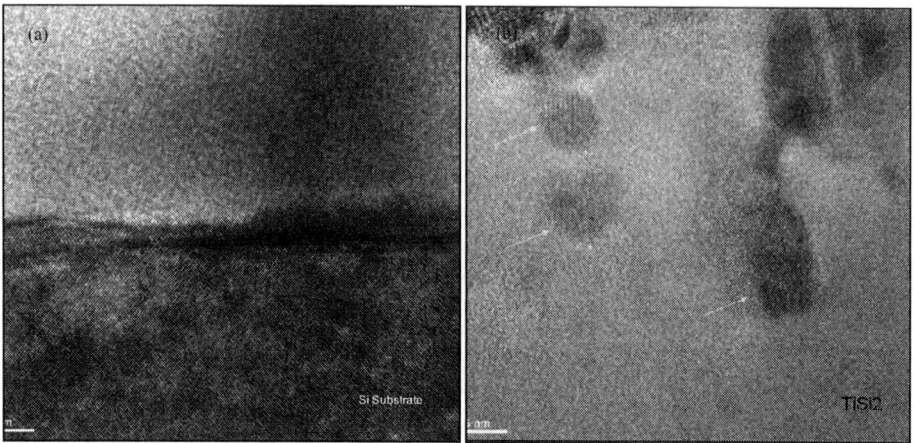

Figure 4: TEM image of (a) the interface layer of the sample and (b) of the grains inside the TiSi layer.

This can be seen in the modification of sheet resistance values and can be linked to different formation processes: the given thermal budget, which first determines the silicide formation, produces, with its further increasing, a modification of the oxide and nitride clusters dimension and/or of the lattice structure. This leads to the formation of TiN and TiN_xO_y compounds' spikes in the silicide film or a modification of the Ti_xSi_y/Si interface with a subsequent degradation of the electrical properties of the Schottky contact.

To conclude, the titanium silicide formation is affected by different types of nitrides and oxides agglomerations on the top and in the bulk material, depending on the annealing process temperature. This lead to a different final result; in fact the residues behave differently to the chemical etch according to their structural and chemical composition.

The SBH of the formed Ti silicide barriers has been measured in order to verify the correct applicability to a Shottky diode. As expected, the correct electrical behaviour occurs in the samples annealed at temperatures higher than 600° and the SBH is about 560 meV.

Acknowledgement

The authors acknowledge the working group of the Politecnico di Torino.

References

[1] S. M. Sze, *Physics of Semiconductor Devices*, John Wiley & Sons, New York, 1981.

[2] J. Nemanich et al,. *Raman Scattering Characterization of Titanium Silicide Formation*, J.Vac.Sci.A Vol. 3, n. 3, p 938, 1985

[3] I. Bertoti, M. Mohai, J.L. Sullivan, S.O. Saied, *Surface characterization of plasma-nitrided titanium: an XPS study*, Applied Surface Science, vol. 84, pp. 357-371, November 1994.

[4] M.O. Aboelfotoh and K.N. Tu, *Schottky-barrier heights of Ti and TiSi2 on n-type and p-type Si(100)*, Physical Review B, Vol.34, n.4, pp. 2311-2318, 1986

[5] M. Wittmer, *Conduction mechanism in PtSi/Si Schottky diodes*, Physical Review B, Vol. 43, n. 5, pp. 4385-4395, 1991

THE COMPOUND OXIDES BASED ON TiO₂ AND NiO THIN FILMS FOR LOW TEMPERATURE GAS DETECTION

I. Kosc[1], I. Hotovy[1], M. Kompitsas[3], R. Grieseler[2], M Wilke[2], V. Rehacek[1], M. Predanocy[1]
T. Kups[2], L. Spiess[2]

[1] Department of Microelectronics, Slovak University of Technology
Ilkovicova 3, SK-812 19 Bratislava, Slovakia
[2] Institute of Materials Science, Ilmenau University of Technology
Ilmenau 98694, Germany
[3] Theoretical and Physical Chemistry Institute, National Hellenic Research Foundation
Vasileos Konstantinou Ave 48, 11635 Athens, Greece
e-mail: ivan.kosc@stuba.sk

The multilayer compound thin films, consisted of metal oxides (TiO₂ and NiO) prepared by dc magnetron sputtering technique, have been studied. The structural, compositional, electrical and gas sensing properties have been investigated by XRD, GDOES and Van der Pauw method considering changes in layout, annealing temperature and addition of Au noble metal catalyst. The Au modified compound oxides exhibit fast response and enhanced sensitivity to hydrogen at low operating temperatures.

1. Introduction

It is well-known that the adsorption of gases on metal oxide films causes electrical conductivity changes, so a measurement of electrical resistance allows us to detect the presence of specific gases. Nanostructured compound materials exhibit unusual and often enhanced properties with respect to single oxides [1]. Under normal conditions, it is well known, that TiO₂ has high resistivity. For usage of the TiO₂ layers at low temperature it is needed to build in material that causes lowering of the resistivity of the whole complex. It seems that NiO is great material for incorporation into complex compound layers based on TiO₂. Under some conditions it is expected that the oxide at lower concentration restrains the growth of the other oxide and stabilizes it, avoiding the well-known grain growth that takes place when the sensor is heated [2]. This circumstance together with fast response time and good sensitivity to concentrations of hydrogen below environmental monitoring limits make investigated compound mixed oxides new promising candidates for gas sensing materials.

2. Thin films preparation and characterization

Two main layouts of compound oxides were prepared (Fig. 1). The difference between

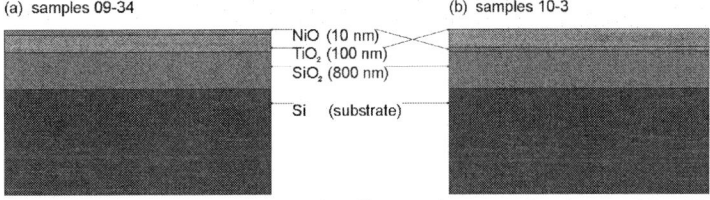

Fig. 1 The cross-section of deposited structures; NiO film on the top – first layout (a), TiO₂ film on the top of the structure – modified layout (b).

the layouts is in sequence of the top two functional layers: TiO_2 and NiO. In the first layout (sample code 09-34) the compound structure was deposited as multilayer complex of films in order $Si/SiO_2/TiO_2/NiO$ (Fig. 1a). Afterwards the sequence of thin films was altered: $Si/SiO_2/NiO/TiO_2$ (Fig. 1b) so as to get the modified layout (sample code 10-3).

Deposition of the TiO_2 and NiO thin films in both layouts undergoes the same settings and parameters. The TiO_2 thin films were prepared by dc reactive magnetron sputtering from a Ti target in a mixture of Ar and O_2. Sputtering power of 600W was used. The apparatus was initially evacuated to a pressure below 0.5 mPa. The relative partial pressure of O_2 in the reactive mixture O_2-Ar was 25%. The 10 nm thick NiO films were also deposited by dc magnetron sputtering in a mixture of Ar and O_2. The total gas pressure was kept at 0.6 Pa.

To obtain enhancement in the structure of deposited layers samples were post-annealed by rapid thermal annealing at 500°C, 600°C, 700°C and 900°C for 10 seconds in a mixture of argon and hydrogen. The relative partial pressure of hydrogen in the mixture Ar-H_2 was 5%. After annealing process representative set of samples from the both layouts was chosen and underwent 30 s of the pulsed laser deposition (PLD) of the gold catalyst.

The glow discharge optical emission spectrometry (GDOES) was used to find the depth profiles of investigated structures. The crystal structure was identified with a Theta-Theta X-ray diffractometer (XRD) D5000 with Goebel mirror oriented into grazing incidence focusing with Cu Kα radiation. The diffraction patterns were recorded for values between 15° and 60°. The electrical measurements were performed using the linear four probe Van der Pauw method at room temperature in order to obtain the resistivity and sheet resistance.

Samples of both layouts were then tested as hydrogen gas sensors in an apparatus that consists of aluminium vacuum chamber. The testing procedure began with initial chamber evacuation down to 1 Pa and thereafter the apparatus was filled with synthetic air to level of atmospheric pressure. The films were tested in the operating temperature range of 180 °C to 200 °C at hydrogen concentrations from 10000 to 1000 ppm. The hydrogen concentration was calculated on the basis of the partial pressures of the sensing gas and air inside the chamber. The pressure was measured by a MKS Baratron gauge. A 1 V bias was applied by two gold-coated electrodes mechanically pressed on the sample surface. The current I through the film was measured with a Keithley Mo. 485 Pico-ammeter. Current changes helped to monitor the gas sensing in real time. The relative response is defined as:

$$R = \frac{Ia - Ig}{Ia} = \frac{Rg - Ra}{Rg} \tag{1}$$

where Ra (Ia) is the resistance (current) in air and Rg (Ig) the resistance (current) in the presence of the gas.

3. Structural and compositional properties observation

GDOES depth profiles yield exact knowledge about the type and position of every element in compound structure. Figure 2 delineates depth profiles of the samples 09-34. Zero level on the x-axis, which corresponds to the surface, shows only small amount of oxygen in the near sub-surface area. Low oxygen content and almost parallel tendency of slope of oxygen and Ti indicates pure Ni covering of sensing film in the case of the first layout. In addition the depth profiles displays dependency of moving and inter-diffusion of incorporated layers in the volume of investigated structure on the parameters of annealing temperature.

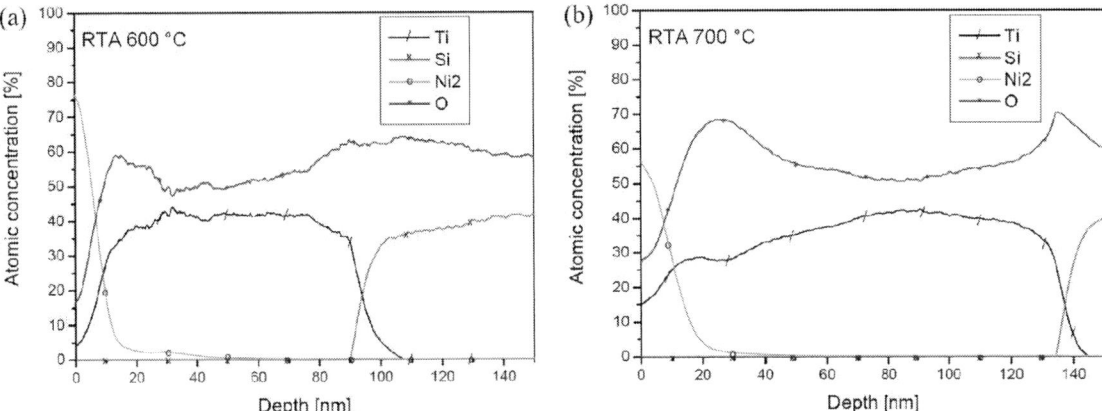

Fig. 2 GDOES depth profiles of samples 09-34 after rapid thermal annealing in chamber heated up to 600 °C (a) and 700 °C (b).

The XRD diagrams (Fig. 3) showed that as-deposited and annealed samples at 500 °C were amorphous, whereas annealed samples at 600 °C and 700 °C were found to be polycrystalline. Diffraction patterns from samples annealed at 600 °C and 700 °C delineate presence of diffraction peaks that appertain to anatase TiO_2 lattice, as referred in JC-PDS 21-1272. In samples 09-34 the rest of the diffraction peaks belongs to cubic Ni lattice, as referred in JC-PDS 4-0850. This fact also indicates presence of pure Ni on the top the 09-34 films.

Sample:	RTA	Rs [Ω/sq]	Thickness [nm]	ρ [Ωcm]
09-34/1	as-dep	2.0×10^6	110	21.7
09-34/3	500 °C	1.9×10^7	100	186.7
09-34/5	600 °C	4.7×10^5	70	3.3
09-34/7	700 °C	5.8×10^3	105	0.1
10-3/1	as-dep	1.3×10^{10}	100	1.3×10^5
10-3/1	500 °C	1.0×10^{10}	100	1.0×10^5
10-3/1	600 °C	9.7×10^9	100	9.7×10^4
10-3/1	800 °C	9.2×10^9	100	9.2×10^4

Tab. 1 Electrical properties of as-dep and annealed samples 09-34 and 10-3.

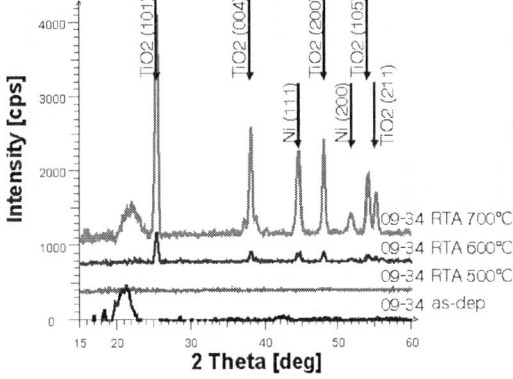

Fig. 3 XRD patterns of as-dep and annealed samples 09-34.

4. Electrical measurements and hydrogen sensing properties

The sheet resistance Rs and resistivity ρ of the investigated thin films were obtained by linear four probe Van der Pauw method. Table 1 depicts decreasing trend of Rs and ρ with raising annealing temperature. According to our previous experiments, Rs of TiO_2 in presence of top Ni layer (09-34) is lowered in three orders of magnitude down in comparison to pure TiO_2 layer. This shift of samples 09-34 can be seen also in comparison to modified samples 10-3. It is assumed that pure Ni on top acts as metal clusters (nanoelectrodes) that causes significant change in conductivity of the sensing film [3].

Compound mixed oxides of the both layouts were tested as potential gas sensors for low temperature (under 200 °C) hydrogen sensing. The best results were obtained for the modified layout (samples 10_3) together with afterwards Au catalyst deposition (Fig. 4). It is known that for pure TiO_2 it is not possible to extract appreciable responses until operating temperature lower than 200 °C [4]. Novel NiO/TiO$_2$ compound oxides together with noble metal catalyst overcome this temperature limitation and provide enhanced sensitivity and very fast response. Over and above these findings indicate possibility of further lowering of the operating temperature and consequential reduction of energy consumption of the final device.

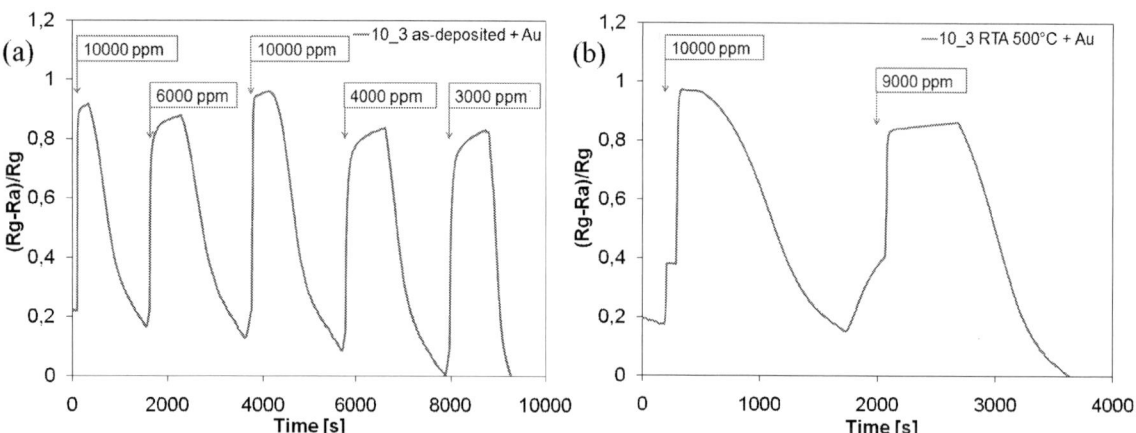

Fig. 4 Relative response of as-dep (a) and annealed at 500 °C (b) compound oxides with Au clusters in time. Operating temperature was for the both measurements 200 °C.

5. Conclusion

Investigation of sputtered and subsequently annealed thin compound films revealed absence of surface NiO phase in the first layout samples (09-34) and presence of pure Ni. This fact influenced electrical Van der Pauw and hydrogen gas measurements. Ni layer situated on the surface caused notable decrease of resistivity of the film, but with and also without of the noble metal catalyst addition, the first layout structures were insensitive to hydrogen in low operating temperatures. We expect full covering of the sensitive layer by pure Ni thereby no response to hydrogen was observed.

On the other side, samples 10-3 with 10 nm thin NiO layer under 100 nm TiO$_2$ layer showed some response to hydrogen, which was afterwards greatly enhanced by Au catalyst. Au modification brought shift of the working temperature down to lower values - under 200 °C desirable for revealing the improved sensitivity and response time of the multilayer compound oxides.

Acknowledgement

This work was supported by the Scientific Grant Agency of the Ministry of Education of the Slovak Republic No. 1/0553/09, by APVV under contact No. VVCE-0049-07 and No. APVV-0655-07, PPP Program project DAAD No. D/08/07742 and training grant from SAIA.

References

[1] K. Galatsis, Y. Li, W. Wlodarski, E. Comini, G. Sberveglieri, C. Cantalini, S. Santucci, M. Passacantando, *Sensors and Actuators* **B83** (2002) 276-280

[2] E. Comini, M. Ferroni, V. Guidi, G. Faglia, G. Martinelli, G. Sberveglieri, *Sensors and Actuators* **B84** (2002) 26-32

[3] A. Tricoli, S. E. Pratsinis, *Nature Nanotechnology*, DOI: 10.1038/NNANO 2009 349

[4] A.M. Taurino, M. Epifani, T. Toccoli , S. Iannotta, P. Siciliano, *Thin Solid Films* **436** (2003) 52-63

RuO$_2$/TiO$_2$ based MIM capacitors for DRAM application

B. Hudec[1], K. Hušeková[1], J. Aarik[2], A. Tarre[2], A. Kasikov[2] and K. Fröhlich[1]

1 – Department of Thin Oxide Films, Institute of Electrical Engineering, Slovak Academy of Sciences, Dúbravská Cesta 9, 841 04 Bratislava, Slovak Republic
2 – Institute of Physics, University of Tartu, Riia 142, 51014 Tartu, Estonia
e-mail: boris.hudec@savba.sk

MIM capacitors with MOCVD-grown RuO$_2$ bottom electrode, ALD-grown TiO$_2$ rutile dielectric and RuO2 and Pt top electrodes were prepared and characterised by the means of C-V and J-V measurements. Dielectric constants were in the range of 140, which correspond to EOT of 0.5 nm and leakage current density as low as $1.5*10^{-6}$ A/cm^2 was achieved. Strong influence of TiO$_2$ stoichiometry on the leakage currents was found and analysed.

1. Introduction

Dynamic Random Access Memories (DRAMs) are widely present in computers today and will probably be used also in the mid-term future. Their main advantages are high speed and relatively easy implementation. One cell of a DRAM matrix consists just of a transistor used for addressing the cell and a capacitor used to store the information in the form of accumulated charge. As DRAM chips are being scaled down, increase in the capacitance density of the cell capacitor is required. This can be achieved by substituting the dielectric material with the one with higher dielectric constant. Hence, SiO$_2$ and Si$_3$N$_4$ used in the past are being substituted by Al$_2$O$_3$, HfO$_2$, ZrO$_2$, Ta$_2$O$_5$ and their mixtures in the present. Capacitor stacks have also changed from MIS (metal-insulator-silicon) structures deposited on doped poly-Si bottom electrode to MIM (metal-insulator-metal) structures using TiN as a bottom electrode in the present. This change was driven by the need to get rid of the SiO$_2$ layer formed on poly-Si during deposition of a dielectric oxide. In order to continue the down-scale process of DRAMs, novel materials have to be studied and implemented. The requirements and trends for the future DRAM capacitors are summarized in International Technology Roadmap for Semiconductors 2009 Edition [1]. For DRAM chips with sub-22 nm node expected in production after year 2016, ITRS 2009 mention requirements like capacitance density >70 fF/μm^2, EOT (equivalent oxide thickness) <0.5 nm and leakage current density <$2*10^{-7}$ A/cm^2 @ 0.8 V bias. As these capacitors are expected to be deposited after the transistors formation (so called back end-of-line), the deposition temperatures are expected to be below 500°C. ITRS proposes that materials like Ru, RuO$_2$, Ir and IrO$_2$ should be used as electrodes and TiO$_2$, SrTiO$_3$ and BaSrTiO$_3$ would be used as dielectrics, fulfilling the above mentioned requirements. However, a lot of research on these materials has to be conducted in order to move from proposals to implementation.

In this work, we have prepared MIM capacitors based on RuO$_2$ bottom electrodes and TiO$_2$-rutile dielectric layers. MIM capacitors with rutile dielectric are promising candidates for future DRAM memories, because of very high permittivity of rutile [2-4] ranging from 90 to 170, dependently on the lattice orientation. Unfortunately, the rutile phase of TiO$_2$ often coexists with low-κ anatase phase that reduces the effective dielectric constant of the whole stack. Therefore high temperature post-deposition annealing (~800 °C) is frequently needed to obtain pure rutile [5]. However, an appropriate seed layer can also result in growth of pure rutile phase. We have successfully used RuO$_2$ bottom electrode as a seed layer and obtained pure rutile phase of TiO$_2$ in as-deposited films [2]. As the band gap of rutile-type TiO$_2$ is

978-1-4244-8574-1/10 $26.00 © 2010 IEEE

relatively small (~3 eV) and the electron affinity of TiO_2 is high (4 eV [6]), the MIM structures often suffer from high leakage currents while a high work function metal should be used as an electrode material to prevent excessive leakage caused by Schottky emission. In addition to Schottky emission, defects existing in the dielectric can contribute to the leakage. In this work, we have prepared MIM capacitors with RuO_2 bottom electrode, TiO_2 dielectric layer and RuO_2 and Pt top electrodes in order to investigate the contribution of different leakage mechanisms.

2. Experimental

Thin RuO_2 films were deposited by MOCVD in a low-pressure, hot-wall quartz reactor operated at a pressure of 2 Torr. Precursor bis(2,2,6,6-tetramethyl-3,5-heptanedionato)(1,5-cyclooctadiene) ruthenium was dissolved in iso-octane (concentration 0.035 M) and injected into the evaporation chamber using solenoid microvalve. The injector was open for 3 ms with a frequency of 0.33 Hz. Oxygen was used as a reactant gas with flow rate of 170 sccm, while argon was used as a carrier gas with a flow rate of 21 sccm. The films were grown at a substrate temperature of 290°C, bottom electrodes on Si(100) substrates covered by a 100 nm SiO_2 layer and top electrodes on RuO_2/TiO_2 bilayers.

The TiO_2 films were grown in a flow-type ALD reactor [7] at a temperature of 400°C on polycrystalline RuO_2 films serving as bottom electrodes and seed layers. In order to synthesize the TiO_2 films, the substrates were exposed to the $TiCl_4$ vapor for 2 s, purged in the flow of pure nitrogen for 2 s, exposed to the H_2O vapor for 2 s, and again purged in the flow of pure nitrogen for 5 s. The ALD cycle was repeated 400 times to obtain 20 nm thick films.

Crystallographic phases were identified by X-ray diffraction (XRD) on Bruker AXS-D8 Discover equipment in grazing incidence (1.5°) mode using Cu Kα radiation. The thickness of the TiO_2 films was determined by X-ray reflectivity (XRR) on samples grown on Si substrate. Capacitance-voltage (C-V) and current density-voltage (J-V) characteristics were measured using an Agilent 4284A LCR meter and Keithley 2400 Source Meter, respectively. Optical lithography and argon ion milling were used to define top electrodes on the samples with RuO_2 top layer. Platinum top electrodes were evaporated by electron beam at room temperature through a shadow mask. On both type of samples, the top electrodes were circle-shaped with a diameter of 50 μm. Bottom electrode was always grounded and top electrode biased. Post-deposition annealing was performed as the last step of processing.

3. Results and discussion

Annealing in oxygen and nitrogen ambient at different times and temperatures was examined in this study in order to evaluate the most appropriate annealing conditions with respect to the leakage currents. As expected, annealing in oxygen resulted in lower leakage currents than annealing in nitrogen (not shown) did. The nitrogen annealing probably led to reduction of TiO_2 and formation of non-stoichiometric oxide, sometimes referred to as TiO_{2-x}, which is considered to be a wide band gap n-type semiconductor rather than an insulator [8]. Impact of oxygen annealing on the leakage currents of the samples with RuO_2 top electrodes is shown in Fig. 1. Two different samples were measured after annealing at different temperatures for 5 minutes (Fig. 1 left) and 20 minutes (Fig. 1 right). In both cases, the annealing at 300°C was evaluated to be the most appropriate. Therefore we have applied post-deposition annealing in oxygen at the pressure of 1 bar for 30 minutes at 300°C to all following samples.

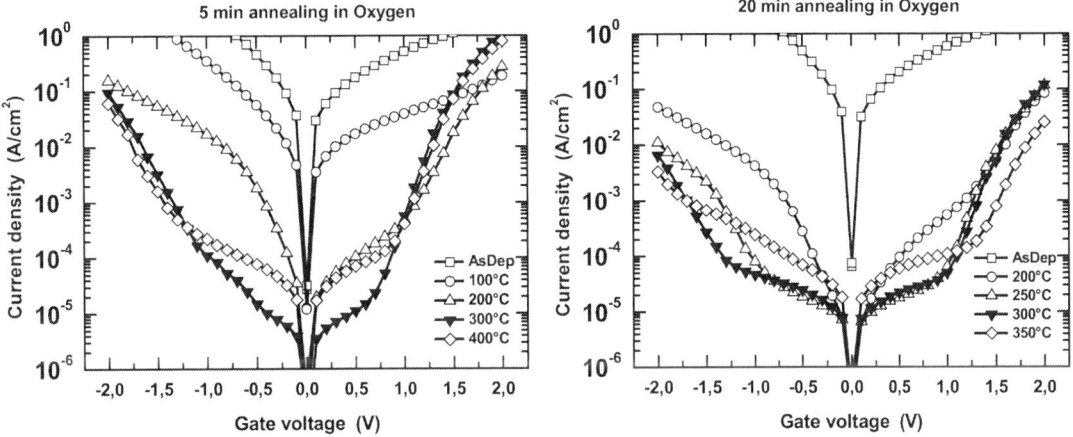

Fig. 1: Current density vs. voltage (J-V) curves of two test samples annealed in O_2 under different temperatures for 5 minutes (left) and 20 minutes (right). It is clear from both experiments that annealing in O_2 for 300°C for at least 20 minutes results in the lowest leakage currents. At higher annealing temperatures the stacks are being degraded. Temperature in the legend stands for temperature of annealing (measurements were done at room temperature) and AsDep means state after deposition of top RuO_2 electrode before annealing.

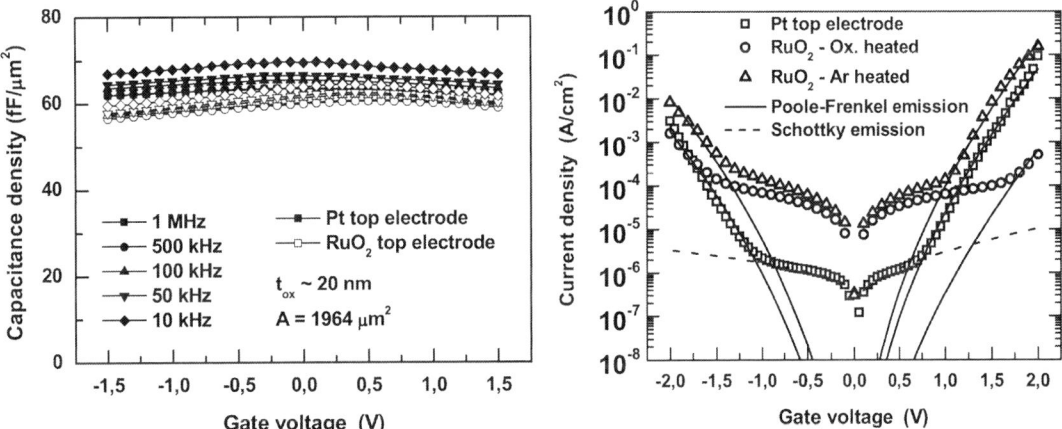

Fig. 2: (left) Capacitance density vs. voltage (C-V) curves of samples with RuO_2 and Pt on top. (right) Current density vs. voltage (J-V) curves of samples with Pt and RuO_2 top electrodes heated in Oxygen and Argon ambient before top RuO_2 electrode deposition. Dashed line represents Schottky emission and solid lines represent Poole-Frenkel emission fits. At 0.8 V, Pt-topped sample exhibits lowest leakage current density.

Electrode material is of crucial importance with regard to electrical properties of the MIM stack. Bottom RuO_2 electrode that serves as a seed layer for rutile growth cannot be replaced. To increase the barrier height for Schottky emission from the top electrode to TiO_2, another material with higher work function could be used for top electrode, however, as the work function of RuO_2 is only 5.0 – 5.2 eV [9, 10]. In our previous study [11], we have prepared similar MIM stacks with Au and Ni top electrodes, as work functions of these metals are both > 5.0 eV [12]. The samples with Ni top electrodes were very leaky. The samples with Au top electrodes exhibited superior leakage, but only half the capacitance compared to samples with RuO_2 top electrodes. As pointed out by Valleé et. al. [12], also the oxygen affinity of the electrode metal is playing a role in electrical properties of the MIM stacks. Thus, the oxygen affinity of Ni that is 2.6 eV higher than that of Au was probably the cause for the leaky behavior of Ni-topped samples. However, the reason for reduced capacitance of Au-topped samples is still unclear.

Pt that was applied as a top electrode material in MIM structures of this study has the work function of 5.65 eV and is chemically stable against oxidation (low oxygen affinity). Thus, we expected a positive impact of Pt top electrodes on the leakage currents. Capacitance density vs. voltage (C-V, Fig.2 left) measurements for the samples with RuO_2 and Pt top electrodes revealed dielectric constants of 135 and 144 with corresponding EOT of 0.55 and 0.50 nm, respectively. Capacitance dispersion of both samples measured in a frequency range of 10 kHz – 1 MHz is at an acceptable level considering DRAM application.

In Fig.2 (right), J-V curves of samples with Pt and RuO_2 electrodes are shown. Two samples with RuO_2 electrodes were measured, one heated in pure oxygen during MOCVD reactor start-up and the other one heated in argon. Schottky emission (SE) and Poole-Frenkel emission (PFE) fits are shown as well. Details of the calculation were reported elsewhere [2]. Schottky emission is an interface-limited mechanism, influenced mainly by the barrier height at the interface between electrode and TiO_2. SE is the dominating mechanism for all 3 samples at lower voltages, causing lower leakage in Pt-topped sample. This was expected, as Pt top electrode exhibits higher work function than RuO_2 does (by ~0.5 eV). As PFE is a bulk-limited mechanism, its presence signifies that defects, probably oxygen vacancies, are present inside TiO_2 layer. This is in agreement with the fact that PFE is suppressed to higher voltages for the sample heated in oxygen before starting the top RuO_2 electrode deposition, compared to the sample heated in argon. PFE is present also in J-V curves of Pt-topped sample, indicating that TiO_2 was non-stoichiometric or contained impurities or other types of defects also in the structures with Pt top electrodes.

3. Summary

MIM structures for DRAM capacitor application were prepared by MOCVD-grown RuO_2 seed layer and bottom electrode and ALD-grown rutile TiO_2 dielectric layer. Samples with RuO_2 and Pt top electrodes exhibited dielectric constants of around 140 and EOT ~ 0.5 nm. The lowest leakage current density of $1.5*10^{-6}$ A/cm^2 @ 0.8 V was achieved for the samples with Pt-top electrode. Leakage currents were very sensitive to TiO_2 stoichiometry and O_2 treatment.

Acknowledgement

This work was financially supported by the Slovak Research and Development Agency project APVV-0133-07 and VEGA project 2/0031/08 and Estonian Science Foundation (Grant No. 7845).

References

[1]   ITRS 2009 Edition, http://www.itrs.net.
[2]   Fröhlich, K. et. al.: J. Vac. Sci. Technol. B 27, 266 (2009).
[3]   Choi, G.-J. et. al.: J. Electrochem. Soc. 156, G71 (2009).
[4]   Cheng, C. H. et. al.: IEEE Elect. Dev. Lett. 29, 845 (2008).
[5]   G. T. Lim and D.-H. Kim, Thin Solid Films, 498, 254, 2006.
[6]   J. Robertson, J. Vac. Sci. Technol. B 18, 1785, 2000.
[7]   J. Aarik, A. Aidla, A.-A. Kiisler, T. Uustare, V. Sammelselg: Thin Solid Films 340, 110 1999.
[8]   P. Kofstad, J. Less Common Metals 13 (1967) 635.
[9]   C. S. Park, G. Bersuker, P. Y. Hung, P. D. Kirsch, R. Jammy: Electrochem. Sol. St. Lett. 13, H105 (2010).
[10] K. Fröhlich, K. Hušeková, D. Machajdík, J. C. Hooker, et. al.: Mat. Sci. Eng. B 109, 117 (2004).
[11] B. Hudec, K. Hušeková, E. Dobročka, T. Lalinský, J. Aarik et. al: IOP Conf. Series: Mat. Sci. Eng. 8, 012024 (2010).
[12] C. Vallée, P. Gonon, C. Jorel and F. El Kamel: Appl. Phys. Lett 96, 233504 (2010).

AUTHOR INDEX

Aarik J.	255, 341	Danneville F.	25
Ahmad N.	305	Daricek M.	203, 251
Alcubilla R.	263	Defrance N.	111
Alexander C.	317	Desplanque L.	25
Allsopp D. W. E.	127	Dobrocka E.	247
Amen R.	127	Domaradzki J.	65, 69
Amir F.	29	Donoval D.	53, 97, 115, 143, 171, 203, 251, 309
Andok R.	85		
Arshad S.	305	Donoval M.	203, 251
Asenov A.	317	Douvry Y.	163
Asgari-Khoshooie A.	45	Drzik M.	175
Balalykin N. I.	77	Dubecky F.	207, 219
Barak V.	85	Durina P.	89
Barbarin E.	333	Dutta G.	325
Basnar B.	163	Dvurechenskii A. V.	9
Basu S.	325	Edward M. J.	127
Behmenburg H.	97, 309	Ezzati N.	287, 291
Bello I.	123, 271	Farajzadeh A.	33
Benkovska J.	135	Fasaki I.	297
Blaho M.	155	Fedor J.	247
Bohacek P.	77, 81, 219	Feltin E.	163
Böhm P.	313	Figueras E.	243
Bollaert S.	25	Fisher M.	329
Bose S.	139	Flickyngerova S.	73, 123
Boura A.	215, 235	Florovic M.	97
Bouzid S.	111	Foit J.	231, 239
Bowen C. R.	127	Fonseca L.	243
Brath T.	61	Fox A.	159
Brezina I.	227	Fröhlich K.	247, 255, 341
Bruncko J.	301	Frolec J.	211
Buc D.	61	Gallo O.	195
Büttgen B.	279	Gaquière Ch.	163
Calaza C.	243	Gaspierik P.	183
Cané C	243, 263	Gawor T.	69
Caplovic L.	61	Gerharz J.	9
Caplovicova M.	61	Gladkov P.	187
Carlin J. F.	163	Gonschorek M.	163
Cester A.	171	Gornik E.	163
Cico K.	247	Gràcia I.	243
Cumeras R.	243	Grandjean N.	163
Daghighi A	33, 45	Grazner R.	195

Gregusova D.	155	Keshtiban P. M.	287, 291
Grieseler R.	337	Khmyrova I.	13
Grützmacher D.	9, 159	Kindl D.	207
Gryglewicz J.	49	Knight R.	17
Grym J.	187	Kobzev A. P.	77, 81
Guastella S.	333	Kodys P.	283
Harmatha L.	135, 195	Kompitstas M.	337
Hascik S.	131	Konecnikova A.	85, 89
Hasenöhrl S.	175	Koptev E. S.	9
Hashizume T.	155	Kordos P.	97, 115, 143, 155, 159
Heuken M.	309		
Hoel V.	111	Kosc I.	297, 337
Horinek F.	203	Kostic I.	85, 89, 131
Horniak M.	203	Kotorova D.	73
Hotovy I.	297, 337	Kovac J.	21, 37, 53, 97, 115, 123, 135, 143, 171, 295, 309
Hotovy J.	37		
Hranai M.	255		
Hrnciar V.	61	Kovac J. jr.	21, 37, 123, 271
Hudec B.	255, 341		
Hudek P.	85	Kovac U.	317
Huran J.	77, 81	Kuball M.	119
Husak M.	93, 211, 215, 223, 235	Kubicova I.	21
		Kucera M.	81
Husekova K.	247, 255, 341	Kulha P.	191
Cheng B.	317	Kups T.	337
Chini A.	41	Kus P.	89
Chlpik J.	301	Kutsay O.	123
Chvala A.	115, 143	Kuzmik J.	163
Ian K. W.	41	Ladziansky M.	207
Iannacci J.	267	Lalinsky T.	131, 135
Isa M. M.	41	Laposa A.	93
Islam A.	321	Lecourt F.	111
Jaeger J. C. De	111, 163	Lee S. T.	271
Jakabovic J.	21, 37, 171	Limiao C.	271
Jakovenko J.	93, 215, 223	Lu B.	105
Janicek V.	239	Maher H.	111
Jarchovsky Z.	187	Malinovsky L.	81
Jelenković E. V.	123	Marek J.	115, 143
Jelinek M.	301	Marso M.	147, 159
Jha S. K.	123	Martincek I.	21
Johander P.	127	Matay L.	85, 89
Jurkovic M.	247	Mazumder S. K.	139
Kaczmarek D.	65, 69	Mazur M.	65, 69
Kadar O.	301	Meneghesso G.	171
Kalna K.	321	Michalka M.	295
Kasikov A.	341	Mikolasek M.	195

Mikulics M.	101, 159	Prusakova L.	179
Milinović V.	183	Pudis D.	21
Milosavljević M.	183	Racko J.	195
Mishra U. K.	119	Ramiączek - Krasowska M.	49
Missous M.	17, 29, 41, 305		
Mitchell C.	29	Rehacek V.	337
Moers J.	9	Remsa J.	301
Molnar M.	143	Rendek K.	53, 203, 251
Mullerova J.	329	Renvoise M.	111
Necas V.	207, 219	Reznak J.	195
Nemec M.	135, 247	Ritomsky A.	85
Netrvalova M.	73, 179, 329	Rodriguez A.	263
Niessner M.	267	Roelens Y.	25
Nikiforov A. I.	9	Roch T.	89
Nohavica D.	187	Romani A.	259
Noskovic J.	89	Roy G.	317
Noudeviwa A.	25	Rufer L.	127
Novak J.	21, 175, 231, 239	Rybar J.	135
		Ryć L.	219
Novotny I.	37, 73, 183	Ryger I.	131
Oleszkiewicz W.	49	Sabaté N.	243, 263
Olivier A.	25	Saguatti D.	41
Ostermaier	163	Salleras M.	243
Osvald J.	167	Sangiorgi E.	259
Oswald C.	283	Santander J.	243, 263
Östling M.	1	Satka A.	53, 295
Palacios T.	105	Sebok J.	135
Partel S.	85	Sedmidubsky D.	101
Paszkiewicz B.	49, 57	Sexton J.	17
Paszkiewicz R.	49, 57, 135	Shirkov G. D.	77
Pei Y.	119	Shvetsov V. N.	81
Pejović M.	123	Schaaf P.	73
Perusko D.	183	Scheirich J.	283
Petrus M.	135	Schrag G.	267
Pidik A.	89	Schrenk W.	163
Piedra D.	105	Schwierz F.	195
Pirri F.	333	Simms R. J. T.	119
Placido F.	65	Skriniarova J.	21, 37, 97, 115, 135
Plecenik	89		
Pogany D.	163	Smith D.	111
Pozzovivo G.	163	Sofer Z.	101
Pravin K. N.	199	Solarikova P.	227
Predanocy M.	61, 297, 337	Soltys J.	175
Priesol J.	295	Song S.	65
Prociow E.	69	Spiess L.	73, 297, 337
Prochazkova O.	275	Sramaty R.	115
		Srnanek R.	123, 309

Srobar F.	275	Yu L.	271
Stafiniak A.	49	Zapien J. A.	123, 271
Stefecka M.	89	Zatko B.	219
Stepina N. P.	9	Zeman M.	329
Stoklas R.	155		
Stopjakova V.	227		
Strang B.	159		
Strasser G.	163		
Stuchlikova L.	135		
Surya C.	123		
Suslik L.	21		
Sutta P.	73, 179, 329		
Szyszka A.	49		
Tajima M.	155		
Tan J.	279		
Tapajna M.	119		
Tarre A.	255, 341		
Tartagni M.	259		
Telek P.	175		
Theuwissen A. J. P.	279		
Tlaczala M.	49, 57, 135		
To C. H.	271		
Tomaska M.	131		
Torres-Herrero N.	263		
Trgala M.	89		
Trifonov T.	263		
Tvarozek V.	73, 183, 227		
Uherek F.	295, 301, 309		
Vallo M.	131		
Valovic A.	77, 81		
Vanko G.	131		
Vavra I.	175		
Vavrinsky E.	203, 227, 251		
Verzellesi G.	41		
Vincze A.	295, 301, 309		
Vittoz S.	127		
Wachutka G.	267, 313		
Waldhoff N.	25		
Wallart X.	25		
Watanabe N.	13		
Wichmann N.	25		
Wilke M.	297, 337		
Wojcieszak D.	65, 69		
Wosko M.	57		
Wrachien N.	171		
Yamase R.	13		